# Gut Microorganisms of Aquatic Animals 2.0

# Gut Microorganisms of Aquatic Animals 2.0

Editor

**Konstantinos Ar. Kormas**

MDPI • Basel • Beijing • Wuhan • Barcelona • Belgrade • Manchester • Tokyo • Cluj • Tianjin

*Editor*
Konstantinos Ar. Kormas
Ichthyology and Aquatic
Environment
University of Thessaly
Volos
Greece

*Editorial Office*
MDPI
St. Alban-Anlage 66
4052 Basel, Switzerland

This is a reprint of articles from the Special Issue published online in the open access journal *Microorganisms* (ISSN 2076-2607) (available at: www.mdpi.com/journal/microorganisms/special_issues/aquaticanimal_gut_microb).

For citation purposes, cite each article independently as indicated on the article page online and as indicated below:

| LastName, A.A.; LastName, B.B.; LastName, C.C. Article Title. *Journal Name* **Year**, *Volume Number*, Page Range. |
|---|

**ISBN 978-3-0365-4712-1 (Hbk)**
**ISBN 978-3-0365-4711-4 (PDF)**

# Contents

# About the Editor

**Konstantinos Ar. Kormas**

Konstantinos Kormas leads the Microbial Communities and Habitats in Aquatic Environments Laboratory (MiCHAEL) of the University of Thessaly, Greece. The team's research interests focus on the diversity and ecological role of prokaryotes and unicellular eukaryotes in the aquatic environment, with special emphasis on the associations and interactions of symbiotic microorganisms with aquatic animals. MiCHAEL's work includes field and laboratory work in marine and freshwater systems, specializing in molecular approaches including genomics and metagenomics. He is currently involved in two H2020 and two national research projects related to animal–microbe interactions. He is the author/co-author of >115 peer-reviewed papers, nine book chapters and one book in Greek ("Ecology of aquatic microorganisms"). His published works have received more than 2600 citations (SCOPUS, excluding self-citations) and his h-index is 30.

microorganisms

MDPI

*Review*

# Methods to Evaluate Bacterial Motility and Its Role in Bacterial–Host Interactions

**Victoria Palma [1], María Soledad Gutiérrez [1,2], Orlando Vargas [1], Raghuveer Parthasarathy [3,4] and Paola Navarrete [1,2,*]**

[1] Laboratory of Microbiology and Probiotics, Institute of Nutrition and Food Technology (INTA), University of Chile, El Líbano 5524, Santiago 7830490, Chile; victoria.palma@ug.uchile.cl (V.P.); soleguti@uchile.cl (M.S.G.); orlandovargasblanco@gmail.com (O.V.)
[2] Millennium Science Initiative Program, Milenium Nucleus in the Biology of the Intestinal Microbiota, National Agency for Research and Development (ANID), Moneda 1375, Santiago 8200000, Chile
[3] Institute of Molecular Biology, University of Oregon, Eugene, OR 97403, USA; raghu@uoregon.edu
[4] Department of Physics and Materials Science Institute, University of Oregon, Eugene, OR 97403, USA
* Correspondence: pnavarre@inta.uchile.cl

**Abstract:** Bacterial motility is a widespread characteristic that can provide several advantages for the cell, allowing it to move towards more favorable conditions and enabling host-associated processes such as colonization. There are different bacterial motility types, and their expression is highly regulated by the environmental conditions. Because of this, methods for studying motility under realistic experimental conditions are required. A wide variety of approaches have been developed to study bacterial motility. Here, we present the most common techniques and recent advances and discuss their strengths as well as their limitations. We classify them as macroscopic or microscopic and highlight the advantages of three-dimensional imaging in microscopic approaches. Lastly, we discuss methods suited for studying motility in bacterial–host interactions, including the use of the zebrafish model.

**Keywords:** bacterial motility; motility methods; bacteria; flagella; bacterial–host interaction; microscopy

**Citation:** Palma, V.; Gutiérrez, M.S.; Vargas, O.; Parthasarathy, R.; Navarrete, P. Methods to Evaluate Bacterial Motility and Its Role in Bacterial–Host Interactions. *Microorganisms* **2022**, *10*, 563. https://doi.org/10.3390/microorganisms10030563

Academic Editor: Konstantinos Ar. Kormas

Received: 30 December 2021
Accepted: 6 February 2022
Published: 4 March 2022



## 1. Introduction

Motility is defined as the movement of cells by some form of self-propulsion [1]. Many bacterial cells are motile as it allows them, for example, to escape from unfavorable conditions and to exploit new resources or opportunities. Combined with chemotaxis, the ability to sense a chemical gradient and direct movement accordingly, it enables bacteria to pursue nutrients and to reach specific niches. In this sense, motility is also involved in the interaction between microorganisms and their host, specifically in colonization or infectious pathogenic processes. Indeed, non-motile mutants are either impaired or completely disabled to colonize and/or cause disease [2].

There are different types of motility, often classified as swimming, swarming, twitching, gliding, and sliding [3,4]. Swimming consists of movement in a liquid environment typically by using flagella, long, thin appendages attached to the cell [1]. Swarming is a coordinated movement of cells that are propelled by flagella through thin liquid films on surfaces and can involve cellular differentiation into a longer and hyper-flagellated phenotype [5]. Other structural molecules can be involved in bacterial movement such as twitching and gliding, both being active ways of moving over a surface. In twitching, type IV pili extend and attach to a solid surface, then retract to allow movement [6]. While twitching is described as intermittent and uneven, gliding is a more organized and smoother cell movement that comprises evolutionarily unrelated machineries which include the use of adhesins that attach to a substratum and either move across the cell or use surface proteins to perform a back-and-forth motion [4,7,8]. Sliding is a passive movement that, instead of requiring an appendage, occurs by bacteria's surfactants (i.e., rhamnolipids) [3]. While dividing, cells are pushed outwards by the growing colony, and surfactants reduce the surface tension decreasing the friction between the surface and bacterial cells, accelerating their spreading [9]. Alternatively, sliding can be attributed to osmolytes (i.e., glycine betaine) secreted by bacteria that draw water to the media surface [9].

Other types of motion are possible [4,10–14]. *Spiroplasma* propagates kinks along its helical body to swim [10], while it is believed that cyanobacteria of the genus *Synechococcus* does so by propagating spicule-like surface extensions along the cell [11,12]. Another example is *Acinetobacter baumannii* 17978, whose type I pili confer surface motility modulated by light [13]. Moreover, some parasitic bacteria can induce actin polymerization to form a tail and move inside the host cell. These motility types and others are included in a recent re-classification based on the structure of the force-producing motor [14].

Different motility types are not mutually exclusive. It has been shown that besides swimming, swarming, and twitching, *Pseudomonas aeruginosa* can also display sliding motility [15], and a recent review discusses different forms of movement observed in *Staphylococci*, including gliding and sliding in *Staphylococcus aureus* [16]. Motility also shows great variability among species and even strains. For example, strains from different serovars of *Salmonella enterica* showed differences up to a factor of 2.7 in swimming speed [17].

Although motility can provide fitness advantages, it also has considerable drawbacks, such as high energetic and metabolic cost [18], and the presence of antigenic structures such as flagella [19]. These costs are a function of the biological context, and therefore realistic assessment of motility requires setting experimental conditions to be as close as possible to the actual environment of interest. We will discuss here common and recent methodological approaches that have been used to study bacterial motility and its role in bacteria–host interactions.

## 2. Macroscopic Techniques

We will distinguish between macroscopic and microscopic methods for studying bacterial motility. The former does not resolve the motions of individual bacteria but rather the spread of a population through some medium. Qualitatively, the link between macroscopic spreading and microscopic motility makes sense—a non-motile species, for example, will have little dispersal, and a vigorously moving species may travel far. Quantitatively, the relationship between macroscopic dispersal and the motility of individual cells is more subtle because the spread of a population is driven by growth (cell division) as well as motility. For example, a bacterium *Escherichia coli* that travels in fairly straight "runs" of a constant speed, $v_{bacteria}$ that persist on average for time $\tau$ before the organism "tumbles" and randomizes its direction, executes a random walk through its three-dimensional world with an effective diffusion coefficient $D$ proportional to the square of its speed [20,21].

$$D = \frac{1}{3} {v_{bacteria}}^2 \tau \quad (1)$$

If the bacteria are also growing exponentially with growth rate $r$, the population will spread with a velocity:

$$v = 2\sqrt{rD} \quad (2)$$

as Fisher, Kolmogorov, and others showed nearly a century ago [22,23]. For typical bacterial swimming speeds and growth rates, the macroscopic dispersal speed (perhaps millimeters per hour) will be one or two orders of magnitude lower than the speed of individual bacteria (perhaps tens of microns per second). Besides considering the expansion described by Fisher (Fisher waves), recent work on bacterial range expansion has taken into consideration phenomena such as intraspecific cooperativity [24] and chemotaxis [25].

The most common macroscopic approach to studying macroscopic motility is by examining bacterial spread through semi-solid agar (soft agar) [26]. Starting from an inoculation stab deep inside the agar, non-motile bacteria will remain near the inoculation zone, while motile bacteria will spread and visibly blur the media (Figure 1a). Because of its simplicity, it is particularly well suited to uncover non-motile or hypermotile strains (Table 1). Some bacteria can form, depending on the environmental conditions, characteristic colony patterns in plates, especially during swarming [5]. Spatial patterns seen using the soft agar method are linked to chemotaxis—directed motion induced by chemicals—as chemoattractants present in the agar that are metabolized by bacteria creating radial concentration gradients that boosts outward expansion [27]. Using low concentrations of the metabolizable chemoattractant would accentuate taxis response [28]. Other methods to study chemotaxis have been described, such as the capillary assay, where a capillary tube filled with a chemical is placed in a bacterial suspension and the accumulation of bacteria towards or away from the chemical is assessed visually [27,28].

In soft agar assays, the agar concentration can be adjusted according to the bacterial species and motility type (Table 2). To assess sliding motility, soft agar assays with flagellum- and/or type IV pili-deficient strains are usually used to discard swarming and/or twitching, respectively [9]. If the motility zone cannot be visualized because of low cell density, for example, in the case of using agar medium low in nutrients, the bacterial density of an agar plug at a standardized distance can be measured to determine if bacteria has reached this position [29]. Labeling can increase the contrast between the spreading bacteria and the culture media. For example, 2,3,5-triphenyltetrazolium chloride (TTC) can be easily incorporated into the media, coloring bacterial growth [30]. Genetically modified bacteria encoding fluorescent proteins (i.e., GFP) or bioluminescent bacteria can also be used. For example, a fluorescent *Pseudomonas* and a bioluminescent *Salmonella* can both be distinguished in a co-swarming experiment [31]. Staining the biosurfactant rhamnolipids produced by bacteria, by adding Red Nile in the medium, showed that its production on agar surfaces was associated with bacterial swarming motility [31].

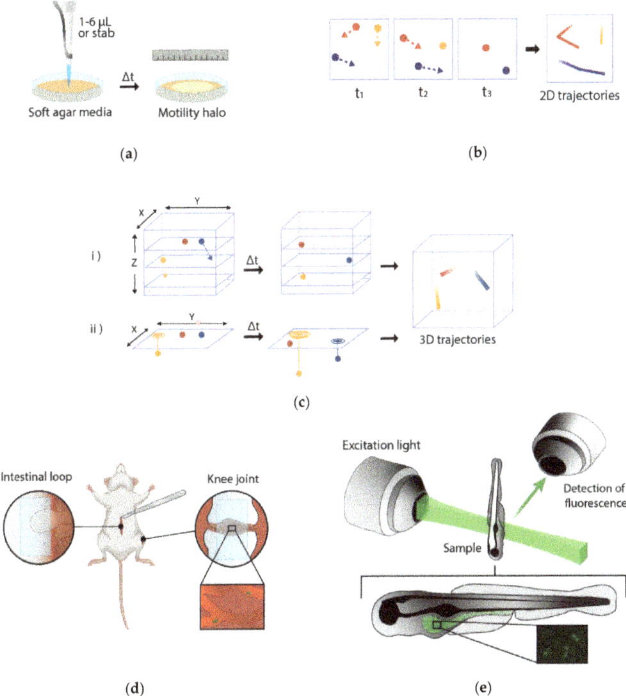


**Figure 1.** Some examples of methods to study bacterial motility and motility tracking. (**a**) Soft agar

assay is the most common macroscopic method used to study motility. After inoculating 1–6 μL or a stab of bacterial culture in soft agar, motile bacteria will spread and blur the media. (**b**) Assessing motility using some common microscopic methods is based on tracking individual bacteria to obtain their 2D trajectories. If a cell leaves the focal plane (orange cell) the track ends. (**c**) Three-dimensional trajectories can be obtained by (**i**) stacks of 2D slices along the z-axis (z-stacking) or by (**ii**) projecting the 2D image in the z-axis according to certain parameters such as depth-dependent shape in the case of defocused imaging methods. (**d**) Intravital microscopy (IVM) aims to visualize phenomena occurring inside live animals. For example, exposing the tissue of an anesthetized mouse by doing small incisions while carefully preserving its physiological conditions, a glass coverslip can be placed in the knee joint [32] or an intestinal loop [33] to visualize the movement of fluorescently labeled bacteria. Bacteria (green) are shown inside knee joint microvasculature. (**e**) The optical transparency of zebrafish larvae allows non-invasive visualization of the in vivo motility of fluorescent bacteria with light-sheet fluorescence microscopy (LSFM) in which a focal plane is illuminated, exciting all points in the plane simultaneously.

**Table 1.** Macroscopic assays to study bacterial motility.

| Macroscopic Assay | Applications | References |
|---|---|---|
| Soft-agar tubes | Easily identification of motile and non-motile bacteria | [26] |
| Soft-agar plates | Quantification of motility level, and identification of a motility type (Table 2) or patterns at a population level | [5,9,26] |
| Using low concentrations of a metabolizable chemoattractant | Assessing chemotactic motility | [27,28] |
| Using fluorescent labelling | Identification of more than two bacteria in co-swarming experiments, increasing contrast with the media, and studying of motility-related compounds | [31] |

**Table 2.** Agar concentration in media according to the type of motility type to assess in a semi-solid (soft) agar assay.

| Motility Type | Agar Concentration | References |
|---|---|---|
| Swimming | ~0.3% | [34] |
| Swarming (temperate) | 0.5–0.8% | [35] |
| Swarming (robust) | >1.5% | [35] |
| Twitching [1] | 1% | [36] |
| Sliding | 0.3–0.4%, or 1–2% has also been used | [37,38] |
| Gliding | ≤7% in *Myxococcus xanthus* | [39] |

[1] The plate is inoculated at the bottom of the media instead of the top.

Environmental factors can also affect motility in agar. Tremblay and Déziel [40] proved that incubation temperature, pH, and drying time of soft agar under laminar flow affected swarming. In fact, even the location of the plates within the laminar flow causes significant differences in the swarming speed. These factors can affect media wetness that causes differences in the thickness of the liquid layer. The wetter the surface, the easier it is for bacteria to overcome frictional forces and move. This makes the reproducibility of these methods difficult to achieve.

## 3. Microscopic Techniques

Direct observation of motile bacterial cells provides the clearest insights into their motility but is challenging due to the length and time scales involved, as well as the potential complexity of the microbe's environments. Bacteria are typically around a micron in size, with speeds up to tens of microns per second for flagella-mediated swimming. Video capture rates of at least 10 frames per second (fps) are therefore needed if cellular positions in adjacent images are to be no more than a body-length apart, facilitating reconstruction of trajectories. Slower rates could capture transitions between straight runs and tumbles, but only rates of 10 fps or higher can capture information about instantaneous speed and angle changes [41]. Moreover, if the bacterial density is too high, bacteria will traverse each other constantly, making the reconstruction process difficult.

Even though bacteria can be tracked using simple bright-field imaging, its discerning from the background can be enhanced by techniques such as dark-field microscopy, differential interference contrast microscopy (DIC), and phase-contrast microscopy (Table 3). In dark-field microscopy, illumination comes from the side so that only light scattered by objects such as bacteria is detected, providing a bright signal on a dark background. This enables, for example, visualization of flagella in addition to bacterial cell bodies when using a high light intensity [1,42]. One-sided dark-field illumination variant is useful to simultaneously determine cell rotation and swimming speed in spirochetes [43]. In DIC microscopy and phase-contrast microscopy, the index of refraction gradients and phase shifts, respectively, are mapped onto intensity differences, enhancing the contrast of relatively transparent objects, making these methods suitable for assessing bacterial movement and orientation [44,45]. Recently, Smith et al. [46] were able to quantify twitching throughout a dense bacterial colony where individual cell tracking was not feasible using DIC microscopy. Substantially, the edge of the colony was observed by microscope and light changes over time were mapped and associated with areas with low and high motility within the field of view, where a higher modulation of light implies higher bacterial motility.

Fluorescent microscopy enables clear identification of labeled cells or even specific bacterial components such as flagella [47] (Table 3). Genetically encoded fluorescent proteins are routinely used in model bacterial strains, such as *E. coli* K12 or *P. aeruginosa* PAO1, and increasingly in non-conventional microbes, such as some *Aeromonas* and *Pleisomonas* isolates from the zebrafish intestinal microbiota [48]. Exogenous labels, such as fluorescent probes, can be simpler to apply but will be diluted as bacteria divide, and one must be aware that they can potentially alter bacterial function. Staining with DAPI, for example, halves the swimming speed of *Pseudomonas* species [49], and fusions of fluorescent proteins to components of the bacterial flagellar motor can alter its dynamics [50].

Microscopy in its forms mentioned so far provides views of a two-dimensional image. The truncated fragments of trajectories as bacteria move in and out of the focal plane still allow measurement of swimming speeds, durations of runs, and other characteristics (Figure 1b). Nonetheless, three-dimensional trajectories obtained through stacks of 2D slices (z-stacking) can be worthwhile, giving a more accurate characterization of motility patterns (Figure 1c). The main disadvantages are the requirement of rapid stack acquisition and the high amount of computational resources needed to process large stacks. On the other hand, methods based on 2D projection allow observing a larger volume in exchange for providing less exact measurements [51]. Berg's classic identifications of *E. coli*'s runs and tumbles tracked a microbe in three dimensions through a feedback loop linking image intensities and stage positions [52]. This is a very precise approach but can only track a single cell.

More recent techniques allow three-dimensional imaging of many bacteria within a field of view. In defocused imaging methods, depth-dependent image shape allows localization along the axis perpendicular to the focal plane ("z") (Figure 1c). This approach has long been used for non-bacterial imaging, e.g., nanoparticles [53], and has been applied to bacteria using fluorescence [54] as well as phase contrast [51] microscopy, with a z-range limit of 200 µm in the latter. Gray values can also be used to determine z-distance in cells close to the focal plane [55].

**Table 3.** Microscopic techniques to study bacterial motility and their main applications.

| Microscopic Techniques | Advantages | Disadvantages | Applications |
|---|---|---|---|
| Bright field microscopy | Simplest, cheapest, and highly accessible | Resolution limited by the wavelength of light, low contrast | Rapidly identification of a motile bacteria |
| Dark field microscopy | Contrast enhancement of unstained samples | Resolution limited by the wavelength of light | Visualization of motile bacteria, flagella |
| Phase contrast microscopy | Contrast enhancement of unstained samples | Resolution limited by the wavelength of light | Visualization of motile bacteria, and bacterial orientation |
| Differential interference contrast microscopy (DIC) | Contrast enhancement of unstained samples, edges of the object are highlighted | Resolution limited by the wavelength of light | Visualization of motile bacteria, and bacterial orientation |
| Confocal microscopy or laser scanning confocal microscopy (LSCM) | High resolution imaging due to reduction of background fluorescence; to collect serial optical sections from thick samples. Contrast and definition are improved | May not be fast enough to capture relevant dynamics; limited to the number of excitation wavelengths available from common lasers; imaging depth limited | Visualization of motile bacteria in thin tissues |
| Spinning disk confocal microscopy | Image acquisition speed is higher than LSCM improving the observation of dynamic processes and reducing photodamage | Imaging depth limited; sensitive camera is needed | Visualization of motile bacteria in thin tissues |
| Multiphoton confocal microscopy | Deeper penetration in tissue (>100 μm) compared to LSCM | Higher phototoxicity and photobleaching in the focal plane compared to LSCM | Visualization of motile bacteria in thick living tissue |
| Light-sheet fluorescent microscopy (LSFM) or selective plane illumination microscopy (SPIM) | High 3D resolution images | Sample mounting may be challenging; reduced resolution in depth compared to confocal microscopy | Visualization of motile bacteria in thick living tissue |
| Light-field-based selective volume illumination microscopy (SVIM) | Captures a 3D volume in a single snapshot | Requires specialized hardware; smaller spatial range than SPIM | Visualization of motile bacteria in thick living tissue in a single snapshot |
| Digital holographic microscopy (DHM) | High imaging speed; high resolution; adjust focus after the image is recorded, since all focus planes are recorded simultaneously by the hologram | Low scattering efficiency of bacteria | Visualization of several free-swimming bacteria |
| Differential dynamic microscopy (DDM) | Great number of bacteria can be processed simultaneously | Unsuited for obtaining specific motility parameters | Quick evaluation of motility responses at a whole-population level |

Another technique for three-dimensional reconstruction that has been applied to bacterial systems is digital holographic microscopy (DHM) [56] (Table 3). DHM reconstructs an image from the interference pattern produced by the specimen, illuminated by a coherent light source, although it does not support three-dimensional fluorescence imaging. While a low scattering efficiency of bacteria is a disadvantage, DHM has high imaging speed and, with recent improvements, a lateral resolution of less than 0.5 μm has been achieved [57,58]. Acres and Nadeau [59] described that DHM 2D projections generally suffice for calculating free-swimming bacteria speeds, but z-stacking is more accurate to study motility near a solid surface.

In light-field microscopy (LFM) a whole volume is illuminated and sampled in one snapshot, instead of using a bidimensional image as an input [60] (Table 3). Then, a microlens array translates depth information into a two-dimensional light field image, which can be computationally transformed back into a three-dimensional image. While LFM employs wide-field illumination, selective volume illumination microscopy (SVIM) is a variant that illuminates only the volume of interest, reducing the background noise and increasing the contrast, allowing a lateral resolution of 3 μm [61] (Table 3). Considering the high number of optimizations available, SVIM has a great potential for visualizing dynamic and complex interactions such as the bacterial flow of *Vibrio fischeri* within the seawater surrounding the light organ of its host, the Hawaiian bobtail squid (*Euprymna scolopes*), as well as the selective colonization of that organ by individual bacteria [61].

Differential dynamic microscopy (DDM) [62,63] relies on light scattering caused by a suspension of particles, instead of tracking (Table 3). The scattering forms a speckle pattern whose intensity will vary at a rate depending on the speed of the particles movement. These fluctuations lead to the differential intensity correlation function from which parameters such as speed and motile fraction can be extracted. While the great number of bacteria that can be processed simultaneously is a considerable advantage, this method is unsuited for obtaining more specific motility parameters. DDM is convenient to quickly evaluate motility responses at a whole-population level, such as the speed recovery after osmotic shocks of different magnitudes [64] and local speed changes caused by a light pattern projection in photokinetic *E. coli* genetically modified to swim smoothly with a light controllable speed [65].

All these techniques and more, under the appropriate conditions, are precise enough to reveal strategies for swimming, chemotaxis, and other behaviors. Lastly, new methods for extracting and assessing image-derived trajectories can be used to produce more accurate characterizations of the bacteria's movement. Accordingly, Liang et al. [66] implemented an unsupervised cluster analysis to fractionate the swimming trajectories of *Azotobacter vinelandii* into run and tumble segments, and then extracted the motility parameters distribution for each segment by fitting mathematical distributions. Other examples are the algorithms developed by Vissers et al. [67] (available on GitLab) to determine the positions, and orientations of individual rod-shaped bacteria, and track and analyze their surface dynamics, discerning between adhering, diffusing, and swimming cells.

Several techniques are available to study the role that bacterial appendages play in motility. However, as they are not in the scope of this review, they will be only briefly presented. Common techniques for visualizing nanomachineries include electron microscopy (EM) and its variations: transmission EM, scanning EM, and cryo-EM [68] are used to observe and study the structure of these bacterial components. Specifically, cryo-EM has recently provided 3D structural models of motility- [4] and chemotaxis-related [69] components with high resolution. However, freezing the cell makes capturing the dynamics of the machinery unachievable. Recent advances in fluorescence microscopy have allowed studying the functionality of these bacterial components. The substitution of amino acid residues of flagellin for cysteines or pilin subunits and subsequent labeling them with maleimide fluorescent dyes has allowed the study of flagellar [70] and pili [68] dynamics in real time. Moreover, a label-free technique, interferometric scattering microscopy (iSCAT), has recently been used to study type IV pili motor dynamics three-dimensionally [71]. These advances are vastly improving our knowledge of how the molecular machinery of bacterial motility operates.

## 4. Study of Bacterial Motility in Bacterial–Host Interactions

The study of bacterial motility inside a host is a more complex affair, which is why many studies simulate host conditions in vitro. Soft agar can, up to some extent, mimic physical, chemical, and nutritional conditions inside and outside the host [29,72,73]. Furthermore, chambers can mimic environments such as xylem vessels [74], enabling the discovery that *Xylella fastidiosa* migrates against the flow via twitching motility, and anaerobiosis, allowing researchers to prove that *Clostridioides difficile* modulates its swimming speed in the presence of a metabolite related to its host colonization [75]. Likewise, vertical diffusion chambers (VDC) were used to study the role of motility in *Campylobacter jejuni* invasion of epithelial cells [76]. An alternative closer to in vivo conditions is tissue culture, which allows investigation of motility behavior in processes such as cell invasion and tumor colonization [77–79]. Lastly, artificial systems that reproduce the successive environmental niches of the human gastrointestinal tract can be used to simulate the host's dynamic conditions [80]. A metagenomic analysis of a gastrointestinal model of the colon developed by The Netherlands Organization for Applied Scientific Research (TIM-2) inoculated with human gut microbes showed that higher iron availability resulted in an enrichment of motility and chemotaxis functions [81]. Meanwhile, an early ex vivo approach in infant mice includes the labeling of motile and non-motile strains of *Vibrio cholerae* with fluorescent antibodies to visualize and compare its distribution in the extracted infected tissue [82].

In vivo real-time imaging is crucial to understand the colonization dynamics of bacteria. Intravital microscopy (IVM) consists of imaging inside live animals and often relies on fluorescence microscopy (Figure 1d; Table 3). The main problem is the thickness of the tissue samples, as off-focus blur and light scattering limit the depth of imaging [83]. Confocal microscopy can suffice; Moriarty et al. [84] reported high-resolution multidimensional visualization of bacterial dissemination inside a living mammal using spinning disk confocal IVM, revealing that dissemination of *Borrelia burgdorferi* in microvasculature of mice is a multi-stage process. Nonetheless, the scattered fluorescence limits the imaging depth of confocal microscopy to tens of microns. On the contrary, with multiphoton fluorescence, which is based on the simultaneous absorption of two or more infrared or near-infrared photons, imaging can be deeper than 100 μm in tissue. This is possible because longer wavelengths can penetrate at higher depths, besides lowering endogenous autofluorescence. Moreover, as excitation occurs only in the focal plane, there is minimal bleaching in the rest of the tissue [85,86]. Because of its advantages, IVM has been widely applied to visualize bacterial motility in colonized organs, such as *B. burgdorferi* in the skin [86] and *V. cholerae* in the intestine [33].

Zebrafish (*Danio rerio*) is a particularly advantageous vertebrate animal model for studying host–bacterial interactions due to their optical transparency at the larval stage, allowing for non-invasive examination of bacterial movement inside a living vertebrate host (Figure 1e). There are considerable similarities between zebrafish and mammals [87]. The gut is anatomically organized in separate sections and the intestinal epithelium is constantly renewing its cells. There is a high degree of orthologue genes [88] and their regulation within the gut is similar. The immune system of teleost fish species shares several traits with the system of mammals including the presence of lymphoid tissues, cell-mediated responses, and mucosal immunity [89].

Another advantage of zebrafish is that larvae hatch at 2–3 days post-fertilization (dpf) and open their mouths at 3 dpf, facilitating the production of germ-free or axenic individuals, great tools to study bacterial–host interactions. Fluorescently labeled bacteria can be inoculated via immersion at this developmental stage and visualized both at a whole population and at a single-cell level [90,91]. Germ-free zebrafish larvae colonized with fluorescent bacteria proved to be useful to examine the relationship between bacterial motility and symbiosis within the intestine [92,93]. In the last few years, the use of the zebrafish model coupled with light-sheet fluorescence microscopy (LSFM, also known as selective plane illumination microscopy) has provided new insights into the field of bacterial dynamics within a living host [94–98]. In this technique, only the focal plane is illuminated, exciting all points in the plane simultaneously, while out-of-focus points are not excited, minimizing photodamage and photobleaching and increasing imaging speed compared to point scanning methods, while achieving much higher resolution than wild field microscopy [96,99]. These characteristics make LSFM very suitable to follow bacterial dynamics inside the whole intestine of zebrafish for several hours. Nevertheless, because of light diffraction, generating a thin plane of excitation light is difficult, causing a loss in resolution compared to confocal and multiphoton imaging.

Combining LSFM, larval zebrafish, and bacteria engineered with inducible switches for a flagellar motor component revealed that the swimming motility of a zebrafish-native

*Vibrio* species was necessary for its persistence inside the host and avoidance of expulsion with intestinal flow [98]. In a separate study, live imaging revealed that sub-lethal doses of the broad-spectrum antibiotic ciprofloxacin promoted its bacterial aggregation and expulsion from the intestine [100].

Finally, transcriptomic approaches can be used to investigate the effect of host or environmental factors [101–103] and phenomena such as macrophage internalization [104] and host cell infection [105] in the transcriptional regulation of genes related to bacterial motility. Employing microarrays, Snyder et al. [106] first assessed an *E. coli* pathotype's transcriptome in vivo from bacteria extracted from infected mice, showing that flagellar genes were downregulated compared to in vitro conditions. Interestingly, this transcriptome was performed from different urine samples taken across 10 days of the infection period. A similar experiment using an *E. coli* expressing a luminescent reporter for the flagellar gene *fliC* showed that its expression was upregulated during the pathogen's ascension through the upper urinary tract, suggesting a major contribution of motility in the colonization of the urinary system [107]. Recently, a comparison by RNA-seq between *Pseudomonas plecoglossicida* infecting spleens of the fish *Larimichthys crocea* and those cultivated in vitro revealed an up-regulation of motility-related and flagellum-related genes during the fish infection [108].

It is important to consider that, as single-cell transcriptomic approaches are difficult to achieve in prokaryotes [109], only homogenized output from a population is usually obtained for bacteria, impeding the study of phenotypically distinct subpopulations that could be present in the sample. Recent works have focused on overcoming these difficulties with strategies including mRNA enrichment methods. Kuchina et al. (2021) modified SPLiT-seq—a technique that uses combinatorial indexing to label the eukaryotic RNA's cellular origin—to optimize its performance in bacteria. This approach was able to assess the fraction of *Bacillus subtilis* PY79 population that expressed flagellin and surfactin while growing in a rich medium [110].

Lastly, proteomic approaches, particularly those based on mass spectrometry (MS)—which measures the mass-to-charge ratio of ionized molecules to identify them—have proven to be a notable tool for assessing abundance changes in bacterial proteins inside a host [111,112]. Proteomic studies using liquid chromatography MS showed that downregulation of *Salmonella enterica* Typhimurium proteins involved in virulence, chemotaxis, and flagellar systems occurs earlier in bacteria inside macrophages compared to bacteria internalized by epithelial cells, suggesting that different host cell types have a different impact on motility adaptations [112].

## 5. Discussion and Concluding Remarks

The crucial role of motility in bacterial survival, host colonization, and/or virulence is a fact. This mini review showed that multiple approaches are available to study motility, from soft agar to a wide variety of microscopic techniques. The optimal choice will depend on the specific questions or requirements of the experiment, such as the number of cells or strains to process, z range needed, and growth conditions. In host–bacterial interactions, in vitro set-ups can provide fair approximations to the host environment, whereas intravital microscopy allows in vivo tracking of bacteria within the host tissue. This approach benefits from techniques that allow a greater depth of imaging, namely, confocal, and multiphoton fluorescence microscopy. Alternatively, the zebrafish model allows direct visualization of bacteria inside the host. Assessing the expression level of motility-related genes is also feasible. All these approaches can be combined to have a wider outlook; for example, coupling semi-solid (soft) agar plates with microscopy visualization. Accordingly, Deforet et al. [55] observed that macroscopically, a *P. aeruginosa* hyperswarmer mutant spreads faster, yet does not swim faster than the wild-type at the single-cell level. Further investigation led to realize if this phenomenon is related to wider turns.

Overall, a considerable number of new methods and advances to study bacterial motility have emerged during the last decade, deepening our understanding of bacterial behavior. Nevertheless, there are several issues that still need improvement, such as protocol standardization in soft agar assays; facilitating the implementation of 3D tracking, mostly achieved by microscopy techniques that are technically demanding and/or require complex set-ups and extending the depth of imaging for bacteria within host tissue in in vivo motility studies.

**Author Contributions:** Conceptualization, V.P. and P.N.; writing—original draft preparation, V.P., M.S.G., O.V., R.P. and P.N.; writing—review and editing, R.P. and P.N.; funding acquisition, P.N. All authors have read and agreed to the published version of the manuscript.

**Funding:** This research was funded by FONDECYT/ANID, grant number 1181499.

**Institutional Review Board Statement:** Not applicable.

**Informed Consent Statement:** Not applicable.

**Acknowledgments:** MSG acknowledges funding by FONDECYT/ANID, grant number 3200998.

**Conflicts of Interest:** The authors declare no conflict of interest. The funders had no role in the writing of the manuscript.

## References

1. Madigan, M.T.; Clark, D.P.; Stahl, D.; Martinko, J.M. Basic Principles of Microbiology. In *Brock Biology of Microorganisms*, 13th ed.; Pearson Prentice Hall: Upper Saddle River, NJ, USA, 2010; pp. 1–84.
2. Josenhans, C.; Suerbaum, S. The Role of Motility as a Virulence Factor in Bacteria. *Int. J. Med. Microbiol.* **2002**, *291*, 605–614. [CrossRef] [PubMed]
3. Henrichsen, J. Bacterial Surface Translocation: A Survey and a Classification. *Bacteriol. Rev.* **1972**, *36*, 478–503. [CrossRef] [PubMed]
4. Wadhwa, N.; Berg, H.C. Bacterial motility: Machinery and mechanisms. *Nat. Rev. Microbiol.* **2021**, 1–13. [CrossRef]
5. Kearns, D.B. A Field Guide to Bacterial Swarming Motility. *Nat. Rev. Microbiol.* **2010**, *8*, 634–644. [CrossRef] [PubMed]
6. Jarrell, K.F.; McBride, M.J. The Surprisingly Diverse Ways That Prokaryotes Move. *Nat. Rev. Microbiol.* **2008**, *6*, 466–476. [CrossRef] [PubMed]
7. Mattick, J.S. Type IV Pili and Twitching Motility. *Annu. Rev. Microbiol.* **2002**, *56*, 289–314. [CrossRef]
8. Nan, B.; Zusman, D.R. Novel mechanisms power bacterial gliding motility. *Mol. Microbiol.* **2016**, *101*, 186–193. [CrossRef]
9. Mattingly, A.E.; Weaver, A.A.; Dimkovikj, A.; Shrout, J.D. Assessing travel conditions: Environmental and host influences on bacterial surface motility. *J. Bacteriol.* **2018**, *200*, 1–17. [CrossRef]
10. Shaevitz, J.W.; Lee, J.Y.; Fletcher, D.A. *Spiroplasma* swim by a processive change in body helicity. *Cell* **2005**, *122*, 941–945. [CrossRef]
11. Wyman, M.; Gregory, R.P.; Carr, N.G. Novel Role for Phycoerythrin in a Marine Cyanobacterium, *Synechococcus* Strain DC2. *Science* **1985**, *230*, 818–820. [CrossRef]
12. Ehlers, K.; Oster, G. On the mysterious propulsion of *Synechococcus*. *PLoS ONE* **2012**, *7*, e36081. [CrossRef] [PubMed]
13. Wood, C.R.; Ohneck, E.J.; Edelmann, R.E.; Actis, L.A. A Light-Regulated Type I Pilus Contributes to *Acinetobacter baumannii* Biofilm, Motility, and Virulence Functions. *Infect. Immun.* **2018**, *86*, e00442-18. [CrossRef]
14. Miyata, M.; Robinson, R.C.; Uyeda, T.Q.P.; Fukumori, Y.; Fukushima, S.; Haruta, S.; Homma, M.; Inaba, K.; Ito, M.; Kaito, C.; et al. Tree of Motility—A Proposed History of Motility Systems in the Tree of Life. *Genes Cells* **2020**, *25*, 6–21. [CrossRef] [PubMed]
15. Murray, T.S.; Kazmierczak, B.I. *Pseudomonas aeruginosa* Exhibits Sliding Motility in the Absence of Type IV Pili and Flagella. *J. Bacteriol.* **2008**, *190*, 2700–2708. [CrossRef] [PubMed]
16. Pollitt, E.J.G.; Diggle, S.P. Defining Motility in the Staphylococci. *Cell. Mol. Life Sci.* **2017**, *74*, 2943–2958. [CrossRef]
17. Zawiah, W.A.N.; Abdullah, W.A.N.; Mackey, B.M. High Phenotypic Variability among Representative Strains of Common *Salmonella enterica* Serovars with Possible Implications for Food Safety. *J. Food Prot.* **2018**, *81*, 93–104. [CrossRef]
18. Macnab, R.M. Flagella and motility. In *Escherichia coli and Salmonella: Cellular and Molecular Biology*, 2nd ed.; Neidhardt, C., Curtiss, R., Ingraham, J.L., Lin, E.C.C., Low, K.B., Magasanik, B., Reznikoff, W.S., Riley, M., Schaechter, M., Umbarger, H.E., Eds.; ASM Press: Washington, DC, USA, 1996; pp. 123–145.
19. Ottemann, K.M.; Miller, J.F. Roles for motility in bacterial–host interactions. *Mol. Microbiol.* **1997**, *24*, 1109–1117. [CrossRef]
20. Lovely, P.S.; Dahlquist, F.W. Statistical Measures of Bacterial Motility. *J. Theor. Biol.* **1974**, *50*, 477–496. [CrossRef]
21. Berg, H.C. *Random Walks in Biology*; Princeton University Press: Princeton, NJ, USA, 1983.
22. Fisher, R.A. The wave of advance of advantageous genes. *Ann. Eugen.* **1937**, *7*, 355–369. [CrossRef]
23. Kolmogorov, A.N.; Petrovsky, N.; Piscounov, N.S. A study of the equation of diffusion with increase in the quantity of matter, and its application to a biological problem. *Mosc. Univ. Bull. Math.* **1937**, *1*, 1–25.
24. Gandhi, S.R.; Yurtsev, E.A.; Korolev, K.S.; Gore, J. Range Expansions Transition from Pulled to Pushed Waves as Growth Becomes More Cooperative in an Experimental Microbial Population. *Proc. Natl. Acad. Sci. USA* **2016**, *113*, 6922–6927. [CrossRef]

25. Cremer, J.; Honda, T.; Tang, Y.; Wong-Ng, J.; Vergassola, M.; Hwa, T. Chemotaxis as a Navigation Strategy to Boost Range Expansion. *Nature* **2019**, *575*, 658–663. [CrossRef]
26. Tittsler, R.P.; Sandholzer, L.A. The Use of Semi-Solid Agar for the Detection of Bacterial Motility. *J. Bacteriol.* **1936**, *31*, 575–580. [CrossRef]
27. Adler, J. A method for measuring chemotaxis and use of the method to determine optimum conditions for chemotaxis by *Escherichia coli*. *J. Gen. Microbiol.* **1973**, *74*, 77–91. [CrossRef]
28. Ditty, J.L.; Parales, R.E. Protocols for the Measurement of Bacterial Chemotaxis to Hydrocarbons. In *Hydrocarbon and Lipid Microbiology Protocols*; McGenity, T.J., Timmis, K.N., Nogales, B., Eds.; Springer: Berlin/Heidelberg, Germany, 2015; pp. 7–42. [CrossRef]
29. Robinson, C.D.; Klein, H.S.; Murphy, K.D.; Parthasarathy, R.; Guillemin, K.; Bohannan, B.J.M. Experimental Bacterial Adaptation to the Zebrafish Gut Reveals a Primary Role for Immigration. *PLoS Biol.* **2018**, *16*, e2006893. [CrossRef]
30. Ball, R.J.; Sellers, W. Improved motility medium. *Appl. Microbiol.* **1966**, *14*, 670–673. [CrossRef]
31. Morris, J.D.; Hewitt, J.L.; Wolfe, L.G.; Kamatkar, N.G.; Chapman, S.M.; Diener, J.M.; Courtney, A.J.; Leevy, W.M.; Shrout, J.D. Imaging and Analysis of *Pseudomonas aeruginosa* Swarming and Rhamnolipid Production. *Appl. Environ. Microbiol.* **2011**, *77*, 8310–8317. [CrossRef]
32. Lee, W.Y.; Sanz, M.J.; Wong, C.H.; Hardy, P.O.; Salman-Dilgimen, A.; Moriarty, T.J.; Chaconas, G.; Marques, A.; Krawetz, R.; Mody, C.H.; et al. Invariant natural killer T cells act as an extravascular cytotoxic barrier for joint-invading Lyme *Borrelia*. *Proc. Natl. Acad. Sci. USA* **2014**, *111*, 13936–13941. [CrossRef]
33. Millet, Y.A.; Alvarez, D.; Ringgaard, S.; von Andrian, U.H.; Davis, B.M.; Waldor, M.K. Insights into *Vibrio cholerae* Intestinal Colonization from Monitoring Fluorescently Labeled Bacteria. *PLoS Pathog.* **2014**, *10*, e1004405. [CrossRef]
34. Partridge, J.D.; Harshey, R.M. Investigating Flagella-Driven Motility in *Escherichia coli* by Applying Three Established Techniques in a Series. *J. Vis. Exp.* **2020**, *2020*, 1–8. [CrossRef]
35. Partridge, J.D.; Harshey, R.M. Swarming: Flexible Roaming Plans. *J. Bacteriol.* **2013**, *195*, 909–918. [CrossRef]
36. Turnbull, L.; Whitchurch, C.B. Motility Assay: Twitching Motility. In *Pseudomonas Methods and Protocols*; Filloux, A., Ramos, J.-L., Eds.; Humana Press Inc.: New York, NY, USA, 2014; pp. 73–86. [CrossRef]
37. Martínez, A.; Torello, S.; Kolter, R. Sliding Motility in Mycobacteria. *J. Bacteriol.* **1999**, *181*, 7331–7338. [CrossRef]
38. Liu, Y.; Kyle, S.; Straight, P.D. Antibiotic Stimulation of a *Bacillus subtilis* Migratory Response. *mSphere* **2018**, *3*, e00586-17. [CrossRef]
39. Tchoufag, J.; Ghosh, P.; Pogue, C.B.; Nan, B.; Mandadapu, K.K. Mechanisms for Bacterial Gliding Motility on Soft Substrates. *Proc. Natl. Acad. Sci. USA* **2019**, *116*, 25087–25096. [CrossRef]
40. Tremblay, J.; Déziel, E. Improving the Reproducibility of *Pseudomonas aeruginosa* Swarming Motility Assays. *J. Basic Microbiol.* **2008**, *48*, 509–515. [CrossRef]
41. Pottash, A.E.; McKay, R.; Virgile, C.R.; Ueda, H.; Bentley, W.E. TumbleScore: Run and Tumble Analysis for Low Frame-Rate Motility Videos. *BioTechniques* **2017**, *62*, 31–36. [CrossRef]
42. Macnab, R.M. Examination of Bacterial Flagellation by Dark Field Microscopy. *J. Clin. Microbiol.* **1976**, *4*, 258–265. [CrossRef]
43. Nakamura, S.; Islam, M.S. Motility of Spirochetes. *Methods Mol. Biol.* **2017**, *1593*, 243–251. [CrossRef]
44. Cheong, F.C.; Wong, C.C.; Gao, Y.; Nai, M.H.; Cui, Y.; Park, S.; Kenney, L.J.; Lim, C.T. Rapid, High-Throughput Tracking of Bacterial Motility in 3D via Phase-Contrast Holographic Video Microscopy. *Biophys. J.* **2015**, *108*, 1248–1256. [CrossRef]
45. Hook, A.L.; Flewellen, J.L.; Dubern, J.-F.; Carabelli, A.M.; Zaid, I.M.; Berry, R.M.; Wildman, R.D.; Russell, N.; Williams, P.; Alexander, M.R. Simultaneous Tracking of *Pseudomonas aeruginosa* Motility in Liquid and at the Solid-Liquid Interface Reveals Differential Roles for the Flagellar Stators. *mSystems* **2019**, *4*, e00390-19. [CrossRef]
46. Smith, B.; Li, J.; Metruccio, M.; Wan, S.; Evans, D.; Fleiszig, S. Quantification of Bacterial Twitching Motility in Dense Colonies Using Transmitted Light Microscopy and Computational Image Analysis. *Bio Protoc.* **2018**, *8*, e2804. [CrossRef] [PubMed]
47. Turner, L.; Ryu, W.S.; Berg, H.C. Real-Time Imaging of Fluorescent Flagellar Filaments. *J. Bacteriol.* **2000**, *182*, 2793–2801. [CrossRef] [PubMed]
48. Wiles, T.J.; Wall, E.S.; Schlomann, B.H.; Hay, E.A.; Parthasarathy, R.; Guillemin, K. Modernized Tools for Streamlined Genetic Manipulation and Comparative Study of Wild and Diverse Proteobacterial Lineages. *mBio* **2018**, *9*, 1–19. [CrossRef]
49. Toepfer, J.A.; Ford, R.M.; Metge, D.; Harvey, R.W. Impact of Fluorochrome Stains Used to Study Bacterial Transport in Shallow Aquifers on Motility and Chemotaxis of *Pseudomonas* Species. *FEMS Microbiol. Ecol.* **2012**, *81*, 163–171. [CrossRef]
50. Heo, M.; Nord, A.L.; Chamousset, D.; van Rijn, E.; Beaumont, H.J.E.; Pedaci, F. Impact of Fluorescent Protein Fusions on the Bacterial Flagellar Motor. *Sci. Rep.* **2017**, *7*, 12583. [CrossRef]
51. Taute, K.M.; Gude, S.; Tans, S.J.; Shimizu, T.S. High-Throughput 3D Tracking of Bacteria on a Standard Phase Contrast Microscope. *Nat. Commun.* **2015**, *6*, 8776. [CrossRef]
52. Berg, H.; Brown, D. Chemotaxis in *Escherichia coli* analysed by Three-dimensional Tracking. *Nature* **1972**, *239*, 500–504. [CrossRef]
53. Speidel, M.; Jonáš, A.; Florin, E.-L. Three-Dimensional Tracking of Fluorescent Nanoparticles with Subnanometer Precision by Use of off-Focus Imaging. *Opt. Lett.* **2003**, *28*, 69. [CrossRef]
54. Wu, M.; Roberts, J.W.; Kim, S.; Koch, D.L.; Delisa, M.P. Collective Bacterial Dynamics Revealed Using a Three-Dimensional Population-Scale Defocused Particle Tracking Technique. *Appl. Environ. Microbiol.* **2006**, *72*, 4987–4994. [CrossRef]

55. Deforet, M.; Van Ditmarsch, D.; Carmona-Fontaine, C.; Xavier, J.B. Hyperswarming Adaptations in a Bacterium Improve Collective Motility without Enhancing Single Cell Motility. *Soft Matter* **2014**, *10*, 2405–2413. [CrossRef]
56. Kim, M.K. Principles and Techniques of Digital Holographic Microscopy. *SPIE Rev.* **2010**, *1*, 018005. [CrossRef]
57. Molaei, M.; Sheng, J. Imaging Bacterial 3D Motion Using Digital In-Line Holographic Microscopy and Correlation-Based de-Noising Algorithm. *Opt. Express* **2014**, *22*, 32119. [CrossRef] [PubMed]
58. Wang, A.; Garmann, R.F.; Manoharan, V.N. Tracking *E. coli* Runs and Tumbles with Scattering Solutions and Digital Holographic Microscopy. *Opt. Express* **2016**, *24*, 23719. [CrossRef] [PubMed]
59. Acres, J.; Nadeau, J. 2D vs. 3D tracking in bacterial motility analysis. *AIMS Biophys.* **2021**, *8*, 385–399. [CrossRef]
60. Levoy, M.; Zhang, Z.; McDowall, I. Recording and Controlling the 4D Light Field in a Microscope Using Microlens Arrays. *J. Microsc.* **2009**, *235*, 144–162. [CrossRef]
61. Truong, T.; Holland, D.B.; Madaan, S.; Andreev, A.; Keomanee-Dizon, K.; Troll, J.; Koo, D.E.S.; McFall-Ngai, M.J.; Fraser, S.E. High-Contrast, Synchronous Volumetric Imaging with Selective Volume Illumination Microscopy. *Commun. Biol.* **2020**, *3*, 74. [CrossRef]
62. Wilson, L.G.; Martinez, V.A.; Tailleur, J.; Bryant, G.; Pusey, P.N.; Poon, W.C.K. Differential Dynamic Microscopy of Bacterial Motility. *Phys. Rev. Lett.* **2011**, *106*, 018101. [CrossRef]
63. Martinez, V.A.; Besseling, R.; Croze, O.A.; Tailleur, J.; Reufer, M.; Schwarz-Linek, J.; Wilson, L.G.; Bees, M.A.; Poon, W.C.K. Differential Dynamic Microscopy: A High-Throughput Method for Characterizing the Motility of Microorganisms. *Biophys. J.* **2012**, *103*, 1637–1647. [CrossRef]
64. Rosko, J.; Martinez, V.A.; Poon, W.C.K.; Pilizota, T. Osmotaxis in *Escherichia coli* through Changes in Motor Speed. *Proc. Natl. Acad. Sci. USA* **2017**, *114*, E7969–E7976. [CrossRef]
65. Frangipane, G.; Dell'arciprete, D.; Petracchini, S.; Maggi, C.; Saglimbeni, F.; Bianchi, S.; Vizsnyiczai, G.; Bernardini, M.L.; Di Leonardo, R. Dynamic density shaping of photokinetic *E. coli*. *eLife* **2018**, *7*, e36608. [CrossRef]
66. Liang, X.; Lu, N.; Chang, L.; Nguyen, T.H.; Massoudieh, A. Evaluation of Bacterial Run and Tumble Motility Parameters through Trajectory Analysis. *J. Contam. Hydrol.* **2018**, *211*, 26–38. [CrossRef]
67. Vissers, T.; Koumakis, N.; Hermes, M.; Brown, A.T.; Schwarz-Linek, J.; Dawson, A.; Poon, W. Dynamical analysis of bacteria in microscopy movies. *PLoS ONE* **2019**, *14*, e0217823. [CrossRef]
68. Ellison, C.K.; Dalia, T.N.; Dalia, A.B.; Brun, Y.V. Real-time microscopy and physical perturbation of bacterial pili using maleimide-conjugated molecules. *Nat. Protoc.* **2019**, *14*, 1803–1819. [CrossRef]
69. Qin, Z.; Zhang, P. Studying bacterial chemosensory array with CryoEM. *Biochem. Soc. Trans.* **2021**, *49*, 2081–2089. [CrossRef]
70. Kühn, M.J.; Schmidt, F.K.; Eckhardt, B.; Thormann, K.M. Bacteria exploit a polymorphic instability of the flagellar filament to escape from traps. *Proc. Natl. Acad. Sci. USA* **2017**, *114*, 6340–6345. [CrossRef]
71. Talà, L.; Fineberg, A.; Kukura, P.; Persat, A. *Pseudomonas aeruginosa* orchestrates twitching motility by sequential control of type IV pili movements. *Nat. Microbiol.* **2019**, *4*, 774–780. [CrossRef]
72. Sun, E.; Liu, S.; Hancock, R.E.W. Surfing Motility: A Conserved yet Diverse Adaptation among Motile Bacteria. *J. Bacteriol.* **2018**, *200*, e00394-18. [CrossRef]
73. Yeung, A.T.Y.; Parayno, A.; Hancock, R.E.W. Mucin Promotes Rapid Surface Motility in *Pseudomonas aeruginosa*. *mBio* **2012**, *3*, e00073-12. [CrossRef]
74. Meng, Y.; Li, Y.; Galvani, C.D.; Hao, G.; Turner, J.N.; Burr, T.J.; Hoch, H.C. Upstream Migration of *Xylella fastidiosa* via Pilus-Driven Twitching Motility. *J. Bacteriol.* **2005**, *187*, 5560–5567. [CrossRef]
75. Courson, D.S.; Pokhrel, A.; Scott, C.; Madrill, M.; Rinehold, A.J.; Tamayo, R.; Cheney, R.E.; Purcell, E.B. Single Cell Analysis of Nutrient Regulation of *Clostridioides (Clostridium) difficile* Motility. *Anaerobe* **2019**, *59*, 205–211. [CrossRef]
76. Mills, D.C.; Gundogdu, O.; Elmi, A.; Bajaj-elliott, M.; Taylor, P.W.; Wren, B.W.; Dorrell, N. Increase in *Campylobacter jejuni* Invasion of Intestinal Epithelial Cells under Low-Oxygen Coculture Conditions That Reflect the In Vivo Environment. *Infect. Immun.* **2012**, *80*, 1690–1698. [CrossRef]
77. Szymanski, C.M.; King, M.; Haardt, M.; Armstrong, G.D. *Campylobacter jejuni* Motility and Invasion of Caco-2 Cells. *Infect. Immun.* **1995**, *63*, 4295–4300. [CrossRef]
78. Higashi, D.L.; Lee, S.W.; Snyder, A.; Weyand, N.J.; Bakke, A.; So, M. Dynamics of *Neisseria gonorrhoeae* Attachment: Microcolony Development, Cortical Plaque Formation, and Cytoprotection. *Infect. Immun.* **2007**, *75*, 4743–4753. [CrossRef]
79. Toley, B.J.; Forbes, N.S. Motility Is Critical for Effective Distribution and Accumulation of Bacteria in Tumor Tissue. *Integr. Biol.* **2012**, *4*, 165–176. [CrossRef]
80. Jubelin, G.; Desvaux, M.; Schüller, S.; Etienne-Mesmin, L.; Muniesa, M.; Blanquet-Diot, S. Modulation of Enterohaemorrhagic *Escherichia coli* Survival and Virulence in the Human Gastrointestinal Tract. *Microorganisms* **2018**, *6*, 115. [CrossRef]
81. Kortman, G.A.M.; Dutilh, B.E.; Maathuis, A.J.H.; Engelke, U.F.; Boekhorst, J.; Keegan, K.P.; Nielsen, F.G.G.; Betley, J.; Weir, J.C.; Kingsbury, Z.; et al. Microbial Metabolism Shifts towards an Adverse Profile with Supplementary Iron in the TIM-2 in Vitro Model of the Human Colon. *Front. Microbiol.* **2016**, *6*, 1481. [CrossRef]
82. Guentzel, M.N.; Field, L.H.; Eubanks, E.R.; Berry, L.J. Use of Fluorescent Antibody in Studies of Immunity to Cholera in Infant Mice. *Infect. Immun.* **1977**, *15*, 539–548. [CrossRef]
83. Masedunskas, A.; Milberg, O.; Porat-Shliom, N.; Sramkova, M.; Wigand, T.; Amornphimoltham, P.; Weigert, R. Intravital microscopy: A practical guide on imaging intracellular structures in live animals. *Bioarchitecture* **2012**, *2*, 143–157. [CrossRef]

84. Moriarty, T.J.; Norman, M.U.; Colarusso, P.; Bankhead, T.; Kubes, P.; Chaconas, G. Real-Time High Resolution 3D Imaging of the Lyme Disease Spirochete Adhering to and Escaping from the Vasculature of a Living Host. *PLoS Pathog.* **2008**, *4*, 17–19. [CrossRef]
85. Schuh, C.D.; Haenni, D.; Craigie, E.; Ziegler, U.; Weber, B.; Devuyst, O.; Hall, A.M. Long Wavelength Multiphoton Excitation Is Advantageous for Intravital Kidney Imaging. *Kidney Int.* **2016**, *89*, 712–719. [CrossRef]
86. Belperron, A.A.; Mao, J.; Bockenstedt, L.K. Two Photon Intravital Microscopy of Lyme Borrelia in Mice. In *Borrelia burgdorferi. Methods in Molecular Biology*; Pal, U., Buyuktanir, O., Eds.; Humana Press: New York, NY, USA, 2018; Volume 1690, pp. 279–290. [CrossRef]
87. López Nadal, A.; Ikeda-Ohtsubo, W.; Sipkema, D.; Peggs, D.; McGurk, C.; Forlenza, M.; Wiegertjes, G.F.; Brugman, S. Feed, Microbiota, and Gut Immunity: Using the Zebrafish Model to Understand Fish Health. *Front. Immunol.* **2020**, *11*, 114. [CrossRef]
88. Howe, K.; Clark, M.D.; Torroja, C.F.; Torrance, J.; Berthelot, C.; Muffato, M.; Collins, J.E.; Humphray, S.; McLaren, K.; Matthews, L.; et al. The Zebrafish Reference Genome Sequence and Its Relationship to the Human Genome. *Nature* **2013**, *496*, 498–503. [CrossRef]
89. Sunyer, J.O. Fishing for Mammalian Paradigms in the Teleost Immune System. *Nat. Immunol.* **2013**, *14*, 320–326. [CrossRef]
90. Van der Sar, A.M.; Musters, R.J.P.; van Eeden, F.J.M.; Appelmelk, B.J.; Vandenbroucke-Grauls, C.M.; Bitter, W. Zebrafish Embryos as a Model Host for the Real Time Analysis of *Salmonella typhimurium* Infections. *Cell. Microbiol.* **2003**, *5*, 601–611. [CrossRef]
91. O'Toole, R.; von Hofsten, J.; Rosqvist, R.; Olsson, P.E.; Wolf-Watz, H. Visualisation of Zebrafish Infection by GFP-Labelled *Vibrio anguillarum*. *Microb. Pathog.* **2004**, *37*, 41–46. [CrossRef]
92. Rawls, J.F.; Mahowald, M.A.; Goodman, A.L.; Trent, C.M.; Gordon, J.I. In Vivo Imaging and Genetic Analysis Link Bacterial Motility and Symbiosis in the Zebrafish Gut. *Proc. Natl. Acad. Sci. USA* **2007**, *104*, 7622–7627. [CrossRef]
93. Caruffo, M.; Navarrete, N.; Salgado, O.; Díaz, A.; López, P.; García, K.; Feijóo, C.G.; Navarrete, P. Potential Probiotic Yeasts Isolated from the Fish Gut Protect Zebrafish (*Danio rerio*) from a *Vibrio anguillarum* Challenge. *Front. Microbiol.* **2015**, *6*, 1093. [CrossRef]
94. Jemielita, M.; Taormina, M.J.; Burns, A.R.; Hampton, J.S.; Rolig, A.S.; Guillemin, K.; Parthasarathy, R. Spatial and Temporal Features of the Growth of a Bacterial Species Colonizing the Zebrafish Gut. *mBio* **2014**, *5*, 1–8. [CrossRef]
95. Wiles, T.J.; Jemielita, M.; Baker, R.P.; Schlomann, B.H.; Logan, S.L.; Ganz, J.; Melancon, E.; Eisen, J.S.; Guillemin, K.; Parthasarathy, R. Host Gut Motility Promotes Competitive Exclusion within a Model Intestinal Microbiota. *PLoS Biol.* **2016**, *14*, e1002517. [CrossRef]
96. Parthasarathy, R. Monitoring Microbial Communities Using Light Sheet Fluorescence Microscopy. *Curr. Opin. Microbiol.* **2018**, *43*, 31–37. [CrossRef]
97. Schlomann, B.H.; Wiles, T.J.; Wall, E.S.; Guillemin, K.; Parthasarathy, R. Bacterial Cohesion Predicts Spatial Distribution in the Larval Zebrafish Intestine. *Biophys. J.* **2018**, *115*, 2271–2277. [CrossRef]
98. Wiles, T.J.; Schlomann, B.H.; Wall, E.S.; Betancourt, R.; Parthasarathy, R.; Guillemin, K. Swimming Motility of a Gut Bacterial Symbiont Promotes Resistance to Intestinal Expulsion and Enhances Inflammation. *PLoS Biol.* **2020**, *18*, e3000661. [CrossRef]
99. Taormina, M.J.; Jemielita, M.; Stephens, W.Z. Investigating Bacterial-Animal Symbioses with Light Sheet Microscopy. *Biol. Bull.* **2012**, *223*, 7–20. [CrossRef]
100. Schlomann, B.H.; Wiles, T.J.; Wall, E.S.; Guillemin, K.; Parthasarathy, R. Sublethal Antibiotics Collapse Gut Bacterial Populations by Enhancing Aggregation and Expulsion. *Proc. Natl. Acad. Sci. USA* **2019**, *116*, 21392–21400. [CrossRef]
101. House, B.; Kus, J.; Prayitno, N.; Mair, R.; Que, L.; Chingcuanco, F.; Gannon, V.; Cvitkovitch, D.G.; Foster, D.B. Acid-Stress-Induced Changes in Enterohaemorrhagic *Escherichia Coli* O157: H7 Virulence. *Microbiology* **2009**, *155*, 2907–2918. [CrossRef]
102. Peano, C.; Chiaramonte, F.; Motta, S.; Pietrelli, A.; Jaillon, S.; Rossi, E.; Consolandi, C.; Champion, O.L.; Michell, S.L.; Freddi, L.; et al. Gene and Protein Expression in Response to Different Growth Temperatures and Oxygen Availability in *Burkholderia thailandensis*. *PLoS ONE* **2014**, *9*, e93009. [CrossRef]
103. Berndt, A.; Müller, J.; Borsi, L.; Kosmehl, H.; Methner, U.; Berndt, A. Reorganisation of the Caecal Extracellular Matrix upon *Salmonella* Infection-Relation between Bacterial Invasiveness and Expression of Virulence Genes. *Vet. Microbiol.* **2009**, *133*, 123–137. [CrossRef]
104. Tolman, J.S.; Valvano, M.A. Global Changes in Gene Expression by the Opportunistic Pathogen *Burkholderia cenocepacia* in Response to Internalization by Murine Macrophages. *BMC Genom.* **2012**, *13*, 63. [CrossRef]
105. Jiang, L.; Ni, Z.; Wang, L.; Feng, L.; Liu, B. Loss of the lac operon contributes to *Salmonella* invasion of epithelial cells through derepression of flagellar synthesis. *Curr. Microbiol.* **2015**, *70*, 315–323. [CrossRef]
106. Snyder, J.A.; Haugen, B.J.; Buckles, E.L.; Lockatell, C.V.; Johnson, D.E.; Donnenberg, M.S.; Welch, R.A.; Mobley, H.L.T. Transcriptome of Uropathogenic *Escherichia coli* during Urinary Tract Infection. *Infect. Immun.* **2004**, *72*, 6373–6381. [CrossRef]
107. Lane, M.C.; Alteri, C.J.; Smith, S.N.; Mobley, H.L.T. Expression of Flagella Is Coincident with Uropathogenic *Escherichia coli* Ascension to the Upper Urinary Tract. *Proc. Natl. Acad. Sci. USA* **2007**, *104*, 16669–16674. [CrossRef]
108. Tang, Y.; Xin, G.; Zhao, L.M.; Huang, L.X.; Qin, Y.X.; Su, Y.Q.; Zheng, W.Q.; Wu, B.; Lin, N.; Yan, Q.P. Novel Insights into Host-Pathogen Interactions of Large Yellow Croakers (*Larimichthys crocea*) and Pathogenic Bacterium *Pseudomonas plecoglossicida* Using Time-Resolved Dual RNA-Seq of Infected Spleens. *Zool. Res.* **2020**, *41*, 314–327. [CrossRef]
109. Chen, Z.; Chen, L.; Zhang, W. Tools for Genomic and Transcriptomic Analysis of Microbes at Single-Cell Level. *Front. Microbiol.* **2017**, *8*, 1831. [CrossRef]

110. Kuchina, A.; Brettner, L.M.; Paleologu, L.; Roco, C.M.; Rosenberg, A.B.; Carignano, A.; Kibler, R.; Hirano, M.; DePaolo, R.W.; Seelig, G. Microbial single-cell RNA sequencing by split-pool barcoding. *Science* **2021**, *371*, eaba5257. [CrossRef]
111. Yang, Y.; Hu, M.; Yu, K.; Zeng, X.; Liu, X. Mass spectrometry-based proteomic approaches to study pathogenic bacteria-host interactions. *Protein Cell* **2015**, *6*, 265–274. [CrossRef]
112. Li, Z.; Liu, Y.; Fu, J.; Zhang, B.; Cheng, S.; Wu, M.; Wang, Z.; Jiang, J.; Chang, C.; Liu, X. *Salmonella* Proteomic Profiling during Infection Distinguishes the Intracellular Environment of Host Cells. *mSystems* **2019**, *4*, e00314-18. [CrossRef]

Review

# Research Progress of the Gut Microbiome in Hybrid Fish

**Xinyuan Cui [1], Qinrong Zhang [1], Qunde Zhang [1], Yongyong Zhang [1], Hua Chen [2], Guoqi Liu [2] and Lifeng Zhu [1,*]**

[1] College of Life Sciences, Nanjing Normal University, Nanjing 210046, China; cuixinyuanwj@163.com (X.C.); z1187438993@163.com (Q.Z.); qundezhang23@163.com (Q.Z.); zyy863098363@163.com (Y.Z.)
[2] Mingke Biotechnology, Hangzhou 310000, China; chenhua@mingkebio.com (H.C.); liuguoqi@mingkebio.com (G.L.)
* Correspondence: zhulf2020@126.com

**Abstract:** Fish, including hybrid species, are essential components of aquaculture, and the gut microbiome plays a vital role in fish growth, behavior, digestion, and immune health. The gut microbiome can be affected by various internal and/or external factors, such as host development, diet, and environment. We reviewed the effects of diet and dietary supplements on intestinal microorganisms in hybrid fish and the difference in the gut microbiome between the hybrid and their hybrids that originate. Then, we summarized the role of the gut microbiome in the speciation and ecological invasion of hybrid fish. Finally, we discussed possible future studies on the gut microbiome in hybrid fish, including the potential interaction with environmental microbiomes, the effects of the gut microbiome on population expansion, and fish conservation and management.

**Keywords:** hybrid fishes; gut microbiome; community and function; speciation; invasion; fish conservation and management

**Citation:** Cui, X.; Zhang, Q.; Zhang, Q.; Zhang, Y.; Chen, H.; Liu, G.; Zhu, L. Research Progress of the Gut Microbiome in Hybrid Fish. *Microorganisms* **2022**, *10*, 891. https://doi.org/10.3390/microorganisms10050891

Academic Editor: Konstantinos Ar. Kormas

Received: 4 April 2022
Accepted: 21 April 2022
Published: 24 April 2022



## 1. Introduction

The host and its microbiome are regarded as a unique biological entity holobiont, including the genome, which is called the hologenome [1]. The combination of complex microbiota and genes in the intestine are collectively referred to as the gut microbiome [2]. Animal hosts maintain a long, close, and complex relationship with their gut microbiome [3]. The gut microbiome plays a vital role in the nervous system development [4], behavior [5], immunity [6], food digestion, and metabolism [7] of the host. Gut microbiota are highly specialized microbial communities with a complex composition that is affected by many interactions among microorganisms, host, diet, and the environment [8]. Host phylogeny and diet are the two main factors shaping the animal gut microbiome [9–14].

Fish comprise nearly 50% of the total vertebrate diversity, and more than 34,000 species have been described to date, constituting a crucial part of the aquatic ecosystem [15,16]. Microorganisms exist in almost every fish organ, including the skin, digestive tract, internal organs, and luminous organs [17]. The fish gut is a complex ecosystem, composed of highly diverse microbiota. The microbiota is influenced by various factors, such as habitat environmental factors, season, host genetics, developmental stage, nutrition level, and diet composition, with the potential major determinant being the habitat environment [16].

Overall, bacteria are the primary microbial colonizers in the gastrointestinal tract of fish [18–21]. The gastrointestinal microbiota of fish mainly consist of aerobic or facultative anaerobic microorganisms and facultative and obligate anaerobes [20,22–24]. Among them, Proteobacteria, Firmicutes, and Bacteroidetes constitute 90% of the gut microbiome of most fish [15]. In addition, Actinobacteria, Fusobacteria, Bacilli, Clostridia, and Verrucomicrobia are the dominant bacterial phyla in fish gut microorganisms [15,25–29]. The gut microbiota of fish participate in various physiological functions. There are several beneficial effects on the host, such as reproduction, development, nutrition, immunity, and stress responses, and the gut microbiota are often referred to as an 'extra organ' [15,30]. Nayak has described the role of fish gastrointestinal microbiota in nutrition, immunity, and health management [20].

Early research on fish gut microflora employed culture-dependent techniques. The emergence of metagenomics and next-generation sequencing techniques has entirely changed fish gut microbiome research by presenting a method that directly analyzes the microbial genome from environmental samples [31,32]. These new research methods have led to a better understanding of the connections between the microorganisms and their respective hosts. The Illumina system, Roche 454 system, and Ion Torrent Personal Genome Machine (PGM) are the primary next-generation sequencing (NGS) platforms used in fish gut microbiome research, and the Illumina system is the most commonly used [15].

The influencing factors and physiological functions of fish intestinal microbiota are two critical issues in NGS analyses [33]. Most studies have explored the effects of various host and environmental factors on the bacterial community composition of gut microbiota. Limited studies have analyzed the beneficial and harmful effects of the gut microbiota on the host [15]. However, there are many valuable bacterial species in the intestines of fish, including *Cetobacterium* spp. and *Lactobacillus* spp. [34]. Hybrid fish are indispensable components of fish species and are essential in aquaculture. We review recent research on the gut microbiome and ecological problems in hybrid fishes and discuss possible future research to improve our understanding of the gut microbiome in fish.

## 2. The Gut Microbiome in Hybrid Fish

### *2.1. Effects of Diet and Dietary Supplements on the Gut Microbiome and Immune Health of Hybrid Fish*

Hybridization is a basic step in the long-term evolution of organisms, which may lead to the production of new species. Heterosis is a complex biological phenomenon where the hybrid offspring show superior natural characteristics, when compared with their parents [35,36]. Heterosis occurs in fish, and hybrid fish have advantages of faster growth performance, higher immunity, improved ecological adaptability, and an enhanced tolerance for transportation. Therefore, as wild catch fisheries can no longer support the world consumption of seafood, fish heterosis has been widely assisting aquaculture since the 1980s [37–41]. However, even the improvements made by heterosis may not be enough for the growing world consumption rate of fish [35].

In addition, different fish species inhabiting the same waters may also naturally hybridize in the wild. Hybrid fish may possess improved ecological adaptability compared to their parents and be more widely distributed in the natural environment with heterosis, due to the survival of the fittest theory [42–44]. In reality, the microbiota in hybrids may provide new favorable physiological functions and promote the utilization of new ecological niches, and the hybrid microbiota may also shape reproductive barriers, which may influence the ecological speciation or the expansion of the population range [45–48]. It has been shown that greater than 30,000 variations of hybrid fish species have formed in the wild, and these large fish populations can produce high diversity in the dietary niches. Therefore, exploring the microbiota of wild hybrid fish is of great significance for understanding the basic biological and ecological processes of speciation, population expansion, and invasion ecology [39].

We, firstly, aimed to provide a whole picture of the diet or dietary supplement effects on the fish gut microbiome (Table 1) [15,20,49,50]. Then, we focused on the relationship between the diet and the hybrid fish gut microbiome. We found that many studies have explored changing the diet or dietary additives on the composition and function of the hybrid fish gut microbiome and their promotion of the growth and health of mixed fish (Table 2), but rare in the comparison between the hybrid and their hybrids' origin.

**Table 1.** The application of diet and dietary supplements in fish.

| Species | Class | Order | Family | Ingredients | Intervention Type | 16s rRNA Sequencing | References |
|---|---|---|---|---|---|---|---|
| *Oreochromis niloticus* | Actinopteri | Cichliformes | Cichlidae | *Rummeliibacillus stabekisii* | Probiotic | Illumina MiSeq, Amplicon: V3–V4 | [51] |
| *Oreochromis niloticus* | Actinopteri | Cichliformes | Cichlidae | *Bacillus subtilis* | Probiotic | Illumina HiSeq, Amplicon: V4 | [52] |
| *Acipenser baerii* | Actinopteri | Acipenseriformes | Acipenseridae | Arabinoxylan-oligosaccharides (A.X.O.S.) + *Lactococcus lactis spp. lactis* or *Bacillus circulans* | Synbiotic | 454 GS FLX Titanium, Amplicon | [53] |
| *Ctenopharyngodon idellus* | Actinopteri | Cypriniformes | Xenocyprididae | Xylo-oligosaccharide | Prebiotic | Illumina MiSeq | [54] |
| *Oreochromis niloticus* | Actinopteri | Cichliformes | Cichlidae | *Lactobacillus rhamnosus* JCM1136 and *Lactococcus lactis* subsp. *lactis* JCM5805 | Probiotic | Illumina MiSeq, Amplicon: V3–V4 | [55] |
| *Dicentrarchus labrax* | Actinopteri | Perciformes | Moronidae | Calcium carbonate | Prebiotic | Illumina MiSeq, Amplicon: V3–V4 | [56] |
| *Ctenopharyngodon idellus* | Actinopteri | Cypriniformes | Xenocyprididae | *Bacillus subtilis* | Probiotic | No | [57] |
| *Ctenopharyngodon idellus* | Actinopteri | Cypriniformes | Xenocyprididae | *Bacillus coagulans*, *Rhodopseudomonas palustris* and *Lactobacillus acidophilus* | Probiotic | No | [58] |
| *Ctenopharyngodon idellus* | Actinopteri | Cypriniformes | Xenocyprididae | *Bacillus subtilis* Ch9 | Probiotic | No | [59] |
| *Ctenopharyngodon idellus* | Actinopteri | Cypriniformes | Xenocyprididae | Exogenous cellulase | Prebiotic | Amplicon: V3 | [60] |
| *Ctenopharyngodon idellus* | Actinopteri | Cypriniformes | Xenocyprididae | Glutathione | Prebiotic | No | [61] |
| *Ctenopharyngodon idellus* | Actinopteri | Cypriniformes | Xenocyprididae | *B. licheniformis* + xylo-oligosaccharide | Synbiotic | No | [62] |
| *Danio rerio* | Actinopteri | Cypriniformes | Danionidae | Gluten formulated diet | Protein | Illumina Miseq, Amplicon: V4 | [63] |
| *Danio rerio* | Actinopteri | Cypriniformes | Danionidae | Protein meal of animal origin (*ragworm Nereis virens*) | Protein | 454 GS FLX Titanium, Amplicon | [64] |
| *Danio rerio* | Actinopteri | Cypriniformes | Danionidae | Chitosan silver nanocomposites (CAgNCs) | Composites | 454 GS FLX Titanium, Amplicon | [65] |
| *Gambusia affinis* | Actinopteri | Cyprinodontiformes | Poeciliidae | Rifampicin | Antibiotic | Illumina HiSeq, Amplicon: V4 | [66] |
| *Oncorhynchus mykiss* | Actinopteri | Salmoniformes | Salmonidae | Dietary plant proteins | Protein | Illumina Miseq, Amplicon: V6–V8 | [67] |
| *Oncorhynchus mykiss* | Actinopteri | Salmoniformes | Salmonidae | *Wickerhamomyces anomalus* + *Saccharomyces cerevisiae* | Synbiotic | Illumina HiSeq, Amplicon | [68] |
| *Oncorhynchus mykiss* | Actinopteri | Salmoniformes | Salmonidae | Microalgae meal (Schizochytrium limacinum) | Prebiotic | Illumina HiSeq, Amplicon | [69] |
| *Oreochromis niloticus* | Actinopteri | Cichliformes | Cichlidae | *Lactobacillus plantarum* CCFM8610 | Probiotic | Illumina MiSeq, Amplicon: V4–V5 | [70] |
| *Oreochromis niloticus* | Actinopteri | Cichliformes | Cichlidae | *Lactobacillus plantarum* CCFM639 | Probiotic | Illumina MiSeq, Amplicon | [71] |
| *Oreochromis niloticus* | Actinopteri | Cichliformes | Cichlidae | *Vibrio sp.* CC8 and *Bacillus cereus* CC27, | Probiotic | No | [72] |
| *Oreochromis niloticus* | Actinopteri | Cichliformes | Cichlidae | *Clostridium butyricum* | Probiotic | Illumina HiSeq, Amplicon- | [73] |
| *Oreochromis niloticus* | Actinopteri | Cichliformes | Cichlidae | *Allium sativum* | Plant | Illumina MiSeq, Amplicon: V4–V5 | [74] |
| *Oreochromis niloticus* | Actinopteri | Cichliformes | Cichlidae | *Bacillus subtilis* and *Bacillus licheniformis* | Probiotic | No | [75] |
| *Oreochromis niloticus* | Actinopteri | Cichliformes | Cichlidae | *Metschnikowia sp.* GXUS03 | Probiotic | No | [76] |
| *Sparus aurata* | Actinopteri | Spariformes | Sparidae | Sodium butyrate | Butyrate | 455 GS FLX Titanium, Amplicon: V1–V3 | [77] |
| *Seriola lalandi* | Actinopteri | Carangiformes | Carangidae | Oxytetracycline, erythromycin and metronidazole | Antibiotic | Illumina MiSeq, Amplicon: V1–V2 | [78] |
| *Piaractus mesopotamicus* | Actinopteri | Characiformes | Serrasalmidae | Florfenicol | Antibiotic | Illumina MiSeq, Shotgun metagenome | [79] |
| *Channa striata* | Actinopteri | Anabantiformes | Channidae | β-glucan, galactooligosaccharides, mannan-oligosaccharide | Prebiotic | T-RFLP fragment sequencing, Amplicon | [80] |
| *Channa striata* | Actinopteri | Anabantiformes | Channidae | *Saccharomyces cerevisiae* and *Lactobacillus acidophilus* | Probiotic | T-RFLP fragment sequencing, Amplicon | [80] |
| *Cyprinus carpio* | Actinopteri | Cypriniformes | Cyprinidae | Chinese yam peel | Plant | Illumina MiSeq, Amplicon: V3–V4 | [81] |

**Table 1.** *Cont.*

| Species | Class | Order | Family | Ingredients | Intervention Type | 16s rRNA Sequencing | References |
|---|---|---|---|---|---|---|---|
| *Lates calcarifer* | Actinopteri | Perciformes | Centropomidae | Sodium diformate | Formate | No | [82] |
| *Oreochromis niloticus* | Actinopteri | Cichliformes | Cichlidae | *Bacillus subtilis* and *Lactobacillus plantarum* | Probiotic | ABI PRISM 377 sequencer (Perkin-Elmer), Amplicon: V6–V8 | [83] |
| *Sparus aurata* | Actinopteri | Spariformes | Sparidae | Poultry by-product meal and Hydrolyzed feather meal | Protein | 455 GS FLX Titanium, Amplicon: V3–V4 | [84] |
| *Sparus aurata* | Actinopteri | Spariformes | Sparidae | Fish protein hydrolysate or Autolysed dried yeast | Protein | Illumina MiSeq, Amplicon: V3–V4 | [85] |
| *Dicentrarchus labrax* | Actinopteri | Perciformes | Moronidae | Galactomannan oligosaccharides and A mixture of garlic and labiatae-plants oils | Prebiotic | Illumina MiSeq, Amplicon: V3–V4 | [86] |
| *Salmo salar* | Actinopteri | Salmoniformes | Salmonidae | *Pediococcus acidilactici* MA18/5M and Short chain fructooligosaccharides | Synbiotic | Amplicon: V3 | [87] |
| *Arapaima gigas* | Actinopteri | Osteoglossiformes | Osteoglossidae | *Lactococcus lactis* subsp. *lactis* and *Enterococcus faecium* | Probiotic | Amplicon: V1–V2 | [88] |
| *Cyprinus carpio* | Actinopteri | Cypriniformes | Cyprinidae | Dietary plant proteins | Protein | Illumina HiSeq, Amplicon: V3–V4 | [89] |
| *Carassius auratus* | Actinopteri | Cypriniformes | Cyprinidae | *Bacillus subtilis* and *Enterococcus faecium* | Probiotic | Amplicon: V3–V4 | [90] |
| *Totoaba macdonaldi* | Actinopteri | Perciformes | Sciaenidae | Commercial dietary prebiotic and probiotic | Synbiotic | Illumina MiSeq, Amplicon: V3–V4 | [91] |
| *Totoaba macdonaldi* | Actinopteri | Perciformes | Sciaenidae | Soy protein concentrate | Protein | Illumina MiSeq, Amplicon: V3–V4 | [92] |

**Table 2.** The studies on the gut microbiome of hybrid fish.

| Host/Parents | Class | Order | Family | NGS Platform | Amplicon Sequencing | Reference |
|---|---|---|---|---|---|---|
| *Culter alburnus* ♀ × *Megalobrama amblycephala* ♂ | Actinopteri | Cypriniformes | Xenocyprididae | Illumina MiSeq | Amplicon: V3–V4 | [39] |
| *Parachondrostoma toxostoma/Chondrostoma nasus* | Actinopteri | Cypriniformes | Leuciscidae | Illumina MiSeq | Amplicon: V4 | [48] |
| *Epinephelus fuscoguttatus* ♀ × *E. lanceolatus* ♂ | Actinopteri | Perciformes | Serranidae | Illumina NovaSeq | Amplicon: V3–V4 | [93] |
| *Epinephelus fuscoguttatus* ♀ × *E. lanceolatus* ♂ | Actinopteri | Perciformes | Serranidae | Unknown | Unknown | [94] |
| *Oreochromis niloticus* ♀ × *O. aureus* ♂ | Actinopteri | Cichliformes | Cichlidae | Unknown | Amplicon: V3 | [95] |
| *Acipenser baerii* × *A. schrenckii* | Actinopteri | Acipenseriformes | Acipenseridae | Illumina HiSeq | Amplicon: V3–V4 | [96] |
| *Pangasianodon gigas* × *Pangasianodon hypophthalmus* | Actinopteri | Siluriformes | Pangasiidae | Unknown | Unknown | [97] |
| *Epinephelus fuscoguttatus* ♀ × *E. lanceolatus* ♂ | Actinopteri | Perciformes | Serranidae | Illumina HiSeq | Amplicon: V3–V4 | [98] |
| *Acipenser baeri Brandt* ♀ × *A. schrenckii Brandt* ♂ | Actinopteri | Acipenseriformes | Acipenseridae | Illumina MiSeq | Amplicon: V3–V4 | [99] |
| *Epinephelus fuscoguttatus* ♀ × *E. lanceolatus* ♂ | Actinopteri | Perciformes | Serranidae | Illumina HiSeq | Amplicon: V3–V4 | [100] |
| *Epinephelus fuscoguttatus* ♀ × *E. lanceolatus* ♂ | Actinopteri | Perciformes | Serranidae | Lon GeneStudio S5™ | Amplicon: V4 | [101] |
| *Epinephelus fuscoguttatus* ♀ × *E. lanceolatus* ♂ | Actinopteri | Perciformes | Serranidae | Illumina | Amplicon: V3–V4 | [102] |
| *Epinephelus fuscoguttatus* ♀ × *E. lanceolatus* ♂ | Actinopteri | Perciformes | Serranidae | Illumina | Amplicon: V3–V4 | [103] |
| *Epinephelus fuscoguttatus* ♀ × *E. lanceolatus* ♂ | Actinopteri | Perciformes | Serranidae | Illumina MiSeq | Amplicon: V3–V4 | [104] |
| *Morone chrysops* × *M. saxatilis* | Actinopteri | Perciformes | Moronidae | Illumina MiSeq | Amplicon: V1–V3 | [105] |
| *Coregonus* | Actinopteri | Salmoniformes | Salmonidae | Illumina MiSeq | Amplicon: V3–V4 | [106] |
| *Hypophthalmichthys nobilis* × *H. molitr* | Actinopteri | Cypriniformes | Xenocyprididae | Illumina MiSeq | Amplicon: V4 | [107] |

**Table 2.** *Cont.*

| Host/Parents | Class | Order | Family | NGS Platform | Amplicon Sequencing | Reference |
|---|---|---|---|---|---|---|
| *Epinephelus fuscoguttatus* ♀ × *E. lanceolatus* ♂ | Actinopteri | Perciformes | Serranidae | Illumina MiSeq | Amplicon: V3–V4 | [108] |
| *Epinephelus fuscoguttatus* ♀ × *E. lanceolatus* ♂ | Actinopteri | Perciformes | Serranidae | Unknown | Amplicon | [109] |
| *Epinephelus fuscoguttatus* ♀ × *E. lanceolatus* ♂ | Actinopteri | Perciformes | Serranidae | Illumina MiSeq | Amplicon | [110] |
| *Morone Chrysops* × *M. Saxatilis* | Actinopteri | Perciformes | Moronidae | Illumina MiSeq | Amplicon: V1–V3 | [111] |
| *Oreochromis niloticus* ♀ × *O. aureus* ♂ | Actinopteri | Cichliformes | Cichlidae | 454 Sequencer F.L.X. | Amplicon: V6–V8 | [112] |
| *Oreochromis niloticus* ♀ × *O. aureus* ♂ | Actinopteri | Cichliformes | Cichlidae | Unknown | Amplicon: V4 | [113] |
| *Tachysurus fulvidraco* ♀ × *Pseudobagrus vachellii* ♂ | Actinopteri | Siluriformes | Bagridae | Illumina MiSeq | Amplicon | [114] |
| *Epinephelus fuscoguttatus* ♀ × *E. lanceolatus* ♂ | Actinopteri | Perciformes | Serranidae | Illumina | Amplicon: V3–V4 | [115] |
| *Acipenser baerii* × *A. schrenckii* | Actinopteri | Acipenseriformes | Acipenseridae | Illumina HiSeq | Amplicon: V3–V4 | [116] |
| *Epinephelus moara* ♀ × *E. lanceolatus* ♂ | Actinopteri | Perciformes | Serranidae | Illumina HiSeq | Amplicon: V3–V4 | [117] |
| *Epinephelus fuscoguttatus* ♀ × *E. lanceolatus* ♂ | Actinopteri | Perciformes | Serranidae | Illumina MiSeq | Amplicon: V3–V4 | [118] |
| *Acipenser baerii* × *A. schrenckii* | Actinopteri | Acipenseriformes | Acipenseridae | Illumina MiSeq | Amplicon: V3–V4 | [119] |
| *Epinephelus fuscoguttatus* ♀ × *E. lanceolatus* ♂ | Actinopteri | Perciformes | Serranidae | Illumina HiSeq | Amplicon: V3–V4 | [120] |
| *Epinephelus fuscoguttatus* ♀ × *E. lanceolatus* ♂ | Actinopteri | Perciformes | Serranidae | Illumina HiSeq | Amplicon | [121] |
| *Epinephelus fuscoguttatus* ♀ × *E. lanceolatus* ♂ | Actinopteri | Perciformes | Serranidae | Unknown | Amplicon | [122] |
| *Epinephelus fuscoguttatus* ♀ × *E. lanceolatus* ♂ | Actinopteri | Perciformes | Serranidae | Illumina HiSeq | Amplicon | [123] |
| *Acipenser baerii* × *A. schrenckii* | Actinopteri | Acipenseriformes | Acipenseridae | Illumina HiSeq | Amplicon: V3–V4 | [124] |

Unknown, the information is unclear in the reference.

### 2.1.1. Antibiotics

Infectious diseases caused by various pathogens have severely harmed the health of aquatic organisms around the world [125]. Antibiotics have been widely used as feed supplements to treat intestinal diseases in fish and have become indispensable in human health [33,126,127]. A short-term (6 days) dietary antibiotic mixture (vancomycin, neomycin sulfate, and metronidazole) can improve the lipid metabolism in hybrid groupers (*Epinephelus fuscoguttatus* ♀ × *E. lanceolatus* ♂) fed medium- and high-lipid diets. However, antibiotic treatments can also strongly alter intestinal microbiota by reducing the relative abundance and diversity of hybrid grouper gut microbiota, resulting in a significant increase in the proportion of Bacteroidetes and a decrease in the proportion of Firmicutes [93]. Long-term antibiotic supplementation can cause several side effects on fish health [127–129]. Presently, the pollution and spread of antibiotic-resistant genes caused by the long-term abuse of antibiotics have become a global problem [130]. Recently, probiotics and prebiotics are an emerging strategic approach for sustainable aquaculture, as they do not cause environmental pollution or public health hazards [51,131,132].

### 2.1.2. Probiotics

Probiotics are beneficial microorganisms that can modulate intestinal microbial composition and improve the host health status [133,134]. Probiotics are commonly used in the aquaculture industry as feed or water additives [20]. The essential probiotic microorganisms employed in aquaculture are lactic acid bacteria (LAB) species [135,136] and *Bacillus spp*. [52,137]. The other general probiotic species used in fish are *Saccharomyces, Clostridium, Enterococcus, Shewanella, Leuconostoc, Lactococcus, Carnobacterium*, and *Aeromonas* [20]. Fish are vulnerable to various pathogenic microorganisms, and innate immunity provides an initial line of defense [138]. The addition of probiotics to the diet plays a vital role in stimulating fish immune responses, and further promotes the innate and adaptive immune system [139]. For an example, *Bacillus subtilis* strain 7k, isolated from the gastrointestinal tract of hybrid hulong grouper (*Epinephelus fuscoguttatus* × *E. lanceolatus*), could be used in grouper culture to stimulate growth, enhance immunity and promote health in the fishes [94]. Studies reveal that *O. mykiss* fed different types of probiotics increased the expression of the TGF-β gene, which regulates fish immunity [140–142]. TGF-β levels increased in juvenile hybrid tilapia (*O. niloticus* ♀ × *Oreochromis aureus* ♂), after consuming a diet supplemented with *Bacillus subtilis* C-3102 [95], and the same occurred in Koi carp (*Cyprinus carpio*) [143]. HWF™ is a paraprobiotic and postbiotic supplementary diet using inactive and beneficial bacteria, and is considered an efficient therapeutic agent in fish. Feeding hybrid sturgeons (*Acipenser baerii* × *Acipensers chrenckii*) with HWF™ improved their growth and immunity by changing the composition and diversity of the gut bacteria, developing their healthy gut microbiota [96].

### 2.1.3. Prebiotics

Prebiotics are an innovative strategy, providing a dietary supplement to improve growth development and the immune system by regulating gut microbiota [144]. Prebiotics are generally non-digestible oligosaccharides added to fish feed as dietary components to promote the proliferation of specific beneficial microorganisms in the intestine and, thus, enhance host health [145]. Previous research has shown that prebiotics can decrease the adherence and colonization of pathogenic microorganisms in the intestinal tract to improve the general immunity of the host by increasing the number of lactic acid bacteria, especially *Bifidobacterium* [20,146,147]. Fructo-oligosaccharides, galactooligosaccharides, mannan-oligosaccharides (MOS), xylooligosaccharides (XOS), inulin, lactulose, and lactosucrose are common prebiotics used in various animals, including humans [20]. The level of gut lactic acid bacteria was significantly increased in hybrid catfish (*Pangasianodon gigas* × *Pangasianodon hypophthalmus*) fed with diets containing 0.6% xylooligosaccharides (XOS) [97]. In addition, several studies have reported that inulin, fructooligosaccharides, xylooligosaccharides, galactooligosaccharides, and arabinoxylan-oligosaccharides can affect growth development, immune health, and the composition and/or diversity of the gut microbiota in different fish species [53,97,148–151]. Indeed, many researchers have reported the effect of prebiotics on the gut microbiota in fish, such as grass carp [54], Siberian sturgeon [53], Nile tilapia [55], and European sea bass [152].

The prebiotic Grobiotic™AE and dietary brewer's yeast can improve the growth performance, immune response, and resistance to *Streptococcus iniae* infection in hybrid striped bass (*Morone chrysops* × *M. saxatilis*) [153]. Dietary supplementation of 4% ESTAQUA® yeast culture (YC) for hybrid grouper (*Epinephelus fuscoguttatus* ♀ × *E. lanceolatus* ♂) could improve the alpha diversity of gut microbiota, growth performance and serum immune responses against *V. harveyi* attacks [98]. N.B.T. is an excellent indicator of the health status and/or immunization effectiveness in fish [56]. Supplementing the diet with raffinose in hybrid sturgeons (*Acipenser baeri* Brandt ♀× *A. schrenckii* Brandt ♂) improved the growth performance and intestinal morphology, modifying the gut microbiota composition and increasing the level of N.B.T. activity [99]. Chitosan oligosaccharide (COS) is a new prebiotic, dietary COS supplementation, which improves the growth performance and health status of *Scopthalmus maximus* [154], *Cyprinus carpio koi* [155], and *Oncorhynchus mykiss* [156]. Dietary COS supplementation improved the intestinal health and immune responses of hybrid groupers (*Epinephelus fuscoguttatus* ♀ × *E. lanceolatus* ♂) when fed a low-fish meal diet [100].

It is worth noting that prebiotic supplementation is only beneficial when a moderate volume is provided; prebiotics at a high concentration can be harmful to the host. Excessive prebiotics may cause an imbalance in the gut microenvironment, which decreases the digestive capacity in fish intestines. A previous study revealed that a high concentration of inulin could damage the enterocytes of *Salvelinus alpinus* [157]. This may explain why 0.4–0.6% COS supplementation was optimum in hybrid groupers [100].

#### 2.1.4. Fishmeal Protein Substitutes

Fishmeal (F.M.) is the most widely utilized high-quality protein source in aquatic feed and has many advantages [158]. However, fishmeal production cannot meet the growing needs of the aquaculture industry due to its rapid development, which is causing a severe impediment to industry development [101,159]. Therefore, using plant proteins is an innovative solution for sustainable aquaculture [160,161].

Cottonseed protein concentrate (CPC) is a new experimental fishmeal (FM) replacement [162]. However, fishmeal replaced with CPC in an inappropriate proportion can have adverse effects on the intestinal health of groupers and leads to intestinal inflammation [163]. A study on pearl gentian groupers (*Epinephelus fuscoguttatus* ♀ × *Epinephelus lanceolate* ♂) revealed that 24% CPC was considered the most appropriate volume for F.M. replacement and growth performance, digestive proteinase activity, intestinal morphology, and intestinal microflora in the pearl gentian grouper reached maximum levels with 24% CPC replacement levels. Subsequently, many physiological parameters are reduced with increasing CPC replacement levels [101]. The substitution of FM with peanut meal (PNM) of up to 50% or CPC up to 60% obviously changed the intestinal microbiota of juvenile hybrid groupers (*E. fuscoguttatus* ♀ × *E. lanceolatus* ♂), which increased intestinal pathogenic bacteria and decreased intestinal beneficial bacteria [102,103]. Similarly, replacing FM with peptides from swine blood (PSB) up to 75% could reduce growth performance for hybrid groupers (*Epinephelus fuscoguttatus* ♀ × *E. lanceolatus* ♂), and increase the abundance of the potentially pathogenic *Pseudomonas* and *Arcobacter* in the gut [104].

Another fishmeal replacement protein is soybean meal (SBM). SBM has been widely considered an inexpensive FM replacement [164]. Nevertheless, anti-nutritional factors in SBM can negatively affect the intestinal morphology of fish [165]. Research reveals that bioprocesses (such as soybean meal ingredients) can reduce the intestinal microorganism diversity in hybrid striped bass (*Morone chrysops* × *M. saxatilis*) [105]. It is challenging to find a suitable fish meal substitute for various fish, and protein substitutes have excellent potential and are important future research topics.

### 2.2. Hybrid Speciation and Gut Microbiome

No living organisms exist in isolation from the microbial world, and microbial symbiosis and speciation profoundly shape the biodiversity composition. Animal hosts and microbiomes are closely interconnected and interact over long evolutionary timeframes. They can even be regarded as a unique biological entity-holobiont and include their entire genome, called the hologenome [1]. Diverse and complex interactions exist between hosts and microorganisms. Microorganisms play essential roles in host physiology, health, and survival. Microorganisms can even alter host reproduction [166], resulting in host embryo death [167–170] and affect the host gametic integrity and embryonic viability, which may be closely related to the formation of new species [45,171]. The microorganisms and their interactions with hosts are potentially important factors in stimulating the formation of new species [172].

Species are reproductively isolated groups composed of potentially interbreeding individuals, and hybrids can suffer from post-mating isolation barriers, such as sterility and/or unviability [173]. The composition and functional effects of animal microbiota are closely related to host evolution, and the survival rate and performance of microorganisms can be reduced when interspecific microbiota transplantation occurs between closely related and different host species pairs. The microbiome compositional relationships (i.e., beta diversity) reflect the evolutionary relationships of the host species [173,174]. Thus, natural selection can drive phylosymbiotic changes within the parental species, which may lead to the evolution of deleterious interactions between hybrids and their microbiomes [173].

Based on the holobiont concept, host-genome–microbiome associations and their role in host adaptability demonstrate that microorganisms may participate in the process of speciation, and symbiotic microorganisms may hinder speciation through isolation, including behavioral isolation, geographical isolation, and reproductive isolation [45]. Microbial symbionts can add new functional genes to the host genome, which assists the host in expanding its dietary niche and obtaining new nutritional opportunities. Unfortunately, hybridization can inhibit symbiotic relationships by destroying the vertical transmission of some microorganisms between the host parents and offspring, which are hybridization disadvantages and hinder species formation, as observed in *Acyrthosiphon pisum* [175], *Sitophilus* [176] and the family *Plataspidae* [177]. In hybrid species, microorganisms can hinder speciation by assisting reproductive isolation. *Wolbachia* is a bacterium that widely exists in the reproductive system of arthropods and may cause hybrid male sterility in *Drosophila paulistorum* [178]. In the two-spotted mite (*Tetranychus urticae*), *Wolbachia* can also cause cytoplasmic incompatibility (CI) in the F1 generation and F2 male offspring deaths from the surviving F1 females in the CI cross [178]. Similarly, different CI Wolbachia in *Nasonia* wasp species can cause high levels of F1 hybrid lethality and the reproductive isolation induced by CI has evolutionary potential in the early stages of the speciation process [179,180].

Similarly, a close interaction exists between the gut microbiome and host, and plays an important role in the speciation of hybrid species. For example, the host gut microbiome may hinder the formation of new species by participating in the death of hybrids in *Nasonia* wasp species [181]. Vertebrates are a vital group for interactions in reproductive isolation and speciation research. Alterations in gut microbiota communities and increases in gut pathology exist in hybrid mice (*Mus musculus* × *Mus domesticus*) [46]. The gut microbiome does not always play negative roles in hybrid species. For example, the hybrid offspring of sika deer (*Cervus nippon*) and elk (*Cervus elaphus*) harbor a high abundance of *Acetitomaculum* bacterial species, which may assist in the absorption and metabolism of nutrients [182,183]. A similar phenomenon was identified in the hybrid offspring of ponies and donkeys, which render a completely different gut microbiota from their parents [184].

In the gut microbiome in hybrid fish research, differences in the gut microbiome between hybrid offspring and parents have been observed. In lake whitefish (*Coregonus clupeaformis*), the gut microbiome is significantly different between the F1 hybrids and their parents, especially the abundance difference between Firmicutes and Proteobacteria [106]. The research also found the interactions of the host-microbiota-environment demonstrated three different evolutionary paths in the gut microbiome [106]. Similarly, the gut characteristics of hybrid fish from herbivorous blunt snout bream (*Megalobrama amblycephala*) and carnivorous topmouth culter (*Culter alburnus*) differ from their parents. The microbial community in the hybrid topmouth culters was markedly distinct from their parents, and varied in the cellulose content in the gut [39]. One study found that the evolutionary characteristics of hybrid fish progeny from *Megalobrama amblycephala* and *Culter alburnus* may be manifested in dietary adaptation and choice; the interactions between gut microbiota and host genetics contributed to hybrid fishes adapting to herbivorous diets more than carnivorous diets [185]. Compared to the parents, the hybrid offspring of two invasive North American carp, *Hypophthalmichthys nobilis* and *Hypophthalmichthys molitrix*, harbor different gut microbiome compositions and display higher alpha diversity than their parents [107].

### 2.3. The Differences in the Gut Microbiome of the Hybrid Fish and Their Hybrids Origin

There are still few studies directly comparing gut microbiome between parental and hybrid progeny. However, it has been shown that existing differences in intestinal microbiota between captive parents and hybrid fishes' offspring exist under a controlled environment [106]. There is no doubt that diet will affect the gut microbiome composition and growth performance of the host, and under the same dietary conditions (Artemia and mixed diet), the taxonomic composition of transient gut microbiota between both whitefish (*Coregonus clupeaformis*) parental species and their reciprocal hybrids showed a slight pattern of differentiation, which, within the Artemia diet group, meant a higher abundance for Firmicutes, but lower for Proteobacteria, was observed in hybrids in comparison with their parents' whitefish, while the opposite result was found in the mixed diet group, where there was a higher abundance of Proteobacteria but it was lower for Firmicutes. In addition, in the abundance composition of some specific bacterial genera, the two reciprocal hybrids, and their parents also showed the opposite pattern, that F1 D♀N♂has more specific bacterial genera than its parents, while F1 N♀D♂with fewer specific bacterial genera than its parents. In the hybridization experiment between whitefish and omul (*Coregonus migratorius*), the researchers found that the hybrid progeny had a lower alpha diversity (e.g., Shannon index) in hindgut microbiota than the parents [186].

Host genetics can strongly affect the gut microbial composition of the hybrid offspring [39]. Compared with carnivorous topmouth culter (*Culter alburnus*, TC) parents, the gut microbiome structure of their two-hybrid progenies is more similar to that of herbivorous blunt snout bream (*Megalobrama amblycephala*, BSB) parents, as the alpha diversity of the two types of hybrids and BSB parent is higher than that of a TC parent, as well as beta diversity analysis, which also showed that there was no significant difference between the two hybrids and the BSB parent. Interestingly, in the composition of gut microbiota, Fusobacteria and Proteobacteria are the most abundant intestinal flora in hybrid fishes, and the proportion of Fusobacteria and Proteobacteria in hybrid offspring is similar to the BSB parent but significantly different from the TC parent. Again, the shared bacterial taxa at the phylum level showed different results; the hybrids of the two types share higher proportions of gut bacterial communities with the BSB parent than the TC parent.

Recently, our study reported a direct comparison of the similarities and differences in gut microbiome (composition and potential function) among bighead carps (*Hypophthalmichthys nobilis*, B), silver carps (*Hypophthalmichthys molitr*, S) and their hybrid offspring (SB and BS) in ponding experiments [107]. The hybrid gut microbiome displays the admixed pattern at the community level and harbors the relatively high alpha diversity (e.g., phylogenetic diversity). For example, the hybrid fish had intermediate abundances of Cyanobacteria and Bacteroidetes in the foregut, while Fusobacteria are significantly enriched in parents in the hindgut. Moreover, the hybrid gut microbiome's predicted function shows the enrichment in the genes coding for putative enzymes involved the diet utilization, which suggests the potential benefits to their local adaptation.

### *2.4. Gut Microbiome Might Promote Ecological Invasion by Hybrid Fish*

Gut microbiota can enhance the adaptability of the host to the environment and improve the successful invasion rate of some invasive species [187]. For invasion success, the species requires a dispersal ability, environmental tolerance, phenotypic plasticity, and associated epigenetics [188,189]. Host shifts can lead to phytophagous insects becoming invasive species [190]. It has already been demonstrated that the gut microbiome plays a vital role in phytophagous insect invasion success [191], and gut bacteria can assist in the successful invasion of insect species by regulating epigenetic factors related to the host [192]. Similarly, some biological mechanisms can enhance the success rate of invasive species, such as genetic diversity [193], reproductive rate [194], food resources [195], and hybridization [44,196].

Therefore, there are complex and close relationships between hybridization, the gut microbiome, and bio-invasion. Bighead carp and silver carp are invasive species, characterized by various hybridization in the Mississippi River Basin [107]. There is higher alpha diversity in the foregut microbiota in the hybrid offspring, and an increasing discrepancy also occurs between the foregut and hindgut. Similarly, the hybrids had a higher proportion of putative genes coding for putative enzymes related to the digestion of filter-feeding phytoplankton (Cyanobacteria, cellulose, and chitin) than their parents. The improved putative enzymes could encourage the utilization of new food resources by the gut microbiota and, therefore, improve survival, environmental adaptation, and invasion by hybrid fish. Therefore, the gut microbiome and host genome may synergistically promote bigheaded carp invasion in the United States [107].

## 3. The Potential Impact of Environmental Microbiota

The current research focuses on fish, not hybrid fish. However, environmental microbiota impacts may also occur in hybrid fish.

### *3.1. Habitat Environmental Microbiome Shapes the Early Gut Microbiome of Juvenile Fish*

The main determinant of fish gut microbiota is the natural environment, and fish intestinal microbiota symbionts are generally obtained from the environment [197] by neutral processes, such as drift and diffusion, which produce most of the microbial diversity [198]. The microorganisms transmitted from the environment to the fish intestine are mainly derived from two paths: the foodborne microorganisms carried by prey and the microorganisms in the water, and most of the environmental microbiota remain temporarily in the fish gut [199]. In most fish species, the ontogeny and colonization of gut microbiota in the early stages of life rely on the horizontal transmission of environmental microbiota [200]. Juvenile zebrafish (*Danio rerio*) acquire gut symbiotic bacteria from the water environment after hatching, which may promote the development and function of their intestines [201]. Similar patterns are observed in wild Atlantic salmon (*Salmo salar*), discus (*Symphysodon aequifasciata*) [200], grass carp (*Ctenopharyngodon idellus*), Mucha perch (*Siniperca chuatsi*), and southern catfish (*Silurus meridionalis*). The composition of the gut microbiota community of juvenile fish was more similar to the habitat water environment than the adults [197,200]. However, fish gut microbiota often differ from their surrounding environment after becoming adults [202]. Therefore, environmental microorganisms play an important role in shaping the gut microbiota in the early juvenile fish stages and, as fish mature, the environmental factors are less influential because the gut microbiota gradually differentiate from the environmental microbiota, showing individual variations [197,203].

### *3.2. Do Fish Specifically Select Proteus from the Water Environment?*

The gut microbiota of fish are mainly Proteobacteria and Firmicutes, whereas amphibians, reptiles, birds, and mammals contain mainly Firmicutes and Bacteroidetes. The excessive reproduction and presence of *Proteus* may be a sign of ecological imbalance in the gut microbial community of mammals [204], as many symbiotic *Proteus* bacteria can translate into pathogens, and infect and promote inflammation in the host under specific conditions. Many studies have demonstrated that, regardless of the fish living environment, the gut microbiome is composed of a common core microbiome [205]. Major environmental microorganisms are rarely observed in fish intestines [49]. *Proteus* dominate the gut microbiota of most fish species [206]. The *Proteus* abundance can increase with the growth and nutritional level of the fish (from herbivorous to carnivorous). Conversely, the abundance of Firmicutes usually decreases with increasing nutritional levels [202]. The gut microbiome not only reflects the microorganisms in its surrounding environment but also characterizes the specific selection of the environmental microbiome by the host in grass carp (*Ctenopharyngodon idellus*) [207], silver Prussian carp (*Carassius auratus gibelio*) [208], and zebrafish (*Danio rerio*) [209]. Notably, the higher proportion of *Proteus* in the fish intestines indicates the fish host has specifically selected *Proteus* from the habitat water or *Proteus* has outperformed the other environmental bacterial taxa in the water. This discrepancy is an urgent problem needing to be explored [16].

## 4. Future Perspectives

The gut microbiome can promote the successful ecological invasion of hybrid fish, which makes them occupy favorable ecological niches and further improves the potential for population expansion. Following Darwin's theory of evolution, this process greatly improves the potential of hybrid fish to evolve into new species in the future (Figure 1). The gut microbiome plays a role in speciation, but its degree of impact remains unclear. Furthermore, the high genomic similarity between bighead and silver carp, and an over 90% embryonic viability in all crosses, indicate that interspecific hybridization between the carps might have promoted their range expansion [44]. In the future, the role of the gut microbiome in population expansion of hybrid species should not be ignored. It is highly significant for us to better combine the genome and metagenome to improve our understanding of the ecological problems of hybrid fish. The fish gut flora and fecal materials discharged into the water may reflect their diet preferences, physiological behaviors, and presence in the river [210], allowing gut microbiota to potentially monitor fish invasion and population expansion, which is an important research issue in fish conservation and management in the future (Figure 1).

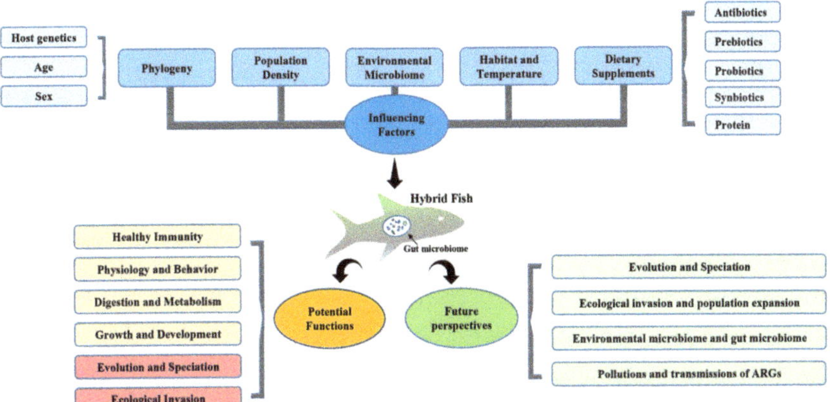


**Figure 1.** Major research progress and future perspectives on hybrid fish gut microbiome.

For a long time, the source of gut microbiota has been an attractive research topic. Environmental microbiome transmission plays an important role in animal gut microbiota, and the differences between terrestrial and aquatic environments cause the gut of aquatic animals to be very different from that of terrestrial organisms, including fish and aquatic mammals. Research shows 13% of the gut microbiota of threespine stickleback (*Gasterosteus aculeatus*) comes from the surrounding water environment and 73% from prey [199]. In addition, in most fish species, the ontogeny and colonization by gut microbiota in the early stages of life mainly occur through the horizontal transmission of environmental microbiota [200]. Juvenile zebrafish (*Danio rerio*) acquire gut symbiotic bacteria from the water environment after hatching, potentially promoting the development and function of the intestines [201]. Similar patterns are observed in wild Atlantic salmon (*Salmo salar*), discus (*Symphysodon aequifasciata*) [200], grass carp (*Ctenopharyngodon idellus*), Mucha perch (*Siniperca chuatsi*), and southern catfish (*Silurus meridionalis*), and the composition of the gut microbiota community of juvenile fish was more similar to the habitat water than the adults [197,200]. In addition, different fish tissue types, such as skin, gills, and intestines, may also be the main determinants of microbiota diversity and composition [48]. Successful hybrid fish invasion depends on the relationships and interactions between an individual's characteristics (age and gender), gut microbiome, environmental microbiome, and post-mating reproductive isolation, associated with environmental microbial transmission. Future research is required to assist our understanding of these interactions (Figure 1). In addition, the aquatic environment can become a reservoir of antibiotic-resistant genes (ARGs), providing an ideal path for the acquisition and dissemination of ARGs [211]. Aquatic animals, such as fish, are direct witnesses and victims of ARG-water pollution. Therefore, wild fish can be recipients and disseminators of ARGs in aquatic environments [130]. At present, there are few studies assessing ARG pollution and transmission in wild hybrid fish, providing great research potential in the future (Figure 1).

**Author Contributions:** L.Z. conceived the ideas and methodology. X.C., Q.Z. (Qinrong Zhang), Q.Z. (Qunde Zhang), Y.Z., H.C. and G.L. collected the data. X.C. and L.Z. wrote the manuscript. All authors have read and agreed to the published version of the manuscript.

**Funding:** Financial support was provided by the Priority Academic Program Development of Jiangsu Higher Education Institutions (PAPD).

**Institutional Review Board Statement:** Not applicable.

**Informed Consent Statement:** Not applicable.

**Acknowledgments:** We thank the team members for the help during literature collection.

**Conflicts of Interest:** The authors declare no conflict of interest.

## References

1. Bordenstein, S.R.; Theis, K.R. Host Biology in Light of the Microbiome: Ten Principles of Holobionts and Hologenomes. *PLoS Biol.* **2015**, *13*, e1002226. [CrossRef] [PubMed]
2. Rosenberg, E.; Zilber-Rosenberg, I. The hologenome concept of evolution after 10 years. *Microbiome* **2018**, *6*, 78. [CrossRef] [PubMed]
3. McFall-Ngai, M.; Hadfield, M.G.; Bosch, T.C.G.; Carey, H.V.; Domazet-Lošo, T.; Douglas, A.E.; Dubilier, N.; Eberl, G.; Fukami, T.; Gilbert, S.F.; et al. Animals in a bacterial world, a new imperative for the life sciences. *Proc. Natl. Acad. Sci. USA* **2013**, *110*, 3229–3236. [CrossRef] [PubMed]
4. Carlson, A.; Xia, K.; Azcarate-Peril, M.A.; Goldman, B.D.; Ahn, M.; Styner, M.A.; Thompson, A.L.; Geng, X.; Gilmore, J.H.; Knickmeyer, R.C. Infant Gut Microbiome Associated With Cognitive Development. *Biol. Psychiatry* **2018**, *83*, 148–159. [CrossRef] [PubMed]
5. Archie, E.A.; Tung, J. Social behavior and the microbiome. *Curr. Opin. Behav. Sci.* **2015**, *6*, 28–34. [CrossRef]
6. Levy, M.; Blacher, E.; Elinav, E. Microbiome, metabolites and host immunity. *Curr. Opin. Microbiol.* **2017**, *35*, 8–15. [CrossRef]
7. Hollister, E.B.; Gao, C.; Versalovic, J. Compositional and functional features of the gastrointestinal microbiome and their effects on human health. *Gastroenterology* **2014**, *146*, 1449–1458. [CrossRef]
8. Kers, J.G.; Velkers, F.; Fischer, E.A.J.; Hermes, G.D.A.; Lamot, D.M.; Stegeman, J.A.; Smidt, H. Take care of the environment: Housing conditions affect the interplay of nutritional interventions and intestinal microbiota in broiler chickens. *Anim. Microbiome* **2019**, *1*, 10. [CrossRef]
9. Youngblut, N.D.; Reischer, G.H.; Walters, W.; Schuster, N.; Walzer, C.; Stalder, G.; Ley, R.E.; Farnleitner, A.H. Host diet and evolutionary history explain different aspects of gut microbiome diversity among vertebrate clades. *Nat. Commun.* **2019**, *10*, 2200. [CrossRef]
10. Ley, R.E.; Lozupone, C.A.; Hamady, M.; Knight, R.; Gordon, J.I. Worlds within worlds: Evolution of the vertebrate gut microbiota. *Nat. Rev. Microbiol.* **2008**, *6*, 776–788. [CrossRef]
11. Sharpton, T.J. Role of the Gut Microbiome in Vertebrate Evolution. *Msystems* **2018**, *3*, e00174-17. [CrossRef] [PubMed]

12. Zhu, L.; Zhang, Y.; Cui, X.; Zhu, Y.; Dai, Q.; Chen, H.; Liu, G.; Yao, R.; Yang, Z. Host Bias in Diet-Source Microbiome Transmission in Wild Cohabitating Herbivores: New Knowledge for the Evolution of Herbivory and Plant Defense. *Microbiol. Spectr.* **2021**, *9*, e0075621. [CrossRef] [PubMed]
13. Wei, F.; Wu, Q.; Hu, Y.; Huang, G.; Nie, Y.; Yan, L. Conservation metagenomics: A new branch of conservation biology. *Sci. China Life Sci.* **2019**, *62*, 168–178. [CrossRef] [PubMed]
14. Ley, R.E.; Hamady, M.; Lozupone, C.; Turnbaugh, P.J.; Ramey, R.R.; Bircher, J.S.; Schlegel, M.L.; Tucker, T.A.; Schrenzel, M.D.; Knight, R.; et al. Evolution of mammals and their gut microbes. *Science* **2008**, *320*, 1647–1651. [CrossRef] [PubMed]
15. Johny, T.K.; Puthusseri, R.M.; Bhat, S.G. A primer on metagenomics and next-generation sequencing in fish gut microbiome research. *Aquac. Res.* **2021**, *52*, 4574–4600. [CrossRef]
16. Kim, P.S.; Shin, N.R.; Lee, J.B.; Kim, M.S.; Whon, T.W.; Hyun, D.W.; Yun, J.H.; Jung, M.J.; Kim, J.Y.; Bae, J.W. Host habitat is the major determinant of the gut microbiome of fish. *Microbiome* **2021**, *9*, 166. [CrossRef]
17. Austin, B. The bacterial microflora of fish. *Sci. World J.* **2002**, *2*, 558–572. [CrossRef]
18. Spanggaard, B.; Huber, I.; Nielsen, J.; Nielsen, T.; Appel, K.F.; Gram, L.J.A. The microflora of rainbow trout intestine: A comparison of traditional and molecular identification. *Aquaculture* **2000**, *182*, 1–15. [CrossRef]
19. Pond, M.J.; Stone, D.M.; Alderman, D.J. Comparison of conventional and molecular techniques to investigate the intestinal microflora of rainbow trout (*Oncorhynchus mykiss*). *Aquaculture* **2006**, *261*, 194–203. [CrossRef]
20. Nayak, S.K. Role of gastrointestinal microbiota in fish. *Aquac. Res.* **2010**, *41*, 1553–1573. [CrossRef]
21. Molinari, L.M.; Scoaris, D.; Pedroso, R.B.; Bittencourt, N.D.L.R.; Filho, B. Bacterial microflora in the gastrointestinal tract of Nile tilapia, Oreochromis niloticus, cultured in a semi-intensive system. *Acta Sci. Biol. Sci.* **2003**, *25*, 267–271.
22. Saha, S.; Roy, R.N.; Sen, S.K.; Ray, A.K. Characterization of cellulase-producing bacteria from the digestive tract of tilapia, *Oreochromis mossambica* (Peters) and grass carp, *Ctenopharyngodon idella* (Valenciennes). *Aquac. Res.* **2010**, *37*, 380–388. [CrossRef]
23. Bairagi, A.; Ghosh, K.S.; Sen, S.K.; Ray, A.K. Enzyme producing bacterial flora isolated from fish digestive tracts. *Aquac. Int.* **2002**, *10*, 109–121. [CrossRef]
24. Clements, K.D. *Gastrointestinal Microbiology: Volume 1 Gastrointestinal Ecosystems and Fermentations*; Mackie, B.A., White, B., Eds.; Springer: Boston, MA, USA, 1997; pp. 156–198.
25. Carda-Diéguez, M.; Mira, A.; Fouz, B. Pyrosequencing survey of intestinal microbiota diversity in cultured sea bass (Dicentrarchus labrax) fed functional diets. *FEMS Microbiol. Ecol.* **2014**, *87*, 451–459. [CrossRef]
26. Desai, A.R.; Links, M.G.; Collins, S.A.; Mansfield, G.S.; Hill, J. Effects of plant-based diets on the distal gut microbiome of rainbow trout (*Oncorhynchus mykiss*). *Aquaculture* **2012**, *350–353*, 134–142. [CrossRef]
27. Ghanbari, M.W.; Kneifel, J.A.; Domig, K.J. A new view of the fish gut microbiome: Advances from next-generation sequencing. *Aquaculture* **2015**, *448*, 464–475. [CrossRef]
28. Givens, C.E.; Ransom, B.; Bano, N.; Hollibaugh, J. Comparison of the gut microbiomes of 12 bony fish and 3 shark species. *Mar. Ecol. Prog. Ser.* **2015**, *518*, 209–223. [CrossRef]
29. Ringø, E.; Sperstad, S.; Myklebust, R.; Refstie, S.; Krogdahl, Å. Characterisation of the microbiota associated with intestine of Atlantic cod (*Gadus morhua* L.): The effect of fish meal, standard soybean meal and a bioprocessed soybean meal. *Aquaculture* **2006**, *261*, 829–841. [CrossRef]
30. Butt, R.L.; Volkoff, H. Gut Microbiota and Energy Homeostasis in Fish. *Front. Endocrinol.* **2019**, *10*, 9. [CrossRef]
31. Handelsman, J.; Rondon, M.R.; Brady, S.F.; Clardy, J.; Goodman, R.M. Molecular biological access to the chemistry of unknown soil microbes: A new frontier for natural products. *Chem. Biol.* **1998**, *5*, R245–R249. [CrossRef]
32. Riesenfeld, C.S.; Schloss, P.D.; Handelsman, J. Metagenomics: Genomic analysis of microbial communities. *Annu. Rev. Genet.* **2004**, *38*, 525–552. [CrossRef] [PubMed]
33. Yukgehnaish, K.; Kumar, P.; Sivachandran, P.; Marimuthu, K.; Arshad, A.; Paray, B.A.; Arockiaraj, J. Gut microbiota metagenomics in aquaculture: Factors influencing gut microbiome and its physiological role in fish. *Rev. Aquac.* **2020**, *12*, 1903–1927. [CrossRef]
34. Méndez-Pérez, R.; García-López, R.; Bautista-López, J.S.; Vázquez-Castellanos, J.F.; Peña-Marín, E.S.; Martínez-García, R.; Domínguez-Rodríguez, V.I.; Adams-Schroeder, R.H.; Baltierra-Trejo, E.; Valdés, C.M.; et al. Gut Microbiome Analysis In Adult Tropical Gars (*Atractosteus tropicus*). *bioRxiv* **2019**. [CrossRef]
35. Zhang, G.; Li, J.; Zhang, J.; Liang, X.; Zhang, X.; Wang, T.; Yin, S. Integrated Analysis of Transcriptomic, miRNA and Proteomic Changes of a Novel Hybrid Yellow Catfish Uncovers Key Roles for miRNAs in Heterosis. *Mol. Cell. Proteom. MCP* **2019**, *18*, 1437–1453. [CrossRef]
36. Liu, S.; Luo, J.; Chai, J.; Ren, L.; Zhou, Y.; Huang, F.; Liu, X.; Chen, Y.; Zhang, C.; Tao, M.; et al. Genomic incompatibilities in the diploid and tetraploid offspring of the goldfish × common carp cross. *Proc. Natl. Acad. Sci. USA* **2016**, *113*, 1327–1332. [CrossRef]
37. Zhang, G.; Tao, P.; Chen, J.; Wang, R.; Zang, X.; Yin, S. The complete mitochondrial genome of the hybrid of *Pelteobagrus fulvidraco* (♀) × *Pelteobagrus vachelli* (♂). *Mitochondrial DNA Part A DNA Mapp. Seq. Anal.* **2016**, *27*, 4191–4192. [CrossRef]
38. Gui, J. Scientific frontiers and hot issues in hydrobiology. *Chin. Sci. Bull.* **2015**, *60*, 2051–2057. [CrossRef]
39. Li, W.; Liu, J.; Tan, H.; Yang, C.; Ren, L.; Liu, Q.; Wang, S.; Hu, F.; Xiao, J.; Zhao, R.; et al. Genetic Effects on the Gut Microbiota Assemblages of Hybrid Fish From Parents With Different Feeding Habits. *Front. Microbiol.* **2018**, *9*, 2972. [CrossRef]
40. Li, X.; Yu, Y.; Li, C.; Yan, Q. Comparative study on the gut microbiotas of four economically important Asian carp species. *Sci. China Life Sci.* **2018**, *61*, 696–705. [CrossRef]

41. Zhong, X.; Wang, X.; Zhou, T.; Jin, Y.; Tan, S.; Jiang, C.; Geng, X.; Li, N.; Shi, H.; Zeng, Q.; et al. Genome-Wide Association Study Reveals Multiple Novel QTL Associated with Low Oxygen Tolerance in Hybrid Catfish. *Mar. Biotechnol.* **2017**, *19*, 379–390. [CrossRef]
42. McGinnity, P.; Prodöhl, P.; Ferguson, A.; Hynes, R.; Maoiléidigh, N.O.; Baker, N.; Cotter, D.; O'Hea, B.; Cooke, D.; Rogan, G.; et al. Fitness reduction and potential extinction of wild populations of Atlantic salmon, Salmo salar, as a result of interactions with escaped farm salmon. *Proc. Biol. Sci.* **2003**, *270*, 2443–2450. [CrossRef] [PubMed]
43. Solberg, M.F.; Kvamme, B.O.; Nilsen, F.; Glover, K.A. Effects of environmental stress on mRNA expression levels of seven genes related to oxidative stress and growth in Atlantic salmon *Salmo salar* L. of farmed, hybrid and wild origin. *BMC Res. Notes* **2012**, *5*, 672. [CrossRef] [PubMed]
44. Wang, J.; Gaughan, S.; Lamer, J.T.; Deng, C.; Hu, W.; Wachholtz, M.; Qin, S.; Nie, H.; Liao, X.; Ling, Q.; et al. Resolving the genetic paradox of invasions: Preadapted genomes and postintroduction hybridization of bigheaded carps in the Mississippi River Basin. *Evol. Appl.* **2020**, *13*, 263–277. [CrossRef] [PubMed]
45. Brucker, R.M.; Bordenstein, S.R. Speciation by symbiosis. *Trends Ecol. Evol.* **2012**, *27*, 443–451. [CrossRef]
46. Wang, J.; Kalyan, S.; Steck, N.; Turner, L.M.; Harr, B.; Künzel, S.; Vallier, M.; Häsler, R.; Franke, A.; Oberg, H.H.; et al. Analysis of intestinal microbiota in hybrid house mice reveals evolutionary divergence in a vertebrate hologenome. *Nat. Commun.* **2015**, *6*, 6440. [CrossRef]
47. Janson, E.M.; Stireman, J.O., 3rd; Singer, M.S.; Abbot, P. Phytophagous insect-microbe mutualisms and adaptive evolutionary diversification. *Evol. Int. J. Org. Evol.* **2008**, *62*, 997–1012. [CrossRef]
48. Guivier, E.; Martin, J.-F.; Pech, N.; Ungaro, A.; Chappaz, R.; Gilles, A. Microbiota Diversity within and between the Tissues of Two Wild Interbreeding Species. *Microb. Ecol.* **2018**, *75*, 799–810. [CrossRef]
49. Talwar, C.; Nagar, S.; Lal, R.; Negi, R.K. Fish Gut Microbiome: Current Approaches and Future Perspectives. *Indian J. Microbiol.* **2018**, *58*, 397–414. [CrossRef]
50. Tran, N.T.; Wang, G.-T.; Wu, S.-G. A review of intestinal microbes in grass carp *Ctenopharyngodon idellus* (Valenciennes). *Aquac. Res.* **2017**, *48*, 3287–3297. [CrossRef]
51. Tan, H.Y.; Chen, S.W.; Hu, S.Y. Improvements in the growth performance, immunity, disease resistance, and gut microbiota by the probiotic Rummeliibacillus stabekisii in Nile tilapia (*Oreochromis niloticus*). *Fish Shellfish Immunol.* **2019**, *92*, 265–275. [CrossRef]
52. Giatsis, C.; Sipkema, D.; Ramiro-Garcia, J.; Bacanu, G.M.; Abernathy, J.; Verreth, J.; Smidt, H.; Verdegem, M. Probiotic legacy effects on gut microbial assembly in tilapia larvae. *Sci. Rep.* **2016**, *6*, 33965. [CrossRef] [PubMed]
53. Geraylou, Z.; Souffreau, C.; Rurangwa, E.; De Meester, L.; Courtin, C.M.; Delcour, J.A.; Buyse, J.; Ollevier, F. Effects of dietary arabinoxylan-oligosaccharides (AXOS) and endogenous probiotics on the growth performance, non-specific immunity and gut microbiota of juvenile Siberian sturgeon (*Acipenser baerii*). *Fish Shellfish Immunol.* **2013**, *35*, 766–775. [CrossRef] [PubMed]
54. Xiong, J.; Wu, Z.; Zhang, P.; Qu, Y.; Fu, S.; Liu, J.; Chen, X. Effects of dietary xylo-oligosaccharide on intestinal microflora of grass carp(Ctenopharyngodon idellus). *Microbiology* **2011**, *38*, 317–321.
55. Xia, Y.; Lu, M.; Chen, G.; Cao, J.; Gao, F.; Wang, M.; Liu, Z.; Zhang, D.; Zhu, H.; Yi, M. Effects of dietary Lactobacillus rhamnosus JCM1136 and Lactococcus lactis subsp. lactis JCM5805 on the growth, intestinal microbiota, morphology, immune response and disease resistance of juvenile Nile tilapia, Oreochromis niloticus. *Fish Shellfish Immunol.* **2018**, *76*, 368–379. [CrossRef]
56. Anderson, D.P.; Moritomo, T.; de Grooth, R. Neutrophil, glass-adherent, nitroblue tetrazolium assay gives early indication of immunization effectiveness in rainbow trout. *Veter Immunol. Immunopathol.* **1992**, *30*, 419–429. [CrossRef]
57. Li, W.; Shen, T.; Chen, N.; Deng, B.; Fu, L.; Zhou, X. Effects of dietary Bacillus subtilis on digestive enzyme activity and intestinal microflora in grass carp Ctenopharyngodon idellus. *J. Dalian Ocean Univ.* **2012**, *27*, 221–225.
58. Wang, Y. Use of probiotics Bacillus coagulans, Rhodopseudomonas palustris and Lactobacillus acidophilus as growth promoters in grass carp (Ctenopharyngodon idella) fingerlings. *Aquac. Nutr.* **2011**, *17*, E372–E378. [CrossRef]
59. Wu, Z.X.; Feng, X.; Xie, L.L.; Peng, X.Y.; Yuan, J.; Chen, X.X. Effect of probiotic Bacillus subtilis Ch9 for grass carp, Ctenopharyngodon idella (Valenciennes, 1844), on growth performance, digestive enzyme activities and intestinal microflora. *J. Appl. Ichthyol.* **2012**, *28*, 721–727. [CrossRef]
60. Zhou, Y.; Yuan, X.C.; Liang, X.F.; Fang, L.; Li, J.; Guo, X.Z.; Bai, X.L.; He, S. Enhancement of growth and intestinal flora in grass carp: The effect of exogenous cellulase. *Aquaculture* **2013**, *416*, 1–7. [CrossRef]
61. Yuan, X.; Zhou, Y.; Liang, X.F.; Guo, X.; Fang, L.; Li, J.; Liu, L.; Li, B. Effect of dietary glutathione supplementation on the biological value of rapeseed meal to juvenile grass carp, Ctenopharyngodon idellus. *Aquac. Nutr.* **2015**, *21*, 73–84. [CrossRef]
62. Yao, D.; Zou, Q.; Liu, W.; Xie, S.; Zhou, A.; Chen, J.; Zou, J. Effects of Bacillus licheniformis and xylo-oligosaccharide on growth performance, intestinal microflora and enzyme activities in grass carp Ctenopharyngodon idella. *J. Dalian Ocean Univ.* **2014**, *29*, 136–140.
63. Koo, H.; Hakim, J.A.; Powell, M.L.; Kumar, R.; Eipers, P.G.; Morrow, C.D.; Crowley, M.; Lefkowitz, E.J.; Watts, S.A.; Bej, A.K. Metagenomics approach to the study of the gut microbiome structure and function in zebrafish Danio rerio fed with gluten formulated diet. *J. Microbiol. Methods* **2017**, *135*, 69–76. [CrossRef] [PubMed]
64. Rurangwa, E.; Sipkema, D.; Kals, J.; Ter Veld, M.; Forlenza, M.; Bacanu, G.M.; Smidt, H.; Palstra, A.P. Impact of a novel protein meal on the gastrointestinal microbiota and the host transcriptome of larval zebrafish Danio rerio. *Front. Physiol.* **2015**, *6*, 133. [CrossRef] [PubMed]

65. Udayangani, R.M.C.; Dananjaya, S.H.S.; Nikapitiya, C.; Heo, G.-J.; Lee, J.; De Zoysa, M. Metagenomics analysis of gut microbiota and immune modulation in zebrafish (Danio rerio) fed chitosan silver nanocomposites. *Fish Shellfish Immunol.* **2017**, *66*, 173–184. [CrossRef]
66. Carlson, J.M.; Leonard, A.B.; Hyde, E.R.; Petrosino, J.F.; Primm, T.P. Microbiome disruption and recovery in the fish Gambusia affinis following exposure to broad-spectrum antibiotic. *Infect. Drug Resist.* **2017**, *10*, 143–154. [CrossRef]
67. Michl, S.C.; Ratten, J.M.; Beyer, M.; Hasler, M.; LaRoche, J.; Schulz, C. The malleable gut microbiome of juvenile rainbow trout (*Oncorhynchus mykiss*): Diet-dependent shifts of bacterial community structures. *PLoS ONE* **2017**, *12*, e0177735. [CrossRef]
68. Huyben, D.; Nyman, A.; Vidaković, A.; Passoth, V.; Moccia, R.; Kiessling, A.; Dicksved, J.; Lundh, T. Effects of dietary inclusion of the yeasts *Saccharomyces cerevisiae* and *Wickerhamomyces anomalus* on gut microbiota of rainbow trout. *Aquaculture* **2017**, *473*, 528–537. [CrossRef]
69. Lyons, P.P.; Turnbull, J.F.; Dawson, K.A.; Crumlish, M. Effects of low-level dietary microalgae supplementation on the distal intestinal microbiome of farmed rainbow trout *Oncorhynchus mykiss* (Walbaum). *Aquac. Res.* **2017**, *48*, 2438–2452. [CrossRef]
70. Zhai, Q.; Yu, L.; Li, T.; Zhu, J.; Zhang, C.; Zhao, J.; Zhang, H.; Chen, W. Effect of dietary probiotic supplementation on intestinal microbiota and physiological conditions of Nile tilapia (*Oreochromis niloticus*) under waterborne cadmium exposure. *Antonie Van Leeuwenhoek* **2017**, *110*, 501–513. [CrossRef]
71. Yu, L.; Qiao, N.; Li, T.; Yu, R.; Chen, W.J.P. Dietary supplementation with probiotics regulates gut microbiota structure and function in Nile tilapia exposed to aluminum. *PeerJ* **2019**, *7*, e6963. [CrossRef]
72. Hortillosa, E.; Amar, M.; Nuñal, S.; Pedroso, F.; Ferriols, V.M.E. Effects of putative dietary probiotics from the gut of milkfish (*Chanos chanos*) on the growth performance and intestinal enzymatic activities of juvenile Nile tilapia (*Oreochromis niloticus*). *Aquac. Res.* **2021**, *53*, 98–108. [CrossRef]
73. Zhang, M.; Dong, B.; Lai, X.; Chen, Z.; Hou, L.; Shu, R.; Huang, Y.; Shu, H. Effects of Clostridium butyricum on growth, digestive enzyme activity, antioxidant capacity and gut microbiota in farmed tilapia (*Oreochromis niloticus*). *Aquac. Res.* **2021**, *52*, 1573–1584. [CrossRef]
74. Foysal, J.; Alam, M.; Momtaz, F.; Chaklader, M.R.; Siddik, M.A.B.; Rahman, M. Dietary supplementation of garlic (*Allium sativum*) modulates gut microbiota and health status of tilapia (*Oreochromis niloticus*) against Streptococcus iniae infection. *Aquac. Res.* **2019**, *50*, 2107–2116. [CrossRef]
75. Tachibana, L.; Telli, G.S.; Dias, D.d.C.; Gonçalves, G.S.; Guimarães, M.C.; Ishikawa, C.M.; Cavalcante, R.B.; Natori, M.M.; Fernandez Alarcon, M.F.; Tapia-Paniagua, S.; et al. Bacillus subtilis and Bacillus licheniformis in diets for Nile tilapia (*Oreochromis niloticus*): Effects on growth performance, gut microbiota modulation and innate immunology. *Aquac. Res.* **2021**, *52*, 1630–1642. [CrossRef]
76. Liao, Q.; Zhen, Y.; Qin, Y.; Jiang, Q.; Lan, T.; Huang, L.; and Shen, P. Effects of dietary Metschnikowia sp. GXUS03 on growth, immunity, gut microbiota and Streptococcus agalactiae resistance of Nile tilapia (*Oreochromis niloticus*). *Aquac. Res.* **2022**, *53*, 1918–1927. [CrossRef]
77. Piazzon, M.C.; Calduch-Giner, J.A.; Fouz, B.; Estensoro, I.; Simó-Mirabet, P.; Puyalto, M.; Karalazos, V.; Palenzuela, O.; Sitjà-Bobadilla, A.; Pérez-Sánchez, J. Under control: How a dietary additive can restore the gut microbiome and proteomic profile, and improve disease resilience in a marine teleostean fish fed vegetable diets. *Microbiome* **2017**, *5*, 164. [CrossRef] [PubMed]
78. Legrand, T.; Catalano, S.R.; Wos-Oxley, M.L.; Wynne, J.W.; Weyrich, L.S.; Oxley, A.P.A. Antibiotic-induced alterations and repopulation dynamics of yellowtail kingfish microbiota. *Anim. Microbiome* **2020**, *2*, 26. [CrossRef] [PubMed]
79. Sáenz, J.S.; Marques, T.V.; Barone, R.S.C.; Cyrino, J.E.P.; Kublik, S.; Nesme, J.; Schloter, M.; Rath, S.; Vestergaard, G. Oral administration of antibiotics increased the potential mobility of bacterial resistance genes in the gut of the fish Piaractus mesopotamicus. *Microbiome* **2019**, *7*, 24. [CrossRef]
80. Munir, M.B.; Marsh, T.L.; Blaud, A.; Hashim, R.; Janti Anak Joshua, W.; Mohd Nor, S.A. Analysing the effect of dietary prebiotics and probiotics on gut bacterial richness and diversity of Asian snakehead fingerlings using T-RFLP method. *Aquac. Res.* **2018**, *49*, 3350–3361. [CrossRef]
81. Meng, X.; Hu, W.; Wu, S.; Zhu, Z.; Lu, R.; Yang, G.; Qin, C.; Yang, L.; Nie, G. Chinese yam peel enhances the immunity of the common carp (*Cyprinus carpio* L.) by improving the gut defence barrier and modulating the intestinal microflora. *Fish Shellfish Immunol.* **2019**, *95*, 528–537. [CrossRef]
82. Reyshari, A.; Mohammadiazarm, H.; Mohammadian, T.; Mozanzadeh, M.T. Effects of sodium diformate on growth performance, gut microflora, digestive enzymes and innate immunological parameters of Asian sea bass (*Lates calcarifer*) juveniles. *Aquac. Nutr.* **2019**, *25*, 1135–1144. [CrossRef]
83. Guimarães, M.C.; Moriñigo, M.Á.; Moyano, F.J.; Tachibana, L. Oral administration of *Bacillus subtilis* and *Lactobacillus plantarum* modulates the gut microbiota and increases the amylase activity of Nile tilapia (*Oreochromis niloticus*). *Aquac. Int.* **2021**, *29*, 91–104. [CrossRef]
84. Psofakis, P.; Meziti, A.; Berillis, P.; Mente, E.; Kormas, K.A.; Karapanagiotidis, I.T. Effects of Dietary Fishmeal Replacement by Poultry By-Product Meal and Hydrolyzed Feather Meal on Liver and Intestinal Histomorphology and on Intestinal Microbiota of Gilthead Seabream (*Sparus aurata*). *Appl. Sci.* **2021**, *11*, 8806. [CrossRef]
85. Rimoldi, S.; Gini, E.; Koch, J.F.A.; Iannini, F.; Brambilla, F.; Terova, G. Effects of hydrolyzed fish protein and autolyzed yeast as substitutes of fishmeal in the gilthead sea bream (*Sparus aurata*) diet, on fish intestinal microbiome. *BMC Vet. Res.* **2020**, *16*, 118.

86. Rimoldi, S.; Torrecillas, S.; Montero, D.; Gini, E.; Makol, A.; Valdenegro, V.V.; Izquierdo, M.; Terova, G. Assessment of dietary supplementation with galactomannan oligosaccharides and phytogenics on gut microbiota of European sea bass (*Dicentrarchus Labrax*) fed low fishmeal and fish oil based diet. *PLoS ONE* **2020**, *15*, e0231494. [CrossRef]
87. Abid, A.; Davies, S.J.; Waines, P.; Emery, M.; Castex, M.; Gioacchini, G.; Carnevali, O.; Bickerdike, R.; Romero, J.; Merrifield, D.L. Dietary synbiotic application modulates Atlantic salmon (*Salmo salar*) intestinal microbial communities and intestinal immunity. *Fish Shellfish Immunol.* **2013**, *35*, 1948–1956. [CrossRef]
88. Pereira, G.d.V.; Pereira, S.A.; Soares, A.; Mouriño, J.L.P.; Merrifield, D. Autochthonous probiotic bacteria modulate intestinal microbiota of Pirarucu, Arapaima gigas. *J. World Aquac. Soc.* **2019**, *50*, 1152–1167. [CrossRef]
89. Xie, M.; Zhou, W.; Xie, Y.; Li, Y.; Zhang, Z.; Yang, Y.; Olsen, R.E.; Ran, C.; Zhou, Z. Effects of Cetobacterium somerae fermentation product on gut and liver health of common carp (*Cyprinus carpio*) fed diet supplemented with ultra-micro ground mixed plant proteins. *Aquaculture* **2021**, *543*, 736943. [CrossRef]
90. Xu, Q.; Yang, Z.; Chen, S.; Zhu, W.; Xiao, S.; Liu, J.; Wang, H.; Lan, S. Effects of Replacing Dietary Fish Meal by Soybean Meal Co-Fermented Using *Bacillus subtilis* and *Enterococcus faecium* on Serum Antioxidant Indices and Gut Microbiota of Crucian Carp *Carassius auratus*. *Fishes* **2022**, *7*, 54. [CrossRef]
91. González-Félix, M.L.; Gatlin, D.M.; Urquidez-Bejarano, P.; de la Reé-Rodríguez, C.; Duarte-Rodríguez, L.; Sánchez, F.; Casas-Reyes, A.; Yamamoto, F.Y.; Ochoa-Leyva, A.; Perez-Velazquez, M. Effects of commercial dietary prebiotic and probiotic supplements on growth, innate immune responses, and intestinal microbiota and histology of *Totoaba macdonaldi*. *Aquaculture* **2018**, *491*, 239–251. [CrossRef]
92. Larios-Soriano, E.; Zavala, R.C.; López, L.M.; Gómez-Gil, B.; Ramírez, D.T.; Sanchez, S.; Canales, K.; Galaviz, M.A. Soy protein concentrate effects on gut microbiota structure and digestive physiology of *Totoaba macdonaldi*. *J. Appl. Microbiol.* **2022**, *132*, 1384–1396. [CrossRef] [PubMed]
93. Xu, J.; Xie, S.; Chi, S.; Zhang, S.; Cao, J.; Tan, B. Short-term dietary antibiotics altered the intestinal microbiota and improved the lipid metabolism in hybrid grouper fed medium and high-lipid diets. *Aquaculture* **2022**, *547*, 737453. [CrossRef]
94. Zhou, S.; Song, D.; Zhou, X.; Mao, X.; Zhou, X.; Wang, S.; Wei, J.; Huang, Y.; Wang, W.; Xiao, S.-M.; et al. Characterization of Bacillus subtilis from gastrointestinal tract of hybrid Hulong grouper (*Epinephelus fuscoguttatus* × *E. lanceolatus*) and its effects as probiotic additives. *Fish Shellfish Immunol.* **2019**, *84*, 1115–1124. [CrossRef] [PubMed]
95. Liu, W.; Wang, X.; Chen, D.; Wang, Y.; Zhang, A.; Zhou, H. Comparison of adhesive gut bacteria composition, immunity, and disease resistance in juvenile hybrid tilapia fed two different Lactobacillus strains. *Fish Shellfish Immunol.* **2013**, *35*, 54–62. [CrossRef]
96. Wu, X.; Teame, T.; Hao, Q.; Ding, Q.; Liu, H.; Ran, C.; Yang, Y.; Zhang, Y.; Zhou, Z.; Duan, M.; et al. Use of a paraprobiotic and postbiotic feed supplement (HWF™) improves the growth performance, composition and function of gut microbiota in hybrid sturgeon (*Acipenser baerii* x *Acipenser schrenckii*). *Fish Shellfish Immunol.* **2020**, *104*, 36–45. [CrossRef]
97. Hahor, W.; Thongprajukaew, K.; Suanyuk, N. Effects of dietary supplementation of oligosaccharides on growth performance, gut health and immune response of hybrid catfish (*Pangasianodon gigas* × *Pangasianodon hypophthalmus*). *Aquaculture* **2019**, *507*, 97–107. [CrossRef]
98. Wang, Q.; Ayiku, S.; Liu, H.; Tan, B.; Dong, X.; Chi, S.; Yang, Q.; Zhang, S.; Zhou, W. Effects of dietary ESTAQUA® yeast culture supplementation on growth, immunity, intestinal microbiota and disease-resistance against Vibrio harveyi in hybrid grouper (♀*Epinephelus fuscoguttatus* × ♂*E. lanceolatus*). *Aquac. Rep.* **2022**, *22*, 100922. [CrossRef]
99. Xu, G.; Xing, W.; Li, T.; Ma, Z.; Liu, C.; Jiang, N.; Luo, L. Effects of dietary raffinose on growth, non-specific immunity, intestinal morphology and microbiome of juvenile hybrid sturgeon (*Acipenser baeri* Brandt ♀ × A. *schrenckii* Brandt ♂). *Fish Shellfish Immunol.* **2018**, *72*, 237–246. [CrossRef]
100. Chen, G.; Yin, B.; Liu, H.; Tan, B.; Dong, X.; Yang, Q.; Chi, S.; Zhang, S. Supplementing Chitosan Oligosaccharide Positively Affects Hybrid Grouper (*Epinephelus fuscoguttatus* & FEMALE; x E. *lanceolatus* & MALE;) Fed Dietary Fish Meal Replacement With Cottonseed Protein Concentrate: Effects on Growth, Gut Microbiota, Antioxidant Function and Immune Response. *Front. Mar. Sci.* **2021**, *8*, 707627. [CrossRef]
101. Chen, G.; Yin, B.; Liu, H.; Tan, B.; Dong, X.; Yang, Q.; Chi, S.; Zhang, S. Effects of fishmeal replacement with cottonseed protein concentrate on growth, digestive proteinase, intestinal morphology and microflora in pearl gentian grouper (♀*Epinephelus fuscoguttatus* × ♂*Epinephelus lanceolatu*). *Aquac. Res.* **2020**, *51*, 2870–2884. [CrossRef]
102. Ye, G.; Dong, X.; Yang, Q.; Chi, S.; Liu, H.; Zhang, H.; Tan, B.; Zhang, S. Dietary replacement of fish meal with peanut meal in juvenile hybrid grouper (*Epinephelus fuscoguttatus* ♀ × *Epinephelus lanceolatus* ♂): Growth performance, immune response and intestinal microbiota. *Aquac. Rep.* **2020**, *17*, 100327. [CrossRef]
103. Ye, G.; Dong, X.; Yang, Q.; Chi, S.; Liu, H.; Zhang, H.; Tan, B.; Zhang, S. Low-gossypol cottonseed protein concentrate used as a replacement of fish meal for juvenile hybrid grouper (*Epinephelus fuscoguttatus* ♀ × *Epinephelus lanceolatus* ♂): Effects on growth performance, immune responses and intestinal microbiota. *Aquaculture* **2020**, *524*, 735309. [CrossRef]
104. He, Y.; Guo, X.; Tan, B.; Dong, X.; Yang, Q.; Liu, H.; Zhang, S.; Chi, S. Partial fishmeal protein replacement with peptides from swine blood modulates the nutritional status, immune response, and intestinal microbiota of hybrid groupers (female *Epinephelus fuscoguttatus*× male *E. lanceolatus*). *Aquaculture* **2021**, *533*, 736154. [CrossRef]
105. Fowler, E.C.; Poudel, P.; White, B.; St-Pierre, B.; Brown, M. Effects of a Bioprocessed Soybean Meal Ingredient on the Intestinal Microbiota of Hybrid Striped Bass, *Morone chrysops* × *M. saxatilis*. *Microorganisms* **2021**, *9*, 1032. [CrossRef]

106. Sevellec, M.; Laporte, M.; Bernatchez, A.; Derome, N.; Bernatchez, L. Evidence for host effect on the intestinal microbiota of whitefish (*Coregonus* sp.) species pairs and their hybrids. *Ecol. Evol.* **2019**, *9*, 11762–11774. [CrossRef]
107. Zhu, L.; Zhang, Z.; Chen, H.; Lamer, J.T.; Wang, J.; Wei, W.; Fu, L.; Tang, M.; Wang, C.; Lu, G. Gut microbiomes of bigheaded carps and hybrids provide insights into invasion: A hologenome perspective. *Evol. Appl.* **2021**, *14*, 735–745. [CrossRef]
108. Deng, Y.; Zhang, Y.; Chen, H.; Xu, L.; Wang, Q.; Feng, J. Gut-Liver Immune Response and Gut Microbiota Profiling Reveal the Pathogenic Mechanisms of Vibrio harveyi in Pearl Gentian Grouper (*Epinephelus lanceolatus* ♂ × *E. fuscoguttatus* ♀). *Front. Immunol.* **2020**, *11*, 607754. [CrossRef]
109. He, Y.; Liang, J.; Dong, X.; Yang, Q.; Liu, H.; Zhang, S.; Chi, S.; Tan, B. Glutamine alleviates β-conglycinin-induced enteritis in juvenile hybrid groupers *Epinephelus fuscoguttatus*♀ × *Epinephelus lanceolatus*♂by suppressing the MyD88/NF-κB pathway. *Aquaculture* **2022**, *549*, 737735. [CrossRef]
110. Liu, X.; Shi, H.; He, Q.; Lin, F.; Wang, Q.; Xiao, S.; Dai, Y.; Zhang, Y.; Yang, H.; Zhao, H. Effect of starvation and refeeding on growth, gut microbiota and non-specific immunity in hybrid grouper (*Epinephelus fuscoguttatus*♀ × *E. lanceolatus*♂). *Fish Shellfish Immunol.* **2020**, *97*, 182–193. [CrossRef]
111. Fowler, E.; St-Pierre, B.; Poudel, P.; White, B.; Brown, M. 183 Effects of Protein Supplementation on the Gut Bacterial Community Composition of Hybrid Striped Bass, Morone Chrysops X M. Saxatilis. *J. Anim. Sci.* **2021**, *99*, 88–89. [CrossRef]
112. Ran, C.; Hu, J.; Liu, W.; Liu, Z.; He, S.; Bui Chau Truc, D.; Nguyen Ngoc, D.; Ooi, E.L.; Zhou, Z. Thymol and Carvacrol Affect Hybrid Tilapia through the Combination of Direct Stimulation and an Intestinal Microbiota-Mediated Effect: Insights from a Germ-Free Zebrafish Model. *J. Nutr.* **2016**, *146*, 1132–1140. [CrossRef] [PubMed]
113. Qin, C.; Zhang, Y.; Liu, W.; Xu, L.; Yang, Y.; Zhou, Z. Effects of chito-oligosaccharides supplementation on growth performance, intestinal cytokine expression, autochthonous gut bacteria and disease resistance in hybrid tilapia *Oreochromis niloticus* ♀ × *Oreochromis aureus* ♂. *Fish Shellfish Immunol.* **2014**, *40*, 267–274. [CrossRef] [PubMed]
114. Zhu, H.; Qiang, J.; He, J.; Tao, Y.; Bao, J.; Xu, P. Physiological parameters and gut microbiome associated with different dietary lipid levels in hybrid yellow catfish (*Tachysurus fulvidraco*♀× *Pseudobagrus vachellii*♂). *Comp. Biochem. Physiol. Part D Genom. Proteom.* **2021**, *37*, 100777. [CrossRef] [PubMed]
115. Ye, G.; Dong, X.; Yang, Q.; Chi, S.; Liu, H.; Zhang, H.; Tan, B.; Zhang, S. A formulated diet improved digestive capacity, immune function and intestinal microbiota structure of juvenile hybrid grouper (*Epinephelus fuscoguttatus* ♀ × *Epinephelus lanceolatus* ♂) when compared with chilled trash fish. *Aquaculture* **2020**, *523*, 735230. [CrossRef]
116. Hao, Q.; Teame, T.; Wu, X.; Ding, Q.; Ran, C.; Yang, Y.; Xing, Y.; Zhang, Z.; Zhou, Z. Influence of diet shift from bloodworm to formulated feed on growth performance, gut microbiota structure and function in early juvenile stages of hybrid sturgeon (*Acipenser baerii* × *Acipenser schrenckii*). *Aquaculture* **2021**, *533*, 736165. [CrossRef]
117. Wang, L.; Tian, Y.; Li, Z.; Li, Z.; Chen, S.; Li, L.; Li, W.; Wang, Q.; Lin, H.; Li, B. Comparison of the gut microbiota composition between asymptomatic and diseased *Epinephelus moara* ♀ × *Epinephelus lanceolatus* ♂with nervous necrosis virus infection. *Aquac. Res.* **2022**, *53*, 633–641. [CrossRef]
118. Hlordzi, V.; Wang, J.; Li, T.; Cui, Z.; Tan, B.; Liu, H.; Yang, Q.; Dong, X.; Zhang, S.; Chi, S. Effects of Lower Fishmeal with Hydrolyzed Fish Protein Powder on the Growth Performance and Intestinal Development of Juvenile Pearl Gentian Grouper (*Epinephelus fuscoguttatus* ♀and *Epinephelus lanceolatus* ♂). *Front. Mar. Sci.* **2022**, *9*, 398. [CrossRef]
119. Huang, X.; Zhong, L.; Fan, W.; Feng, Y.; Xiong, G.; Liu, S.; Wang, K.; Geng, Y.; Ouyang, P.; Chen, D.; et al. Enteritis in hybrid sturgeon (*Acipenser schrenckii*♂ × *Acipenser baeri*♀) caused by intestinal microbiota disorder. *Aquac. Rep.* **2020**, *18*, 100456. [CrossRef]
120. Long, S.; You, Y.; Dong, X.; Tan, B.; Zhang, S.; Chi, S.; Yang, Q.; Liu, H.; Xie, S.; Yang, Y.; et al. Effect of dietary oxidized fish oil on growth performance, physiological homeostasis and intestinal microbiome in hybrid grouper (♀*Epi-nephelus fuscoguttatus* × ♂*Epinephelus lanceolatus*). *Aquac. Rep.* **2022**, *24*, 101130. [CrossRef]
121. Long, S.; Dong, X.; Yan, X.; Liu, H.; Tan, B.; Zhang, S.; Chi, S.; Yang, Q.; Liu, H.; Yang, Y.; et al. The effect of oxidized fish oil on antioxidant ability, histology and transcriptome in intestine of the juvenile hybrid grouper (♀*Epinephelus fuscoguttatus* × ♂*Epinephelus lanceolatus*). *Aquac. Rep.* **2022**, *22*, 100921. [CrossRef]
122. Yan, X.; Chen, Y.; Dong, X.; Tan, B.; Liu, H.; Zhang, S.; Chi, S.; Yang, Q.; Liu, H.; Yang, Y. Ammonia Toxicity Induces Oxidative Stress, Inflammatory Response and Apoptosis in Hybrid Grouper (♀*Epinephelus fuscoguttatus* × ♂*E. lanceolatu*). *Front. Mar. Sci.* **2021**, *8*, 100880. [CrossRef]
123. Wei, H.; Li, R.; Yang, Q.; Tan, B.; Ray, G.W.; Dong, X.; Chi, S.; Liu, H.; Zhang, S. Effects of Zn on growth performance, immune enzyme activities, resistance to disease and intestinal flora for juvenile pearl gentian grouper(*Epinephelus lanceolatus* ♂ × *Epinephelus fuscoguttatus* ♀) under low fishmeal diet. *Aquac. Rep.* **2021**, *21*, 100880. [CrossRef]
124. Wu, X.; Hao, Q.; Teame, T.; Ding, Q.; Liu, H.; Ran, C.; Yang, Y.; Xia, L.; Wei, S.; Zhou, Z.; et al. Gut microbiota induced by dietary GWF®contributes to growth promotion, immune regulation and disease resistance in hybrid sturgeon (*Acipenserbaerii* x *Acipenserschrenckii*): Insights from a germ-free zebrafish model. *Aquaculture* **2020**, *520*, 734966. [CrossRef]
125. Limbu, S.M.; Chen, L.-Q.; Zhang, M.-L.; Du, Z.-Y. A global analysis on the systemic effects of antibiotics in cultured fish and their potential human health risk: A review. *Rev. Aquac.* **2021**, *13*, 1015–1059. [CrossRef]
126. Sx, A.; Yin, P.; Tian, L.; Liu, Y.; Tan, B.; Niu, J. Interactions between dietary lipid levels and chronic exposure of legal aquaculture dose of sulfamethoxazole in juvenile largemouth bass *Micropterus salmoides*. *Aquat. Toxicol.* **2020**, *229*, 105670.

127. Limbu, S.M.; Ma, Q.; Zhang, M.-L.; Du, Z.-Y. High fat diet worsens the adverse effects of antibiotic on intestinal health in juvenile Nile tilapia (*Oreochromis niloticus*). *Sci. Total Environ.* **2019**, *680*, 169–180. [CrossRef]
128. Zhou, L.; Limbu, S.M.; Qiao, F.; Du, Z.Y.; Zhang, M.J.Z. Influence of Long-Term Feeding Antibiotics on the Gut Health of Zebrafish. *Zebrafish* **2018**, *15*, 340–348. [CrossRef]
129. Nakano, T.; Hayashi, S.; Nagamine, N. Effect of excessive doses of oxytetracycline on stress-related biomarker expression in coho salmon. *Environ. Sci. Pollut. Res. Int.* **2018**, *25*, 7121–7128. [CrossRef]
130. Zhou, Z.C.; Lin, Z.J.; Shuai, X.Y.; Zheng, J.; Meng, L.X.; Zhu, L.; Sun, Y.J.; Shang, W.C.; Chen, H. Temporal variation and sharing of antibiotic resistance genes between water and wild fish gut in a peri-urban river. *J. Environ. Sci.* **2021**, *103*, 12–19. [CrossRef]
131. Carbone, D.; Faggio, C. Importance of prebiotics in aquaculture as immunostimulants. Effects on immune system of *Sparus aurata* and *Dicentrarchus labrax*. *Fish Shellfish Immunol.* **2016**, *54*, 172–178. [CrossRef]
132. Song, S.K.; Beck, B.R.; Kim, D.; Park, J.; Kim, J.; Kim, H.D.; Ringø, E. Prebiotics as immunostimulants in aquaculture: A review. *Fish Shellfish Immunol.* **2014**, *40*, 40–48. [CrossRef]
133. Taverniti, V.; Guglielmetti, S. The immunomodulatory properties of probiotic microorganisms beyond their viability (ghost probiotics: Proposal of paraprobiotic concept). *Genes Nutr.* **2011**, *6*, 261–274. [CrossRef]
134. Xia, Y.; Wang, M.; Gao, F.; Lu, M.; Chen, G. Effects of dietary probiotic supplementation on the growth, gut health and disease resistance of juvenile Nile tilapia (*Oreochromis niloticus*). *Anim. Nutr.* **2020**, *6*, 69–79. [CrossRef]
135. Beck, B.R.; Kim, D.; Jeon, J.; Lee, S.M.; Kim, H.K.; Kim, O.J.; Lee, J.I.; Suh, B.S.; Do, H.K.; Lee, K.H.J.F.; et al. The effects of combined dietary probiotics *Lactococcus lactis* BFE920 and *Lactobacillus plantarum* FGL0001 on innate immunity and disease resistance in olive flounder (*Paralichthys olivaceus*). *Fish Shellfish Immunol.* **2015**, *42*, 177–183. [CrossRef]
136. Liu, L.; Wu, R.; Zhang, J.; Shang, N.; Li, P. D-Ribose Interferes with Quorum Sensing to Inhibit Biofilm Formation of *Lactobacillus paraplantarum* L-ZS9. *Front. Microbiol.* **2017**, *8*, 1860. [CrossRef]
137. Chai, P.C.; Song, X.L.; Chen, G.F.; Xu, H.; Huang, J. Dietary supplementation of probiotic Bacillus PC465 isolated from the gut of *Fenneropenaeus chinensis* improves the health status and resistance of *Litopenaeus vannamei* against white spot syndrome virus. *Fish Shellfish Immunol.* **2016**, *54*, 602–611. [CrossRef]
138. Staykov, Y.; Spring, P.; Denev, S.; Sweetman, J.W. Effect of a mannan oligosaccharide on the growth performance and immune status of rainbow trout (*Oncorhynchus mykiss*). *Aquac. Int.* **2007**, *15*, 153–161. [CrossRef]
139. Foligné, B.; Dewulf, J.; Breton, J.; Claisse, O.; Lonvaud-Funel, A.; Pot, B. Probiotic properties of non-conventional lactic acid bacteria: Immunomodulation by Oenococcus oeni. *Int. J. Food Microbiol.* **2010**, *140*, 136–145. [CrossRef]
140. Panigrahi, A.; Kiron, V.; Satoh, S.; Hirono, I.; Kobayashi, T.; Sugita, H.; Puangkaew, J.; Aoki, T. Immune modulation and expression of cytokine genes in rainbow trout *Oncorhynchus mykiss* upon probiotic feeding. *Dev. Comp. Immunol.* **2007**, *31*, 372–382. [CrossRef]
141. Yang, M.; Wang, X.; Chen, D.; Wang, Y.; Zhang, A.; Zhou, H. TGF-β1 exerts opposing effects on grass carp leukocytes: Implication in teleost immunity, receptor signaling and potential self-regulatory mechanisms. *PLoS ONE* **2012**, *7*, e35011. [CrossRef]
142. Kadowaki, T.; Yasui, Y.; Takahashi, Y.; Kohchi, C.; Soma, G.; Inagawa, H. Comparative immunological analysis of innate immunity activation after oral administration of wheat fermented extract to teleost fish. *Anticancer Res.* **2009**, *29*, 4871–4877. [PubMed]
143. He, S.; Liu, W.; Zhou, Z. Evaluation of probiotic strain Bacillus subtilis C-3102 as a feed supplement for koi carp (*Cyprinus carpio*). *J. Aquac. Res. Dev.* **2011**, *S1*, 005. [CrossRef]
144. Gibson, G.R.; Probert, H.M.; Loo, J.V.; Rastall, R.A.; Roberfroid, M.B. Dietary modulation of the human colonic microbiota: Updating the concept of prebiotics. *Nutr. Res. Rev.* **2004**, *17*, 259–275. [CrossRef] [PubMed]
145. Gibson, G.R.; Roberfroid, M.B. Dietary Modulation of the Human Colonic Microbiota: Introducing the Concept of Prebiotics. *J. Nutr.* **1995**, *125*, 1401–1412. [CrossRef]
146. Hoseinifar, S.H.; Zoheiri, F.; Dadar, M.; Rufchaei, R.; Ringø, E. Dietary galactooligosaccharide elicits positive effects on non-specific immune parameters and growth performance in Caspian white fish (*Rutilus frisii kutum*) fry. *Fish Shellfish Immunol.* **2016**, *56*, 467–472. [CrossRef]
147. Watzl, B.; Girrbach, S.; Roller, M. Inulin, oligofructose and immunomodulation. *Br. J. Nutr.* **2005**, *93* (Suppl. 1), S49–S55. [CrossRef]
148. Hoseinifar, S.H.; Khalili, M.; Rostami, H.K.; Esteban, M. Immunology, Dietary galactooligosaccharide affects intestinal microbiota, stress resistance, and performance of Caspian roach (*Rutilus rutilus*) fry. *Fish Shellfish Immunol.* **2013**, *35*, 1416–1420. [CrossRef]
149. Hoseinifar, S.H.; Sharifian, M.; Vesaghi, M.J.; Khalili, M.; Esteban, M. The effects of dietary xylooligosaccharide on mucosal parameters, intestinal microbiota and morphology and growth performance of Caspian white fish (*Rutilus frisii kutum*) fry—ScienceDirect. *Fish Shellfish Immunol.* **2014**, *39*, 231–236. [CrossRef]
150. Ali, S.S.R.; Ambasankar, K.; Nandakumar, S.; Praveena, P.E.; Syamadayal, J. Effect of dietary prebiotic inulin on growth, body composition and gut microbiota of Asian seabass (*Lates calcarifer*). *Anim. Feed Sci. Technol.* **2016**, *217*, 87–94.
151. Guerreiro, I.; Serra, C.R.; Enes, P.; Couto, A.; Salvador, A.; Costas, B.; Oliva-Teles, A. Effect of short chain fructooligosaccharides (scFOS) on immunological status and gut microbiota of gilthead sea bream (*Sparus aurata*) reared at two temperatures. *Fish Shellfish Immunol.* **2016**, *49*, 122–131. [CrossRef]
152. Lp, A.; Yúfera, M.; Navarro-Guillén, C.; Moyano, F.J.; Soverini, M.; D'Amico, F.; Candela, M.; Fontanillas, R.; Gatta, P.P.; Bonaldo, A. Effects of calcium carbonate inclusion in low fishmeal diets on growth, gastrointestinal pH, digestive enzyme activity and gut bacterial community of European sea bass (*Dicentrarchus labrax* L.) juveniles. *Aquaculture* **2019**, *510*, 283–292.

153. Li, P.; Gatlin, D.M. Dietary brewers yeast and the prebiotic Grobiotic™AE influence growth performance, immune responses and resistance of hybrid striped bass (*Morone chrysops* × M. *saxatilis*) to Streptococcus iniae infection. *Aquaculture* **2004**, *231*, 445–456. [CrossRef]
154. Cui, L.; Xu, W.; Ai, Q.; Wang, D.; Mai, K. Effects of dietary chitosan oligosaccharide complex with rare earth on growth performance and innate immune response of turbot, *Scophthalmus maximus* L. *Aquac. Res.* **2013**, *44*, 683–690. [CrossRef]
155. Lin, S.; Mao, S.; Yong, G.; Lin, L.; Li, L.; Yu, P. Effects of dietary chitosan oligosaccharides and *Bacillus coagulans* on the growth, innate immunity and resistance of koi (*Cyprinus carpio koi*). *Aquaculture* **2012**, *342–343*, 36–41. [CrossRef]
156. Luo, L.; Cai, X.; He, C.; Xue, M.; Wu, X.; Cao, H. Immune response, stress resistance and bacterial challenge in juvenile rainbow trouts *Oncorhynchus mykiss* fed diets containing chitosan-oligosaccharides. *Curr. Zool.* **2009**, *55*, 416–422. [CrossRef]
157. Olsen, R.E.; Myklebust, R.; Kryvi, H.; Mayhew, T.M.; Ringø, E. Damaging effect of dietary inulin on intestinal enterocytes in Arctic charr (*Salvelinus alpinus* L.). *Aquac. Res.* **2001**, *32*, 931–934. [CrossRef]
158. Tacona, A.G.J.; Metianb, M. Global overview on the use of fish meal and fish oil in industrially compounded aquafeeds: Trends and future prospects. *Aquaculture* **2008**, *285*, 146–158. [CrossRef]
159. Hardy, R.W.; Tacon, A.G.J. Fish meal: Historical uses, production trends and future outlook for sustainable supplies. *Responsible Mar. Aquac.* **2002**, 311–325.
160. Hardy, R.W. Hardy, Utilization of plant proteins in fish diets: Effects of global demand and supplies of fishmeal. *Aquac. Res.* **2010**, *41*, 770–776. [CrossRef]
161. Daniel, A. Studies, A review on replacing fish meal in aqua feeds using plant protein sources. *Int. J. Fish. Aquat. Stud.* **2018**, *6*, 164–179.
162. Robinson, E.H.; Li, M.H. Use of Plant Proteins in Catfish Feeds: Replacement of Soybean Meal with Cottonseed Meal and Replacement of Fish Meal with Soybean Meal and Cottonseed Meal. *J. World Aquac. Soc.* **1994**, *25*, 271–276. [CrossRef]
163. Li, Y.; Ai, Q.; Mai, K.; Xu, W.; Deng, J.; Cheng, Z. Comparison of high-protein soybean meal and commercial soybean meal partly replacing fish meal on the activities of digestive enzymes and aminotransferases in juvenile Japanese seabass, *Lateolabrax japonicus* (Cuvier, 1828). *Aquac. Res.* **2014**, *45*, 1051–1060. [CrossRef]
164. Laporte, J.; Trushenski, J. Production performance, stress tolerance and intestinal integrity of sunshine bass fed increasing levels of soybean meal. *J. Anim. Physiol. Anim. Nutr.* **2012**, *96*, 513–526. [CrossRef] [PubMed]
165. Baeverfjord, G.T.; Krogdahl, A.J. Development and regression of soybean meal induced enteritis in Atlantic salmon, *Salmo salar* L., distal intestine: A comparison with the intestines of fasted fish. *J. Fish Dis.* **2010**, *19*, 375–387. [CrossRef]
166. Theodosopoulos, A.N.; Hund, A.K.; Taylor, S.A. Parasites and Host Species Barriers in Animal Hybrid Zones. *Trends Ecol. Evol.* **2019**, *34*, 19–30. [CrossRef]
167. Dittmer, J.; Bouchon, D. Feminizing Wolbachia influence microbiota composition in the terrestrial isopod *Armadillidium vulgare*. *Sci. Rep.* **2018**, *8*, 6998. [CrossRef]
168. Zchori-Fein, E.; Faktor, O.; Zeidan, M.; Gottlieb, Y.; Czosnek, H.; Rosen, D. Parthenogenesis-inducing microorganisms in *Aphytis* (Hymenoptera: Aphelinidae). *Insect Mol. Biol.* **1995**, *4*, 173–178. [CrossRef]
169. Nadal-Jimenez, P.; Griffin, J.S.; Davies, L.; Frost, C.L.; Marcello, M.; Hurst, G.D.D. Genetic manipulation allows in vivo tracking of the life cycle of the son-killer symbiont, Arsenophonus nasoniae, and reveals patterns of host invasion, tropism and pathology. *Environ. Microbiol.* **2019**, *21*, 3172–3182. [CrossRef]
170. Perlmutter, J.I.; Bordenstein, S.R. Microorganisms in the reproductive tissues of arthropods. *Nat. Rev. Microbiol.* **2020**, *18*, 97–111. [CrossRef]
171. Gebiola, M.; Kelly, S.E.; Hammerstein, P.; Giorgini, M.; Hunter, M.S. Darwin's corollary and cytoplasmic incompatibility induced by *Cardinium* may contribute to speciation in *Encarsia* wasps (Hymenoptera: Aphelinidae). *Evol. Int. J. Org. Evol.* **2016**, *70*, 2447–2458. [CrossRef]
172. Gatenby, J.B. Symbionticism and the Origin of Species. *Nature* **1928**, *121*, 164–165. [CrossRef]
173. Miller, A.K.; Westlake, C.S.; Cross, K.L.; Leigh, B.A.; Bordenstein, S.R. The microbiome impacts host hybridization and speciation. *PLoS Biol.* **2021**, *19*, e3001417. [CrossRef] [PubMed]
174. Brooks, A.W.; Kohl, K.D.; Brucker, R.M.; van Opstal, E.J.; Bordenstein, S.R. Phylosymbiosis: Relationships and Functional Effects of Microbial Communities across Host Evolutionary History. *PLoS Biol.* **2016**, *14*, e2000225. [CrossRef] [PubMed]
175. Tsuchida, T.; Koga, R.; Fukatsu, T. Host plant specialization governed by facultative symbiont. *Science* **2004**, *303*, 1989. [CrossRef] [PubMed]
176. Nardon, P.; Grenier, A.M. Serial endosymbiosis theory and weevil evolution: The role of symbiosis. In *Symbiosis as a Source of Evolutionary Innovation: Speciation and Morphogenesis*; Margulis, L., Fester, R., Eds.; MIT Press: Cambridage, MA, USA, 1991; pp. 153–169.
177. Hosokawa, T.; Kikuchi, Y.; Nikoh, N.; Shimada, M.; Fukatsu, T. Strict host-symbiont cospeciation and reductive genome evolution in insect gut bacteria. *PLoS Biol.* **2006**, *4*, e337. [CrossRef] [PubMed]
178. Miller, W.J.; Ehrman, L.; Schneider, D. Infectious speciation revisited: Impact of symbiont-depletion on female fitness and mating behavior of Drosophila paulistorum. *PLoS Pathog.* **2010**, *6*, e1001214. [CrossRef]
179. Breeuwer, J.A.; Werren, J.H. Microorganisms associated with chromosome destruction and reproductive isolation between two insect species. *Nature* **1990**, *346*, 558–560. [CrossRef]

180. Bordenstein, S.R.; O'Hara, F.P.; Werren, J.H. Wolbachia-induced incompatibility precedes other hybrid incompatibilities in Nasonia. *Nature* **2001**, *409*, 707–710. [CrossRef]
181. Brucker, R.M.; Bordenstein, S.R. The hologenomic basis of speciation: Gut bacteria cause hybrid lethality in the genus Nasonia. *Science* **2013**, *341*, 667–669. [CrossRef]
182. Li, Z.; Wright, A.G.; Si, H.; Wang, X.; Qian, W.; Zhang, Z.; Li, G. Changes in the rumen microbiome and metabolites reveal the effect of host genetics on hybrid crosses. *Environ. Microbiol. Rep.* **2016**, *8*, 1016–1023. [CrossRef]
183. Petri, R.M.; Schwaiger, T.; Penner, G.B.; Beauchemin, K.A.; Forster, R.J.; McKinnon, J.J.; McAllister, T.A. Characterization of the core rumen microbiome in cattle during transition from forage to concentrate as well as during and after an acidotic challenge. *PLoS ONE* **2013**, *8*, e83424. [CrossRef] [PubMed]
184. Edwards, J.E.; Schennink, A.; Burden, F.; Long, S.; van Doorn, D.A.; Pellikaan, W.F.; Dijkstra, J.; Saccenti, E.; Smidt, H. Domesticated equine species and their derived hybrids differ in their fecal microbiota. *Anim. Microbiome* **2020**, *2*, 8. [CrossRef] [PubMed]
185. Li, W.; Zhou, Z.; Li, H.; Wang, S.; Ren, L.; Hu, J.; Liu, Q.; Wu, C.; Tang, C.; Hu, F.; et al. Successional Changes of Microbial Communities and Host-Microbiota Interactions Contribute to Dietary Adaptation in Allodiploid Hybrid Fish. *Microb. Ecol.* **2022**. [CrossRef] [PubMed]
186. Belkova, N.L.; Sidorova, T.V.; Glyzina, O.Y.; Yakchnenko, V.M.; Sapozhnikova, Y.P.; Bukin, Y.S.; Baturina, O.A.; Sukhanova, L.V. Gut microbiome of juvenile coregonid fishes: Comparison of sympatric species and their F1 hybrids. *Fundam. Appl. Limnol.* **2017**, *189*, 279–290. [CrossRef]
187. Bahrndorff, S.; Alemu, T.; Alemneh, T.; Nielsen, J.L. The Microbiome of Animals: Implications for Conservation Biology. *Int. J. Genom.* **2016**, *2016*, 5304028. [CrossRef]
188. Chown, S.L.; Hodgins, K.A.; Griffin, P.C.; Oakeshott, J.G.; Byrne, M.; Hoffmann, A.A. Biological invasions, climate change and genomics. *Evol. Appl.* **2015**, *8*, 23–46. [CrossRef]
189. Simberloff, D.; RejmÁNek, M. *Encyclopedia of Biological Invasions*, 1st ed.; Simberloff, D., Rejmánek, M., Eds.; University of California Press: Oakland, CA, USA, 2011. [CrossRef]
190. Lefort, M.C.; Boyer, S.; De Romans, S.; Glare, T.; Armstrong, K.; Worner, S. Invasion success of a scarab beetle within its native range: Host range expansion versus host-shift. *PeerJ* **2014**, *2*, e262. [CrossRef]
191. Lefort, M.C.; Boyer, S.; Glare, T.R. A response to Pennisi—How do gut microbiomes help herbivores, a hint into next-generation biocontrol solutions. *Rethink. Ecol.* **2017**, *1*, 9–13. [CrossRef]
192. Kim, D.; Thairu, M.W.; Hansen, A.K. Novel Insights into Insect-Microbe Interactions-Role of Epigenomics and Small RNAs. *Front. Plant Sci.* **2016**, *7*, 1164. [CrossRef]
193. Dlugosch, K.M.; Parker, I.M. Founding events in species invasions: Genetic variation, adaptive evolution, and the role of multiple introductions. *Mol. Ecol.* **2008**, *17*, 431–449. [CrossRef]
194. Clark, J.S.; Lewis, M.; Horvath, L. Invasion by extremes: Population spread with variation in dispersal and reproduction. *Am. Nat.* **2001**, *157*, 537–554. [CrossRef] [PubMed]
195. Ficetola, G.F.; Siesa, M.E.; de Bernardi, F.; Padoa-Schioppa, E. Complex impact of an invasive crayfish on freshwater food webs. *Biodivers. Conserv.* **2012**, *21*, 2641–2651. [CrossRef]
196. Figueroa, M.E.; Castillo, J.M.; Redondo, S.; Luque, T.; Castellanos, E.M.; Nieva, F.J.; Luque, C.J.; Rubio-Casal, A.E.; Davy, A.J. Facilitated invasion by hybridization of Sarcocornia species in a salt-marsh succession. *J. Ecol.* **2003**, *91*, 616–626. [CrossRef]
197. Llewellyn, M.S.; Boutin, S.; Hoseinifar, S.H.; Derome, N. Teleost microbiomes: The state of the art in their characterization, manipulation and importance in aquaculture and fisheries. *Front. Microbiol.* **2014**, *5*, 207. [CrossRef]
198. Burns, A.R.; Stephens, W.Z.; Stagaman, K.; Wong, S.; Rawls, J.F.; Guillemin, K.; Bohannan, B.J. Contribution of neutral processes to the assembly of gut microbial communities in the zebrafish over host development. *ISME J.* **2016**, *10*, 655–664. [CrossRef]
199. Smith, C.C.; Snowberg, L.K.; Caporaso, J.G.; Knight, R.; Bolnick, D.I. Dietary input of microbes and host genetic variation shape among-population differences in stickleback gut microbiota. *ISME J.* **2015**, *9*, 2515–2526. [CrossRef]
200. Sylvain, F.E.; Derome, N. Vertically and horizontally transmitted microbial symbionts shape the gut microbiota ontogenesis of a skin-mucus feeding discus fish progeny. *Sci. Rep.* **2017**, *7*, 5263. [CrossRef]
201. Bates, J.M.; Mittge, E.; Kuhlman, J.; Baden, K.N.; Cheesman, S.E.; Guillemin, K. Distinct signals from the microbiota promote different aspects of zebrafish gut differentiation. *Dev. Biol.* **2006**, *297*, 374–386. [CrossRef]
202. Yan, Q.; Li, J.; Yu, Y.; Wang, J.; He, Z.; Van Nostrand, J.D.; Kempher, M.L.; Wu, L.; Wang, Y.; Liao, L.; et al. Environmental filtering decreases with fish development for the assembly of gut microbiota. *Environ. Microbiol.* **2016**, *18*, 4739–4754. [CrossRef]
203. Stephens, W.Z.; Burns, A.R.; Stagaman, K.; Wong, S.; Rawls, J.F.; Guillemin, K.; Bohannan, B.J. The composition of the zebrafish intestinal microbial community varies across development. *ISME J.* **2016**, *10*, 644–654. [CrossRef]
204. Shin, N.R.; Whon, T.W.; Bae, J.W. Proteobacteria: Microbial signature of dysbiosis in gut microbiota. *Trends Biotechnol.* **2015**, *33*, 496–503. [CrossRef] [PubMed]
205. Roeselers, G.; Mittge, E.K.; Stephens, W.Z.; Parichy, D.M.; Cavanaugh, C.M.; Guillemin, K.; Rawls, J.F. Evidence for a core gut microbiota in the zebrafish. *ISME J.* **2011**, *5*, 1595–1608. [CrossRef] [PubMed]
206. Sullam, K.E.; Essinger, S.D.; Lozupone, C.A.; O'Connor, M.P.; Rosen, G.L.; Knight, R.; Kilham, S.S.; Russell, J.A. Environmental and ecological factors that shape the gut bacterial communities of fish: A meta-analysis. *Mol. Ecol.* **2012**, *21*, 3363–3378. [CrossRef] [PubMed]

207. Wu, S.; Wang, G.; Angert, E.R.; Wang, W.; Li, W.; Zou, H. Composition, diversity, and origin of the bacterial community in grass carp intestine. *PLoS ONE* **2012**, *7*, e30440. [CrossRef]
208. Wu, S.G.; Tian, J.Y.; Gatesoupe, F.J.; Li, W.X.; Zou, H.; Yang, B.J.; Wang, G.T. Intestinal microbiota of gibel carp (*Carassius auratus gibelio*) and its origin as revealed by 454 pyrosequencing. *World J. Microbiol. Biotechnol.* **2013**, *29*, 1585–1595. [CrossRef]
209. Wong, S.; Stephens, W.Z.; Burns, A.R.; Stagaman, K.; David, L.A.; Bohannan, B.J.; Guillemin, K.; Rawls, J.F. Ontogenetic Differences in Dietary Fat Influence Microbiota Assembly in the Zebrafish Gut. *Mbio* **2015**, *6*, e00687-15. [CrossRef]
210. Ye, L.; Amberg, J.; Chapman, D.; Gaikowski, M.; Liu, W.T. Fish gut microbiota analysis differentiates physiology and behavior of invasive Asian carp and indigenous American fish. *ISME J.* **2014**, *8*, 541–551. [CrossRef]
211. Yang, Y.; Song, W.; Lin, H.; Wang, W.; Du, L.; Xing, W. Antibiotics and antibiotic resistance genes in global lakes: A review and meta-analysis. *Environ. Int.* **2018**, *116*, 60–73. [CrossRef]

*Article*

# Changes in Lake Sturgeon Gut Microbiomes Relative to Founding Origin and in Response to Chemotherapeutant Treatments

**Shairah Abdul Razak [1,2], John M. Bauman [3], Terence L. Marsh [4] and Kim T. Scribner [1,5,*]**

1 Department of Fisheries & Wildlife, Michigan State University, East Lansing, MI 48824, USA; shairah@ukm.edu.my
2 Department of Applied Physics, Faculty of Science & Technology, Universiti Kebangsaan Malaysia, Bangi 43600, Malaysia
3 Michigan Department of Natural Resources Fisheries Division, Escanaba Customer Service Center, Gladstone, MI 49837, USA; baumanj@michigan.gov
4 Department of Microbiology and Molecular Genetics, Michigan State University, East Lansing, MI 48824, USA; marsht@msu.edu
5 Department of Integrative Biology, Michigan State University, East Lansing, MI 48824, USA
* Correspondence: scribne3@msu.edu

**Abstract:** Antibiotics, drugs, and chemicals (collectively referred to as chemotherapeutants) are widely embraced in fish aquaculture as important tools to control or prevent disease outbreaks. Potential negative effects include changes in microbial community composition and diversity during early life stages, which can reverse the beneficial roles of gut microbiota for the maintenance of host physiological processes and homeostatic regulation. We characterized the gut microbial community composition and diversity of an ecologically and economically important fish species, the lake sturgeon (*Acipenser fulvescens*), during the early larval period in response to weekly treatments using chemotherapeutants commonly used in aquaculture (chloramine-T, hydrogen peroxide, and $NaCl_2$ followed by hydrogen peroxide) relative to untreated controls. The effects of founding microbial community origin (wild stream vs. hatchery water) were also evaluated. Gut communities were quantified using massively parallel next generation sequencing based on the V4 region of the 16S rRNA gene. Members of the phylum Firmicutes (principally unclassified *Clostridiales* and *Clostridium_sensu_stricto*) and Proteobacteria were the dominant taxa in all gut samples regardless of treatment. The egg incubation environment (origin) and its interaction with chemotherapeutant treatment were significantly associated with indices of microbial taxonomic diversity. We observed large variation in the beta diversity of lake sturgeon gut microbiota between larvae from eggs incubated in hatchery and wild (stream) origins based on nonmetric dimensional scaling (NMDS). Permutational ANOVA indicated the effects of chemotherapeutic treatments on gut microbial community composition were dependent on the initial source of the founding microbial community. Influences of microbiota colonization during early ontogenetic stages and the resilience of gut microbiota to topical chemotherapeutic treatments are discussed.

**Keywords:** chemotherapeutants; environmental variation; founder effects; gut microbiome; lake sturgeon

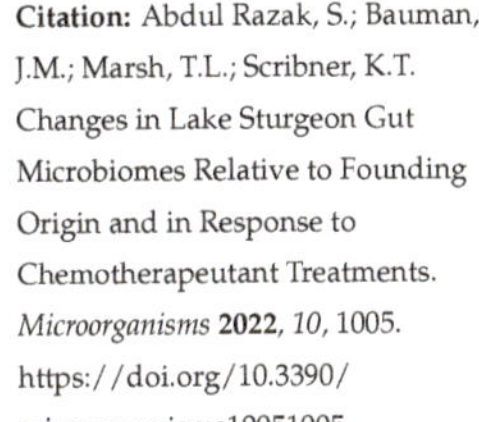

**Citation:** Abdul Razak, S.; Bauman, J.M.; Marsh, T.L.; Scribner, K.T. Changes in Lake Sturgeon Gut Microbiomes Relative to Founding Origin and in Response to Chemotherapeutant Treatments. *Microorganisms* **2022**, *10*, 1005. https://doi.org/10.3390/microorganisms10051005

Academic Editor: Konstantinos Ar. Kormas

Received: 2 April 2022
Accepted: 19 April 2022
Published: 10 May 2022



## 1. Introduction

Developing therapeutic regimes that limit stress-induced microbial infection or that reduces the occurrence of high mortality events in aquaculture is essential to successful fish production [1–3]. In aquaculture systems, stress in fish increases as a result of unfavorable rearing conditions (e.g., water quality, water source) or common production practices (e.g., handling, disease treatment), and interferes with physiological processes that aid in the defense against pathogens [3,4]. In response to a growing need for approved therapeutic regimes, fish culture managers have experimented with a variety of external (topical) disinfectant treatment strategies (hereafter referred to as "chemotherapeutants"). Indirect effects of these compounds, for example, associated with changes in gut microbial community composition and diversity, have not been rigorously evaluated in fishes [5]. However, disruptive effects of antimicrobial compounds on gut microbial communities are widely recognized in humans [6,7]. Given the propensity of larval fishes to internalize water and associated microbial communities via ingestion or respiration [5], one can postulate similar disruption to the gut microbiome of fish.

Common aquaculture treatment strategies include the use of chemotherapeutants (1) to treat infected fish as a function of visual detection of disease or in response to high mortality events, or (2) to administer regimented chemotherapeutant prophylactic treatments to reduce stress and prevent incidences of high mortality associated with pathogen infection [8]. Chemotherapeutant prophylactics used to reduce stress and prevent most prevalent disease-causing bacteria among cold-, cool-, and warm-water fish include chloramine-t (CT), hydrogen peroxide ($H_2O_2$), and sodium chloride ($NaCl_2$) [8]. CT is an external disinfectant found to effectively treat fish with or by prophylaxis to prevent external bacterial infections [9,10], particularly those associated with flavobacteriosis [8]. Similarly, hydrogen peroxide is an oxidative external disinfectant that has been used in aquaculture since the 1930s [11], and has been shown to reduce or eliminate infections, improving survival across multiple species at multiple life periods [12–15]. For example, $H_2O_2$ has been used to control mortality associated with finfish egg Saprolegniosis, as well as mortality of larval and juvenile fish infected with external pathogens, such as *Flavobacterium* [8]. $NaCl_2$ is one of the most commonly used chemotherapeutants for the control and treatment of external pathogens [16,17] as well as for osmoregulatory aid [8,18,19]. In addition, $NaCl_2$ use is believed to be associated with the 'shedding' of the mucosal layers, which exposes potential pathogens to treatment [20]. The toxicity and effectiveness of chemotherapeutants utilized in aquaculture differs by fish species, treatment regime, treatment concentration, as well as the life period during which treatments are administered [16,21–23]. Given that approved chemotherapeutants were initially and most commonly assessed using salmonids, and largely associated with external infections [8,9,15,16], further research is needed to evaluate the applicability of common chemotherapeutants when internalized and for other fish species, including those of conservation concern, such as lake sturgeon (*Acipenser fulvescens*).

Community ecological theory (e.g., [24]) can play an important role in studies of microbial communities and aquatic animal health. Theoretical and empirical studies emphasize the effects of processes associated with patterns in diversity, abundance, and species composition. One established theory in community ecology involves drift or neutral stochasticity on random compositional variation associated with initial colonization [25–27]. Other processes associated with community compositional changes involve response to disturbance [28,29]. Disturbance can be defined as a "single disruptive event or set of events that significantly changes ecological community structure and function" [28,30]. Some microbial communities might experience irreversible changes in taxonomic composition and function, for example, certain populations may be extirpated. Other communities may be resilient, where compositional changes are transitory, and community composition and diversity returns to pre-disturbance levels. Due to high functional redundancy in microbial communities [31], changes in community composition may also occur due to changes to and/or loss of minor populations. However, there may not be appreciable change to community function as roles of newly added constituents maintain the role(s) of original community members [30,32–34].

Widespread use of chemicals, drugs, and antibiotics is an example of a disturbance to microbiota, and is a rising concern in aquaculture [35–37]. With recent expansion and rapid growth in demand for aquaculture products in conservation and food production [38], chemical and antibiotic applications are increasingly used in aquaculture to control pathogens [39,40]. While short-term benefits are often realized, there is potential for damaging impacts of these practices, including disruption of co-adapted microbial communities. Further, large amounts of chemotherapeutants are passed into aquatic environments [29–41], including reduction in abundance of susceptible members of microbial communities.

Chemotherapeutants and antimicrobial compounds used in prophylactic treatments have been shown to be effective at reducing or preventing mortalities caused by pathogens [41]. However, some compounds are indiscriminate in their effects, and may also eradicate symbiotic and commensal gut microbial communities [42]. Downstream effects of antibiotic or chemical treatments on microbiomes are likely to have important consequences to fish hosts, and these effects are currently under-studied.

Few studies have documented changes in a fish-associated gut microbial community in response to chemical or antibiotic exposure to externally (topically) applied chemotherapeutants. The effect of ingested antimicrobial compounds on the gut microbiome was widely reported over a considerable period of time in several important aquaculture species, including rainbow trout (*Onchorynkuss mykiss*), using culture methods [43] or molecular-based methods [44]; hybrid tilapia (*Oreochromis niloticus* × *O. aureus*) [45]; and gibel carp (*Carassius auratus gibelito*) [46] (see review in [47]). Collectively, these studies reported that the taxonomic composition and diversity of gut microbial communities were impacted by antimicrobial treatments. These studies, however, focused mainly on describing gut microbiomes in fish at the juvenile stage. Fish at earlier life stages are more prone to pathogen infection [48], and thus may be more frequently exposed to antimicrobial compounds and chemotherapeutants. To evaluate the suitability of prophylactic treatments on fish larvae without compromising fish normal function, more studies are warranted pertaining to the influence of chemotherapeutants utilized in fish culture on gut microbiota.

In this study we characterized microbial community composition and diversity of the larval lake sturgeon gut using 16S rRNA-based next generation sequencing. Lake sturgeon are a species of conservation concern throughout most of their historic range. Where restoration goals to enhance lake sturgeon populations can be met by stocking, streamside rearing facilities (SRFs) are widely used [49]. SRFs utilize a natal water source and are believed to improve the probability of imprinting, compared to traditional hatcheries, which use non-natal well-water for rearing [49,50]. However, the use of SRFs pose challenges, which include increased exposure to temperature fluctuations and spatially and temporally variable surface water (e.g., stream) and hatchery microbial communities [51], including fish pathogens, during early development when mortality is high.

The objective of this study was to quantify and compare gut community diversity and taxonomic composition of larval lake sturgeon raised in an SRF as a function of different chemotherapeutant prophylactics and founding origin. Samples originated from individuals hatched from different egg sources (hatchery vs. wild stream) that were used to quantify the effects of four chemotherapeutants applied prophylactically. We hypothesized that colonization of the gastrointestinal tract would occur during early life stages [52], and that microbial communities associated with different egg incubation environments (hatchery vs. stream) would be reflected in different egg surface community composition and serve as innocula for the gut prior to initiation of chemotherapeutant treatments [5]. We further hypothesized that lake sturgeon larvae treated topically with chemotherapeutants would exhibit decreased GI tract microbial taxonomic diversity and different community composition relative to individuals from a control (no chemotherapeutant) treatment. Detailed effects of microbial founding source and chemotherapeutant treatment will provide insight into the consequences of these effects on host microbe compositional resiliency.

## 2. Methods

### 2.1. Study Site

Use of SRFs and natal water sources, such as the Black River Streamside Rearing Facility (BR-SRF), have been widely advocated in the Great Lakes basin as the preferred method for culturing lake sturgeon [49]. This study was conducted from 26 June to 30 July 2013 at the BR-SRF that is supplied with ambient river water (~680 L/min) from the Kleber Reservoir, located near primary spawning areas for lake sturgeon in the upper Black River in Cheboygan County, Michigan. The mean water temperature recorded during this study was 22.7 °C (min-max 19.9–26.3 °C).

### 2.2. Study Fish

Fish from different egg sources (hereafter called 'origins') were employed in this study. The first interaction of bacterial communities and fish progeny occur during early ontogenies even prior to larval hatch at the egg developmental stages [5]. Our previous data indicated that microbial colonization of egg surfaces and the egg microbial succession process is influenced by the community in surrounding water [51,53]. In the context of this study, eggs fertilized and incubated in the hatchery using water pumped from upstream was expected to differ chemically and in terms of biological (e.g., microbial) communities from eggs naturally fertilized and deposited on stream substrate in the natural spawning areas, owing to differences in substrate, groundwater, and surface water influences.

#### 2.2.1. Hatchery-Produced Gamete Collection, Fertilization and Incubation

The purpose of using hatchery-produced larvae was to quantify and compare the effects of different chemotherapeutant prophylactics on gut microbial community diversity and taxonomic composition of a progeny source produced using direct gamete takes, which is commonly utilized in finfish aquaculture including for lake sturgeon [54]. Gametes were collected from two male and two female lake sturgeon spawning in the upper Black River (designated as hatchery family A and B or HA and HB, respectively). Gametes were retained in coolers in the field with an ice pack and transported in plastic bags in river water to the BR-SRF for fertilization to maintain ambient river water temperature. Fertilization took place within four hours of collection. Egg de-adhesion procedures began by applying a Fuller's Earth solution (Sigma Aldrich, St. Louis, MO, USA) and gently mixing for 50 min using 50 micron-filtered river water. Subsequently, Fuller's Earth was rinsed from the eggs in 50 micron-filtered river water and at 15 min a 50 ppm iodiphor disinfection treatment was administered. Following a 10 min rinse in 50 micron-filtered river water to remove residual iodiphor using ambient river water, eggs were transferred to Aquatic Eco-Systems (Pentair, Inc., Delevan, WI, USA) J32 Mini Egg-hatching jars for incubation. Beginning two days post-fertilization, eggs were treated daily using a 500 ppm, 15 min bath treatment of hydrogen peroxide until 24 h prior to hatch. After hatch and during the free-embryo period (~7–10 days), lake sturgeon seek refuge in available substrate [55]. Therefore, free-embryos were raised in 10 L polycarbonate tanks (Aquatic Habitats, Inc., Speonk, NY, USA) with a single layer of 2.54 $cm^3$ sinking Bio-Balls (Pentair, Inc., Delevan, WI, USA; #CBB1-S) covering the tank bottom. Free-embryo lake sturgeon were raised until endogenous yolk resources were absorbed and fish began a 'swim-up' drift behavior (approx. 7–10 days post-hatch). At the onset of exogenous feeding the Bio-Balls were removed and live brine shrimp were provided at 28% body weight three times daily [56].

#### 2.2.2. Field Collection and Incubation of Wild Harvested Eggs and Larval Production

The purpose of using wild, naturally produced larvae for this study was to quantify the effects of different chemotherapeutant prophylactic treatments on gut microbial community diversity and taxonomic composition of an additional progeny source utilized in sturgeon aquaculture [54]. Naturally produced, fertilized eggs were collected from stream substrate in the Upper Black River at two spawning site locations approximately three days post-fertilization. Eggs were transported to the BR-SRF in river water and incubated, separated by capture location (wild site B, and site C and designated as WB and WC), in Aquatic Eco-Systems (Pentair) J32 Mini Egg-hatching jars. Eggs were treated daily using a 500 ppm, 15 min bath treatment of hydrogen peroxide until 24 h prior to hatch. After hatch and during the free-embryo period, lake sturgeon were reared in the BR-SRF under conditions described above as originally developed for Michigan State University Animal Use and Care standard operating procedures and subsequently published [56].

*2.3. Experimental Treatments*

Details concerning the experimental design including descriptions of facilities and equipment used to conduct the experiment and background to the major independent variables (hatchery or wild sample source and chemotherapeutant treatments) are provided in Figure 1 and below. At twelve days after initiation of exogenous feeding, we transferred 400 fish from each hatchery origin family (HA and HB) and each stream spawning origin group (WB and WC) into four 1.2 m diameter tanks, which were divided into eight partitions (50 fish per partition). Filtered (50 micron) river water was used in all tanks to remove large particulates and aquatic invertebrates and fish. Each partition was randomly assigned to one of four weekly chemotherapeutant treatment types, each with two replicates (Figure 1). The study began at fourteen days post-exogenous feeding after a two-day tank acclimation period, and continued for thirty-five days to quantify and compare the effects of different prophylactic chemotherapeutants on gut microbial community diversity and taxonomic composition. Chemotherapeutants administered in this study included those commonly utilized in traditional hatcheries and SRFs. Weekly prophylactic treatments in this study included: (1) 60 min, 15 parts per million (ppm) CT bath; (2) 15 min, 60 ppm $H_2O_2$; (3) 3 parts per thousand (ppt) $NaCl_2$ bath for 15 min followed 24 h later by a 15 min, 60 ppm $H_2O_2$ bath; and (4) a control (no chemical treatment). Fish were fed three times daily as described above, except on treatment days when feeding was delayed until all treatments had been performed. Each week, all fish from each treatment type (including no treatment controls) were transferred using a small aquarium dip net that was unique to each tank and section, to 10 L polycarbonate tanks equipped with one aerator in each tank. Fish were administered respective treatments, briefly rinsed in 50 micron-filtered river water, and placed back into their rearing tank. All treatments were administered on the same day, once per week except treatment 3, which included an additional treatment the following day with $H_2O_2$. Controls were handled in the same manner as all other treatment groups; however, similar to treatment 1, were held for 60 min in their 'treatment' tank before being rinsed and returned to their rearing tanks. Mortalities were removed from the tanks each day and recorded to quantify survival at the end of the study. The duration of this experiment lasted thirty-five days (forty-nine days post-exogenous feeding) to encompass the period of high mortality documented in SRFs.

Sampling for microbiota analysis took place following the end of the five-week treatment period the day following the last chemotherapeutant exposure. From each partition ($n = 2$), four fish were randomly collected ($n = 4$), and were euthanized with an overdose of MS-222 (Sigma-Aldrich, St. Louis, MO, USA) according to Michigan State University IACUC-approved animal use and care protocols. All fish ($n = 128$) were preserved in 80% ethanol and transported to MSU until dissections were performed within one month of collection.

*2.4. Fish Dissection, DNA Isolation, PCR Validation*

The distal gut (spiral valve) of each lake sturgeon larvae was recovered from fish following aseptic techniques. The distal gut was defined as the section that includes the end of the intestine through the distal end of the spiral valve. The spiral valve serves as the primary region of digestion and absorption, and thus may provide an area of abundant nutrients where a microbial community can flourish [57,58]. Exterior surfaces were swabbed with 100% ethanol before dissections of the whole digestive tract using sterile instruments. Dissections were performed with slight modification, as previously described by [59]. The intact alimentary tracts were cut from the body cavity, and the excised gut was immediately transferred into filtered-sterilized 80% ethanol solution for DNA isolation. All dissected samples were stored in −20 °C for <1 wk until DNA extractions were performed following the dissection.

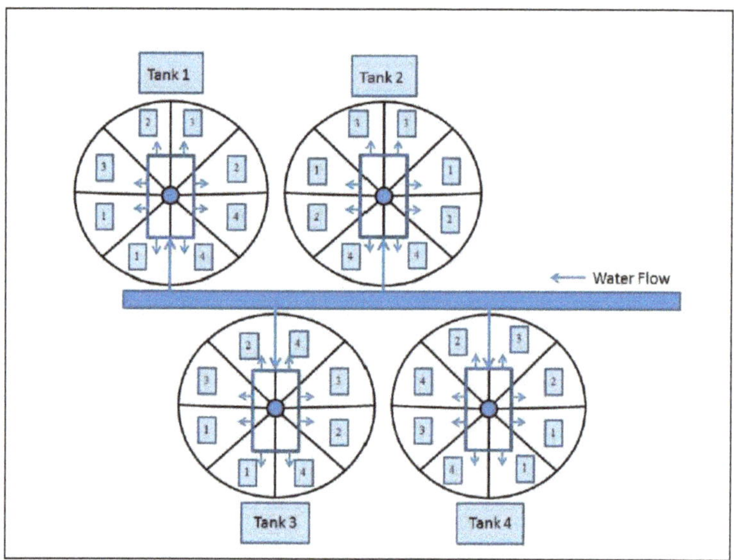


**Figure 1.** Schematic design of the larval chemotherapeutant study. Each 1.2 m diameter of tank held 400 fish from hatchery and wild naturally produced fish, which were divided into eight equal sized partitions (50 fish per partition). There were four tanks. Each partition was randomly assigned to one of four weekly treatment types, each with two replicates. Chemotherapeutant treatments included: (1) 60 min, 15 ppm CT bath; (2) 15 min, 60 ppm $H_2O_2$; (3) 3 parts per thousand (ppt) NaCl- bath for 15 min followed 24 h later by a 15 min, 60 ppm $H_2O_2$ bath labeled as NaCl/$H_2O_2$; and (4) a control (no chemical treatment) labeled as CTRL. Arrows indicate directions of water flow.

Each gut sample was first centrifuged at 12,000× *g* rpm for 15 min at 4 °C to pellet bacteria that may have leached from the sample before DNA was extracted. The combined gut and pelleted bacteria were extracted using The MoBio PowerSoil® DNA Isolation Kit (Carlsbad, CA, USA) including a bead-beating step, following protocols for low-biomass samples, as suggested by the manufacturer. The integrity of each DNA sample was assessed based on amplification of 1.4k bp of the 16S rRNA gene (amplicon based on 27F and 1389R primers) followed by gel agarose electrophoresis (1% agarose in TAE buffer). DNA concentrations were quantified by absorbance at 260 nm in a Microplate spectrophotometer (BioTek®, Winooski, VT, USA).

### *2.5. 16S rRNA Amplicon Sequencing and Sequence Pipeline Analyses*

Gut microbiota from lake sturgeon larvae were surveyed using high-throughput sequencing of the V4 region of the 16S rRNA gene. In total, 152 DNA samples (over four treatments, sampled at three time periods, including four positive controls, water samples, and technical replicates; see Figure 1 and description in Section 2.3) that had been validated to contain sufficient bacterial DNA (as shown by the presence of amplicon bands in electrophoresis) were submitted for sequencing at Michigan State University Research Technology Support Facility, (RTSF—(https://rtsf.natsci.msu.edu/genomics/ (East Lansing, MI, USA, accessed on 20 August 2014)). All sequencing procedures, including the construction of the Illumina sequencing library, emulsion PCR, and MiSeq paired-end sequencing v2 platforms of the V4 region (~250 bp; primer 515F and 806R) followed standard Illumina (San Diego, CA, USA) protocols. Michigan State Genomics RTSF provided standard Illumina quality control, including base calling by Illumina Real Time Analysis v1.18.61, demultiplexing, adaptor and barcode removal, and RTA conversion to FastQ format by Illumina Bcl2Fastq v1.8.4. Raw sequence reads were deposited to the NCBI

Sequence Reads Archive (SRA) under BioProject accession number PRJNA820564 (accessed on 28 March 2022).

Details of the microbial sequence data analyses pipeline and computing workflow were made following the suggested settings of mothur's operation protocol (https://www.mothur.org/wiki/MiSeq_SOP, accessed on 28 March 2022). Briefly, paired-end sequence merging, quality filtering, "denoising", chimera checking, and pre-cluster steps were conducted using an open-source workflow based on methods implemented by program mothur v.1.42 [60]. Sequence pipeline analyses were performed in mothur v.1.42 to accomplish reference-based OTU clustering (method = opticluster). Taxonomic assignment was performed by first aligning sequences data using the SILVA 132 bacterial reference database followed by clustering sequences defined with 97% identity and later classified using Ribosomal Database Project (RDP) 16 (V5.4) training set. Given the length of retained sequences, Operational Taxonomic Unit (OTU) criteria representing sequences that are not more than 3% different from each other, and our desire to compare data presented here to previous gut microbiome research (e.g., [52]), we chose to define taxonomic variation based on OTUs rather than Amplicon Sequence Variant (ASVs). Any sequence singletons that were detected were removed prior to downstream analyses. Rarefaction analyses were performed to evaluate the coverage for each sample based on the selected sequence depth. To minimize effects of under-sampling while maintaining as broad a dataset as possible, the final OTU table (Supplemental Table S1) was rarefied to a depth of 10,000 sequences per sample. Nine DNA samples with low sequence depth were discarded prior to downstream analyses. The community matrix describing sequence counts for all OTUs for all treatments associated with this study can also be found on GitHub at https://github.com/ScribnerLab/Chemotherapeutants.git (doi.org/10.5281/zenodo.6418537, accessed on 21 April 2022).

### 2.6. *Statistical Analysis of Bacterial Community Profiles and Ecological Statistical Analyses*

#### 2.6.1. Alpha Diversity

Measures of microbial community diversity including inverse Simpson (1/D) diversity indices and OTU richness for each sample from larvae from each chemotherapeutant treatment and origin (wild and hatchery egg sources) were calculated from community matrices derived from program mothur based on sequence data. All statistical analyses were carried out in the R program (v3.0.2).

Diversity indices (inverse Simpson and OTU richness) were first evaluated using a Kruskal–Wallis non-parametric test for control groups (no chemotherapeutant added) to determine whether there were statistical differences that existed in unperturbed microbial community alpha [α] diversity measures in the lake sturgeon larval GI tracts as a function of egg origin (wild vs. hatchery). The test was performed instead of parametric tests that assume a normal distribution. Next, the effects of chemotherapeutants and egg origin on measures of microbial gut community diversity were estimated based on a generalized linear model (GLM) using suitable probability distributions (inverse Simpson = Gamma distribution; Richness = Quasipoisson distribution) in R program (v3.0.2) using glm(). The GLM method has been shown to have high efficiency when estimating parameters, yielding interpretable estimates that also avoid transformation bias [53,54]. $p$-values $< 0.05$ indicated significance of the effect of variable on alpha diversity measures. Relative abundance estimates of bacterial phyla in all fish gut and water-associated microbial community samples at the end of the fifth and final treatment was determined using packages dplyr and reshape2 in program R (v3.0.2).

#### 2.6.2. Beta Diversity

We included several packages implemented in program R to estimate (beta [β]) diversity measures quantifying bacterial community compositional differences between samples and ecological statistics at the bacterial OTU level. Briefly, vegan [61] was used to produce a Bray–Curtis (BC) [62] dissimilarity matrix, and to perform non-metric dimensional scaling (NMDS) ordination as a means of characterizing differences in microbial community composition among samples. We used the nmds function to perform non-metric dimensional scaling (NMDS) ordination to visualize community compositional differences based on sample BC dissimilarity. The ggplot and ggplots2 packages [63] were used to create ordination plots to visually compare sample gut community composition as a function of different treatments, between sampling origins, and water samples.

Next, we performed multivariate hypothesis testing to quantify differences in community composition among samples originating from different groups based on locations of egg origins and exposed to different chemotherapeutant treatments using the adonis function [61] in program R (v3.0.2). Two different fish families (hatchery origin) and two river spawning locations (wild origin) were treated as replicates. Analyses focused on the effects of chemotherapeutant treatments and origin. Permutational multivariate analyses of variance (PERMANOVA) was conducted on BC dissimilarity matrices of fish associated microbial community composition [64,65]. Under the null hypotheses, the centroids of the groups (fish from either hatchery and wild groups that were exposed to different chemotherapeutant treatments) were expected to be equivalent for all groups under random allocation (i.e., based on permutation) of individual sample units to the groups.

Analyses investigated whether host origin and/or chemotherapeutant treatment had a significant effect on microbial community structure. NMDS and PERMANOVA were performed on fish gut communities within each origin group to determine whether chemical treatments had effects on fish gut microbiota. Under the null hypothesis, chemotherapeutant treatments were not expected to significantly affect fish gut community taxonomic composition within an origin group, in part because eggs from both hatchery and wild origins were exposed to peroxide during incubation that was believed to reduce and taxonomically homogenize samples for all treatments and both origins. PERMANOVA analyses that indicated significant treatment effects were then analyzed using post hoc tests using betadisper and permutest functions followed by a Tukey test to determine which treatment(s) differ significantly in larval lake sturgeon gut bacterial taxonomic composition.

2.6.3. Differential Abundance of OTUs and Biomarker Identification across Treatments

To determine the operational taxonomic units (OTUs) that most likely explained differences in microbial larval lake sturgeon gut community composition between fish from different origins and among different chemotherapeutant treatment groups, we next employed linear discriminant analysis (LDA) effect size (LEfSe) methods [66]. In general, the LEfSe algorithm identifies genomic features (i.e., bacterial OTUs) that were differentially abundant in different experimental groups (origin groups and treatments), then ranks them based on that abundance differential. The larger the difference in relative abundance between groups, the higher the importance of that OTU.

The algorithm first identified features (OTUs) that were statistically different among origin groups based on the nonparametric factorial Kruskal–Wallis (KW) rank sum test. Additional tests assessed the consistency of differences using unpaired Wilcoxon rank sum tests. In the final step, LEfSe used LDA to rank each differentially abundant taxon in order of the difference in abundance based on an LDA Score (log-scale). Results represent a scale indicating "importance" of an OTU in origin group differences in microbiota composition [66].

To run LEfSe, a tabular file was generated from a shared file that contained no singletons in the program mothur v.1.39.5. The tabular file consisted of taxonomic relative abundance in gut community samples from the four different origin samples that were all exposed to four chemotherapeutant treatments. This tabular file was transferred using an online bioinformatics toolkit developed by the Huttenhower lab to perform LEfSe analyses (https://huttenhower.sph.harvard.edu/galaxy/, accessed 20 November 2014).

## 3. Results

### 3.1. Diversity of Gut Microbial Community Composition

A total of 144 samples were retained after quality filtering was performed in the sequence pipeline analyses. Comparisons of lake sturgeon larvae gut microbial community composition at the level of phyla indicated that three major phyla dominated more than 65% of total community abundance across all fish samples (*Firmicutes* 16%, *Proteobacteria* 36.5%, and *Actinobacteria* 15.1%). Phyla detected in the remainder of the gut community included *Acidobacteria*, *Bacteroidetes*, and *Verrucomicrobia* that collectively comprised 30% of gut communities.

The relative abundance of the most dominant phylum, *Firmicutes*, was fairly consistent across treatments for fish samples from all hatchery and wild origin groups and wild groups (HA, HD and WB, WC, respectively). One exception was WB larvae exposed to salt (mean 58%) and WC fish exposed to peroxide (mean 50%) that were relatively low compared to other treatments (Figure 2a). When comparing the abundance of *Firmicutes* across all groups, fish from hatchery family D (HD) had a lower percentage of *Firmicutes* (mean range from 51–66%). *Proteobacteria* relative abundance was likewise relatively uniform across treatments (13–28% of total abundance) with the exception of WB fish that were treated with chloramine-T, CT (6%). *Actinobacteria* were present at 1% in fish that were not exposed to any chemotherapeutant (control) and only in fish from HA and WC origin groups. At the genus level, Firmicutes were represented by two genera, *Clostridium_sensu_stricto* & unclassified genera from family *Clostridiaceae*. We found that *Clostridium_sensu_stricto* were the most dominant genus (mean range: 30–51% of the total community) for all fish of hatchery origin (except for HA fish exposed to peroxide), whereas all fish of wild origin had unclassified taxa from *Clostridiaceae* family (mean range: 29–62%) as the most abundant genus across any treatment (Figure 2b). Genera from phylum *Proteobacteria* including several unclassified taxa from *Betaproteobacteria*, unclassified taxa from *Enterobacteriaceae*, unclassified taxa from *Rhodobacteriaceae*, and *Deefgea* all were present at lower percentages of abundance with more amounts of variation across fish groups and treatments (Figure 2b). The only genus in the phylum *Actinobacteria* that was detected among dominant taxa was the genus *Zhihengliuella*, present in HA control fish (mean 2.2%) and WC control fish (mean 1.4%).

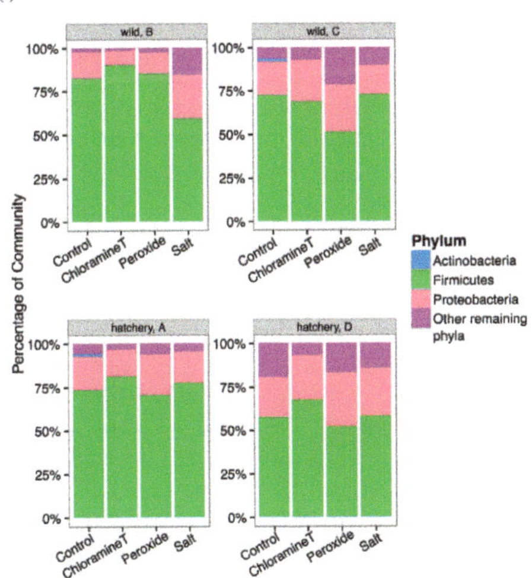


**Figure 2.** *Cont.*

(b)

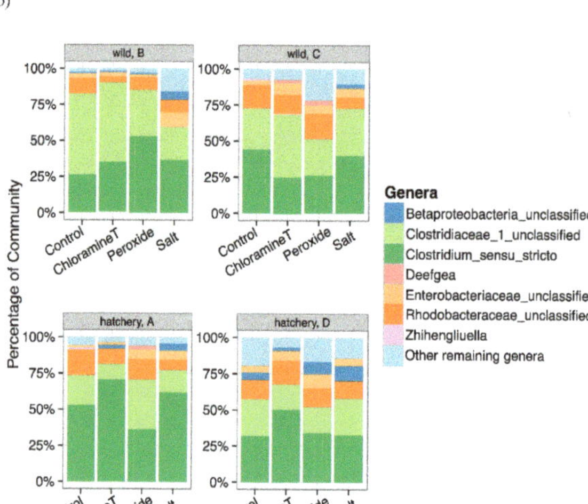


**Figure 2.** Taxonomic composition of bacterial communities identified from the lake sturgeon larval GI tracts (**a**) at the phyla level and (**b**) at the genera level. (**a**) Relative abundance (percentage) of dominant bacterial phyla found in the gut microbiota of lake sturgeon larvae separated based on sample family/group to display variation in communities across prophylactic treatments. Three predominant phyla were present in gut microbial communities (*Firmicutes*, *Proteobacteria*, *Actinobacteria*). The other phyla were characterized as Others; (**b**) relative abundance (percentage) of dominant bacterial taxa found in fish gut samples, separated by family/group and treatment. Among the most abundant taxa included *Unclassified Betaproteobacteria*, *Unclassified Clostridiaceae_1*, *Clostridium_sensu_stricto*, and *Unclassified Enterobacteriaceae*.

Figure 3a,b revealed results of GLM tests comparing inverse Simpson indices and a number of observed taxa among chemotherapeutant treatments and origin groups. As opposed to our initial hypothesis, fish in the control treatment (CT) had less diverse gut communities (both inverse Simpson and richness) with the exception of fish in family HD. Fish exposed to salt treatment were characterized by higher inverse Simpson and greater taxa richness than communities from samples exposed to other chemotherapeutant treatments in wild family, WB. For individuals from wild family WC, we found that fish exposed to peroxide had a greater number of taxa relative to fish from wild family WB from the control group (Figure 3b). Our analyses did not quantify family effects since families (hatchery origin) and stream locations (wild origin) served as replicates for each origin. We observed large heterogeneity among samples collected following different treatments and between egg origins (Figure 3a,b). For example, differences between communities sampled from individuals from the control and salt treatment groups associated with hatchery family HD and wild family WB were observed for Simpson's inverse diversity and were higher for samples in the peroxide and salt treatments. To summarize, Kruskal–Wallis tests for taxa richness and inverse Simpson among fish from control groups indicated that no significant difference existed between groupings based on egg origins ($p > 0.05$). Statistical analyses based on the generalized linear model (GLM) indicated that gut communities of individuals exposed to certain treatments (salt and peroxide) had significantly different levels of taxa richness, but not on the inverse Simpson indices (Supplementary Table S2).

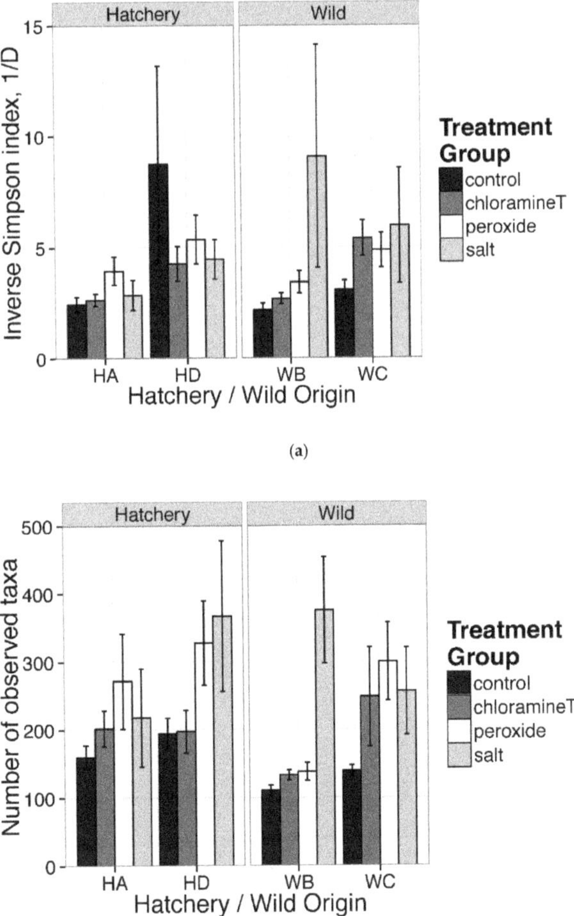


**Figure 3.** Estimates of alpha diversity: (**a**) inverse Simpson index; (**b**) number of observed taxa (OTU richness) for lake sturgeon gut microbial communities from all samples of treatments and families within location of origins. Each bar indicates mean with S.E. for each treatment from each family within origins.

### *3.2. Differences in Gut Microbial Community Composition between Fish Group Origin and among Chemotherapeutant Treatments*

Non-metric dimensional scaling (NMDS) ordination of BC dissimilarities in microbial taxonomic composition of gut communities was performed to visualize community compositional relationships among larval gut samples associated with fish from different egg origin and exposed to different chemotherapeutant treatments. Four NMDS plots were generated, including Figure 4a: all fish gut microbiota; Figure 4b: gut microbiota for fish in control treatment groups only; Figure 4c: gut microbiota community relationships among chemotherapeutant treatments for fish originating from a hatchery (two families, HA and HD); and Figure 4d: gut microbiota for fish among chemotherapeutant treatments originated from the stream substrate (wild groups from two spawning locations; WB and WC). All ordination plots were characterized by stress values ~0.2 indicating that data were well represented in 2D NMDS plots. Community membership across samples of similar origin (either from the wild, or from the hatchery production) were clustered together regardless of treatment groups as denoted by the ordination pattern suggesting influence of egg origins on fish gut microbiome (Figure 4a). Baseline community membership in fish without any chemotherapeutic treatment (control group) was visualized in Figure 4b, revealing that fish from eggs collected from the wild (WB and WC) exhibited considerably higher inter-sample variation in community composition relative to the variation among fish originating from hatchery crosses (HA and HD) across all chemotherapeutant treatments.

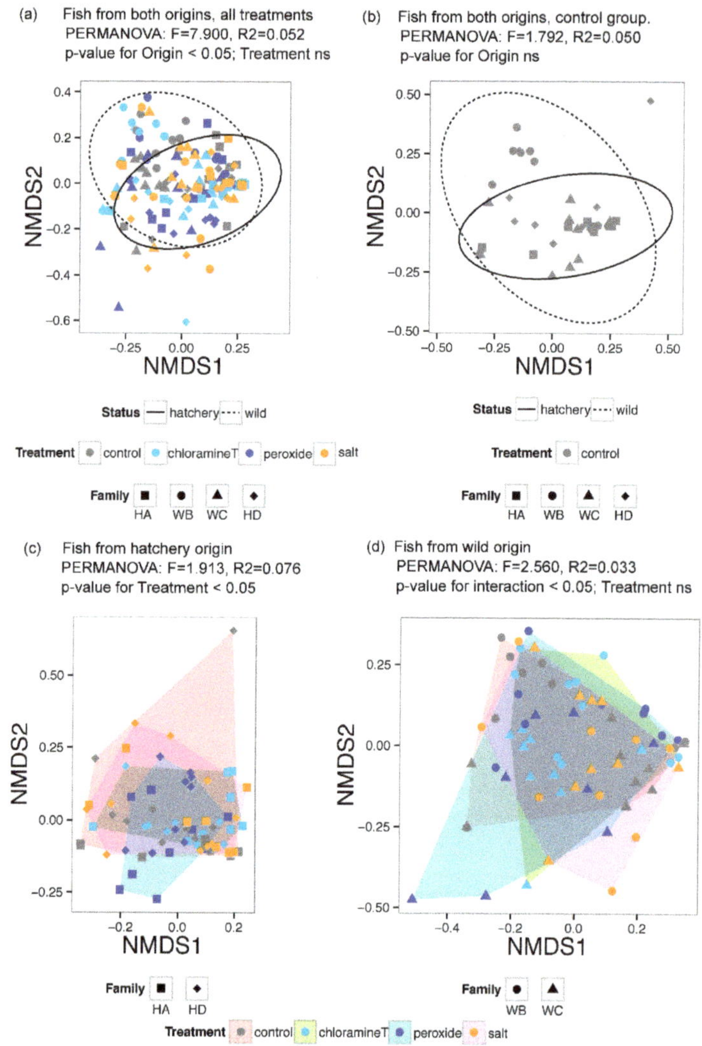


**Figure 4.** Non-metric dimensional scaling (NMDS) ordination of Bray–Curtis differences in larval lake sturgeon gut microbial taxonomic composition (**a**,**b**) and results of accompanying PERMANOVA analyses (**c**,**d**) characterizing relationships among larval gut samples associated with fish from different groups associated with egg origin and those exposed to different chemotherapeutant treatments.

To quantitatively test for gut community compositional differences among chemotherapeutant treatment and origins, PERMANOVA was performed. Comparisons of microbial OTU beta diversity across samples from the controlled groups indicated that gut microbial communities from control groups were not significantly influenced by the egg origin (Table 1). Subsequently, the effects of chemotherapeutant treatments on fish gut microbiomes were investigated across all samples taking into consideration both the effects of chemotherapeutant treatment and where the fish originated from (hatchery vs. wild). No influence of chemotherapeutant treatment was detected, but the effects of egg origin were significant (Table 2).

**Table 1.** PERMANOVA showing variability among fish gut microbiota across control groups only.

| | Df | Sum Sq | Mean Sq | F-Model | $R^2$ | Pr (>F) |
|---|---|---|---|---|---|---|
| Origin (O) | 1 | 0.416 | 0.416 | 1.792 | 0.050 | 0.107 |
| Residuals | 34 | 7.887 | 0.232 | | 0.950 | |
| Total | 35 | 8.302 | | | 1.000 | |

**Table 2.** PERMANOVA showing variability among fish gut microbiota across all samples. Results revealed that origin effect (O) significantly influencing gut microbial communities composition for at least one sample across treatments and origins (PERMANOVA test permutation = 1000).

| | Df | Sum Sq | Mean Sq | F-Model | $R^2$ | Pr (>F) |
|---|---|---|---|---|---|---|
| Treatment (T) | 3 | 0.897 | 0.299 | 1.389 | 0.028 | 0.126 |
| Origin (O) | 1 | 1.699 | 1.699 | 7.900 | 0.052 | $p < 0.01$ |
| Residuals | 139 | 29.900 | 0.215 | | 0.920 | |
| Total | 143 | 32.496 | | | 1.000 | |

Additional analyses of chemotherapeutant effects on fish gut microbiome composition were investigated separately based on fish origin (hatchery vs. wild). Chemotherapeutant treatments had significant effects on larval gut microbiomes between individuals from different hatchery families (HA, HD) as indicated by PERMANOVA test results (Table 3; $p = 0.012$). However, the effect of chemotherapeutant treatment was not evident between fish from eggs collected in different regions of the stream (WB, WC), although a significant interaction was observed between origin group and treatment (Table 4).

**Table 3.** PERMANOVA showing variability among fish gut microbiota across all samples originating from the hatchery.

| | Df | Sum Sq | Mean Sq | F-Model | $R^2$ | Pr (>F) |
|---|---|---|---|---|---|---|
| Treatment (T) | 3 | 1.108 | 0.369 | 1.913 | 0.076 | 0.012 |
| Family (F) | 1 | 0.402 | 0.402 | 2.084 | 0.028 | 0.057 |
| Treatment (T) × Family (F) | 3 | 0.737 | 0.246 | 1.272 | 0.050 | 0.191 |
| Residuals | 64 | 12.357 | 0.193 | | 0.846 | |
| Total | 71 | 14.603 | | 1.000 | | |

**Table 4.** PERMANOVA showing variability among fish gut microbiota across all samples originated from the stream substrate (wild).

| | Df | Sum Sq | Mean Sq | F-Model | $R^2$ | Pr (>F) |
|---|---|---|---|---|---|---|
| Treatment (T) | 3 | 0.784 | 0.261 | 1.241 | 0.048 | 0.209 |
| Family (F) | 1 | 0.539 | 0.539 | 2.560 | 0.033 | 0.022 |
| Treatment (T) × Family (F) | 3 | 1.390 | 0.463 | 2.202 | 0.086 | 0.007 |
| Residuals | 64 | 13.471 | 0.210 | | 0.832 | |
| Total | 71 | 16.184 | | 1.000 | | |

To better understand the effects of different chemotherapeutants in gut communities from hatchery fish, post hoc tests, betadisper and permutest, were conducted followed by Tukey's test. The adjusted *p*-value from Tukey's test indicated that none of the communities associated with different treatments differed statistically, although betadisper revealed that the distance of each point to the centroid for salt and peroxide differed.

*3.3. Identification of Bacterial Taxa Influenced by Chemotherapeutant Treatments*

Given findings of effects of origin and treatment on microbial community beta diversity, we used LEfSe to identify which taxonomic groups showed the largest differences in relative abundance when fish from the same origin were exposed to treatments (Figure 5). We first compared microbial communities from fish from the control groups from hatchery and wild origins at the genus level (all vs. all). We found taxa associated with phylum Actinobacteria, including genus *Methylocystis* from phylum *Firmicutes*, and genus *Salinicoccus* from phylum *Proteobacteria* differed in abundance (LDA score higher than 2.0, $p < 0.05$, see Figure 5a) for the comparison between fish communities in the control group between both egg origins. These three genera were present in higher abundance in hatchery fish samples as opposed to wild fish.

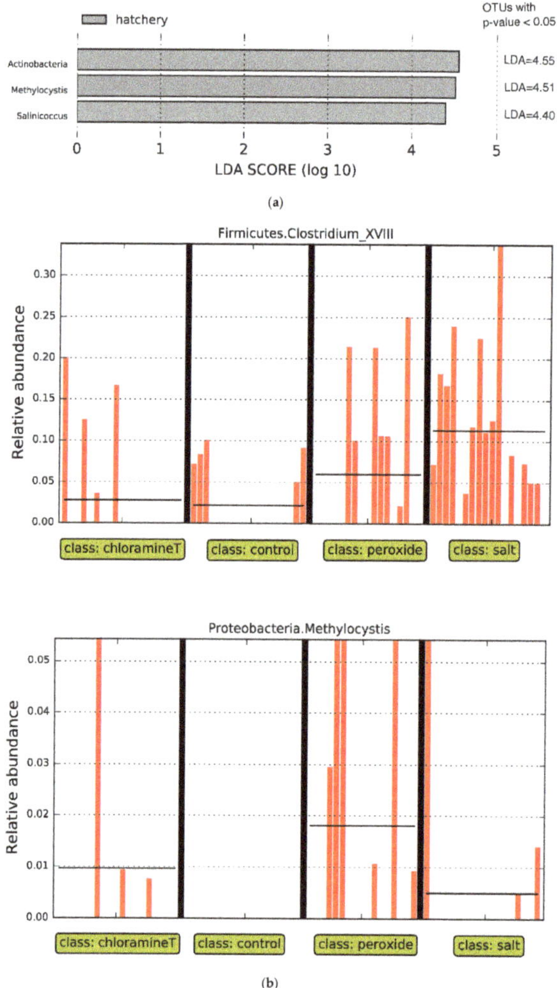


**Figure 5.** *Cont.*

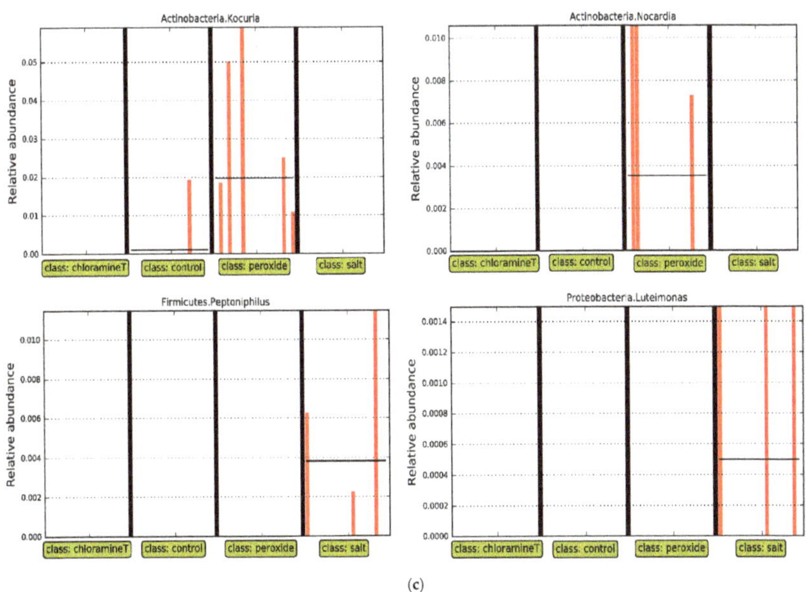


(c)

**Figure 5.** LEfSe analyses for (**a**) fish exposed to control treatment only, comparing hatchery and wild origins; (**b**) fish from all chemotherapeutant treatments against control across fish samples originated from wild; (**c**) fish from all chemotherapeutant treatments against control across fish samples originated from hatchery.

We likewise compared communities of fish from the control treatment within each origin (wild and hatchery, respectively) to other chemotherapeutant treatment groups (one vs. all). LEfSe analyses performed with fish from the wild group detected two differentially abundant taxa associated with genus *Clostridium_XVIII* (Phylum *Firmicutes*) and *Methylocystis* (Phylum *Proteobacteria*). Both genera were present in high abundance in the guts of fish exposed to the peroxide treatment, and for genus *Clostridium_XVIII*. The taxa were also abundant in the guts of fish from the salt treatment (LDA score higher than 2.0, $p < 0.05$, see Figure 5b). For hatchery origin fish, LEfSe analyses on fish that were treated with peroxide revealed the presence of genera *Kocuria* and *Nocardia* (both from phylum *Actinobacteria*) in high abundance, while fish exposed to the salt treatment had *Peptoniphlus* and *Luteimonas* that were in higher abundance compared to individuals from other treatments (see Figure 5c).

## 4. Discussion

### 4.1. General Findings and Relevance to Aquaculture

Understanding interactions between microbes and the host surfaces they colonize is important to aquatic animal health and aquacultural production [5,35]. Potentially harmful changes can occur to beneficial gut microbes from the over-utilization of chemotherapeutants [34,67], which can result in ecological drift or selective community alteration that can favor increases in the abundance of undesirable taxa [68–70]. Additional adverse effects related to antibiotic use include pathogen resistance, suppression of the immune system, increased rates of allergies, autoimmunity, and other immune-inflammatory conditions [34]. Microbial community changes anatomically and ontogenetically in response to spatial and temporal environmental variation, and changes related to perturbations have also been described in fish taxa [5], but are less well investigated. In well studied humans, microbiomes within individual hosts usually vary in composition across anatomical sites, and microbial taxonomic composition can vary over time in response to factors such as diet, physical activities, and medication intake [32,71,72].

In this study, we found little evidence for the influence of commonly used chemotherapeutant treatments applied topically in water baths to larval lake sturgeon prophylactically on gut microbiome composition. Data did indicate greater influence of microbial founder effects (hatchery vs. wild stream origin), which may be explained by exposure to environmental sources during earlier life stages or influences of genetic effects [73,74]. Results could also indicate genetic or maternal effects reflecting different family membership of fish from different origin groups. We provide an interpretation of origin and chemotherapeutant treatment results and discuss implications for aquatic animal health in aquaculture generally.

### 4.2. *Effects of Chemotherapeutant Treatments on Larval Gut Microbial Communities*

All chemotherapeutants used in our study are commonly used for the treatment of external pathogens rather than orally administered to fish. In fish aquaculture, prophylactic treatments are widely used to control pathogenic bacteria disease outbreaks that commonly occur in hatcheries during vulnerable early life stages. Chloramine-T and peroxide are widely used to control and eliminate infection associated with flavobacteriosis [8]. Overall, our results indicate that chemotherapeutant treatments during larval stages did not result in large changes in the composition of the intestinal microbiota, at least during the short observation and experimental period (five weekly exposures). Although GLM suggested that taxa richness may be significantly influenced by certain treatments applied, such as salt and peroxide, the same treatments did not have a significant effect on the inverse Simpson indices. PERMANOVA and least square means tests revealed that the chemotherapeutant treatments employed in our study had only a minor effect on intestinal gut microbiome in lake sturgeon larvae; although effects varied among fish with different backgrounds associated with families and their sampling origin.

There are several potential explanations for the comparatively small effects of chemotherapeutant treatments on larval gut microbial communities. One explanation is that the externally administered treatment did not enter the digestive tract in significant enough concentrations or duration to alter the gut community. When larval fish are provided chemotherapeutants prior to feeding, rather than during feeding, microbial compositional stasis suggested that the chemicals may not enter the gastrointestinal tract. Alternatively, the effect of the treatment may not have been evident due to the short treatment duration (15–60 min bath immersion) and weekly periodicity of chemotherapeutant treatments. Exposure to chemotherapeutants, consistent with our methodology, may not have been of sufficient concentration to result in quantifiable changes in gut community composition. In addition, fish were returned into their tank partition after treatments, and that may have allowed rapid recolonization of gut microbiota from the surrounding water. Further, chemotherapeutant treatments were administered at seven-day intervals, potentially allowing community recovery. The microbial communities may have exhibited resiliency to chemotherapeutant treatment; returning to a similar compositional state during the several day period between the timing of treatment and sampling for gut interrogation. Further studies are warranted to quantify the amount of any compound entering the gut during the treatment period to ascertain causal relationships.

In the LEfSe analyses, three out of thousands of microbial taxa appeared to be tied to differences between untreated fish in a hatchery and the wild. After fish were exposed to chemical treatments, different taxa were reported to be differentially abundant. Those taxa, however, are not among the dominant taxa. It is unclear how treatment differentially affected the relative abundance of these taxa. Results could indicate that the gut microbiota were either resistant or exhibited resilience in community composition, where treatment-based changes were short-lived and communities rapidly returned to their original state [30]. The communities could also have had different compositional taxonomy, yet were still able to maintain function (functional redundancy). Navarrete and colleagues [75] focused on determining the effects of a dietary inclusion of *Thymus vulgaris* essential oil (TVEO) on microbiota composition, compared with a control diet without TVEO over a 5 week period. Their study indicated high similarities between gut microbiota in treated and non-treated fish, and TVEO induced negligible changes in gut microbiota profiles. Essential oils include volatile liquid fractions produced by plants that contain the substances usually responsible for defenses against pathogens and pests due to their antibacterial, antiviral, antifungal, and insecticidal activities [76]. We conclude, based on LEfSe results (Figure 5b,c), that gut microbiota composition in lake sturgeon was persistent and stable throughout the trial, producing relatively consistent molecular profiles.

We detected three major phyla *Firmicutes*, *Proteobacteria*, and *Actinobacteria*, that dominated the lake sturgeon larval gut community across all samples (Figure 2). The most predominant taxa that were detected from phyla *Proteobacteria* (such as *Enterobacteriaceae*, *Rhodobacteriaceae*) are Gram-negative bacteria. Many studies have shown that Gram-negative bacteria are resistant to commercially available antibiotics partly due to their thick cell wall structure compared to Gram-positive bacteria [77,78]. *Enterobacteriaceae* include a group of bacteria known as extended-spectrum beta-lactamase (ESBL) *Enterobacteriaceae* that can confer resistance to antibiotics via production of the β-lactamase enzyme, which can inactivate certain β-lactam antibiotics [79].

Another major phylum, *Firmicutes* that were detected in fish guts across all families and treatments was primarily represented by Unclassified *Clostridiaceae1* and taxa *Clostridium sensu stricto*. Although *Clostridia* are Gram-positive, these bacteria have been identified as part of commensal gut microbiota that plays major roles in the maintenance of the gut homeostasis. Several features associated with *Clostridium* spp. could explain why this taxon can thrive in the gut and can likewise be resistant to prophylactic treatments administered in our study. In humans, *Clostridium* spp. are involved in defenses inside the intestinal microecosystem along with gut-associated lymphoid tissue (GALT), and confer resistance against pathogen infections. This taxon is thought to have immunological tolerance [80]. In addition, cultured *Clostridium* spp. exhibit the ability to form endospores, which offers this bacteria ecological advantages for survival under adverse conditions [80,81].

Comprehensive studies on adverse effects of antibiotic use to the gut microbiomes were reported in other fish species [5,37,39] and in humans [33]. Exposure to antibiotics can have profound effects on resident microbial communities inside human guts [34,72]. Several studies reported changes in density or gut microbiome composition, for instance in human infants who receive antibiotics [82]. Dethlefsen et al. [6,7] documented the pervasive effects of an orally administered antibiotic to adult gut microbiomes, associated with decreases in taxa richness and evenness and can lead to community changes in composition and function [33].

Relatively few studies have been conducted addressing the effects of chemotherapeutants administered topically in water baths on fish gut microbial communities as conducted in this study. Most studies have been conducted on salmonids or tilapia [43,45,75] and gibel carp (*Carassius auratus gibelito*) [46], and have focused on the effects of antibiotics administered orally to address infection levels of known pathogenic bacteria. Navarrete et al. [44] reported that gut microbiomes of salmonids exposed to the antibiotic oxytetracycline (OTC) that was orally administered were characterized by lower taxonomic diversity and were primarily composed of *Aeromonas*. The results were consistent with findings from another study conducted to evaluate the effects of orally administered antibiotics to gibel carp [46]. Importantly, the results from studies using orally administered antibiotics different from our data.

### *4.3. Sources of Heterogeneity Associated with Microbial Community Origin*

A prerequisite for developing a strategy for microbial pathogen control is a knowledge of resident aquatic microflora associated with fish larvae, and how interactions between larvae and microflora occur. De Schryver & Vadstein [83] suggested that the primary means by which pathogens could be controlled is the water surrounding animals.

Fish produced from wild eggs show greater community diversity compared to artificially produced fish in the hatchery (Figure 4b). Thus, the initial inoculation location on the egg chorion surface likely determined their community during later life stages, as community successional changes occurred [5]. Alternatively, gut microbial communities in wild fish may have exhibited greater resilience to treatments and maintained their gut compositional similarity. In contrast, hatchery fish originated from eggs that have been artificially produced in hatchery facilities; therefore, they had limited contact with their respective natural habitat like the wild eggs, except their egg surfaces reflect aquatic communities where their parents spawned (in the hatchery). This could also suggests that domestication selection, in terms of hatchery gut community establishment, occurs in fish produced in a hatchery, affecting the community structure of their gut microbiome.

In fishes, microbial binding to host cell surfaces is often mediated through the interactions of bacterial carbohydrate binding proteins (lectins) with host cell surface carbohydrates [84,85]. Stream substrates are extremely variable and likely harbor different microbial communities than are present in stream water used in stream-side or traditional (often ground water) hatchery facilities. Different microbial communities have been characterized from naturally spawned lake sturgeon eggs in the WC and WB areas of the upper Black River previously (Marsh unpubl. data). Larvae hatching from eggs deposited on stream substrates typically remain in close proximity to egg surfaces for long periods when gill surfaces likely acquire and internalize egg surface-bound microbial taxa. If this period is indeed the point at which larvae internalize egg surface-bound microbial taxa, then this occurs prior to the full development of alimental structures [5]. Thus, differences in founding microbial communities between hatchery and wild sources are probable. This source of heterogeneity and subsequent successional changes in community diversity and composition can be important for later life stages of fishes.

Several studies of gut microbiota in fish with different genetic backgrounds have documented that host genotype (broadly defined) may contribute to compositional heterogeneity among individuals in fish gut microbiota, at least to some extent. Abdul Razak et al. [53] studied catfish gut microbiome assembly and quantified changes in gut microbiome development from eggs to stock-out juveniles released into nursery ponds. The study identified host genotype (families), dietary factors, and environmental (rearing pond) effects. Significant differences in alpha diversity were evident at the egg stage, yet the differences diminished as fish matured. The authors found evidence of significant interactions between family and stocking pond environment on larval gut microbiota composition, as was also found in this study.

Another study [86] demonstrated evidence of host effects on the intestinal microbiota of captive and wild whitefish. Whitefish (*Coregonus* spp.) species pairs and their reciprocal hybrids were reared in captivity under a controlled environment. Analyses revealed significant effects of the host genetic background on the taxonomic composition of the transient microbiota. Navarrete et al. [87] assessed the relative effects of a host (genotype) and diet to gut microbiome composition of rainbow trout (*Onchorhynchus mykiss*). Full-sibling fish from four non-related families were fed two diet regimes in comparison to the control group. Results showed that some relative abundance of several bacterial taxa differed among trout families, indicating that the host genotypes may influence gut microbiota composition. In addition, the authors reported that the effect of diet on microbiota composition was dependent on the trout family. Studies on other organisms, such as chickens, also showed that under a common diet and husbandry practices, gut microbiota composition differed between two lines (high weight, HW and low weight, LW) [88]. Findings from Blekhman et al. [89] indicate that human gut microbial variation are driven by host genetic variation involving genes that have been previously associated with microbiome-related complex diseases. They also showed that host genomic regions associated with microbiomes have high levels of genetic differentiation among human populations, suggesting host-genomic adaptation to environment-specific microbiomes.

This finding could be possibly true for fish as well where variation in gut microbiome is attributed to genetic background.

## 5. Conclusions

Findings in this study detail observed differences in microbial founding sources (water borne and substrate specific egg microbial incubation environments) and chemotherapeutic treatments to developing microbial communities during early ontogenetic stages. These results provide an insight into the consequences of prophylactic treatments and host-microbe interactions. Our study serves as a baseline providing information on the indirect effects of chemotherapeutant intervention that could either positively or negatively affect the normal gut microbiota. Results of minor effects associated with use of chemotherapeutants prophylactically suggest that topical use at the ontogenetic stage and concentration used may not have negative indirect effects on resident gut microbial communities. Thought should be given to the selection of locations to collect gametes to bring into culture. Future work could profitably focus on identifying microbial taxa that colonize the external surfaces of the fish (gill plate, gills, ventral area between pectoral fins, etc.) to see how external treatments impact the colonization of external microbes. Further studies are also warranted that would compare the effects of treatments when administered following pathogen infection.

**Supplementary Materials:** The following supporting information can be downloaded at: https://www.mdpi.com/article/10.3390/microorganisms10051005/s1, Supplementary Materials including the final microbial community matrix describing sequence counts associated with each microbial taxa and sample is provided in Supplemental Table S1. Output of GLM analyses of sample alpha diversity is presented in Supplementary Table S2.

**Author Contributions:** K.T.S. and T.L.M. obtained funding for the project. All authors contributed to the formulation of the experimental design. S.A.R. and J.M.B. performed the experiment and collected the data. S.A.R. and T.L.M. performed the statistical analysis. All authors participated in the manuscript preparation. All authors have read and agreed to the published version of the manuscript.

**Funding:** Funding was provided by the Great Lakes Fishery Trust (Project 1202), the Michigan Department of Natural Resources, Department of Microbiology and Molecular Genetics, Michigan State University (MSU), and Michigan State University Ag-Bio Research. S.A.R. was supported under Universiti Kebangsaan Malaysia (UKM) and Ministry of Higher Education (MOHE), Government of Malaysia, and a Dissertation Completion Fellowship from the College of Agriculture and Natural Resources from MSU.

**Institutional Review Board Statement:** Not applicable.

**Informed Consent Statement:** Not applicable.

**Data Availability Statement:** Sequences have been deposited in the NCBI Sequence Reads Archive (SRA) under BioProject accession number PRJNA820564 (https://www.ncbi.nlm.nih.gov/sra/PRJNA820564, accessed 28 March 2022). The community matrix is provided in the Supplemental materials (Supplemental Table S1). The community matrix describing sequence counts for all OTUs for all treatments associated with this study can also be found on GitHub at https://github.com/ScribnerLab/Chemotherapeutants.git (doi.org/10.5281/zenodo.6418537, accessed on 22 April 2022).

**Acknowledgments:** We thank the technical staff at the Black River Lake Sturgeon Stream Side Facility who assisted with egg ad larval care, particularly S. Valentine and N. Barton. Brian Mauer contributed to discussions pertaining to statistical analyses. Members of the Scribner lab provided valuable comments on earlier manuscript drafts.

**Conflicts of Interest:** The authors declare no conflict of interest.

## References

1. Post, G. *Textbook of Fish Health*; T.F.H. Publications: Austin, TX, USA, 1987.
2. Subasinghe, R.P.; Phillips, M.J. Aquatic animal health management: Opportunities and challenges for rural, small-scale aquaculture and enhanced-fisheries development: Workshop introductory remarks. In *Primary Aquatic Animal Health Care in Rural, Small-Scale, Aquaculture Development*; Arthur, J.R., Phillips, M.J., Eds.; FAO: Rome, Italy, 2002; pp. 1–15.

3. Conte, F.S. Stress and the welfare of cultured fish. *Appl. Anim. Behav. Sci.* **2004**, *86*, 205–223. [CrossRef]
4. Davis, M.W.; Ottmar, M.L. Wounding and reflex impairment may be predictors for mortality in discarded or escaped fish. *Fish. Res.* **2006**, *82*, 1–6. [CrossRef]
5. Llewellyn, M.S.; Boutin, S.; Hoseinifar, S.H.; Derome, N. Teleost microbiomes: The state of the art in their characterization, manipulation and importance in aquaculture and fisheries. *Front. Microbiol.* **2014**, *5*, 1–15. [CrossRef] [PubMed]
6. Dethlefsen, L.; Huse, S.; Sogin, M.L.; Relman, D.A. The pervasive effects of an antibiotic on the human gut microbiota, as revealed by deep 16S rRNA sequencing. *PLoS Biol.* **2008**, *6*, e280. [CrossRef]
7. Dethlefsen, L.; Relman, D.A. Incomplete recovery and individualized responses of the human distal gut microbiota to repeated antibiotic perturbation. *Proc. Natl. Acad. Sci. USA* **2011**, *108*, 4554–4561. [CrossRef]
8. Bowker, J.; Trushenski, J.; Straus, D.; Gaikowski, M.; Goodwin, A. *Guide to Using Drugs, Biologics, and Other Chemicals in Aquaculture*; American Fisheries Society Fish Culture Section: Bethesda, MD, USA, 2016.
9. Thorburn, M.A.; Moccia, R.D. Use of Chemotherapeuties on Trout Farms in Ontario. *J. Aquat. Anim. Health* **1993**, *5*, 85–91. [CrossRef]
10. Gaikowski, M.P.; Larson, W.J.; Gingerich, W.H. Survival of cool and warm freshwater fish following chloramine-T exposure. *Aquaculture* **2008**, *275*, 20–25. [CrossRef]
11. Tort, L.; Balasch, J.C.; Mackenzie, S. Fish immune system. A crossroads between innate and adaptive responses. *Immunologia* **2003**, *22*, 277–286.
12. Speare, D.J.; Arsenault, G.J. Effects of intermittent hydrogen peroxide exposure on growth and columnaris disease prevention of juvenile rainbow trout (*Oncorhynchus mykiss*). *Can. J. Fish. Aquat. Sci.* **1997**, *54*, 2653–2658. [CrossRef]
13. Rach, J.J.; Gaikowski, M.P.; Ramsay, R.T. Efficacy of Hydrogen Peroxide to Control Mortalities Associated with Bacterial Gill Disease Infections on Hatchery-Reared Salmonids. *J. Aquat. Anim. Health* **2000**, *12*, 119–127. [CrossRef]
14. Rach, J.J.; Gaikowski, M.P.; Ramsay, R.T. Efficacy of Hydrogen Peroxide to Control Parasitic Infestations on Hatchery-Reared Fish. *J. Aquat. Anim. Health* **2000**, *12*, 267–273. [CrossRef]
15. Rach, J.J.; Schleis, S.M.; Gaikowski, M.; Johnson, A. Efficacy of Hydrogen Peroxide in Controlling Mortality Associated with External Columnaris on Walleye and Channel Catfish Fingerlings. *N. Am. J. Aquac.* **2003**, *65*, 300–305. [CrossRef]
16. Schelkle, B.; Doetjes, R.; Cable, J. The salt myth revealed: Treatment of gyrodactylid infections on ornamental guppies. *Poecilia reticulata. Aquaculture* **2011**, *311*, 74–79. [CrossRef]
17. Noga, E.J. *Fish Disease: Diagnosis and Treatment*; Wiley-Blackwell: Ames, IA, USA, 2011.
18. Swann, L.; Fitzgerald, S. Use and Application of Salt in Aquaculture. North Central Regional Aquaculture Center. 1992. Available online: http://agrilife.org/fisheries2/files/2013/09/The-Use-and-Application-of-Salt-in-Aquaculture.pdf (accessed on 10 January 2022).
19. Durborow, R.M.; Francis-floyd, R. Medicated Feed for Food Fish. Sounthern Regional Aquaculture Center. 1996. Available online: https://www.ncrac.org/files/biblio/SRAC0473.pdf (accessed on 10 January 2022).
20. Piper, R.G.; McElwain, I.B.; Orme, L.E.; McCraren, J.P.; Fowler, L.G.; Leonard, R.J. *Fish Hatchery Management*; U.S. Fish and Wildlife Service: Washington, DC, USA, 1982.
21. Sanchez, J.G.; Speare, D.J.; Macnair, N.; Johnson, G. Effects of a Prophylactic Chloramine-T Treatment on Growth Performance and Condition Indices of Rainbow Trout. *J. Aquat. Anim. Health* **1996**, *8*, 278–284. [CrossRef]
22. Gaikowski, M.P.; Rach, J.J.; Ramsay, R.T. Acute toxicity of hydrogen peroxide treatments to selected lifestages of cold-, cool-, and warmwater fish. *Aquaculture* **1999**, *178*, 191–207. [CrossRef]
23. Magondu, E.W.; Rasowo, J.; Oyoo-Okoth, E.; Charo-Karisa, H. Evaluation of sodium chloride (NaCl) for potential prophylactic treatment and its short-term toxicity to African catfish *Clarias gariepinus* (Burchell 1822) yolk-sac and swim-up fry. *Aquaculture* **2011**, *319*, 307–310. [CrossRef]
24. Vellend, M. Conceptual synthesis in community ecology. *Q Rev. Biol.* **2010**, *85*, 183–206. [CrossRef]
25. Sloan, W.T.; Lunn, M.; Woodcock, S.; Head, I.M.; Nee, S.; Curtis, T.P. Quantifying the roles of immigration and chance in shaping prokaryote community structure. *Environ. Microbiol.* **2006**, *8*, 732–740. [CrossRef]
26. Burns, A.R.; Stephens, W.Z.; Stagaman, K.; Wong, S.; Rawls, J.F.; Guillemin, K.; Bohannan, B.J.M. Contribution of neutral processes to the assembly of gut microbial communities in the zebrafish over host development. *ISME J.* **2016**, *10*, 655–664. [CrossRef]
27. Venkataraman, A.; Bassis, C.M.; Beck, J.M.; Young, V.B.; Curtis, J.L.; Huffnagle, G.B.; Schmidt, T.M. Application of a Neutral Community Model to Assess Structuring of the Human Lung Microbiome. *mBio* **2015**, *6*, e02284-14. [CrossRef]
28. Leibold, M.A.; Holyoak, M.; Mouquet, N.; Amarasekare, P.; Chase, J.M.; Hoopes, M.F.; Holt, R.D.; Shurin, J.B.; Law, R.; Tilman, D.; et al. The metacommunity concept: A framework for multi-scale community ecology. *Ecol. Lett.* **2004**, *7*, 601–613. [CrossRef]
29. Nemergut, D.R.; Schmidt, S.K.; Fukami, T.; O'Neill, S.P.; Bilinski, T.M.; Stanish, L.F.; Knelman, J.E.; Darcy, J.L.; Lynch, R.C.; Wickey, P.; et al. Patterns and processes of microbial community assembly. *Microbiol. Mol. Biol. Rev.* **2013**, *77*, 342–356. [CrossRef] [PubMed]
30. Christian, N.; Whitaker, B.K.; Clay, K. Microbiomes: Unifying animal and plant systems through the lens of community ecology theory. *Front. Microbiol.* **2015**, *6*, 869. [CrossRef] [PubMed]
31. Louca, S.; Polz, M.F.; Mazel, F.; Albright, M.B.N.; Huber, J.A.; O'Connor, M.I.; Ackermann, M.; Hahn, A.S.; Srivastava, D.S.; Crowe, S.A.; et al. Function and functional redundancy in microbial systems. *Nat. Ecol. Evol.* **2018**, *2*, 936–943. [CrossRef]
32. Cho, I.; Blaser, M.J. The human microbiome: At the interface of health and disease. *Nat. Rev. Genet.* **2012**, *13*, 260–270. [CrossRef]

33. Francino, M.P. Antibiotics and the human gut microbiome: Dysbioses and accumulation of resistances. *Front. Microbiol.* **2016**, *6*, 1543. [CrossRef]
34. Langdon, A.; Crook, N.; Dantas, G. The effects of antibiotics on the microbiome throughout development and alternative approaches for therapeutic modulation. *Genome Med.* **2016**, *8*, 39. [CrossRef]
35. Verschuere, L.; Rombaut, G.; Sorgeloos, P.; Verstraete, W. Probiotic Bacteria as Biological Control Agents in Aquaculture. *Microbiol. Mol. Biol. Rev.* **2000**, *64*, 655–671. [CrossRef]
36. Serrano, P.H. *Responsible Use of Antibiotics in Aquaculture*; Technical Paper; Food & Agriculture Org.: Rome, Italy, 2005.
37. Dfoirdt, T.; Sorgeloos, P.; Bossier, P. Alternatives to antibiotics for the control of bacterial disease in aquaculture. *Curr. Opin. Microbiol.* **2011**, *14*, 251–258. [CrossRef]
38. Diana, J.S. Aquaculture Production and Biodiversity Conservation. *Bioscience* **2009**, *59*, 27–38. [CrossRef]
39. Cabello, F.C. Heavy use of prophylactic antibiotics in aquaculture: A growing problem for human and animal health and for the environment. *Environ. Microbiol.* **2006**, *8*, 1137–1144. [CrossRef] [PubMed]
40. Mortazavi, A.L. Poppin' the Prophylactics: An Analysis of Antibiotics in Aquaculture. *Columbia Univ. J. Glob. Health* **2014**, *4*, 23–27.
41. Burka, J.F.; Hammell, K.L.; Horsberg, T.E.; Johnson, G.R.; Rainnie, D.J.; Speare, D.J. Drugs in salmonid aquaculture—A review. *J. Vet. Pharmacol. Ther.* **2003**, *20*, 333–349. [CrossRef] [PubMed]
42. Romero, J.; Ringø, E.; Merrifield, D.L. The Gut Microbiota of Fish. In *Aquaculture Nutrition: Gut Health, Probiotics and Prebiotics*; Wiley: Chichester, UK, 2014.
43. Austin, B.; Al-Zahrani, A.M.J. The effect of antimicrobial compounds on the gastrointestinal microflora of rainbow trout, Salmo gairdneri Richardson. *J. Fish. Biol.* **1988**, *33*, 1–14. [CrossRef]
44. Navarrete, P.; Mardones, P.; Opazo, R.; Espejo, R.; Romero, J. Oxytetracycline treatment reduces bacterial diversity of intestinal microbiota of Atlantic salmon. *J. Aquat. Anim. Health* **2008**, *20*, 177–183. [CrossRef]
45. He, S.; Zhou, Z.; Liu, Y.; Cao, Y.; Meng, K.; Shi, P.; Yao, B.; Ringø, E. Effects of the antibiotic growth promoters flavomycin and florfenicol on the autochthonous intestinal microbiota of hybrid tilapia (*Oreochromis niloticus* ♀× *O. aureus* ♂). *Arch. Microbiol.* **2010**, *192*, 985–994. [CrossRef]
46. Liu, Y.; Zhou, Z.; Wu, N.; Tao, Y.; Xu, L.; Cao, Y.; Zhang, Y.; Yao, B. Gibel carp Carassius auratus gut microbiota after oral administration of trimethoprim/sulfamethoxazole. *Dis. Aquat. Organ.* **2012**, *99*, 207–213. [CrossRef]
47. Wang, A.R.; Ran, C.; Ringø, E.; Zhou, Z.G. Progress in fish gastrointestinal microbiota research. *Rev Aquac.* **2018**, *10*, 626–640. [CrossRef]
48. Vadstein, O.; Bergh, Ø.; Gatesoupe, F.-J.; Galindo-Villegas, J.; Mulero, V.; Picchietti, S.; Scapigliati, G.; Makridis, P.; Olsen, Y.; Dierckens, K.; et al. Microbiology and immunology of fish larvae. *Rev. Aquac.* **2012**, *5*, S1–S25. [CrossRef]
49. Holtgren, J.M.; Ogren, S.A.; Paquet, A.J.; Fajfer, S. Design of a portable streamside rearing facility for lake sturgeon. *N. Amer. J. Aqua.* **2007**, *69*, 317–323. [CrossRef]
50. Flagg, T.A.; Nash, C.E. (Eds.) *A Conceptual Framework for Conservation Hatchery Strategies for Pacific Salmonids*; NOAA Technical Memorandum NMFS; NOAA: Washington, DC, USA, 1999.
51. Fujimoto, M.; Crossman, J.A.; Scribner, K.T.; Marsh, T.L. Microbial Community Assembly and Succession on Lake Sturgeon Egg Surfaces as a Function of Simulated Spawning Stream Flow Rate. *Microb. Ecol.* **2013**, *66*, 500–511. [CrossRef] [PubMed]
52. Abdul Razak, S.; Scribner, K.T. Ecological and Ontogenetic Components of Larval Lake Sturgeon Gut Microbiota Assembly, Successional Dynamics, and Ecological Evaluation of Neutral Community Processes. *Appl. Environ. Microbiol.* **2020**, *86*, e02662-19. [CrossRef] [PubMed]
53. Abdul Razak, S.; Griffin, M.J.; Mischke, C.C.; Bosworth, B.G.; Waldbieser, G.C.; Wise, D.J.; Marsh, T.L.; Scribner, K.T. Biotic and abiotic factors influencing channel catfish egg and gut microbiome dynamics during early life stages. *Aquaculture* **2019**, *498*, 556–567. [CrossRef]
54. Crossman, J.A.; Scribner, K.T.; Yen, D.T.; Davis, C.A.; Forsythe, P.S.; Baker, E.A. Gamete and larval collection methods and hatchery rearing environments affect levels of genetic diversity in early life stages of lake sturgeon (*Acipenser fulvescens*). *Aquaculture* **2011**, *310*, 312–324. [CrossRef]
55. Hastings, R.P.; Bauman, J.M.; Baker, E.A.; Scribner, K.T. Post-hatch dispersal of lake sturgeon (*Acipenser fulvescens*, Rafinesque, 1817) yolk-sac larvae in relation to substrate in an artificial stream. *J. Appl. Ichthyol.* **2013**, *29*, 1208–1213. [CrossRef]
56. Bauman, J.M.; Woodward, B.M.; Baker, E.A.; Marsh, T.L.; Scribner, K.T. Effects of Family, Feeding Frequency, and Alternate Food Type on Body Size and Survival of Hatchery-Produced and Wild-Caught Lake Sturgeon Larvae. *N. Am. J. Aquac.* **2016**, *78*, 136–144. [CrossRef]
57. Callman, J.L.; Macy, J.M. The predominant anaerobe from the spiral intestine of hatchery-raised sturgeon (*Acipenser transmontanus*), a new Bacteroides species. *Arch. Microbiol.* **1984**, *140*, 57–65. [CrossRef]
58. Buddington, R.K.; Christofferson, J.P. Digestive and feeding characteristics of the chondrosteans. *Environ. Biol. Fishes* **1985**, *14*, 31–41. [CrossRef]
59. Milligan-Myhre, K.; Charette, J.R.; Phennicie, R.T.; Stephens, W.Z.; Rawls, J.F.; Guillemin, K.; Kim, C.H. Study of host-microbe interactions in zebrafish. In *Methods in Cell Biology*, 3rd ed.; Elsevier Inc.: Amsterdam, The Netherlands, 2011.

60. Kozich, J.J.; Westcott, S.L.; Baxter, N.T.; Highlander, S.K.; Schloss, P.D. Development of a dual-index sequencing strategy and curation pipeline for analyzing amplicon sequence data on the miseq illumina sequencing platform. *Appl. Environ. Microbiol.* **2013**, *79*, 5112–5120. [CrossRef]
61. Oksanen, J.; Guillaume Blanchet, F.; Kindt, R.; Legendre, P.; Minchin, P.R.; O'hara, R. *Package 'Vegan'*. Community Ecology Package, Version 2(9). 2015. Available online: https://github.com/vegandevs/vegan (accessed on 10 January 2022).
62. Bray, J.R.; Curtis, J.T. An Ordination of the Upland Forest Communities of Southern Wisconsin. *Ecol. Monogr.* **1957**, *27*, 325–349. [CrossRef]
63. Wickham, H. *Ggplot2Applied Spatial Data Analysis with R*; Springer: New York, NY, USA, 2009.
64. Anderson, M. A new method for non parametric multivariate analysis of variance. *Austral. Ecol.* **2001**, *26*, 32–46.
65. Anderson, M.J. Distance-based tests for homogeneity of multivariate dispersions. *Biometrics* **2006**, *62*, 245–253. [CrossRef] [PubMed]
66. Segata, N.; Izard, J.; Waldron, L.; Gevers, D.; Miropolsky, L.; Garrett, W.S.; Huttenhower, C. Metagenomic biomarker discovery and explanation. *Genome Biol.* **2011**, *12*, R60. [CrossRef]
67. Becattini, S.; Taur, Y.; Pamer, E.G. Antibiotic-Induced Changes in the Intestinal Microbiota and Disease. *Trends Mol. Med.* **2016**, *22*, 458–478. [CrossRef] [PubMed]
68. Manichanh, C.; Reeder, J.; Gibert, P.; Varela, E.; Llopis, M.; Antolin, M.; Guigo, R.; Knight, R.; Guarner, F. Reshaping the gut microbiome with bacterial transplantation and antibiotic intake. *Genome Res.* **2010**, *20*, 1411–1419. [CrossRef] [PubMed]
69. Costello, E.K.; Stagaman, K.; Dethlefsen, L.; Bohannan, B.J.M.; Relman, D.A. The Application of Ecological Theory toward an Understanding of the Human Microbiome. *Science* **2012**, *336*, 1255–1262. [CrossRef]
70. Panda, S.; El Khader, I.; Casellas, F.; López Vivancos, J.; García Cors, M.; Santiago, A.; Cuenca, S.; Guarner, F.; Manichanh, C. Short-term effect of antibiotics on human gut microbiota. *PLoS ONE* **2014**, *9*, e95476. [CrossRef]
71. Huttenhower, C.; Gevers, D.; Knight, R.; Abubucker, S.; Badger, J.H.; Chinwalla, A.T.; Creasy, H.H.; Earl, A.M.; FitzGerald, M.G.; Fulton, R.S.; et al. Structure, function and diversity of the healthy human microbiome. *Nature* **2012**, *486*, 207–214.
72. Dave, M.; Higgins, P.D.; Middha, S.; Rioux, K.P. The human gut microbiome: Current knowledge, challenges, and future directions. *Transl. Res.* **2012**, *160*, 246–257. [CrossRef]
73. Goodrich, J.K.; Waters, J.L.; Poole, A.C.; Sutter, J.L.; Koren, O.; Blekhman, R. Human genetics shape the gut microbiome. *Cell* **2014**, *159*, 789–799. [CrossRef]
74. Wilkins, D.; Leung, M.H.Y.; Lee, P.K.H. Indoor air bacterial communities in Hong Kong households assemble independently of occupant skin microbiomes. *Environ. Microbiol.* **2016**, *18*, 1754–1763. [CrossRef] [PubMed]
75. Navarrete, P.; Toledo, I.; Mardones, P.; Opazo, R.; Espejo, R.; Romero, J. Effect of Thymus vulgaris essential oil on intestinal bacterial microbiota of rainbow trout, *Oncorhynchus mykiss* (Walbaum) and bacterial isolates. *Aquac. Res.* **2010**, *41*, e667–e678. [CrossRef]
76. Romero, J.; Feijoó, C.G.; Navarrete, P. Antibiotics in Aquaculture—Use, Abuse and Alternatives. In *Health and Environment in Aquaculture*; Carvalho, D.E., Gianmarco, S.D., Reinaldo, J.S., Eds.; Books on Demand: Norderstedt, Germany, 2012; pp. 160–198.
77. Slama, T.G. Gram-negative antibiotic resistance: There is a price to pay. *Crit. Care* **2008**, *12*, S4. [CrossRef] [PubMed]
78. Vasoo, S.; Barreto, J.N.; Tosh, P.K. Emerging Issues in Gram-Negative Bacterial Resistance: An Update for the Practicing Clinician. *Mayo Clin. Proc.* **2015**, *90*, 395–403. [CrossRef] [PubMed]
79. Jacoby, G.A.; Munoz-Price, L.S. The New β-Lactamases. *N. Engl. J. Med.* **2005**, *352*, 380–391. [CrossRef] [PubMed]
80. Lopetuso, L.R.; Scaldaferri, F.; Petito, V.; Gasbarrini, A. Commensal Clostridia: Leading players in the maintenance of gut homeostasis. *Gut Pathog.* **2013**, *5*, 23. [CrossRef]
81. Gupta, R.S.; Gao, B. Phylogenomic analyses of clostridia and identification of novel protein signatures that are specific to the genus Clostridiumsensu stricto (cluster I). *Int. J. Syst. Evol. Microbiol.* **2009**, *59*, 285–294. [CrossRef]
82. Palmer, C.; Bik, E.M.; DiGiulio, D.B.; Relman, D.A.; Brown, P.O. Development of the human infant intestinal microbiota. *PLoS Biol.* **2007**, *5*, e177. [CrossRef]
83. De Schryver, P.; Vadstein, O. Ecological theory as a foundation to control pathogenic invasion in aquaculture. *ISME J.* **2014**, *8*, 2360–2368. [CrossRef]
84. Gisbert, E.; Sarasquete, M.C.; Williot, P.; Castelló-Orvay, F. Histochemistry of the development of the digestive system of Siberian sturgeon during early ontogeny. *J. Fish Biol.* **1999**, *55*, 596–616. [CrossRef]
85. Berois, N.; Arezo, M.J.; Papa, N.G. Gamete interactions in teleost fish: The egg envelope. Basic studies and perspectives as environmental biomonitor. *Biol. Res.* **2011**, *44*, 119–124. [CrossRef] [PubMed]
86. Sevellec, M.; Laporte, M.; Bernatchez, A.; Derome, N.; Bernatchez, L. Evidence for host effect on the intestinal microbiota of whitefish (*Coregonus* sp.) species pairs and their hybrids. *Ecol. Evol.* **2019**, *9*, 11762–11774. [CrossRef] [PubMed]
87. Navarrete, P.; Magne, F.; Araneda, C.; Fuentes, P.; Barros, L.; Opazo, R.; Espejo, R.; Romero, J. PCR-TTGE analysis of 16S rRNA from rainbow trout (*Oncorhynchus mykiss*) gut microbiota reveals host-specific communities of active bacteria. *PLoS ONE* **2012**, *7*, e31335.
88. Zhao, L.; Wang, G.; Siegel, P.; He, C.; Wang, H.; Zhao, W.; Zhai, Z.; Tian, F.; Zhao, J.; Zhang, H.; et al. Quantitative Genetic Background of the Host Influences Gut Microbiomes in Chickens. *Sci. Rep.* **2013**, *3*, 1163. [CrossRef] [PubMed]
89. Blekhman, R.; Goodrich, J.K.; Huang, K.; Sun, Q.; Bukowski, R.; Bell, J.T.; Spector, T.D.; Keinan, A.; Ley, R.E.; Gevers, D.; et al. Host genetic variation impacts microbiome composition across human body sites. *Genome Biol.* **2015**, *16*, 191. [CrossRef]

Article

# Gut Microbial Composition of Pacific Salmonids Differs across Oregon River Basins and Hatchery Ancestry

Nicole S. Kirchoff [1,*], Trevan Cornwell [2], Staci Stein [2], Shaun Clements [2] and Thomas J. Sharpton [1,3]

[1] Department of Microbiology, Oregon State University, Corvallis, OR 97331, USA; thomas.sharpton@oregonstate.edu
[2] Oregon Department of Fish and Wildlife, Corvallis, OR 97330, USA; trevan.cornwell@oregonstate.edu (T.C.); staci.stein@oregonstate.edu (S.S.); shaun.clements@oregonstate.edu (S.C.)
[3] Department of Statistics, Oregon State University, Corvallis, OR 97331, USA
* Correspondence: nicole.kirchoff@uconn.edu

**Abstract:** The gut microbiome may represent a relatively untapped resource in the effort to manage and conserve threatened or endangered fish populations, including wild and hatchery-reared Pacific salmonids. To clarify this potential, we defined how steelhead trout gut microbiome composition varies across watersheds and as a function of ancestry. First, we measured this variation across watersheds using wild steelhead trout sampled from nine locations spanning three river basins. While gut microbial composition differs across basins, there exist bacterial clades that are ubiquitous across all populations. Correlating the phylogenetic composition of clades with geographic distance reveals 395 clades of bacteria whose ecological distribution implicates their co-diversification with steelheads. Second, we quantified how microbiome composition varies between first generation hatchery-reared steelhead and traditional hatchery-reared steelhead. Despite being subject to the same hatchery management strategies, fish bred from wild parents carry distinct microbiomes from those bred from hatchery broodstock, implicating the role of genotype on microbiome composition. Finally, we integrated all data from both studies to reveal two distinct, yet robust clusters of community composition. Collectively, our study documents for the first time how the steelhead gut microbiome varies by geography or broodstock and uncovers microbial taxa that may indicate the watershed or hatchery from which an individual was sourced.

**Keywords:** steelhead trout; gut microbiome; hatcheries; aquaculture

**Citation:** Kirchoff, N.S.; Cornwell, T.; Stein, S.; Clements, S.; Sharpton, T.J. Gut Microbial Composition of Pacific Salmonids Differs across Oregon River Basins and Hatchery Ancestry. *Microorganisms* **2022**, *10*, 933. https://doi.org/10.3390/microorganisms10050933

Academic Editor: Konstantinos Ar. Kormas

Received: 4 March 2022
Accepted: 23 April 2022
Published: 29 April 2022

**Publisher's Note:** MDPI stays neutral with regard to jurisdictional claims in published maps and institutional affiliations.



## 1. Introduction

Steelhead trout (*Oncorhynchus mykiss)* is an economically, culturally, and ecologically important fish. However, climate change, overfishing, and habitat destruction threaten and endanger steelhead populations. Even efforts to preserve access to steelhead through the hatchery production of fish are met with rising challenges, as fewer hatchery-reared adults return to spawn compared to their wild counterparts. Simply put, the management and conservation of Pacific salmonids faces grave challenges and may benefit from new tools that aid outcomes.

The gut microbiome is an increasingly considered but relatively untapped resource in the management and conservation of wildlife, including fisheries. Ample evidence shows that anthropogenic-caused land-use changes, climate change, environmental contamination, as well as captivity disrupts gut microbial communities [1]. This disruption is known to involve the elimination or reduction of microorganisms that are important to host health and fitness. For example, red colobus monkeys living in fragmented forests have fewer bacteria that can degrade tannins, a toxic xenobiotic present in many of their food sources [2]. The loss of these bacteria may impact their ability to digest their preferred diet and thus impact their survival. The augmentation or supplementation of microbes important to host survival and health may mitigate anthropogenic disturbances and aid conservation efforts.

Therefore, learning more about the gut microbiome of vulnerable animals will embolden potential microbial related mitigation efforts with the mission of aiding threatened hosts and their microbial consortia. Knowledge of the steelhead gut microbiome is critical if we wish to use gut microbial manipulation to improve the conservation efforts related to these fish.

Despite the importance of the fish gut microbiome to their host, little is known about the steelhead gut microbiome, especially with respect to the diversity of the microbiome across distinct watersheds, wild populations, and hatchery broodstocks [3]. This paucity of knowledge challenges efforts to link the gut microbiome to management and conservation practices. Previous studies have focused on characterizing the non-anadromous member of the *O. mykiss* species, rainbow trout [4–6]. Additionally, previous rainbow trout gut microbiome studies have mostly been conducted in laboratory or aquaculture facilities and not in wild or hatchery populations. Thus, we were interested in characterizing the wild and hatchery steelhead gut microbiome as well as determining how the gut microbiome varies across river systems, thus informing conservation efforts regarding the necessity of location-based gut microbial interventions.

In order to characterize the steelhead gut microbiome and evaluate the gut microbial composition based on location and broodstock ancestry we conducted two studies. The first study investigated the differences between the gut microbiome of steelhead from several different locations in western Oregon. The second study investigated differences in the gut microbiome between traditional hatchery broodstock and hatchery steelhead with wild parents. We found that the steelhead gut microbiome presents geographical effects and varies based on a wild broodstock or hatchery broodstock host background, which suggests that host genotype contributes to gut microbial differences. Additionally, we reveal bacterial clades that demonstrate a phylogenetic composition in the steelhead gut that is associated with geography and that the steelhead gut microbiome has two predominant microbiome types.

## 2. Materials and Methods

### 2.1. Sample Locations

For the comparison of fish across river basins, we sampled ten wild-born, juvenile steelhead from each of nine freshwater systems within three Oregon river basins (Figure 1). We sampled Gravel Creek, Sunshine Creek, and Cedar Creek in the Siletz Basin; Fall Creek, Tobe Creek, and East Fork Lobster Creek in the Alsea Basin; and Alder Creek, East Fork Beaver Creek, and Elk Creek in the Nestucca River Basin.

For the comparison of wild broodstock versus traditional hatchery broodstock fish, we collected traditional juvenile steelhead from two different hatcheries as well as corresponding first hatchery generation juvenile steelhead. Specifically, thirty wild broodstock juvenile fish and thirty hatchery broodstock fish were sampled from Cedar Creek Hatchery in the Nestucca River basin and North Fork Alsea Hatchery in the Alsea River basin, respectively (Figure 1).

### 2.2. Sample Collection

For both studies, samples were collected from already scheduled steelhead sacrifices. Fish were collected with backpack electroshockers from several Oregon river basins between October 2016 and March 2017 in accordance with Oregon Department of Fish and Wildlife permits. Fish were sacrificed with a buffered tricaine methanesulfonate (i.e., MS-222) overdose, weighed, cut from anal vent to gills, and gut digesta from stomach to intestines were squeezed into 50 mL conical tubes. To preserve the DNA content, intestinal samples were first placed on ice in the field and then placed into a −20 °C freezer within four hours of sampling. Within 24 h of sampling, samples were finally moved into an −80 °C freezer.

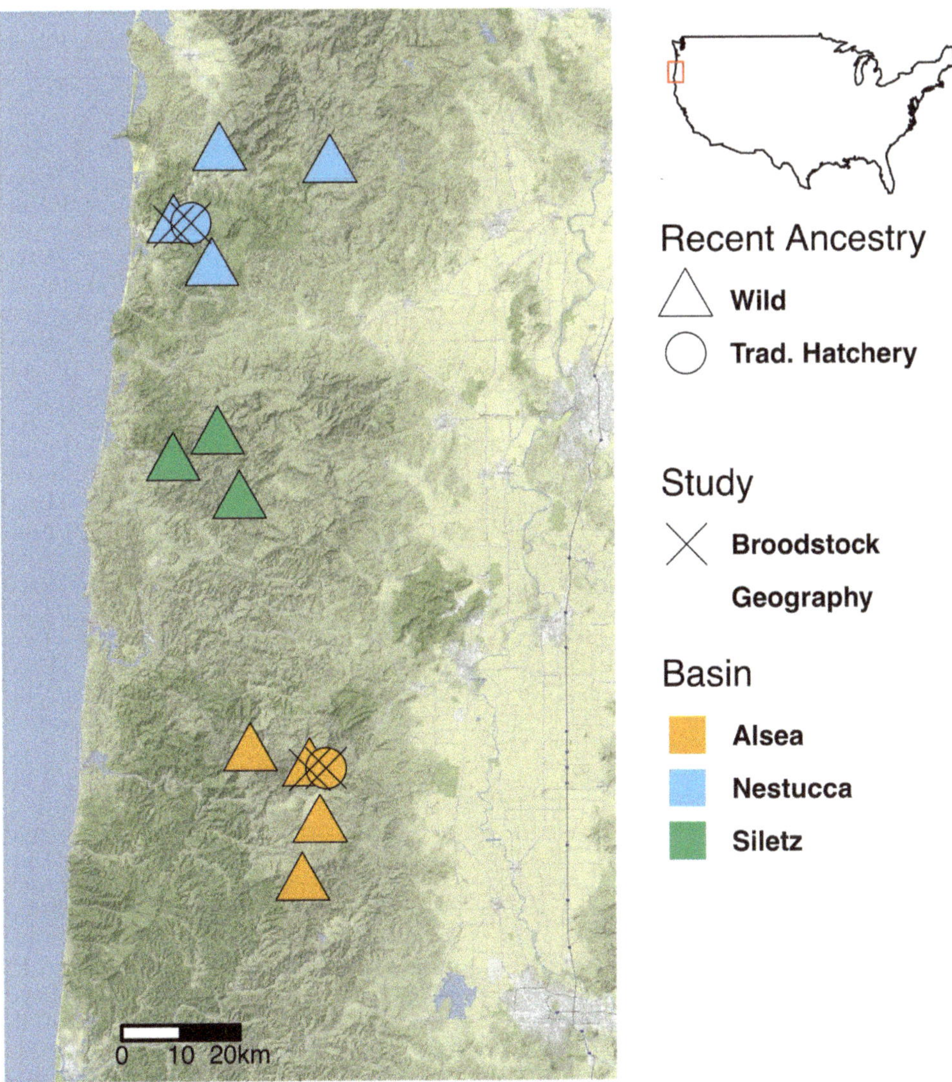


**Figure 1.** A map displaying sampling locations of wild steelhead intestinal samples. This is an image of the western Oregon coast between Portland, OR and Eugene, OR. Point shape, color, and inclusion of an "X" indicate recent ancestry, study origin, and river basin origin, respectively. The Oregon coast is located at the position of the red box on the border of the United States. Nestucca River basin samples are from the north sites in blue, Siletz River basin samples from the middle sites in green, and Alsea River basin samples from the southern sites in orange.

### *2.3. Microbiome Profiling and Analyses*

DNA extraction was conducted using the Qiagen DNeasy PowerSoil kit (QIAGEN, Germantown, MD, USA) with an addition of a 10 min incubation step at 65 °C, as explained previously [7]. The 16S V4 rRNA gene was amplified using Caporaso (515F/806R) primers according to previous protocols [8,9]. DNA was then quantified using a Qubit dsDNA HS kit (Thermo Fisher, Waltham, MA, USA), then pooled and cleaned with the QIAquick PCR Purification Kit (QIAGEN). Amplicons were sequenced at the Center for Quantitative Life

Sciences at Oregon State University with an Illumina MiSeq (v3 chemistry) generating 300 bp paired end reads. Sequences were generated for each study on distinct flow cells.

### *2.4. Bioinformatics and Statistical Analyses*

We generated an amplicon sequence variant (ASV) table by running FASTQ sequence files through the DADA2 (v 1.9.0) pipeline [10]. Separately for each study, forward reads were truncated at 240 base pairs, chimeras were removed, and bacterial taxonomy was assigned with the SILVA rRNA database (release 128) and the Ribosomal Database Project's naïve Bayesian classifier [11]. We then created a phylogenetic tree using V4 rRNA gene sequence alignments via FastTree (v 2.1.10) [12]. We used the R (v 3.6.2) phyloseq package (v 1.3) to rarify sequence abundances for each sample within a study [13,14]. Pairwise Bray–Curtis dissimilarities for each gut microbial sample were calculated to compare abundance-weighted bacterial community compositions across sample location, steelhead weight, and management strategy using the vegan package (2.5–6) [15]. Monophyletic bacterial clades within taxonomic phylotypes were identified with the ClaaTU algorithm [16].

The non-metric multi-dimensional scaling (NMDS) plot was generated in R also using the vegan package to visualize the similarity of compositional abundance with a method that is robust to data sparsity [15]. Beta dispersion was calculated and compared with a Tukey HSD test using the vegan and stats packages, respectively. The coin package (v1.3-1) was used to conduct Kruskal–Wallis tests comparing bacterial cladal abundances across early life history categories and geographic location [17]. Multiple test correction was performed with the p.adjust() function in the stats package (v 3.6.2) with a false discovery rate cut-off of 0.05 [13]. Weighted pairwise UniFrac values were also calculated with vegan to determine the phylogenetic distance between bacterial clades present in the steelhead gut microbiome [15]. Additionally, we computed the straight line geographic distances between steelhead sample sites using the geosphere package (v 1.5-10) [18]. Hierarchical clustering (median, ward-d2) and dendrogram visualization was conducted using the stats package [13].

### *2.5. Combination of Both Studies*

Data from both the geography and hatchery broodstock vs. wild broodstock studies were pooled and bioinformatically and statistically analyzed together. The combined FASTQ files were re-processed through DADA2 quality filtering, and forward reads were cut at 240 base pairs [10]. The phyloseq package was used to normalize the library size and randomly subsample (i.e., rarefy) to a maximum of 1576 reads for each sample (median reads per sample = 11,919), and 16S classification was conducted with the SILVA rRNA gene database. Phylogenetic tree inferences were conducted in FastTree, as in the two studies above [12]. Partitioning around medoids (PAM) cluster analysis was performed in R with the cluster package (v 2.1.0) [19].

## 3. Results

### *3.1. Wild Juvenile Steelhead Trout Gut Microbial Communities Are Structured by Geography and Host Fitness*

To determine if the composition of the steelhead gut microbiome associates with steelhead geography, we rarefied to 13,635 bacterial reads and evaluated the beta diversity of the gut microbiome across locations (Supplementary Table S1). The bacterial community composition of the steelhead gut is significantly different across Oregon river basins, though the effect sizes are weak (PERMANOVA, Bray–Curtis, $R^2 = 0.06$, $p = 0.001$) (Figure 2). This associative pattern is retained when comparing the beta diversity of individual sample sites, and moreover, the model improvably fits the data (PERMANOVA, Bray–Curtis, $R^2 = 0.19$, $p < 0.05$). These results indicate that a steelhead's gut microbiome is related to their geographic location, but the steelhead gut microbiome has a stronger association with the exact river or stream the fish inhabited.

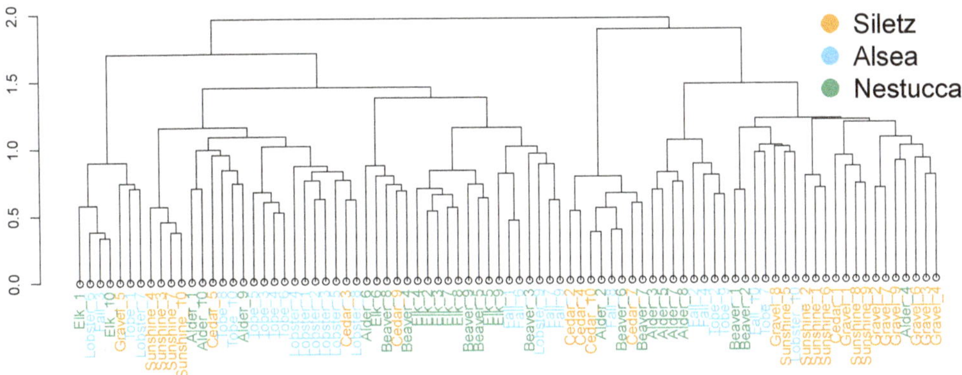


**Figure 2.** Steelhead gut microbiome samples from three different river basins in Oregon roughly group together. Dendrogram showing hierarchal clustering (Ward's method with Ward's clustering criterion) comparing Bray–Curtis dissimilarities between samples. Samples are colored by river basin origin. Samples do not neatly separate into three groups, but the samples tend to cluster into smaller groups with like colors. The differences between steelhead gut microbial composition are confirmed statistically (PERMANOVA, $R^2 = 0.05$, $p < 0.01$).

To discern which taxa may drive these river basin-specific patterns in community composition, we leveraged a phylogenetic approach that aggregates observed counts of ASVs among lineages that constitute monophyletic clades and applied Kruskal–Wallis tests to focus on clades whose aggregated abundances differ between river basins. In so doing, we resolved 21 bacterial clades that stratify the Alsea River basin from the Siletz and Nestucca basins. For example, a *Ferruginibacter* clade is more abundant in Alsea than in the Siletz and Nestucca basins (Figure 3A). We also found 36 clades whose abundances in the Nestucca basin differ from those in the Siletz and Alsea basins, including a clade of Sphingomonadaceae that is more abundant in the Nestucca basin (Figure 3B). Finally, we discovered four bacterial clades that differ in terms of abundance in the Siletz basin compared to the Nestucca and Alsea basins that includes one clade of *Novosphingobium*, two clades of *Aeromonas*, and one clade of *Flavobacterium* that are more abundant in the Siletz basin (Figure 3C).

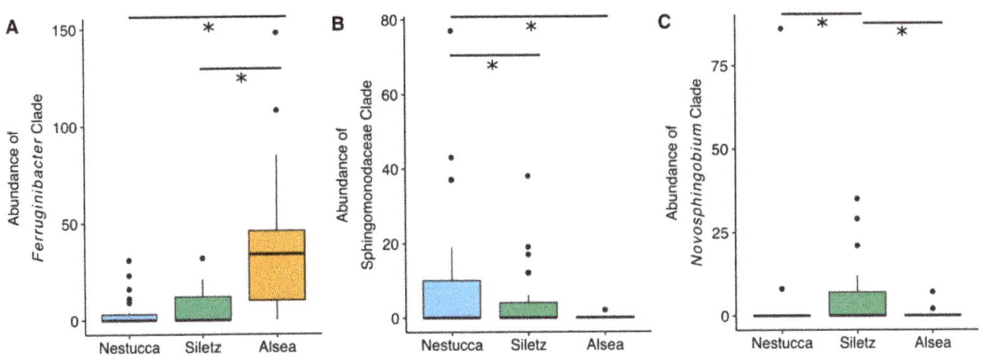


**Figure 3.** Examples of bacterial clades that are more abundant in each of the three river basins sampled in the geography study. Boxplots visualizing the abundance of steelhead gut microbiome bacterial clades across three western Oregon river basins. Asterisk (*) indicates a statistically significant result using Kruskal-Wallis tests and false discovery rate multiple test correction. (**A**) shows one clade's abundances from the genus *Ferruginibacter*, (**B**) from the family Sphingomonadcaeae, and (**C**) from the genus *Novosphingobium*.

Despite these differences across location, we also resolved several microbial clades that were common to all locations. In particular, we identified 1489 clades that are significantly more prevalent across samples than expected by chance (FDR < 0.05). Thirty-six of these conserved clades were present in every steelhead gut sample and they encompass taxa such as *Flavobacterium*, *Hyphomicrobium*, and *Singulisphaera*. Such microbes may manifest these ubiquitous distributions because they are common in the environment, apt at colonizing the salmonid gut, or specifically selected for by the host.

Given the pattern of variation in the salmonid gut microbiome that we observed across locations, we next sought to determine if any salmonid gut microbial clades manifest phylogenetic compositions that are statistically structured by the geography of their host, which may imply population-level co-diversification. To discern such associations, we correlated the pairwise-weighted phylogenetic beta diversity and geographic distances of steelhead gut bacterial clades. This analysis revealed 395 monophyletic clades of bacteria whose phylogenetic compositional differences across samples correlates with the geographic distance spanning sampling locations (Supplementary Table S2). The gut microbial clades that display this phylogenetic distance by geographic distance structure include members of the families Sphingomonadaceae and Rhodobacteraceae. For example, forty-one Sphingomonadaceae clades have a weighted UniFrac value that is significantly correlated with geographic distance between sample site (Mantel test < 0.01) (Figure 4). These patterns indicate that the gut bacterial phylogeny of some clades is related to the geographic location of their host. However, our analysis was based on a limited number of sampling locations and relied on a test of correlation that may be subject to relatively high type I error rates.

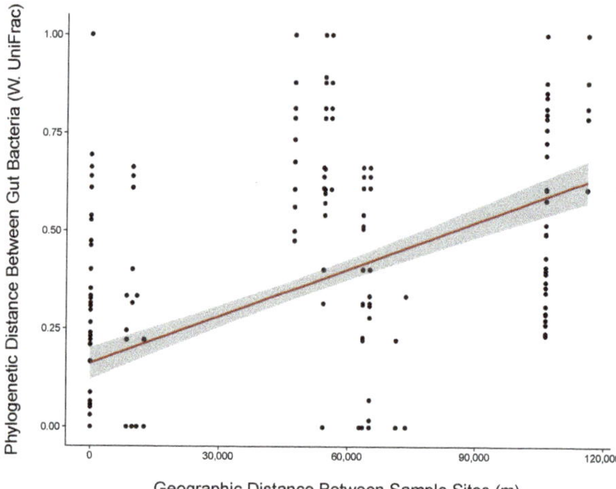


**Figure 4.** Sphingomonodaceae cladal abundance from north to south geography sampling sites shows a relationship between the phylogenetic composition of the clade and geography. Scatter plot representing a Sphingomonadaceae clade that has a significant correlation between weighted UniFrac and physical straight-line distance between coordinates of sample sites (Mantel test < 0.01). The red line represents the slope of all the data points and shows the positive relationship between geographic distance and phylogenetic distance. The shading represents the 95% confidence interval. This significant trend indicates that sampling sites that are geographically closer together tend to host bacteria with a more similar phylogenetic history. Forty other Sphingomonadaceae clades also display weighted UniFrac values that correlate with geographic distance, and Sphingomonadaceae was the taxon with the most significant clades after this analysis.

Some of the variation in the composition of the gut microbiome observed here could hold implications for salmonid fitness. For example, larger sized salmonids have greater reproductive success (i.e., the number of offspring that survive to maturity) compared to their smaller siblings [20]. Accordingly, a larger animal size is related to greater fitness (i.e., reproductive success) in steelhead trout. We thus determined whether the composition of the gut microbiome links to this salmonid fitness indicator through a test of association. In particular, we compared the steelhead gut bacterial structure to the weight of all fish and found that the gut microbiome is associated with steelhead weight (PERMANOVA, Bray–Curtis, $R^2 = 0.1273$, $p = 0.03$).

### 3.2. *Juvenile Steelhead Trout Gut Microbiome Varies as a Function of Hatchery Broodstock and Hatchery Location*

Traditional hatchery broodstock are subject to several genetic bottlenecks after each successive generation compared to wild broodstock fish (i.e., F1 hatchery populations with wild parents) that only experience one generation in a hatchery facility. Despite this fact, it remains generally unknown how hatchery broodstock origins impact the composition of the gut microbiome compared to their wild broodstock counterparts. Addressing this question is critical given the fact that traditional hatchery broodstock fish are less likely to survive than their wild born counterparts for reasons we do not fully understand.

After subsampling bacterial reads to 1237 reads, our analyses indicated that traditional hatchery broodstock fish carry different gut microbiome assemblages relative to their wild broodstock counterparts (PERMANVOA, Bray–Curtis, $R^2 = 0.07$, $p = 0.001$) (Figure 5) (Supplementary Table S3). A total of 665 bacterial clades are differentially abundant across fish ancestry (Supplemetary Table S4). For instance, all 13 of the significant clades from the genus *Peptoniphilus* are more abundant in the gut microbiome of first-generation steelhead (Figure 6A). All four *Pleurocapsa* clades are more abundant in the guts of traditional hatchery broodstock steelhead (Figure 6B). Additionally, there appear to be hatchery-specific effects on the interindividual variation of the microbiome. For example, the NMDS plot of beta diversity shows that traditional North Fork Alsea Hatchery samples are more tightly gathered than the North Fork Alsea Hatchery wild broodstock samples. Thus, we compared the beta dispersion of the steelhead gut microbial samples and found that North Fork Alsea Hatchery wild broodstock samples are more dispersed than the traditional hatchery broodstock samples (Tukey HSD of beta dispersion <0.001). This differentiation in dispersion could contribute to the observed differences in beta diversity. Furthermore, we determined that the gut microbial structure of steelhead is also associated with their creek of origin, irrespective of their hatchery or wild broodstock status (PERMANOVA $R^2 = 0.29$, $p < 0.01$), suggesting that specific aquatic environments play a role in shaping steelhead gut microbial structure. The contribution of geographic origin may also explain the overlap of North Fork Alsea Hatchery samples visible in the NMDS plot that is not seen between Cedar Creek Hatchery traditional broodstock samples and Cedar Creek Hatchery wild broodstock samples, as both North Fork Alsea Hatchery broodstocks were established using fish from the Alsea River and the Cedar Creek hatchery fish were established using fish from two different locations. There are 1664 bacterial clades with different abundances between the North Fork Alsea Hatchery and Cedar Creek Hatchery locations (Supplementary Table S5). All 29 of the clades assigned to the genera *Flavobacterium* are more abundant in the Cedar Creek Hatchery location (Figure 7A). Furthermore, most of the 33 clades assigned to the genus *Bacteroides* are also more abundant in the Cedar Creek Hatchery samples, but eight of the clades are more abundant in the North Fork Alsea Hatchery location. We visualized the abundance distributions of one of the *Flavobacterium* clades and one of the *Bacteroides* clades (Figure 7B). Collectively, our results indicate that broodstock generation and watersheds impact the assembly of the steelhead gut microbiome.

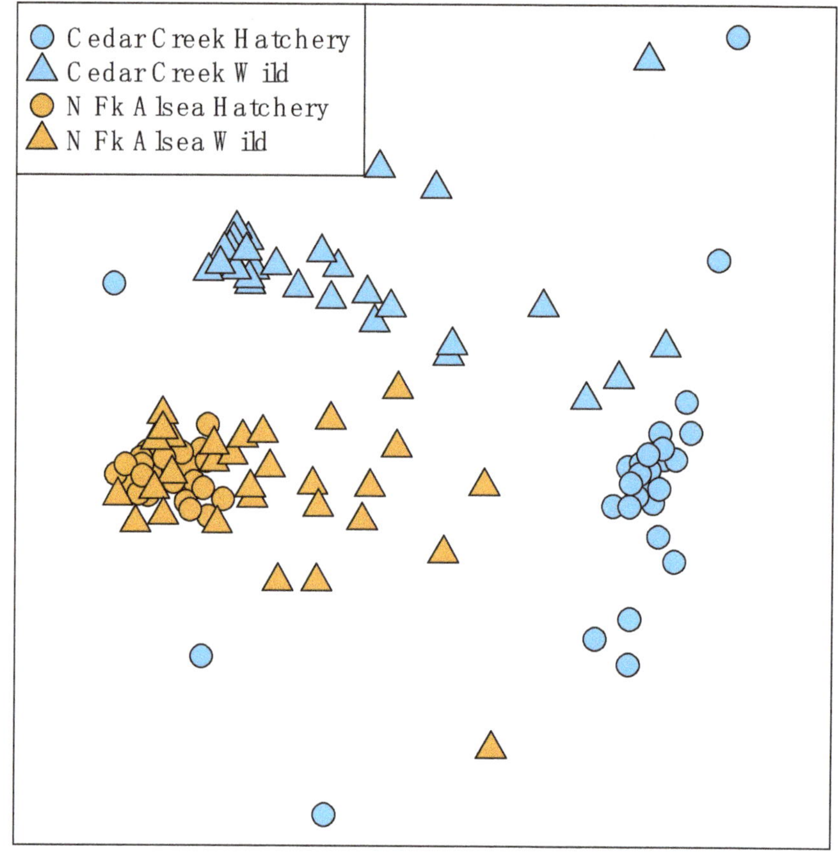


**Figure 5.** The gut microbiomes of traditional hatchery-reared steelhead differ compositionally compared to their wild broodstock counterparts. NMDS plot showcasing the differences between hatchery broodstock and wild broodstock gut microbial samples (PERMANOVA $R^2 = 0.29$, $p < 0.01$) as well as differences between Cedar Creek Hatchery and North Fork Alsea Hatchery locations (PERMANOVA $R^2 = 0.07$, $p < 0.001$). Stress = 0.13. Visually, there is separation between wild steelhead gut microbial composition and hatchery steelhead gut microbial composition within their respective river basins.

### 3.3. Combination of Both Studies

After combining all our available wild-born, wild broodstock, and hatchery broodstock gut microbiome samples, we found that the river basin, broodstock history, and weight remained associated with the beta diversity of the steelhead gut microbiome (PERMANOVAbasin, Bray–Curtis, $R^2 = 0.1223$, $p = 0.001$; PERMANOVAbroodstock, Bray–Curtis, $R^2 = 0.0479$, $p = 0.0001$; PERMANOVAweight, Bray–Curtis, $R^2 = 0.1170$, $p = 0.0001$). Additionally, the dimensions displayed in the NMDS show two all-encompassing potential clusters that we confirmed with a partitioning around medoids (PAM) cluster-based analysis (Figure 8). Despite their separation, these clusters are not explained by any of our measured variables, suggesting some other variables underlie this observed structure in the diversity of the steelhead gut microbiome.

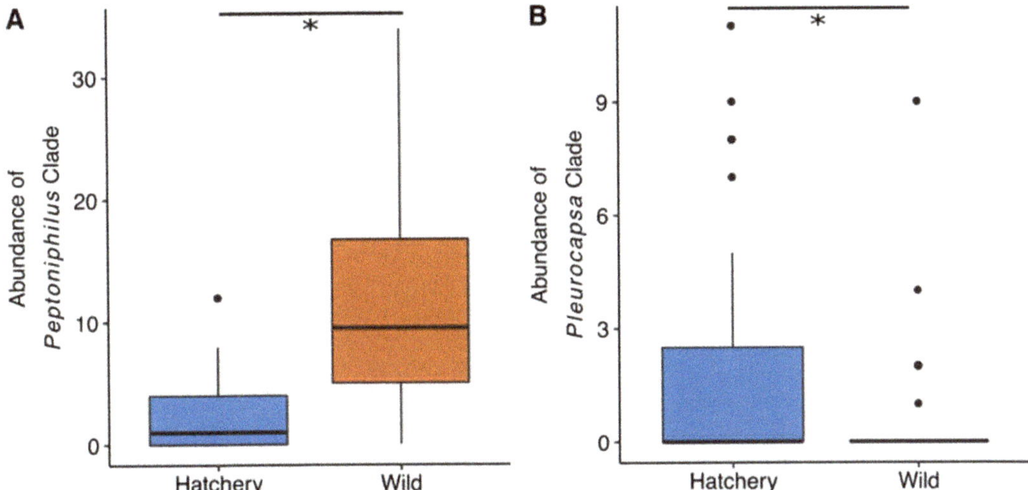


**Figure 6.** Examples of gut bacterial clades that are more abundant in either traditional hatchery or wild broodstock fish. Asterisk (*) indicates a statistically significant result using Kruskal-Wallis tests and false discovery rate multiple test correction. (**A**) This *Peptonophilus* clade example is more abundant in wild broodstock fish guts. (**B**) The *Pluerocapsa* clade example is more abundant in traditional hatchery broodstock fish.

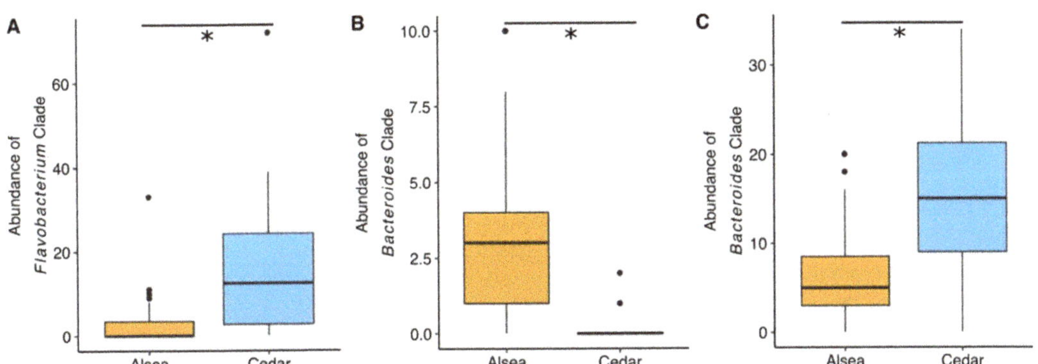


**Figure 7.** Examples of gut bacterial clades that are more abundant in all Cedar Creek Hatchery samples compared to all East Fork Alsea Hatchery samples. Asterisk (*) indicates a statistically significant result using Kruskal-Wallis tests and false discovery rate multiple test correction. (**A**) A clade from the genus *Flavobacterium* that is more abundant in Cedar Creek Hatchery fish guts. Different *Bacteriodes* clades that are (**B**) more abundant in East Fork Alsea Hatchery fish guts and (**C**) more abundant in Cedar Creek Hatchery fish guts.

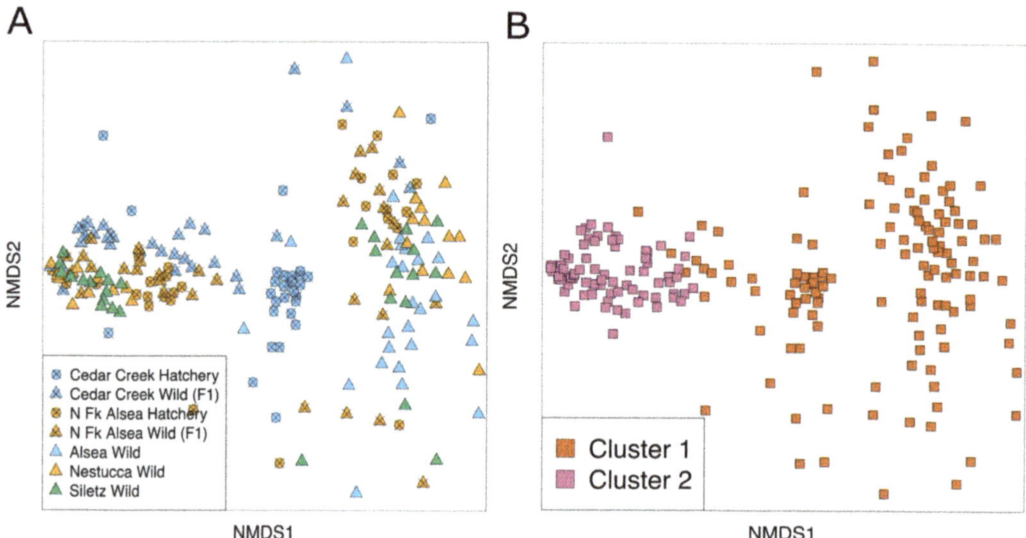


**Figure 8.** NMDS plots and PAM cluster analysis reveal two clusters that may represent two steelhead gut microbiome types. (**A**) shows the combined NMDS visualization of all wild-born and hatchery-reared steelhead gut microbiome samples. The samples aggregate into two groups separated by space in the ordination. The coloring based on river basins and the shapes based on management type indicate that neither of these variables separates out into these two clusters. (**B**) colors each microbiome sample based on the PAM cluster designation and shows that PAM cluster assignment corresponds with the two speculated clusters.

## 4. Discussion

Pacific salmonid fisheries have the task of keeping up with consumer demands as wild and hatchery population numbers decline. Understanding how the Pacific salmonid gut microbiome varies based on broodstock ancestry or geographic location will provide insight into how gut bacteria may be manipulated to improve fish health and survival. This study defines how the steelhead trout gut microbiome varies across three river basins and as a function of their broodstock background. In particular, this study reveals geographic, geographic by phylogenetic lineage, and ancestry effects on the steelhead gut microbiome. Additionally, this study found an association between the steelhead gut microbial community and weight, which may have fitness implications for these fish. We document several bacterial clades that stratify groups with differing gut microbial diversity. Finally, a combined analysis revealed two predominant types of steelhead gut microbiome composition. This work clarifies how geographic location and broodstock affect the steelhead gut microbiome and informs our understanding of how the gut microbiome manifests in declining fish populations, which may lead to improved management practices or conservation efforts.

This study highlights the existence of geographic effects that influence the composition of the gut microbiome. These observations generally agree with prior studies of wildlife gut microbiomes in terrestrial systems [21,22], and a recent meta-analysis revealed differences in the gut microbiome of over 85 species of fish based on the five Korean water sources they were sampled from [23]. Another study, though, found that the wild gut microbiome of Atlantic salmon did not associate with geographic location [24]. However, our study was conducted with a larger sample size of Pacific salmonids, suggesting a larger effect size many be needed to reveal geographic patterns in salmonids or that differences between Atlantic and Pacific salmonids—such as differences in physiology, ecology, or geography—may account for these distinct results.

Importantly, cryptic variation in host physiology or genetic background may shape the gut microbial composition in this study observing wild-born gut microbes. Salmonids show evidence of subpopulations and genetic differences even within the same river system, and genetic differences have been seen in trout with spawning habitats as low as 2 km apart [25,26]. Given that the host genotype plays a role in shaping gut microbial composition in other fish hosts, the differences seen in the gut microbiome across geographic locations may be related to the accompanying differences in host genetics [27]. However, the steelhead genetics of the wild samples were not explicitly documented in this study, and future work should attempt to correlate gut microbial members with genetic differences in wild steelhead trout.

We uncovered that specific bacterial clades were more abundant in one of the three river basins, which supports the hypothesis that gut microbes may be useful for assessing salmonid biogeography. For example, a clade belonging to the genera *Ferruginibacter* was more abundant in the Alsea Basin, a Sphingomondaceae clade was more abundant in the Nestucca Basin, and *Novosphingobium* was more abundant in the Siletz Basin. In a previous study, *Novosphingobium* abundance varied by geographic location in the gut of another fish species, suggesting that members of this bacterial genera typically show geographic patterns within the fish gut microbiome [28]. While the function of *Ferruginibacter* in the fish gut is unknown, bacteria from this genus are often isolated from freshwater sediment, suggesting that these bacteria are dispersing from sediment to fish gut or from fish gut to sediment [29]. Regardless, given the cross-sectional nature of our study, it is not clear if these geographic associations are maintained over the course of a fish's lifespan, a topic that should be explored in future work.

An additional analysis revealed several clades that display correlations between geographic distance and phylogenetic distance in the steelhead gut microbiome, suggesting that these gut bacterial members co-diversified with their hosts. Alternatively, these bacterial clades may manifest a geographic distribution in the environment and then occupy the fish host. Clades demonstrating this geographic and phylogenetic correlation include clades from the bacterial families Rhodobacteraceae and Sphingomonodaceae. Rhodobacteraceae may play a role in fish health, as this family was previously found to be more abundant in the guts of healthy shrimp compared to diseased shrimp [30], but future studies should explicitly test its role in steelhead health. Also, bacterial clades from Sphingomonodaceae produce sphingolipids, which are organic compounds that can modulate *O. mykiss* mucosal homeostasis and B cell abundance [31]. Although we did not sample mucosal-associated bacteria, mucosal membranes and digesta share some microbial members and microbes in the digesta and lumen can still produce compounds that affect host immune responses [32]. Additionally, Sphingomonadaceae possess sphingolipids in their cell membranes that improve chances of successful colonization and survival in the gut, which can be advantageous for both commensal and pathogenic organisms [33,34]. This speculative role of Rhodobacteraceae and Sphingomonadaceae may be the reason for a potentially prolonged association between these bacterial families and steelhead that gave rise to this geographic lineage sorting.

Despite differences in gut microbial structure across basins, we discovered bacterial clades that are prevalent in all steelhead guts of our first study. Bacterial clades from the genera *Flavobacterium*, *Hyphomicrobium*, and *Singulisphaera* represent such core taxa. The ubiquitous presence of these bacteria suggest steelhead physiology selects for these specific clades, as they may have critical functions within the steelhead gut, or that these bacterial clades are also commonly found in the surrounding aquatic meta-communities. The function of *Hyphomicrobium* and *Singulishpaera* in the gut are unknown, but they have been found in aquatic systems as well as other fish guts [35,36]. Several members of the *Flavobacterium* genus are pathogenic to fish, although some *Flavobacterium* are commensal [37]. The pathogenicity of the *Flavobacterium* clades in this study is unclear but could have widespread consequences, as these clades were found in every fish of our first study. Given their ubiquitous distribution, future investigations should seek to discern the physiological impacts of these taxa.

In addition, this study clarifies the impact of hatchery broodstock on the gut microbiome. Previous work suggested that hatcheries elicit strong selective pressure on Pacific salmonids that differentiate fish reared in hatcheries for several generations from fish reared in hatcheries for one generation, who are both different from wild-born salmonids. For example, the relative fitness levels and rates of reproductive success of fish with greater hatchery ancestry are significantly lower than those of fish with wild ancestry [38]. Also, the expression of several genes from the first generation of hatchery steelhead trout (i.e., previously wild trout) are heritably altered after a single generation in a hatchery environment [39]. The differences in the diversity of traditional hatchery and first-generation hatchery gut microbiomes suggest this selective pressure is also applied to steelhead gut microbial communities. Other heavily managed animals are known to have different gut microbial communities compared to their wild counterparts, as is the case with animals in captivity [40].

The fact that the first-generation hatchery stock and traditional hatchery stock were reared in the same hatchery environment suggests that the differences in the gut microbiome between these two groups is due to differences in genetics. Genotype has previously affected gut microbial composition in fish and other hosts [27,41]. Future conservation efforts may use the identification of specific clades that stratify or are indicative of a hatchery or wild steelhead gut microbiome to identify a fish as early generation or traditional hatchery-reared. Additionally, our resolution of clades that differentiate traditional hatchery and first-generation hatchery fish microbiomes may help hatcheries develop management practices that ultimately normalize the composition of hatchery-reared microbiomes closer to their wild counterparts. This study only focused on the gut microbiome of juvenile steelhead as this is the life stage steelhead are contained in hatcheries, and more mature steelhead undergo a great deal of physiological changes in preparation to travel out to the ocean, which may induce changes in the gut microbiome.

A combined analysis using all steelhead gut microbiome samples from our two studies revealed two robust clusters that demonstrate densely populated areas in the multidimensional space of steelhead gut microbiome beta diversity. These two clusters may be evidence of two different steelhead gut microbiome types. However, an unknown covariate that we did not measure, such as the sex of the fish, may be responsible for the clustering. Future studies may find these two clusters are robust among other populations of steelhead gut microbiomes and they should focus on measuring more variables that may be causing these clusters. If these two clusters are robust, future researchers should consider that the effectiveness of microbial interventions may be different based on the steelhead microbiome type measured in future studies. Therefore, microbiome type should be another variable considered when studying the steelhead gut microbiome.

Our results indicate that the steelhead gut microbiome varies as a function of geography and broodstock ancestry. Additionally, several steelhead gut bacterial clades show geographic lineage sorting across western Oregon, and a collective analysis showed two gut microbiome types. Given the declining populations of wild salmonids and the comparatively poor fitness of successive generations of hatchery-reared supplementation stock, characterizing the gut microbial communities across these populations is critical in learning 1) how the steelhead gut microbiome plays a role in the health and fitness of these fish and 2) how we can use steelhead gut microbiota or microbial interventions to improve conservation and supplementation efforts.

**Supplementary Materials:** The following supporting information can be downloaded at: https://www.mdpi.com/article/10.3390/microorganisms10050933/s1, Table S1: Number of reads before and after rarefaction for the geography study; Table S2: Table of clades that were correlated with phylogenetic distance and geographic distance. The table includes mantel test $p$-values and associated bacterial taxa identification; Table S3: Number of reads before and after rarefaction for the broodstock study; Table S4: Gut microbial clades significantly different across traditional hatchery broodstock and wild broodstock fish; Table S5: Gut microbial clades significantly different between North Fork Alsea Hatchery and Cedar Creek Hatchery.

**Author Contributions:** Conceptualization, N.S.K., S.C. and T.J.S.; data curation, N.S.K., T.C. and S.S.; methodology, N.S.K., S.C. and T.J.S.; resources, T.C., S.S., S.C. and T.J.S.; formal analysis, N.S.K. and T.J.S.; writing—original draft preparation, N.S.K. and T.J.S.; writing—review and editing, N.S.K., T.C. and T.J.S.; visualization, N.S.K.; supervision, T.J.S.; funding acquisition, N.S.K. and T.J.S. All authors have read and agreed to the published version of the manuscript.

**Funding:** This work was supported by The National Science Foundation (grant 1557192), institutional funds to T.J.S., and an Oregon Department of Fish and Wildlife fellowship to N.S.K.

**Institutional Review Board Statement:** Animal handling, sampling, and necropsy were conducted by the Oregon Department of Fish and Wildlife as part of their routine management and conservation efforts in accordance with state permits. Microbiome samples were collected from animals that were otherwise subject to these efforts; no animals were handled, sampled, or sacrificed specifically for this study.

**Data Availability Statement:** DNA and R code generated during this research is located on the Sharpton Lab Repository (http://files.cqls.oregonstate.edu/Sharpton_Lab/Papers/Kirchoff_Microorganisms_2022/ accessed on 22 April 2022).

**Acknowledgments:** We would like to thank Mark Dasenko at the Center for Quantitative Life Sciences for sequencing samples and providing advice on sample preparation. We thank Jerri Bartholomew for helpful discussions. We would also like to thank the employees at North Fork Alsea Hatchery and Cedar Creek Hatchery for allowing us to sample fish and Jamie Anthony for providing additional historical information about the two hatcheries.

**Conflicts of Interest:** The authors declare no conflict of interest.

## References

1. Trevelline, B.K.; Fontaine, S.S.; Hartup, B.K.; Kohl, K.D. Conservation Biology Needs a Microbial Renaissance: A Call for the Consideration of Host-Associated Microbiota in Wildlife Management Practices. *Proc. R. Soc. B Biol. Sci.* **2019**, *286*, 20182448. [CrossRef] [PubMed]
2. Barelli, C.; Albanese, D.; Donati, C.; Pindo, M.; Dallago, C.; Rovero, F.; Cavalieri, D.; Tuohy, K.M.; Hauffe, H.C.; De Filippo, C. Habitat Fragmentation Is Associated to Gut Microbiota Diversity of an Endangered Primate: Implications for Conservation. *Sci. Rep.* **2015**, *5*, 14862. [CrossRef] [PubMed]
3. Talwar, C.; Nagar, S.; Lal, R.; Negi, R.K. Fish Gut Microbiome: Current Approaches and Future Perspectives. *Indian J. Microbiol.* **2018**, *58*, 397–414. [CrossRef] [PubMed]
4. Brown, R.M.; Wiens, G.D.; Salinas, I. Analysis of the Gut and Gill Microbiome of Resistant and Susceptible Lines of Rainbow Trout (*Oncorhynchus mykiss*). *Fish Shellfish Immunol.* **2019**, *86*, 497–506. [CrossRef] [PubMed]
5. Desai, A.R.; Links, M.G.; Collins, S.A.; Mansfield, G.S.; Drew, M.D.; Van Kessel, A.G.; Hill, J.E. Effects of Plant-Based Diets on the Distal Gut Microbiome of Rainbow Trout (*Oncorhynchus mykiss*). *Aquaculture* **2012**, *350–353*, 134–142. [CrossRef]
6. Kim, D.-H.; Brunt, J.; Austin, B. Microbial Diversity of Intestinal Contents and Mucus in Rainbow Trout (*Oncorhynchus mykiss*). *J. Appl. Microbiol.* **2007**, *102*, 1654–1664. [CrossRef]
7. Gaulke, C.A.; Barton, C.L.; Proffitt, S.; Tanguay, R.L.; Sharpton, T.J. Triclosan Exposure Is Associated with Rapid Restructuring of the Microbiome in Adult Zebrafish. *PLoS ONE* **2016**, *11*, e0154632. [CrossRef] [PubMed]
8. Caporaso, J.G.; Lauber, C.L.; Walters, W.A.; Berg-Lyons, D.; Lozupone, C.A.; Turnbaugh, P.J.; Fierer, N.; Knight, R.; Gordon, J.I. Global Patterns of 16S RRNA Diversity at a Depth of Millions of Sequences per Sample. *Proc. Natl. Acad. Sci. USA* **2011**, *108*, 4516–4522. [CrossRef]
9. Caporaso, J.G.; Lauber, C.L.; Walters, W.A.; Berg-Lyons, D.; Huntley, J.; Fierer, N.; Owens, S.M.; Betley, J.; Fraser, L.; Bauer, M.; et al. Ultra-High-Throughput Microbial Community Analysis on the Illumina HiSeq and MiSeq Platforms. *ISME J.* **2012**, *6*, 1621–1624. [CrossRef]
10. Callahan, B.J.; McMurdie, P.J.; Rosen, M.J.; Han, A.W.; Johnson, A.J.A.; Holmes, S.P. DADA2: High-Resolution Sample Inference from Illumina Amplicon Data. *Nat. Methods* **2016**, *13*, 581–583. [CrossRef]
11. Wang, Q.; Garrity, G.M.; Tiedje, J.M.; Cole, J.R. Naïve Bayesian Classifier for Rapid Assignment of RRNA Sequences into the New Bacterial Taxonomy. *Appl Env. Microbiol* **2007**, *73*, 5261–5267. [CrossRef] [PubMed]
12. Price, M.N.; Dehal, P.S.; Arkin, A.P. FastTree 2—Approximately Maximum-Likelihood Trees for Large Alignments. *PLoS ONE* **2010**, *5*, e9490. [CrossRef] [PubMed]
13. R Core Team. R: The R Project for Statistical Computing. Available online: https://www.r-project.org/ (accessed on 13 April 2022).

14. McMurdie, P.J.; Holmes, S. Phyloseq: An R Package for Reproducible Interactive Analysis and Graphics of Microbiome Census Data. *PLoS ONE* **2013**, *8*, e61217. [CrossRef]
15. Oksanen, J.; Blanchet, F.G.; Friendly, M.; Kindt, R.; Legendre, P.; McGlinn, D.; Minchin, P.R.; O'Hara, R.B.; Simpson, G.L.; Solymos, P.; et al. Vegan: Community Ecology Package. 2020. Available online: http://CRAN.R-project.org/package=vegan (accessed on 13 April 2022).
16. Gaulke, C.A.; Arnold, H.K.; Humphreys, I.R.; Kembel, S.W.; O'Dwyer, J.P.; Sharpton, T.J. Ecophylogenetics Clarifies the Evolutionary Association between Mammals and Their Gut Microbiota. *mBio* **2018**, *9*, e01348-18. [CrossRef] [PubMed]
17. Hothorn, T.; Hornik, K.; van de Wiel, M.A.; Zeileis, A. Implementing a Class of Permutation Tests: The Coin Package. *J. Stat. Softw.* **2008**, *28*, 1–23. [CrossRef]
18. Hijmans, R.J.; Karney, C.; Williams, E.; Vennes, C. Geosphere: Spherical Trigonometry. 2021. Available online: https://cran.r-project.org/web/packages/geosphere/geosphere.pdf (accessed on 13 April 2022).
19. Maechler, M.; Rousseeuw, P.; Struyf, A.; Hubert, M.; Hornik, K. Cluster: Cluster Analysis Basics and Extensions. 2021. Available online: http://CRAN.R-project.org/package-cluster (accessed on 13 April 2022).
20. Auld, H.L.; Noakes, D.L.G.; Banks, M.A. Advancing Mate Choice Studies in Salmonids. *Rev. Fish Biol. Fish.* **2019**, *29*, 249–276. [CrossRef]
21. Gupta, V.K.; Paul, S.; Dutta, C. Geography, Ethnicity or Subsistence-Specific Variations in Human Microbiome Composition and Diversity. *Front. Microbiol.* **2017**, *8*, 1162. [CrossRef]
22. Yatsunenko, T.; Rey, F.E.; Manary, M.J.; Trehan, I.; Dominguez-Bello, M.G.; Contreras, M.; Magris, M.; Hidalgo, G.; Baldassano, R.N.; Anokhin, A.P.; et al. Human Gut Microbiome Viewed across Age and Geography. *Nature* **2012**, *486*, 222–227. [CrossRef]
23. Kim, P.S.; Shin, N.-R.; Lee, J.-B.; Kim, M.-S.; Whon, T.W.; Hyun, D.-W.; Yun, J.-H.; Jung, M.-J.; Kim, J.Y.; Bae, J.-W. Host Habitat Is the Major Determinant of the Gut Microbiome of Fish. *Microbiome* **2021**, *9*, 166. [CrossRef]
24. Llewellyn, M.S.; McGinnity, P.; Dionne, M.; Letourneau, J.; Thonier, F.; Carvalho, G.R.; Creer, S.; Derome, N. The Biogeography of the Atlantic salmon (*Salmo salar*) Gut Microbiome. *ISME J* **2016**, *10*, 1280–1284. [CrossRef]
25. Skaala, Ø.; Nævdal, G. Genetic Differentiation between Freshwater Resident and Anadromous Brown Trout, Salmo Trutta, within Watercourses. *J. Fish Biol.* **1989**, *34*, 597–605. [CrossRef]
26. Verspoor, E. Genetic Diversity among Atlantic Salmon (*Salmo salar* L.) Populations. *ICES J. Mar. Sci.* **1997**, *54*, 965–973. [CrossRef]
27. Dąbrowska, K.; Witkiewicz, W. Correlations of Host Genetics and Gut Microbiome Composition. *Front Microbiol* **2016**, *7*, 1357. [CrossRef] [PubMed]
28. Walter, J.M.; Bagi, A.; Pampanin, D.M. Insights into the Potential of the Atlantic Cod Gut Microbiome as Biomarker of Oil Contamination in the Marine Environment. *Microorganisms* **2019**, *7*, 209. [CrossRef] [PubMed]
29. Lim, J.H.; Baek, S.-H.; Lee, S.-T. *Ferruginibacter alkalilentus* Gen. Nov., Sp. Nov. and *Ferruginibacter lapsinanis* Sp. Nov., Novel Members of the Family "Chitinophagaceae" in the Phylum Bacteroidetes, Isolated from Freshwater Sediment. *Int. J. Syst. Evol. Microbiol.* **2009**, *59*, 2394–2399. [CrossRef]
30. Yao, Z.; Yang, K.; Huang, L.; Huang, X.; Qiuqian, L.; Wang, K.; Zhang, D. Disease Outbreak Accompanies the Dispersive Structure of Shrimp Gut Bacterial Community with a Simple Core Microbiota. *AMB Express* **2018**, *8*, 120. [CrossRef]
31. Sepahi, A.; Cordero, H.; Goldfine, H.; Esteban, M.Á.; Salinas, I. Symbiont-Derived Sphingolipids Modulate Mucosal Homeostasis and B Cells in Teleost Fish. *Sci. Rep.* **2016**, *6*, 39054. [CrossRef]
32. Duarte, M.E.; Kim, S.W. Intestinal Microbiota and Its Interaction to Intestinal Health in Nursery Pigs. *Anim. Nutr.* **2022**, *8*, 169–184. [CrossRef]
33. An, D.; Na, C.; Bielawski, J.; Hannun, Y.A.; Kasper, D.L. Membrane Sphingolipids as Essential Molecular Signals for Bacteroides Survival in the Intestine. *Proc. Natl. Acad. Sci. USA* **2011**, *108*, 4666–4671. [CrossRef]
34. Wang, J.; Chen, Y.-L.; Li, Y.-K.; Chen, D.-K.; He, J.-F.; Yao, N. Functions of Sphingolipids in Pathogenesis During Host–Pathogen Interactions. *Front. Microbiol.* **2021**, *12*, 1926. [CrossRef]
35. Martineau, C.; Mauffrey, F.; Villemur, R. Comparative Analysis of Denitrifying Activities of *Hyphomicrobium nitrativorans*, *Hyphomicrobium denitrificans*, and *Hyphomicrobium zavarzinii*. *Appl. Environ. Microbiol.* **2015**, *81*, 5003–5014. [CrossRef] [PubMed]
36. Ray, C.; Bujan, N.; Tarnecki, A.; Davis, A.D.; Browdy, C.; Arias, C.R. Analysis of the Gut Microbiome of Nile Tilapia *Oreochromis niloticus* L. Fed Diets Supplemented with Previda® and Saponin. *J. FisheriesSciences.Com* **2017**, *11*, 36. [CrossRef]
37. Loch, T.P.; Faisal, M. Emerging Flavobacterial Infections in Fish: A Review. *J. Adv. Res.* **2015**, *6*, 283–300. [CrossRef] [PubMed]
38. Christie, M.R.; Ford, M.J.; Blouin, M.S. On the Reproductive Success of Early-Generation Hatchery Fish in the Wild. *Evol. Appl.* **2014**, *7*, 883–896. [CrossRef]
39. Christie, M.R.; Marine, M.L.; Fox, S.E.; French, R.A.; Blouin, M.S. A Single Generation of Domestication Heritably Alters the Expression of Hundreds of Genes. *Nat. Commun.* **2016**, *7*, 10676. [CrossRef]
40. Gibson, K.M.; Nguyen, B.N.; Neumann, L.M.; Miller, M.; Buss, P.; Daniels, S.; Ahn, M.J.; Crandall, K.A.; Pukazhenthi, B. Gut Microbiome Differences between Wild and Captive Black Rhinoceros—Implications for Rhino Health. *Sci. Rep.* **2019**, *9*, 7570. [CrossRef]
41. Blekhman, R.; Goodrich, J.K.; Huang, K.; Sun, Q.; Bukowski, R.; Bell, J.T.; Spector, T.D.; Keinan, A.; Ley, R.E.; Gevers, D.; et al. Host Genetic Variation Impacts Microbiome Composition across Human Body Sites. *Genome Biol.* **2015**, *16*, 191. [CrossRef]

Article

# Effects of a Bioprocessed Soybean Meal Ingredient on the Intestinal Microbiota of Hybrid Striped Bass, *Morone chrysops x M. saxatilis*

**Emily Celeste Fowler [1], Prakash Poudel [1], Brandon White [2], Benoit St-Pierre [1,*] and Michael Brown [2,*]**

[1] Department of Animal Science, South Dakota State University, Brookings, SD 57007, USA; Emily.Fowler@sdstate.edu (E.C.F.); Prakash@Himalayandogchew.com (P.P.)
[2] Department of Natural Resource Management, South Dakota State University, Brookings, SD 57007, USA; brandon@prairieaquatech.com
* Correspondence: Benoit.St-Pierre@sdstate.edu (B.S.-P.); Michael.Brown@sdstate.edu (M.B.)

**Abstract:** The hybrid striped bass (*Morone chrysops x M. saxatilis*) is a carnivorous species and a major product of US aquaculture. To reduce costs and improve resource sustainability, traditional ingredients used in fish diets are becoming more broadly replaced by plant-based products; however, plant meals can be problematic for carnivorous fish. Bioprocessing has improved nutritional quality and allowed higher inclusions in fish diets, but these could potentially affect other systems such as the gut microbiome. In this context, the effects of bioprocessed soybean meal on the intestinal bacterial composition in hybrid striped bass were investigated. Using high-throughput sequencing of amplicons targeting the V1–V3 region of the 16S rRNA gene, no significant difference in bacterial composition was observed between fish fed a control diet, and fish fed a diet with the base bioprocessed soybean meal. The prominent Operational Taxonomic Unit (OTU) in these samples was predicted to be a novel species affiliated to *Peptostreptococcaceae*. In contrast, the intestinal bacterial communities of fish fed bioprocessed soybean meal that had been further modified after fermentation exhibited lower alpha diversity ($p < 0.05$), as well as distinct and more varied composition patterns, with OTUs predicted to be strains of *Lactococcus lactis*, *Plesiomonas shigelloides*, or *Ralstonia pickettii* being the most dominant. Together, these results suggest that compounds in bioprocessed soybean meal can affect intestinal bacterial communities in hybrid striped bass.

**Keywords:** hybrid striped bass; microbiome; bacteria; bioprocessed soybean meal

**Citation:** Fowler, E.C.; Poudel, P.; White, B.; St-Pierre, B.; Brown, M. Effects of a Bioprocessed Soybean Meal Ingredient on the Intestinal Microbiota of Hybrid Striped Bass, *Morone chrysops x M. saxatilis*. *Microorganisms* **2021**, *9*, 1032. https://doi.org/10.3390/microorganisms9051032

Academic Editor: Konstantinos Ar. Kormas

Received: 31 March 2021
Accepted: 9 May 2021
Published: 11 May 2021



## 1. Introduction

As a result of the growing market demand for seafood and the depletion of wild fish populations, the aquaculture industry has considerably expanded over the last few decades [1]. Of the various fish species available for production, the hybrid striped bass has proven to be well suited for aquaculture because of its high growth performance, survival, and disease resistance, as well as its ability to be reared under a number of different culture systems and conditions [2]. The hybrid striped bass is the result of crossing female white bass (*Morone chrysops*) with male striped bass (*M. saxatilis*), and its higher performance compared to its parent species is attributed to hybrid vigor [2]. The success of the hybrid striped bass has been well illustrated by the rapid expansion of its global production, starting at five metric tons in 1986, increasing by 36.8-fold to 184 metric tons in 1987, then peaking at 6203 metric tons by 2005 [3]. World production levels then fluctuated between 3764 metric tons and 5884 metric tons between 2006 and 2016 [3]. Hybrid striped bass has become one of the leading aquaculture industries in the United States, behind channel catfish (*Ictalurus punctatus*), Atlantic salmon (*Salmo salar*), and rainbow trout (*Oncorhynchus mykiss*) [4].

As with other intensive animal production systems, minimizing operating costs represents one of the main challenges faced by aquaculture producers, with purchasing of dietary ingredients generally representing a major expense. In the case of hybrid striped bass production, for instance, nutrition costs have been estimated to make up approximately 40% of total variable costs [5]. As fishmeal remains an important source of dietary protein in aquaculture diets, its increasing market price and decreasing availability have been particularly problematic [6,7]. In response to this challenge, fishmeal is being replaced by more economically and environmentally sustainable sources of dietary protein in the formulation of aquaculture diets; inclusion of fishmeal in salmonid diets, for example, has decreased from 50% in the 1990s to 15% by 2012 [8].

Of the available alternatives to fishmeal, plant-based protein ingredients such as soybean meal have become an attractive substitute because of their availability and lower cost [9]. However, the inclusion of soybean meal in carnivorous fish diets needs to be limited because of the presence of anti-nutritional factors, as these reduce digestibility and increase digestive tract inflammation, which is associated with intestinal enteritis [10]. Another concern of higher inclusion of plant-based protein sources in carnivorous fish diets is their higher carbohydrate content; carbohydrate levels need to be low enough to avoid negative effects on digestion and gut physiology. One effective solution to these problems has been the production of soy protein concentrate, a feed ingredient generated by the extraction of carbohydrates from soybean byproducts using ethanol [11]. Another approach involves the use of bioprocessing, a biotechnological strategy that aims to generate value-added products by treatment of a substrate with biocatalysts such as enzymes, microorganisms (bacteria and yeast or other fungi), or cells cultured from plants or animals. In the case of feed ingredients, microbial utilization of a substrate can effectively neutralize undesirable compounds, such as non-starch polysaccharides, protease inhibitors, lectins, saponins, phytic acid, phytoestrogens, and allergens. While bioprocessing has permitted higher inclusion of soybean meal in fish and livestock diets [12], efforts are still ongoing to increase the effectiveness of the procedure, optimize production scale-up, as well as improve the quality of the final product and/or custom tailor its composition to better suit specific areas of animal production.

For carnivorous fish, such as the hybrid striped bass, one unintentional consequence of including plant-based protein ingredients in diets may be changing the composition of the gut microbiome. Gut microbial communities have been shown to be important for the health and nutrition of a wide variety of host species, including fish [13,14]. Indeed, they promote the development and regulation of immune defenses, compete against pathogenic bacteria, and produce short-chain fatty acids from substrates that host enzymes are unable to digest [15–17]. In fish, the gut environment can be colonized as early as the larval stage, a process that can be modulated by factors such as diet, season, stage of development, and habitat [18]. Generally, the most abundant phyla in the fish gastrointestinal tract tend to be Proteobacteria followed by Firmicutes and Bacteroidetes, but this can vary depending on the trophic level of the host [19,20].

Considering the importance of the hybrid striped bass in the aquaculture industry, the composition of its gut microbiome in healthy individuals has remained mostly unexplored. Indeed, in contrast to salmon [21–25], trout [26–32], and catfish [33–38], no published studies are currently available on hybrid striped bass or its parental strains using DNA sequencing-based methods; culture-independent approaches using high throughput Next-Generation sequencing platforms have been established as the gold standard for the analysis of microbiomes. Using a culturing approach, *Aeromonas hydrophila* was identified as a dominant species in the gut of striped bass [39–42] and hybrid striped bass [43]. Considering the limited scope of culture-dependent techniques for analysis of gut microbial environments and that *A. hydrophila* has been recognized as a pathogen for a number of freshwater fish species [44], it can be concluded that the gut bacterial communities of the hybrid striped bass and of its parental fish species have yet to be investigated.

In this context, the study described in the present report aimed to determine and compare the intestinal bacterial community composition of hybrid striped bass fed diets that included bioprocessed soybean meal or modified bioprocessed soybean meal. These products were selected as test ingredients because of their lower cost compared to soy protein concentrate, as well as their potential to provide additional biotic properties. Together, results show that the composition of intestinal bacterial communities of hybrid striped bass fed bioprocessed soybean meal did not differ from the composition of a control diet that did not include the bioprocessed soybean meal. However, three diets that each included a different product variant of the bioprocessed soybean meal resulted in bacterial compositions that were very different.

## 2. Materials and Methods

### 2.1. Diet Formulations

The Control (CON) diet used in this study did not include soybean meal ingredients. It was designed from a documented formulation from the Agricultural Research Service digestibility database [45]. The treatment diets, which contained bioprocessed soybean meal or a subsequent product variant that was further modified after initial fermentation, were designed to replace 54.4% of wheat middling, 47.8% of the poultry meal, and 66.7% of the feather meal in a control diet (CON) (Table 1). Production of bioprocessed soybean meal consists of growing *Aureobasidium pullulans* on a pasteurized slurry of soybean meal in water for 4 to 5 days. The fungus converts sugars and oligosaccharides into fungal cell mass while also neutralizing anti-nutritional factors and other undesirable compounds. After completion of the fungal treatment, solids are recovered by centrifugation and then dried. The bioprocessed soybean meal ingredients resulting from post-fermentation treatments that were used in this study included three different fractions of the product (BP-F1, BP-F2, and BP-F3), bioprocessed soybean meal after an enzymatic treatment (BP-E), as well as bioprocessed soybean meal after an additional rinse or wash step (BP-W). The diets used in this study were formulated to be isocaloric and isonitrogenous, but some differences were detected by the proximate composition of nutrients (Table 2). While crude fat concentrations were similar among diets (8.13–8.99%), the CON diet had the lowest crude protein concentration (45.49%), compared to a range of 47.02–47.92% for the other diets, and it concomitantly had a higher nitrogen-free extract content (32.69% vs. 30.64–31.80%).

**Table 1.** Experimental diet formulations used in the 105-day growth trial. All values are shown as g/(100 g dry matter).

| Ingredient | Diet | | | | |
|---|---|---|---|---|---|
| | BP-F1 | BP-E | BP-W | BP | CON |
| BP-SBM Fraction #1 [a] | 25.00 | 0.00 | 0.00 | 0.00 | 0.00 |
| BP-SBM Fraction #2 [a] | 0.00 | 0.00 | 0.00 | 0.00 | 0.00 |
| BP-SBM Fraction #3 [a] | 0.00 | 0.00 | 0.00 | 0.00 | 0.00 |
| BP-SBM + Enzyme [a] | 0.00 | 25.00 | 0.00 | 0.00 | 0.00 |
| BP-SBM Base + Wash [a] | 0.00 | 0.00 | 25.00 | 0.00 | 0.00 |
| BP-SBM Base [a] | 0.00 | 0.00 | 0.00 | 25.00 | 0.00 |
| Blood Meal [b] | 2.50 | 2.50 | 2.50 | 2.50 | 2.50 |
| Wheat Midds [c] | 10.00 | 10.00 | 10.00 | 10.00 | 21.92 |
| Whole Cleaned Wheat [d] | 16.67 | 16.67 | 16.67 | 16.67 | 15.00 |
| Poultry Meal [e] | 12.00 | 12.00 | 12.00 | 12.00 | 23.00 |
| Feather Meal [e] | 2.50 | 2.50 | 2.50 | 2.50 | 7.50 |
| Fish Meal [f] | 10.00 | 10.00 | 10.00 | 10.00 | 10.00 |
| Vitamin Premix [g] | 1.25 | 1.25 | 1.25 | 1.25 | 1.25 |
| Lysine [h] | 1.00 | 1.00 | 1.00 | 1.00 | 1.75 |
| Methionine [h] | 0.50 | 0.50 | 0.50 | 0.50 | 0.50 |
| Choline Chloride [i] | 0.58 | 0.58 | 0.58 | 0.58 | 0.58 |
| Mineral Premix [j] | 0.75 | 0.75 | 0.75 | 0.75 | 0.75 |
| Stay C [k] | 0.25 | 0.25 | 0.25 | 0.25 | 0.25 |
| Fish Oil [l] | 6.50 | 6.50 | 6.50 | 6.50 | 4.50 |

**Table 1.** *Cont.*

| Ingredient | Diet | | | | |
|---|---|---|---|---|---|
| | BP-F1 | BP-E | BP-W | BP | CON |
| Dicalcium phosphate [m] | 0.50 | 0.50 | 0.50 | 0.50 | 0.50 |
| Defatted SBM [n] | 10.00 | 10.00 | 10.00 | 10.00 | 10.00 |
| Totals | 100.0 | 100.0 | 100.0 | 100.0 | 100.0 |

BP-SBM: bioprocessed soybean meal.[a] South Dakota State University, Brookings, SD; [b] Mason City By-Products, Mason City, IA; [c] Consumer Supply Distributing, Sioux City, IA; [d] Ag First Farmer's Cooperative, Brookings, South Dakota; [e] Tyson Foods, Springdale, AR; [f] Special Select, Omega Protein, Houston, TX; [g] ARS 702 premix, Nelson and Sons, Murray, UT; [h] Pure Bulk, Roseburg, OR; [i] BalChem Corporation, New Hampton, NJ; [j] ARS 640 trace mix, Nelson and Sons, Murray, UT; [k] DSM Nutritional Products, Parsippany, NJ; [l] Viginia Prime Gold, Omega Protein, Houson, TX; [m] Feed Products Inc., St. Louis, MO; [n] South Dakota Soybean Processors, Volga, South Dakota.

**Table 2.** Proximate composition of diets used in growth study. All values are shown as g/(100 g dry matter).

| Diet | Ash | Fat | Fiber | Protein | NFE |
|---|---|---|---|---|---|
| BP-F1 | 7.71 | 8.79 | 4.65 | 47.32 | 31.53 |
| BP-F2 | 7.89 | 8.53 | 3.86 | 47.92 | 31.80 |
| BP-F3 | 7.63 | 8.97 | 4.47 | 47.25 | 31.68 |
| BP-E | 7.71 | 8.99 | 5.33 | 47.32 | 30.64 |
| BP-W | 7.77 | 8.49 | 5.62 | 47.02 | 31.10 |
| BP | 8.07 | 8.13 | 5.27 | 47.65 | 30.88 |
| CON | 8.88 | 8.42 | 4.52 | 45.49 | 32.69 |

NFE: Nitrogen Free Extract.

Dry ingredients were ground using a Fitzpatrick Commutator (Elmhurst, IL, USA) equipped with a 1.27 mm screen prior to blending. Milled ingredients were transferred to a ribbon mixer (Patterson Equipment, Toronto, ON, Canada), then blended for five minutes. The resulting homogenous feedstuff was extruded with an Extru-Tech E325 single-screw extruder (Sabetha, KS, USA), which was equipped with 2.5 mm die inserts to produce 3.2 mm diameter floating pellets. Extruded pellets were then dried with a conveyor oven drier (Colorado Mill Equipment, Canon City, CO, USA), screen sifted using a Rotex screener (Rotex Inc., Cincinnati, OH, USA), then lipid-coated with a Phlauer vacuum coater (A & J Mixing, Oakville, ON, Canada). Finally, diets were bagged for storage at room temperature until use. Feeds were manufactured at Prairie AquaTech (Brookings, SD, USA).

### 2.2. Feeding Trial

The feeding trial was run at the fish-holding laboratory in the Northern Plains Biostress Facility at South Dakota State University. Naïve, juvenile hybrid striped bass ($n$ = 560; 17.83 ± 0.11 g; Keo Fish Farm, Keo, AR, USA) were randomly distributed at a density of 20 fish per tank, with 4 replicate tanks randomly assigned to each dietary treatment. All fish were fed the same ration of a fishmeal-based holding diet prior to the start of the trial, and then each tank was switched to its assigned experimental diets upon the start of the trial. Feed was hand fed and offered to satiety to each tank three times per day (08:00, 12:00, and 16:00) for 105 consecutive days, and the amount of feed consumed was recorded. The trial was conducted with a 4682 L recirculating aquaculture system (RAS), consisting of 28 tanks, each with a 114 L capacity. Tanks were each equipped with a "recirculating" drain which withdrew water from the subsurface, and a "sludge" drain which was affixed to the lowest point in the bottom at the center of the tank. Each tank also contained forced air diffusers fed by a blower, as well as half covers to minimize disturbance. The RAS was also equipped with a pump, bead filter, bag filter, UV filter, biofilter, solids settling sump, clarifying sump, water inlet float valve, and heater/chiller unit. Water temperature was maintained between 25 °C and 27 °C, dissolved oxygen was held at levels greater than 5 mg/L, and the range in pH was 7–8. Temperature, dissolved oxygen, and pH were monitored daily, while ammonia ($NH_3$) and nitrite ($NO_2^-$) were monitored on average three times per week.

At the end of the study, randomly selected individual fish were euthanized using lethal levels of buffered MS-222, according to a protocol approved by the Institutional Animal Care and Use Committee (IACUC) (Approval Number 16-089A). Liver and viscera were collected by dissection from three randomly selected fish from each tank to calculate the hepatosomatic index (HSI = (liver weight/body weight) $\times$ 100) and viscerosomatic index (VSI = (viscera weight/body weight) $\times$ 100). Each fish used for microbiota analysis was randomly selected from an individual tank. Distal intestines were cut from the vent; then feces were recovered into a sterile tube by gently running forceps along the outside of the intestine. Collected feces were flash-frozen in liquid nitrogen, then stored at −80 °C until they were processed for bacterial community composition.

### 2.3. *Microbial DNA Isolation and PCR Amplification*

Microbial genomic DNA was isolated from intestinal samples by a repeated bead beating plus column method [46], which included the use of the QIAamp DNA Mini Kit (Qiagen, Hilden, Germany). One dissected intestine was used as starting material for each microbial genomic DNA preparation. Bead beating was performed twice for each DNA preparation, for a duration of 3 min at 3500 rpm for each repetition. The V1–V3 region of the bacterial 16S rRNA gene was targeted using the 27F forward [47] and 519R reverse [48] primer pair by PCR with the Phusion Taq DNA polymerase (Thermo Scientific, Waltham, MA, USA) under the following conditions: hot start (4 min, 98 °C), followed by 35 cycles of denaturation (10 s, 98 °C), annealing (30 s, 50 °C) and extension (30 s, 72 °C), then ending with a final extension period (10 min, 72 °C). A total of 5–30 ng of purified microbial genomic DNA was used per PCR reaction in a total reaction volume of 50 µL. PCR products were separated by agarose gel electrophoresis, and amplicons of the expected size (~500bp) were excised for gel purification using the QiaexII Gel extraction kit (Qiagen, Hilden, Germany). A negative control reaction (no DNA) was included for each series of PCR reactions; amplicon DNA from experimental samples was not recovered from sets of PCR reactions whose negative control showed detectable amplified DNA. For each sample, approximately 400 ng of amplified DNA were submitted to Molecular Research DNA (MRDNA, Shallowater, TX, USA), which performed all subsequent steps for Next-Generation sequencing, including indexing and library preparation, to generate overlapping paired-end reads with the Illumina MiSeq (2 $\times$ 300) platform.

### 2.4. *Computational Analysis of PCR Generated 16S rRNA Amplicon Sequences*

Unless specified, sequence data analysis was performed using custom-written Perl scripts. Raw bacterial 16S rRNA gene V1–V3 amplicon sequences were provided by Molecular Research DNA (MRDNA, Shallowater, TX, USA) as assembled contigs from overlapping MiSeq (2 $\times$ 300) paired-end reads from the same flow cell clusters. Reads were then selected to meet the following criteria: the presence of both intact 27F (forward) and 519R (reverse) primer nucleotide sequences, a length between 400 and 580 nt, and a minimum quality threshold of no more than 1% of nucleotides with a Phred quality score lower than 15 [49,50].

Following quality screens, sequence reads were aligned, then clustered into Operational Taxonomic Units (OTUs) at a genetic distance cutoff of 5% sequence dissimilarity [49,50]. OTUs were screened for DNA sequence artifacts using the following methods. Chimeric sequences were first identified with the 'chimera.uchime' [51] and 'chimera.slayer' [52] commands from the MOTHUR (version 1.44.1) open-source software package [53]. Secondly, the integrity of the 5′ and 3′ ends of OTUs was evaluated using a database alignment search-based approach; when compared to their closest match of equal or longer sequence length from the NCBI 'nt' database, as determined by BLAST [54], OTUs with more than five nucleotides missing from the 5′ or 3′ end of their respective alignments were discarded as artifacts. Single read OTUs were subjected to an additional screen, where only sequences that had a perfect or near-perfect match to a sequence in the NCBI 'nt' database were kept for analysis, i.e., that the alignment had to span the entire sequence of the OTU, and a maximum of 1% of dissimilar nucleotides was tolerated.

After removal of sequence chimeras and artifacts, OTUs were subjected to taxonomic assignments as follows: two general taxonomic level assignments (Phylum and Family) for all OTUs using RDP Classifier [55], and closest relative identification for select OTUs using BLAST queries [54]. Alpha diversity indices (Observed OTUs, Chao, Ace, Shannon, and Simpson) were determined using the 'summary.single' command from MOTHUR (version 1.44.1) [53] on a dataset subsampled to 5000 reads for each sample. Principle Coordinate Analysis (PCoA) for beta diversity was performed using the same rarefied dataset, by determining Bray–Curtis distances with the 'summary.shared' command followed by the 'pcoa' command in MOTHUR (version 1.44.1) [53].

### 2.5. Statistical Analysis

Normal distribution of fish performance data was first confirmed using the Shapiro-Wilk test, then an Analysis of Variance (ANOVA), with a Tukey's HSD post hoc test for multiple comparisons, was performed for statistical analysis using JMP (Version 12, SAS Institute Inc., Cary, NC, USA). Comparisons of abundance for bacterial taxonomic groups and OTUs amongst different dietary treatments were performed in R (Version R-3.6.2) using the non-parametric test Kruskal–Wallis (command 'kruskal.test'), followed by the Wilcoxon test (command 'pairwise.wilcox.test') for multiple pairwise comparisons, which included the Benjamini-Hochberg correction to control for false discovery rate. For alpha diversity indices, normal distribution of data was first confirmed using the Shapiro Wilk test (command 'shapiro.test'), then comparison across the different diet groups was performed using ANOVA followed by Tukey's range test for multiple comparisons; these tests were conducted using R (Version R-3.6.2). Statistical significance was set at $p \leq 0.05$.

PERMANOVA (permutational multivariate analysis) was performed in R (Version R-3.6.2) using the command 'adonis', followed by the command ' pairwise.adonis' to identify pairs of sample groups that were different. For all analyses, tests resulting in $p \leq 0.05$ were considered significant. Analysis by LDA Effect Size (LEfSe) [56] was performed using a publicly available online implementation of the program (https://huttenhower.sph.harvard.edu/galaxy/ accessed on 16 October 2020).

### 2.6. Next-Generation Sequencing Data Accessibility

Raw sequence data are available from the NCBI Sequence Read Archive under Bioproject PRJNA718291.

## 3. Results

### 3.1. Feeding Trial Performance

Overall, all fish grew well across the seven dietary treatments, with 100% survival for the duration of the 105-day trial. While no difference in biomass gain per fish was detected among dietary treatments, the respective feed conversion ratios for all diets that included bioprocessed soybean meal were found to be improved since they were lower than for the CON diet ($p < 0.05$; Table 3).

### 3.2. Taxonomic Composition Analysis

A combined total of 15 samples from five of the dietary treatments were selected for investigating the intestinal bacterial composition of hybrid striped bass in response to the inclusion of bioprocessed soybean meal. In addition to samples with or without the inclusion of bioprocessed soybean meal (BP vs. CON), samples from three diets with modified bioprocessed soybean meal were also analyzed: BP-F1 and BP-W, which had the lowest FCR means, and BP-E, which had the highest digestibility of the seven diets (digestibility data not shown). A combined total of 302,427 high-quality sequence reads, ranging between 5561 and 87,780 sequence reads per sample (Supplementary Table S1), from the V1–V3 region of the 16S rRNA gene were generated from the five diets. Taxonomy-based composition analyses revealed that Firmicutes and Proteobacteria were the most abundant phyla across all samples (Table 4, Figure 1), with the former showing the highest representation across all diets except diet BP-F1. The respective abundances of the five main families from the phylum Firmicutes were all found to vary across dietary treatments ($p \leq 0.05$). *Peptostreptococcaceae*, *Peptoniphilaceae*, and *Clostridiaceae* were numerically more abundant in the BP and CON samples. In contrast, *Leuconostocaceae* showed their highest representation in the BP-F1, BP-E, and BP-W groups, while *Streptococcaceae* were at higher levels in the BP-E and BP-W groups. Of the three main families of Proteobacteria identified in this study, only *Enterobacteriaceae* were found to vary across dietary treatments ($p \leq 0.05$), with the highest levels observed in the BP-F1 and BP-E groups.

**Table 3.** Mean of performance indices for each dietary treatment over a 15 week trial.

| Diet | Gain [1] | Consumption [2] | FCR [3] | VSI [4] | HIS [5] |
|---|---|---|---|---|---|
| BP-F1 | 128.2 [a] | 185.0 [ab] | 1.44 [c] | 8.85 [a] | 1.49 [bc] |
| BP-F2 | 119.9 [a] | 178.5 [b] | 1.49 [b] | 8.61 [a] | 1.44 [bc] |
| BP-F3 | 110.8 [a] | 179.6 [b] | 1.62 [b] | 9.09 [a] | 1.40 [c] |
| BP-E | 105.6 [a] | 163.5 [b] | 1.55 [b] | 9.28 [a] | 1.56 [abc] |
| BP-W | 126.6 [a] | 183.8 [ab] | 1.45 [bc] | 8.57 [a] | 1.44 [bc] |
| BP | 119.7 [a] | 184.0 [ab] | 1.54 [bc] | 9.32 [a] | 1.64 [ab] |
| CON | 109.8 [a] | 211.2 [a] | 1.93 [a] | 8.60 [a] | 1.76 [a] |

Significant differences ($p < 0.05$) are indicated by different superscripts within a given column. [1]: Average weight (g)/fish; [2]: Average total consumption (g, dry)/fish; [3]: Feed Conversion Ratio; [4]: Viscerosomatic index (VSI); [5]: Hepatosomatic index (HSI).

**Table 4.** Mean relative abundance (%) of main bacterial phyla and families identified in the intestine of hybrid striped bass.

| Taxonomic Affiliation | CON | BP-F1 | BP-E | BP-W | BP | *p* Value |
|---|---|---|---|---|---|---|
| Firmicutes | 95.32 | 12.83 | 61.53 | 61.30 | 88.92 | 0.08 |
| *Peptostreptococcaceae* # | 72.66 [c] | 6.90 [abc] | 1.67 [a] | 6.22 [ab] | 66.54 [bc] | 0.02 |
| *Streptococcaceae* # | 4.09 [ab] | 2.29 [a] | 52.09 [b] | 49.18 [b] | 6.40 [ab] | 0.02 |
| *Leuconostocaceae* # | 0.10 [b] | 1.05 [a] | 4.64[a] | 2.10 [a] | 0.17 [ab] | 0.03 |
| *Peptoniphilaceae* # | 1.90 [b] | 0.22 [ab] | 0.07[a] | 0.23 [ab] | 2.35 [b] | 0.02 |
| *Clostridiaceae* 1 # | 3.27 [d] | 0.44 [abc] | 0.06 [b] | 0.25 [abc] | 1.38 [cd] | 0.01 |
| unclass. Clostridiales x | 8.89 | 0.78 | 0.29 | 0.84 | 7.94 | - |
| Other Firmicutes x | 4.42 | 1.16 | 2.70 | 2.47 | 4.15 | - |
| Proteobacteria | 2.00 | 85.23 | 37.04 | 37.03 | 7.43 | 0.06 |
| *Enterobacteriaceae* # | 1.12 [c] | 31.99 [a] | 31.28 [ac] | 1.02 [bc] | 6.10 [a] | 0.03 |
| *Sphingomonadaceae* | 0.30 | 18.04 | 2.46 | 15.29 | 0.25 | 0.26 |
| *Burkholderiaceae* | 0.24 | 34.43 | 3.01 | 20.06 | 0.26 | 0.11 |
| Other Proteobacteria x | 0.35 | 0.77 | 0.29 | 0.65 | 0.83 | - |
| Bacteriodetes | 2.25 | 0.79 | 0.49 | 0.74 | 3.31 | 0.06 |
| *Porphyromonadaceae* | 1.25 | 0.28 | 0.18 | 0.29 | 1.63 | 0.06 |
| Other Bacteroidetes x | 1.00 | 0.52 | 0.31 | 0.45 | 1.67 | - |
| Other Bacteria x$ | 0.42 | 1.14 | 0.94 | 0.93 | 0.34 | - |

Mean relative abundance of taxonomic groups is presented as a percentage (%) of the total number of analyzed reads per sample. Please see Supplementary Table S2 for standard errors of the means. # Taxa showing a statistically significant difference by the Kruskal–Wallis sum rank test ($p < 0.05$). Different superscripts in the same row indicate that groups are significantly different by the Wilcoxon test for multiple pairwise comparisons. x Statistical test not performed because of group heterogeneity. $ Other bacteria include Actinobacteria, Spirochaetes, Fusobacteria, Acidobacteria, Planctomycetes, as well as unclassified bacteria.

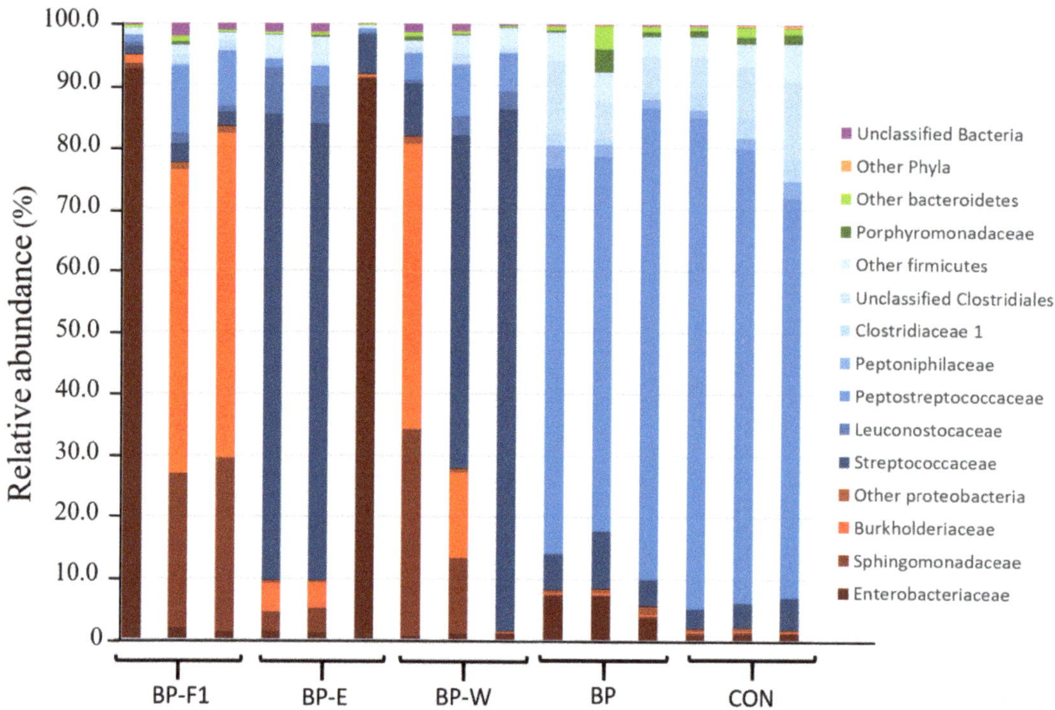


**Figure 1.** Taxonomic profiles at the phylum and family levels of intestinal bacterial communities of hybrid striped bass. Families belonging to the same phylum are represented by different shades of the same color: Firmicutes (blue), Bacteroidetes (green), and Proteobacteria (red).

### 3.3. Alpha and Beta Diversity

Since taxonomic profiling indicated differences in composition associated with diets, OTU-level analyses were performed to gain further insight (Table 5). Based on the alpha diversity indices Observed OTUs, Ace, and Chao, dietary treatments appeared to fall into two distinct groups, with the group consisting of treatments BP-F1, BP-E, and BP-W having a lower number of OTUs compared to the group with BP and CON ($p \leq 0.05$). Clustering of treatments into separate groups was consistent with PCoA (Figure 2) and supported by the PERMANOVA test ($p = 0.001$).

**Table 5.** Observed OTUs and alpha-diversity indices in five dietary treatment groups. Values are shown as means.

| Index | CON | BP-F1 | BP-E | BP-W | BP | *p*-Value |
|---|---|---|---|---|---|---|
| Observed OTUs # | 266.67 [b] | 172.67 [a] | 134.00 [a] | 169.67 [a] | 301.67 [b] | <0.001 |
| Ace # | 799.75 [b] | 397.48 [a] | 305.62 [a] | 377.39 [a] | 725.13 [b] | <0.001 |
| Chao # | 550.36 [b] | 284.72 [a] | 269.85 [a] | 306.63 [a] | 545.64 [b] | <0.001 |
| Shannon | 2.81 | 1.94 | 1.59 | 2.17 | 3.03 | 0.078 |
| Simpson | 0.24 | 0.40 | 0.53 | 0.32 | 0.20 | 0.287 |

# Taxa showing a statistically significant difference by ANOVA ($p < 0.05$). Please see Supplementary Table S3 for standard errors of the means. Different superscripts in the same row indicate that groups are significantly different by the Tukey's range test for multiple pairwise comparisons.

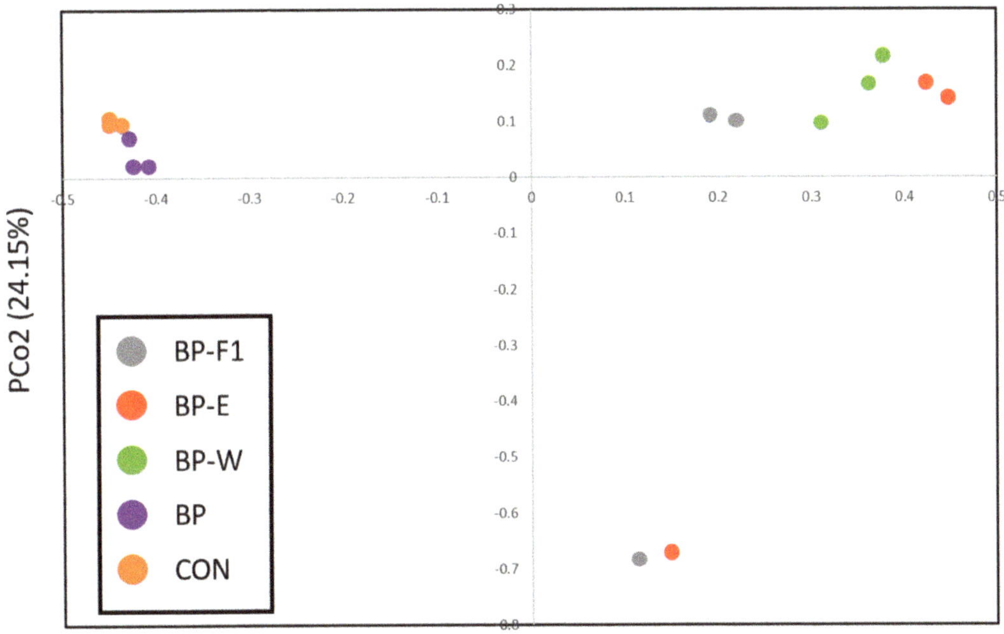


**Figure 2.** Comparison of intestinal bacterial communities in hybrid striped bass for five dietary treatments (BP-F1, BP-E, BP-W, BP, and CON). Principal Coordinate Analysis (PCoA) was performed based on the Bray–Curtis distance matrix. The x and y axes correspond to Principal Components 1 (PCo1) and 2 (PCo2), respectively, which together explained 67.82% of the variance. The PERMANOVA test supported the separation of samples into different groups ($p = 0.001$), but differences between groups could not be resolved by pairwise comparisons ($p > 0.05$).

### 3.4. OTU Composition Analysis

Of the 1132 OTUs that were identified across all samples, the most abundant OTUs, defined as representing at least 1.0% of sequences in at least one set of samples, were further analyzed (Figure 3; Table 6). Eleven of these OTUs, one assigned to Proteobacteria and ten affiliated to Firmicutes, were found to vary across dietary treatments ($p \leq 0.05$), and they exhibited composition patterns that were consistent with their respective taxonomic groups. For instance, SD_McMs-00002 was the most highly represented OTU of the family *Enterobacteriaceae*, representing 79.0–95.8% of sequence reads from this taxonomic group across all samples, and it accordingly was most abundant in diets BP-F1 and BP-E. Five of the OTUs of interest were assigned to *Peptostreptococcaceae* (SD_McMs-00001, SD_McMs-00011, SD_McMs-00012, SD_McMs-00014, and SD_McMs-00015), and their highest representation was in samples from diets BP and CON. SD_McMs-00001 was the most abundant OTU from this group, representing 2.86–4.42-times the combined read abundances from the other four *Peptostreptococcaceae* OTUs in each sample. Of the remaining Firmicutes OTUs that varied across dietary treatments, three were assigned to *Streptococcaceae* (SD_McMs-00007, SD_McMs-00010, and SD_McMs-00016). SD_McMs-00007 and SD_McMs-00016 were most closely related to *Lactococcus lactis*, and they were at their highest representation in diets BP-E and BP-W, while SD_McMs-00010 was most closely related to *Streptococcus dysgalactiae*, and it was most abundant in diets BP and CON. LEfse analysis was also used to identify biomarkers for the dietary treatments tested in the study; these consisted of 55 OTUs for CON, 38 OTUs for BPP, 17 OTUs for BP-E, four OTUs for BP-F1, and two OTUs for BP-W (Supplementary Figure S1). The eleven abundant

OTUs identified as significantly different by Kruskal–Wallis (Table 6) were identified as biomarkers by LEfse.

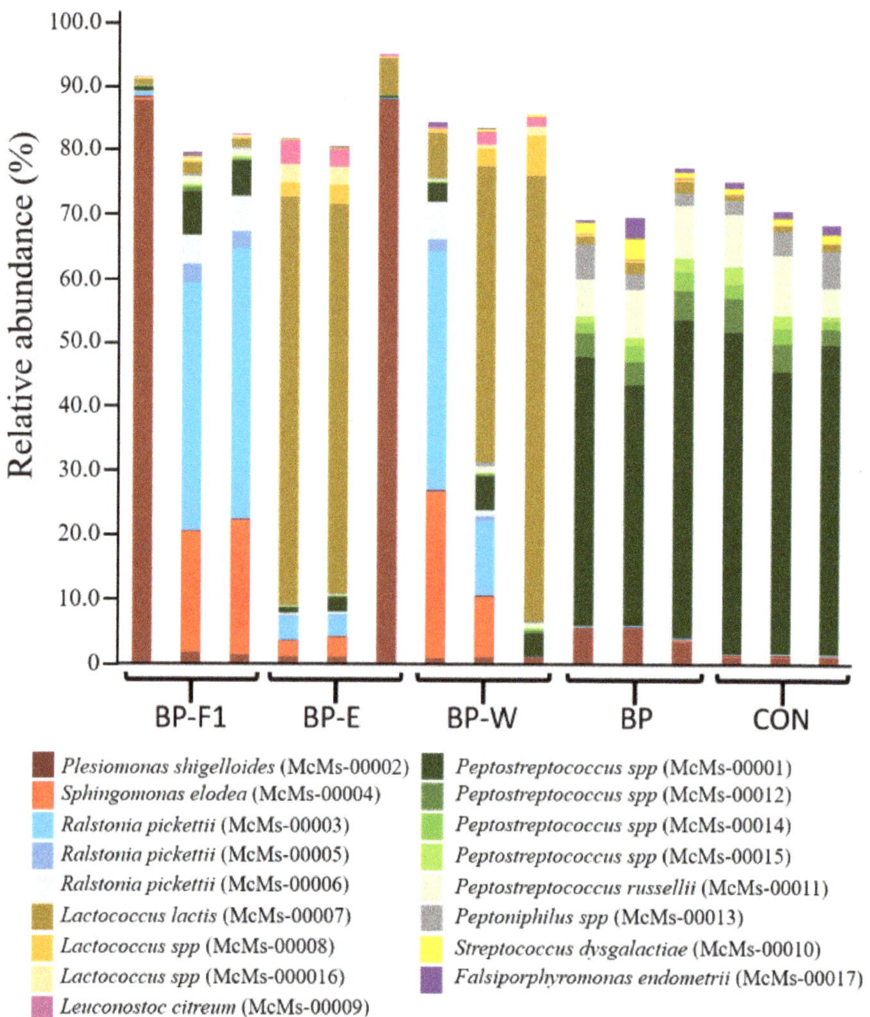


**Figure 3.** Histogram showing the relative abundance of the most highly represented intestinal OTUs in hybrid striped bass for five dietary treatments. OTUs showing 97% sequence identity or greater to their closest relative are represented by the full species name, while OTUs showing less than 97% identity to their closest relative are represented by their assigned genus.

**Table 6.** Mean relative abundance of the main bacterial OTUs identified in hybrid striped bass. Abundance is presented as a percentage (%) of the total number of analyzed reads per sample.

| OTUs | CON | BP-F1 | BP-E | BP-W | BP | *p*-Value | Closest Taxon (id%) |
|---|---|---|---|---|---|---|---|
| Proteobacteria | | | | | | | |
| SD_McMs-00002 # | 1.03 $^{ab}$ | 30.28 $^{a}$ | 29.96 $^{ab}$ | 0.84 $^{b}$ | 4.82 $^{a}$ | 0.05 | *Pl. shigelloides* (99%) |
| SD_McMs-00003 | 0.18 | 27.22 | 2.44 | 16.41 | 0.21 | 0.09 | *R. pickettii* (99%) |
| SD_McMs-00004 | 0.23 | 13.54 | 1.90 | 11.97 | 0.18 | 0.29 | *Sp. elodea* (99%) |
| SD_McMs-00005 | 0.02 | 1.92 | 0.14 | 0.92 | 0.01 | 0.16 | *R. pickettii* (98%) |
| SD_McMs-00006 | 0.03 | 3.39 | 0.26 | 2.24 | 0.03 | 0.24 | *R. pickettii* (99%) |
| Firmicutes | | | | | | | |
| SD_McMs-00001 # | 47.61 $^{a}$ | 4.29 $^{abc}$ | 1.15 $^{c}$ | 3.91 $^{bc}$ | 43.13 $^{ab}$ | 0.02 | *Ps. russellii* (91%) |
| SD_McMs-00007 # | 0.86 $^{a}$ | 1.35 $^{ab}$ | 43.47 $^{b}$ | 41.01 $^{b}$ | 1.59 $^{ab}$ | 0.02 | *La. lactis* (100%) |
| SD_McMs-00008 | 0.16 | 0.16 | 1.72 | 3.19 | 0.34 | 0.08 | *La. lactis* (96%) |
| SD_McMs-00009 # | 0.03 $^{a}$ | 0.10 $^{ab}$ | 2.25 $^{b}$ | 1.20 $^{b}$ | 0.10 $^{ab}$ | 0.02 | *Le. citreum* (100%) |
| SD_McMs-00010 # | 0.98 $^{c}$ | 0.05 $^{a}$ | 0.10 $^{ac}$ | 0.29 $^{abc}$ | 1.97 $^{bc}$ | 0.02 | *St. dysgalactiae* (100%) |
| SD_McMs-00011 # | 7.41 $^{a}$ | 0.86 $^{ab}$ | 0.11 $^{b}$ | 0.75 $^{ab}$ | 7.20 $^{a}$ | 0.02 | *Ps. russellii* (99%) |
| SD_McMs-00012 # | 3.99 $^{a}$ | 0.31 $^{ab}$ | 0.07 $^{b}$ | 0.27 $^{ab}$ | 3.98 $^{a}$ | 0.02 | *Ps. russellii* (91%) |
| SD_McMs-00013 # | 3.97 $^{a}$ | 0.28 $^{b}$ | 0.11 $^{b}$ | 0.37 $^{ab}$ | 3.30 $^{ab}$ | 0.02 | *Pe. stercorisuis* (89%) |
| SD_McMs-00014 # | 1.98 $^{bc}$ | 0.16 $^{ac}$ | 0.04 $^{a}$ | 0.17 $^{ab}$ | 2.32 $^{b}$ | 0.02 | *Ps. russellii* (94%) |
| SD_McMs-00015 # | 1.81 $^{a}$ | 0.17 $^{ab}$ | 0.04 $^{b}$ | 0.10 $^{ab}$ | 1.51 $^{a}$ | 0.02 | *Ps. russellii* (94%) |
| SD_McMs-00016 # | 0.02 $^{c}$ | 0.31 $^{ab}$ | 1.96 $^{b}$ | 0.72 $^{ab}$ | 0.04 $^{ac}$ | 0.03 | *La. lactis* (96%) |
| Bacteroidetes | | | | | | | |
| SD_McMs-00017 | 1.18 | 0.25 | 0.16 | 0.27 | 1.42 | 0.06 | *F. endometrii* (99%) |

# OTUs showing a statistically significant difference by the Kruskal-Wallis sum rank test ($p < 0.05$). Different superscripts in the same row indicate that groups are significantly different by the Wilcoxon test for multiple pairwise comparisons. Please see Supplementary Table S4 for standard errors of the means and Supplementary Table S1 for a complete list of OTUs and their respective abundances. Abbreviations: *F.*: *Falsiporphyromonas*; *La.*: *Lactococcus*; *Le.*: *Leuconostoc*; *Pe.*: *Peptoniphilus*; *Ps.*: *Peptostreptococcus*; *Pl.*: *Plesiomonas*; *R.*: *Ralstonia*; *Sp.*: *Sphingomonas*; *St.*: *Streptococcus*.

## 4. Discussion

As a result of challenges such as cost and availability, the inclusion of fishmeal as a primary ingredient in aquaculture has become difficult to sustain [6]. Fishmeal use in aquaculture and livestock diets is also cause for ethical and social sustainability concerns, as increased demand for fishmeal would not only risk reducing the supply of fish available as food for humans but also promote overexploitation of species that are not for human consumption [57]. While other animal protein sources, such as poultry meal and feather meal, have served as suitable alternatives in aquaculture diets, there remains a critical need to find more economical replacement ingredients. While lower cost and availability make plant-based protein ingredients attractive alternatives, the presence of anti-nutritional factors such as non-starch polysaccharides, protease inhibitors, lectins, saponins, phytic acid, phytoestrogens, and allergens limits the extent to which they can be included in fish diets [58].

Thus, even if soybean meal provides a well-balanced amino acid profile, a favorable protein content, and lower amounts of anti-nutrients relative to other plant-based protein sources [59,60], its inclusion in diets of carnivorous species still needs to be restricted. Bioprocessing, i.e., modification of plant-based primary ingredients by microbial metabolism, has provided a solution to this problem. Indeed, fermentation of soybean meal into 'bioprocessed' soybean meal results in a product with an enhanced nutritional profile, as it is highly digestible and has a high protein content with increased lysine and methionine concentrations, while its anti-nutritional factor levels are greatly reduced [61].

One possible effect of the high inclusion of bioprocessed soybean meal in aquaculture diets is its potential impact on the composition of intestinal microbial communities. Considering their contributions to the health and nutrition of their host, alterations in the composition of symbiotic gut microbial communities could have unintended consequences on aquaculture production. In this context, the study presented in this report aimed at investigating the potential effects of bioprocessed soybean meal on the intestinal bacterial composition of aquaculture-raised hybrid striped bass. In addition to gaining further insight on an alternative feed ingredient with great potential for aquaculture, this report is also the first to provide insight on the gut microbiome of the hybrid striped bass using a culture-independent method.

The first observation was the absence of major differences in intestinal bacterial composition between fish fed a diet with the bioprocessed soybean meal (BP) and fish fed a diet without bioprocessed soybean meal (CON). Based on its limited sequence identity to its closest valid relative (*Peptostreptococcus russellii*, 91%), the main OTU (SD_McMs-00001) in these samples was predicted to correspond to a currently uncharacterized or uncultured bacterial species. Because of its taxonomic affiliation, SD_McMs-00001 would be predicted to utilize proteins as a main substrate. Indeed, strains of *P. russellii* were originally isolated from swine manure, and they were reported to produce elevated amounts of ammonia when grown in culture with various peptide-based ingredients [62]. This activity was interpreted as *P. russellii* playing an active role in the digestion and fermentation of proteinaceous material [63,64], which is consistent with the high protein content in carnivorous fish diets and the dietary treatments used in this study. Bacteria affiliated to the genus *Peptostreptococcus* have been identified as one of the main bacterial groups in rainbow trout [31] and proposed as an indicator taxon of fast-growing fish for this host [65]. This bacterial group was also reported to be well represented in aquaculture raised Arctic char (*Salvelinus alpinus*) [66]. Interestingly, many species of the genus *Peptostreptococcus* can increase the production of indoleacrylic acid and decrease the susceptibility of epithelial injury in mice [60]. Furthermore, research in humans has revealed that increased production of indoleacrylic acid could provide relief to inflammatory bowel disease [67]. Together, these previously published studies suggest that diets resulting in increased *P. russellii* may improve fish health by benefiting the host's intestinal mucus layer.

Unexpectedly, the main intestinal OTUs in fish fed different post-fermentation modified versions of the bioprocessed soybean meal, i.e., diets BP-F1, BP-E, and BP-W, were very different from BP. The most abundant OTU in four of the six combined samples for BP-E and BP-W samples, SD_McMs-00007, was predicted to be a strain of *Lactococcus lactis* based on their high nucleotide sequence identity. *L. lactis* is known for its broad use in the food industry [68] because of its basic ability to utilize proteins and ferment carbohydrates into lactate. Strains of this species have been isolated from a number of distinct sources, including drain water and human vaginal samples [69,70], indicating that *L. lactis* is suited to many different types of environments. While it is not considered a typical resident of the gastrointestinal tract, *L. lactis* is capable of surviving in the gut environment [71,72], where it may interact with the mucus layer [73]. This species has been used as a probiotic in red sea bream (*Pagrus major*), resulting in a higher final weight, percent weight gain, and specific growth rate [74], as well as in olive flounder (*Paralichythys olivaceus*), where it was found to increase levels of growth-promoting metabolites, such as short-chain fatty acids, citrulline, taurine and vitamins [75].

The other two main OTUs in samples from diets BP-F1, BP-E, and BP-W may have represented potential pathogens. SD_McMs-00002 was 99% identical to *Plesiomonas shigelloides*, a bacterial species predominantly found in freshwater fish [40,76,77]. Notably, *P. shigelloides* was reported as a pathogen in cultured tilapia, with infected fish suffering tissue damage in the liver, spleen, kidney, heart, and intestine [78]. Since *P. shigelloides* has been reported as pathogenic in another report [76], future investigations will be required to gain further insight into the biological roles of SD_McMs-00002 and other related strains to determine whether this species represents a pathogen or a commensal in the hybrid striped bass. Similarly, SD_McMs-00003 was very closely related to *Ralstonia pickettii* (99%), also a bacterial species of the phylum Proteobacteria. While it has been proposed as a normal resident of the fish gastrointestinal tract [79], and it has been identified in soil and water samples [80], *R. pickettii* was also reported as a low virulence pathogen in certain cases of invasive infections in humans [81].

Based on the results presented in this report, the inclusion of bioprocessed soybean meal did not dramatically alter the intestinal bacterial composition of hybrid striped bass in the context of an aquaculture-based diet that consisted of a combination of fishmeal and alternative protein sources. However, inclusion of bioprocessed soybean meal that had been further processed after fermentation resulted in different and less consistent intestinal bacterial composition patterns. It will be of interest to further investigate these different effects of post-fermentation treated bioprocessed soybean meal on gut bacterial communities of aquaculture-raised fish in order to determine if the benefits of these alternative feed ingredients on performance are worth potential risks to gut microbiome function.

**Supplementary Materials:** The following are available online at https://www.mdpi.com/article/10.3390/microorganisms9051032/s1.

**Author Contributions:** Conceptualization M.B.; methodology M.B. and B.S.-P.; investigation, B.W., E.C.F. and P.P.; formal analysis, E.C.F., B.W. and B.S.-P.; data curation, M.B. and B.S.-P.; writing—original draft preparation, E.C.F.; writing—review and editing, M.B. and B.S.-P.; funding acquisition, M.B. All authors have read and agreed to the published version of the manuscript.

**Funding:** Funding for this research was provided by the South Dakota State University Agricultural Experiment Station, the South Dakota Soybean Research and Promotion Council, and the United Soybean Board.

**Institutional Review Board Statement:** The study was conducted according to the guidelines of the Declaration of Helsinki, and approved by the Institutional Animal Care and Use Committee (IACUC) of South Dakota State University (Approval Number 16-089A, issued on 11/22/2016).

**Informed Consent Statement:** Not applicable.

**Data Availability Statement:** Raw sequence data are available from the NCBI Sequence Read Archive under Bioproject PRJNA718291.

**Conflicts of Interest:** The authors declare no conflict of interest.

## References

1. Anguiano, M.; Pohlenz, C.; Buentello, A.; Gatlin, D.M., 3rd. The effects of prebiotics on the digestive enzymes and gut histomorphology of red drum (*Sciaenops ocellatus*) and hybrid striped bass (*Morone chrysops* × *M. saxatilis*). *Br. J. Nutr.* **2013**, *109*, 623–629. [CrossRef] [PubMed]
2. Garber, A.F.; Sullivan, C.V. Selective breeding for the hybrid striped bass (*Morone chrysops*, Rafinesque × *M. saxatilis*, Walbaum) industry: Status and perspectives. *Aquac. Res.* **2006**, *37*, 319–338. [CrossRef]
3. FAO 2016–2021. Cultured Aquatic Species Information Programme. In *Morone Hybrid*; Harrell, R.M., Ed.; FAO Fisheries Division: Rome, Italy, 2012.
4. Quagrainie, K. *Profitability of Hybrid Striped Bass Cage Aquaculture in the Midwest*; Purdue University: West Lafayette, IN, USA, 2015.
5. Lougheed, M.; Nelson, B. *Hybrid Striped Bass Production, Markets and Marketing*; Michigan State University, North Central Regional Aquaculture Center: East Lansing, MI, USA, 2001.
6. Tacon, A.G.J.; Metian, M. Global overview on the use of fish meal and fish oil in industrially compounded aquafeeds: Trends and future prospects. *Aquaculture* **2008**, *285*, 146–158.
7. Ytrestøyl, T.; Aas, T.S.; Åsgård, T. Utilisation of feed resources in production of Atlantic salmon (*Salmo salar*) in Norway. *Aquaculture* **2015**, *448*, 365–374.
8. Asche, F.; Oglend, A.; Tveteras, S. Regime shifts in the fish meal/soybean meal price ratio. *J. Agric. Econ.* **2013**, *64*, 97–111. [CrossRef]
9. Yaghoubi, M.; Mozanzadeh, M.T.; Marammazi, J.G.; Safari, O.; Gisbert, E. Dietary replacement of fish meal by soy products (soybean meal and isolated soy protein) in silvery-black porgy juveniles (*Sparidentex hasta*). *Aquaculture* **2016**, *464*, 50–59. [CrossRef]
10. Sindelar, S.C. Utilization of Soybean Products as Fish-Meal Protein Replacements in Yellow Perch *Perca flavescens* Feeds. Master's Thesis, South Dakota State University, Brookings, SD, USA, 2014.
11. Perkins, E.G. Composition of soybeans and soybean products. In *Practical Handbook of Soybean Processing and Utilization*; Erickson, D.R., Ed.; AOCS Press: Champaign, IL, USA, 1995; pp. 9–28.
12. Bruce, T.J.; Neiger, R.D.; Brown, M.L. Gut histology, immunology and the intestinal microbiota of rainbow trout, *Oncorhynchus mykiss* (Walbaum), fed process variants of soybean meal. *Aquac. Res.* **2018**, *49*, 492–504. [CrossRef]
13. Lathrop, S.K.; Bloom, S.M.; Rao, S.M.; Nutsch, K.; Lio, C.W.; Santacruz, N.; Peterson, D.A.; Stappenbeck, T.S.; Hsieh, C.S. Peripheral education of the immune system by colonic commensal microbiota. *Nature* **2011**, *478*, 250–254. [CrossRef]

14. Ley, R.E.; Lozupone, C.A.; Hamady, M.; Knight, R.; Gordon, J.I. Worlds within worlds: Evolution of the vertebrate gut microbiota. *Nat. Rev. Microbiol.* **2008**, *6*, 776–788. [CrossRef]
15. Ciarlo, E.; Heinonen, T.; Herderschee, J.; Fenwick, C.; Mombelli, M.; Le Roy, D.; Roger, T. Impact of the microbial derived short chain fatty acid propionate on host susceptibility to bacterial and fungal infections in vivo. *Sci. Rep.* **2016**, *6*, 37944. [CrossRef]
16. Gómez, G.D.; Balcázar, J.L. A review on the interactions between gut microbiota and innate immunity of fish. *FEMS Immunol. Med. Microbiol.* **2008**, *52*, 145–154. [CrossRef]
17. Llewellyn, M.S.; Boutin, S.; Hoseinifar, S.H.; Derome, N. Teleost microbiomes: The state of the art in their characterization, manipulation and importance in aquaculture and fisheries. *Front. Microbiol.* **2014**, *5*, 207. [CrossRef]
18. Talwar, C.; Nagar, S.; Lal, R.; Negi, R.K. Fish gut microbiome: Current approaches and future perspectives. *Indian J. Microbiol.* **2018**, *58*, 397–414. [CrossRef]
19. Sullam, K.E.; Essinger, S.D.; Lozupone, C.A.; O'Connor, M.P.; Rosen, G.L.; Knight, R.; Kilham, S.S.; Russell, J.A. Environmental and ecological factors that shape the gut bacterial communities of fish: A meta-analysis. *Mol. Ecol.* **2012**, *21*, 3363–3378. [CrossRef]
20. Liu, H.; Guo, X.; Gooneratne, R.; Lai, R.; Zeng, C.; Zhan, F.; Wang, W. The gut microbiome and degradation enzyme activity of wild freshwater fishes influenced by their trophic levels. *Sci. Rep.* **2016**, *6*, 1–12. [CrossRef] [PubMed]
21. Uren Webster, T.M.; Rodriguez-Barreto, D.; Castaldo, G.; Gough, P.; Consuegra, S.; Garcia de Leaniz, C. Environmental plasticity and colonisation history in the Atlantic salmon microbiome: A translocation experiment. *Mol. Ecol.* **2020**, *29*, 886–898. [CrossRef] [PubMed]
22. Egerton, S.; Wan, A.; Murphy, K.; Collins, F.; Ahern, G.; Sugrue, I.; Busca, K.; Egan, F.; Muller, N.; Whooley, J. Replacing fishmeal with plant protein in Atlantic salmon (*Salmo salar*) diets by supplementation with fish protein hydrolysate. *Sci. Rep.* **2020**, *10*, 1–16. [CrossRef] [PubMed]
23. Huyben, D.; Roehe, B.K.; Bekaert, M.; Ruyter, B.; Glencross, B. Dietary lipid: Protein ratio and n-3 long-chain polyunsaturated fatty acids alters the gut microbiome of Atlantic salmon under hypoxic and normoxic conditions. *Front. Microbiol.* **2020**, *11*, 3385. [CrossRef]
24. Fogarty, C.; Burgess, C.M.; Cotter, P.D.; Cabrera-Rubio, R.; Whyte, P.; Smyth, C.; Bolton, D.J. Diversity and composition of the gut microbiota of Atlantic salmon (*Salmo salar*) farmed in Irish waters. *J. Appl. Microbiol.* **2019**, *127*, 648–657. [CrossRef]
25. Gupta, S.; Fečkaninová, A.; Lokesh, J.; Koščová, J.; Sørensen, M.; Fernandes, J.; Kiron, V. Lactobacillus dominate in the intestine of Atlantic salmon fed dietary probiotics. *Front. Microbiol.* **2019**, *9*, 3247. [CrossRef]
26. Brown, R.M.; Wiens, G.D.; Salinas, I. Analysis of the gut and gill microbiome of resistant and susceptible lines of rainbow trout (Oncorhynchus mykiss). *Fish Shellfish. Immunol.* **2019**, *86*, 497–506. [CrossRef]
27. Giang, P.T.; Sakalli, S.; Fedorova, G.; Tilami, S.K.; Bakal, T.; Najmanova, L.; Grabicova, K.; Kolarova, J.; Sampels, S.; Zamaratskaia, G. Biomarker response, health indicators, and intestinal microbiome composition in wild brown trout (Salmo trutta m. fario L.) exposed to a sewage treatment plant effluent-dominated stream. *Sci. Total Environ.* **2018**, *625*, 1494–1509. [CrossRef]
28. Lyons, P.; Turnbull, J.; Dawson, K.A.; Crumlish, M. Phylogenetic and functional characterization of the distal intestinal microbiome of rainbow trout *Oncorhynchus mykiss* from both farm and aquarium settings. *J. Appl. Microbiol.* **2017**, *122*, 347–363. [CrossRef]
29. Michl, S.C.; Beyer, M.; Ratten, J.-M.; Hasler, M.; LaRoche, J.; Schulz, C. A diet-change modulates the previously established bacterial gut community in juvenile brown trout (*Salmo trutta*). *Sci. Rep.* **2019**, *9*, 1–12. [CrossRef]
30. Michl, S.C.; Ratten, J.-M.; Beyer, M.; Hasler, M.; LaRoche, J.; Schulz, C. The malleable gut microbiome of juvenile rainbow trout (*Oncorhynchus mykiss*): Diet-dependent shifts of bacterial community structures. *PLoS ONE* **2017**, *12*, e0177735. [CrossRef]
31. Rimoldi, S.; Terova, G.; Ascione, C.; Giannico, R.; Brambilla, F. Next generation sequencing for gut microbiome characterization in rainbow trout (*Oncorhynchus mykiss*) fed animal by-product meals as an alternative to fishmeal protein sources. *PLoS ONE* **2018**, *13*, e0193652. [CrossRef]
32. Valdés, N.; Gonzalez, A.; Garcia, V.; Tello, M. Analysis of the microbiome of rainbow trout (*Oncorhynchus mykiss*) exposed to the pathogen *Flavobacterium psychrophilum* 10094. *Microbiol. Resour. Announc.* **2020**, *9*. [CrossRef]
33. Bledsoe, J.W.; Waldbieser, G.C.; Swanson, K.S.; Peterson, B.C.; Small, B.C. Comparison of channel catfish and blue catfish gut microbiota assemblages shows minimal effects of host genetics on microbial structure and inferred function. *Front. Microbiol.* **2018**, *9*, 1073. [CrossRef]
34. Burgos, F.A.; Ray, C.L.; Arias, C.R. Bacterial diversity and community structure of the intestinal microbiome of Channel Catfish (*Ictalurus punctatus*) during ontogenesis. *Syst. Appl. Microbiol.* **2018**, *41*, 494–505. [CrossRef]
35. Zhang, H.; Wang, H.; Hu, K.; Jiao, L.; Zhao, M.; Yang, X.; Xia, L. Effect of Dietary supplementation of *Lactobacillus casei* YYL3 and *L. plantarum* YYL5 on growth, immune response and intestinal microbiota in channel catfish. *Animals* **2019**, *9*, 1005. [CrossRef]
36. Zhang, Z.; Li, D.; Refaey, M.M.; Xu, W. High spatial and temporal variations of microbial community along the southern catfish gastrointestinal tract: Insights into dynamic food digestion. *Front. Microbiol.* **2017**, *8*, 1531. [CrossRef]
37. Zhang, Z.; Li, D.; Refaey, M.M.; Xu, W.; Tang, R.; Li, L. Host age affects the development of southern catfish gut bacterial community divergent from that in the food and rearing water. *Front. Microbiol.* **2018**, *9*, 495. [CrossRef]
38. Zhong, L.; Li, D.; Wang, M.; Chen, X.; Bian, W.; Zhu, G. Dynamics of the bacterial community in a channel catfish nursery pond with a cage–pond integration system. *Can. J. Microbiol.* **2018**, *64*, 954–967. [CrossRef]
39. MacFarlane, R.D.; McLaughlin, J.J.; Bullock, G.L. Quantitative and qualitative studies of gut flora in striped bass from estuarine and coastal marine environments. *J. Wildl. Dis.* **1986**, *22*, 344–348. [CrossRef]

40. Nedoluha, P.C.; Westhoff, D. Microbiology of striped bass grown in three aquaculture systems. *Food Microbiol.* **1997**, *14*, 255–264. [CrossRef]
41. Nedoluha, P.C.; Westhoff, D. Microbiological analysis of striped bass (*Morone saxatilis*) grown in flow-through tanks. *J. Food Prot.* **1995**, *58*, 1363–1368. [CrossRef]
42. Nedoluha, P.C.; Westhoff, D. Microbiological analysis of striped bass (*Morone saxatilis*) grown in a recirculating system (†). *J. Food Prot.* **1997**, *60*, 948–953. [CrossRef]
43. Nedoluha, P.C.; Westhoff, D. Microbiological flora of aquacultured hybrid striped bass. *J. Food Prot.* **1993**, *56*, 1054–1060. [CrossRef]
44. Aly, S.M.; Abd-El-Rahman, A.M.; John, G.; Mohamed, M.F. Characterization of some bacteria isolated from *Oreochromis niloticus* and their potential use as probiotics. *Aquaculture* **2008**, *277*, 1–6. [CrossRef]
45. Barrows, F.; Gaylord, T.; Sealey, W.; Rawles, S. Database of Nutrient Digestibility's of Traditional and Novel Feed Ingredients for Trout and Hybrid Striped Bass. USDA-ARS; 2021. Available online: http://www.ars.usda.gov/Main/docs.htm (accessed on 16 October 2020).
46. Yu, Z.; Morrison, M. Improved extraction of PCR-quality community DNA from digesta and fecal samples. *Biotechniques* **2004**, *36*, 808–812. [CrossRef]
47. Edwards, U.; Rogall, T.; Blöcker, H.; Emde, M.; Böttger, E.C. Isolation and direct complete nucleotide determination of entire genes. Characterization of a gene coding for 16S ribosomal RNA. *Nucleic Acids Res.* **1989**, *17*, 7843–7853. [CrossRef] [PubMed]
48. Lane, D.J.; Pace, B.; Olsen, G.J.; Stahl, D.A.; Sogin, M.L.; Pace, N.R. Rapid determination of 16S ribosomal RNA sequences for phylogenetic analyses. *Proc. Natl. Acad. Sci. USA* **1985**, *82*, 6955–6959. [CrossRef] [PubMed]
49. Opdahl, L.J.; Gonda, M.G.; St-Pierre, B. Identification of uncultured bacterial species from Firmicutes, Bacteroidetes and CANDIDATUS Saccharibacteria as candidate cellulose utilizers from the rumen of beef cows. *Microorganisms* **2018**, *6*, 17. [CrossRef] [PubMed]
50. Poudel, P.; Levesque, C.L.; Samuel, R.; St-Pierre, B. Dietary inclusion of Peptiva, a peptide-based feed additive, can accelerate the maturation of the fecal bacterial microbiome in weaned pigs. *BMC Vet. Res.* **2020**, *16*, 60. [CrossRef]
51. Edgar, R.C.; Haas, B.J.; Clemente, J.C.; Quince, C.; Knight, R. UCHIME improves sensitivity and speed of chimera detection. *Bioinformatics* **2011**, *27*, 2194–2200. [CrossRef]
52. Haas, B.J.; Gevers, D.; Earl, A.M.; Feldgarden, M.; Ward, D.V.; Giannoukos, G.; Ciulla, D.; Tabbaa, D.; Highlander, S.K.; Sodergren, E. Chimeric 16S rRNA sequence formation and detection in Sanger and 454-pyrosequenced PCR amplicons. *Genome Res.* **2011**, *21*, 494–504. [CrossRef]
53. Schloss, P.D.; Westcott, S.L.; Ryabin, T.; Hall, J.R.; Hartmann, M.; Hollister, E.B.; Lesniewski, R.A.; Oakley, B.B.; Parks, D.H.; Robinson, C.J.; et al. Introducing mothur: Open-Source, platform-independent, community-supported software for describing and comparing microbial communities. *Appl. Environ. Microbiol.* **2009**, *75*, 7537–7541. [CrossRef]
54. Altschul, S.F.; Madden, T.L.; Schäffer, A.A.; Zhang, J.; Zhang, Z.; Miller, W.; Lipman, D.J. Gapped BLAST and PSI-BLAST: A new generation of protein database search programs. *Nucleic Acids Res.* **1997**, *25*, 3389–3402. [CrossRef]
55. Wang, Q.; Garrity, G.M.; Tiedje, J.M.; Cole, J.R. Naive Bayesian classifier for rapid assignment of rRNA sequences into the new bacterial taxonomy. *Appl. Environ. Microbiol.* **2007**, *73*, 5261–5267. [CrossRef]
56. Segata, N.; Izard, J.; Walron, L.; Gevers, D.; Miropolsky, L.; Garrett, W.; Huttenhower, C. Metagenomic biomarker discovery and explanation. *Genome Biology* **2011**, *12*, 1–8. [CrossRef]
57. Naylor, R.L.; Goldburg, R.J.; Primavera, J.H.; Kautsky, N.; Beveridge, M.C.M.; Clay, J.; Folke, C.; Lubchenco, J.; Mooney, H.; Troell, M. Effects of aquaculture on world fish supplies. *Nature* **2000**, *405*, 1017–1024. [CrossRef]
58. Francis, G.; Makkar, H.P.S.; Becker, K. Antinutritional factors present in plant-derived alternate fish feed ingredients and their effects in fish. *Aquaculture* **2001**, *199*, 197–227. [CrossRef]
59. Dersjant-Li, Y. The use of soy protein in aquafeeds. In Proceedings of the Avances en nutrición acuícola VI. Memorias del VI Simposium Internacional de Nutrición Acuícola., Cancún, Quintana Roo, México, 3–6 September 2002.
60. Gatlin, D.M.; Barrows, F.T.; Brown, P.; Dabrowski, K.; Gaylord, T.G.; Hardy, R.W.; Herman, E.; Hu, G.; Krogdahl, A.; Nelson, R.; et al. Expanding the utilization of sustainable plant products in aquafeeds: A review. *Aquac. Res.* **2007**, *38*, 551–579. [CrossRef]
61. Suman, G.; Nupur, M.; Anuradha, S.; Pradeep, B. Single-cell protein production: A review. *Int. J. Curr. Microbiol. Appl. Sci.* **2015**, *4*, 11.
62. Whitehead, T.R.; Cotta, M.A.; Falsen, E.; Moore, E.; Lawson, P.A. *Peptostreptococcus russellii* sp. nov., isolated from a swine-manure storage pit. *Int. J. Syst. Evol. Microbiol.* **2011**, *61*, 1875–1879. [CrossRef]
63. Attwood, G.T.; Klieve, A.V.; Ouwerkerk, D.; Patel, B.K. Ammonia-hyperproducing bacteria from New Zealand ruminants. *Appl. Environ. Microbiol.* **1998**, *64*, 1796–1804. [CrossRef]
64. Whitehead, T.R.; Cotta, M.A. Isolation and identification of hyper-ammonia producing bacteria from swine manure storage pits. *Curr. Microbiol.* **2004**, *48*, 20–26. [CrossRef]
65. Chapagain, P.; Arivett, B.; Cleveland, B.M.; Walker, D.M.; Salem, M. Analysis of the fecal microbiota of fast- and slow-growing rainbow trout (*Oncorhynchus mykiss*). *BMC Genom.* **2019**, *20*, 788. [CrossRef]
66. Nyman, A.; Huyben, D.; Lundh, T.; Dicksved, J. Effects of microbe-and mussel-based diets on the gut microbiota in Arctic charr (Salvelinus alpinus). *Aquac. Rep.* **2017**, *5*, 34–40. [CrossRef]

67. Wlodarska, M.; Luo, C.; Kolde, R.; d'Hennezel, E.; Annand, J.W.; Heim, C.E.; Krastel, P.; Schmitt, E.K.; Omar, A.S.; Creasey, E.A.; et al. Indoleacrylic acid produced by commensal *Peptostreptococcus* species suppresses inflammation. *Cell Host Microbe* **2017**, *22*, 25–37.e6. [CrossRef]
68. Cavanagh, D.; Fitzgerald, G.F.; McAuliffe, O. From field to fermentation: The origins of Lactococcus lactis and its domestication to the dairy environment. *Food Microbiol.* **2015**, *47*, 45–61. [CrossRef] [PubMed]
69. Gao, Y.; Lu, Y.; Teng, K.L.; Chen, M.L.; Zheng, H.J.; Zhu, Y.Q.; Zhong, J. Complete genome sequence of *Lactococcus lactis* subsp. *lactis* CV56, a probiotic strain isolated from the vaginas of healthy women. *J. Bacteriol.* **2011**, *193*, 2886–2887. [CrossRef] [PubMed]
70. Kato, H.; Shiwa, Y.; Oshima, K.; Machii, M.; Araya-Kojima, T.; Zendo, T.; Shimizu-Kadota, M.; Hattori, M.; Sonomoto, K.; Yoshikawa, H. Complete genome sequence of *Lactococcus lactis* IO-1, a lactic acid bacterium that utilizes xylose and produces high levels of L-lactic acid. *J. Bacteriol.* **2012**, *194*, 2102–2103. [CrossRef] [PubMed]
71. Kimoto, H.; Kurisaki, J.; Tsuji, N.; Ohmomo, S.; Okamoto, T. *Lactococci* as probiotic strains: Adhesion to human enterocyte-like Caco-2 cells and tolerance to low pH and bile. *Lett. Appl. Microbiol.* **1999**, *29*, 313–316. [CrossRef]
72. Meyrand, M.; Guillot, A.; Goin, M.; Furlan, S.; Armalyte, J.; Kulakauskas, S.; Cortes-Perez, N.G.; Thomas, G.; Chat, S.; Pechoux, C. Surface proteome analysis of a natural isolate of *Lactococcus lactis* reveals the presence of pili able to bind human intestinal epithelial cells. *Mol. Cell Proteom.* **2013**, *12*, 3935–3947. [CrossRef]
73. Mercier-Bonin, M.; Chapot-Chartier, M.-P. Surface Proteins of *Lactococcus lactis*: Bacterial resources for muco-adhesion in the gastrointestinal tract. *Front. Microbiol.* **2017**, *8*. [CrossRef]
74. Dawood, M.A.O.; Koshio, S.; Ishikawa, M.; Yokoyama, S.; El Basuini, M.F.; Hossain, M.S.; Nhu, T.H.; Dossou, S.; Moss, A.S. Effects of dietary supplementation of *Lactobacillus rhamnosus* or/and *Lactococcus lactis* on the growth, gut microbiota and immune responses of red sea bream, *Pagrus major*. *Fish Shellfish Immunol.* **2016**, *49*, 275–285. [CrossRef]
75. Nguyen, T.L.; Chun, W.-K.; Kim, A.; Kim, N.; Roh, H.J.; Lee, Y.; Yi, M.; Kim, S.; Park, C.-I.; Kim, D.-H. Dietary probiotic effect of *Lactococcus lactis* WFLU12 on low-molecular-weight metabolites and growth of olive flounder (*Paralichythys olivaceus*). *Front. Microbiol.* **2018**, *9*. [CrossRef]
76. Girijakumari Nisha, R.; Rajathi, V.; Manikandan, R.; Marimuthu Prabhu, N. Isolation of Plesiomonas shigelloides from Infected Cichlid fishes using 16S rRNA characterization and its control with probiotic *Pseudomonas* sp. *Acta Sci. Vet.* **2014**, *42*, 1–7.
77. Roeselers, G.; Mittge, E.K.; Stephens, W.Z.; Parichy, D.M.; Cavanaugh, C.M.; Guillemin, K.; Rawls, J.F. Evidence for a core gut microbiota in the zebrafish. *ISME J.* **2011**, *5*, 1595–1608. [CrossRef]
78. Liu, Z.; Ke, X.; Lu, M.; Gao, F.; Cao, J.; Zhu, H.; Wang, M. Identification and pathological observation of a pathogenic Plesiomonas shigelloides strain isolated from cultured tilapia (Oreochromis niloticus). *Acta Microbiol. Sin.* **2015**, *55*, 96–106.
79. Bai, Z.; Ren, T.; Han, Y.; Rahman, M.M.; Hu, Y.; Li, Z.; Jiang, Z. Influences of dietary selenomethionine exposure on tissue accumulation, blood biochemical profiles, gene expression and intestinal microbiota of *Carassius auratus*. *Comp. Biochem. Physiol. Part C Toxicol. Pharmacol.* **2019**, *218*, 21–29. [CrossRef]
80. Ryan, M.P.; Pembroke, J.T.; Adley, C.C. *Ralstonia pickettii* in environmental biotechnology: Potential and applications. *J. Appl. Microbiol.* **2007**, *103*, 754–764. [CrossRef]
81. Makaritsis, K.P.; Neocleous, C.; Gatselis, N.; Petinaki, E.; Dalekos, G.N. An immunocompetent patient presenting with severe septic arthritis due to *Ralstonia pickettii* identified by molecular-based assays: A case report. *Cases J.* **2009**, *2*, 8125. [CrossRef]

*Article*

# Does the Composition of the Gut Bacteriome Change during the Growth of Tuna?

**Elsa Gadoin [1], Lucile Durand [1], Aurélie Guillou [1], Sandrine Crochemore [1], Thierry Bouvier [1], Emmanuelle Roque d'Orbcastel [1], Laurent Dagorn [1], Jean-Christophe Auguet [1], Antoinette Adingra [2], Christelle Desnues [3] and Yvan Bettarel [1,*]**

[1] MARBEC, University Montpellier, CNRS, Ifremer, IRD, 34095 Montpellier, France; elsa.gadoin@umontpellier.fr (E.G.); durandlucile.ld@gmail.com (L.D.); aurelie.guillou@ird.fr (A.G.); Sandrine.CROCHEMORE@cnrs.fr (S.C.); thierry.bouvier@cnrs.fr (T.B.); emmanuelle.roque@ifremer.fr (E.R.d.); laurent.dagorn@ird.fr (L.D.); Jean-christophe.AUGUET@cnrs.fr (J.-C.A.)
[2] Centre de Recherches Océanologiques, Abidjan, Côte d'Ivoire; adingraantoinette_ama@yahoo.fr
[3] Mediterranean Insitute of Oceanodraphy (MIO), 13009 Marseille, France; Christelle.DESNUES@univ-amu.fr
* Correspondence: yvan.bettarel@ird.fr

**Abstract:** In recent years, a growing number of studies sought to examine the composition and the determinants of the gut microflora in marine animals, including fish. For tropical tuna, which are among the most consumed fish worldwide, there is scarce information on their enteric bacterial communities and how they evolve during fish growth. In this study, we used metabarcoding of the 16S rDNA gene to (1) describe the diversity and composition of the gut bacteriome in the three most fished tuna species (skipjack, yellowfin and bigeye), and (2) to examine its intra-specific variability from juveniles to larger adults. Although there was a remarkable convergence of taxonomic richness and bacterial composition between yellowfin and bigeyes tuna, the gut bacteriome of skipjack tuna was distinct from the other two species. Throughout fish growth, the enteric bacteriome of yellowfin and bigeyes also showed significant modifications, while that of skipjack tuna remained relatively homogeneous. Finally, our results suggest that the gut bacteriome of marine fish may not always be subject to structural modifications during their growth, especially in species that maintain a steady feeding behavior during their lifetime.

**Keywords:** tuna; microbiome; enteric bacteria; fish; barcoding; gut

**Citation:** Gadoin, E.; Durand, L.; Guillou, A.; Crochemore, S.; Bouvier, T.; d'Orbcastel, E.R.; Dagorn, L.; Auguet, J.-C.; Adingra, A.; Desnues, C.; et al. Does the Composition of the Gut Bacteriome Change during the Growth of Tuna? *Microorganisms* **2021**, 9, 1157. https://doi.org/10.3390/microorganisms9061157

Academic Editor: Konstantinos Ar. Kormas

Received: 13 April 2021
Accepted: 24 May 2021
Published: 27 May 2021



## 1. Introduction

In recent years, the number of microbiome studies on marine organisms was increasing, particularly for corals and fish, about which we now know that their microbial associates play a considerable role in their health and fitness [1–4]. Among the different biological compartments that harbor microorganisms, the digestive tract is certainly the one that has received the greatest deal of attention; mostly because enteric bacteria are involved in a wide range of important functions for the host, including digestion, production of useful metabolites, protection against pathogens, promotion of the immune system, behavior, to name a few [5–8]. Previous studies have shown that for humans, the gut microbiome of fish include, from the larval stage onwards, a wide range of taxa mostly dependent on several factors such as life stage [9], sex [10], phylogeny [11], trophic level [12], diet [7], season [13], habitat [14] and captive-state [10].

At various growth stages, the life traits of fish may evolve and could therefore result in major changes in the composition of enteric bacterial communities [6,15,16]. This is the case, for example, in wild migratory species that undergo several ontogenetic transformations during their development and need to adapt their metabolism and diet to major environmental changes, such as the transition from fresh water to salt water [17]. A handful of studies revealed that the gut microbiome changes throughout fish growth, as shown in the farmed olive flounders *Paralichthys olivaceus* [18] and Chinook salmon *Oncorhynchus tshawytsha* [19]. However, these studies have been mostly conducted on experimental reared fish, hence it is timely to evaluate whether such patterns also occur in wild fish, especially for species that represent an important food source for humans.

Tunas are teleostean species living in tropical and temperate waters. They are top predators playing a fundamental role in the marine trophic food chain and ecosystem resilience. They are among the most fished species in the world, exploited in all oceans by both industrial and small-scale fisheries [20].

Despite their ecological importance, but also in terms of ecosystem services provided by tunas, little is currently known about the composition, diversity and role of their gut bacteriome, including its possible changes during fish growth. Knowing that the ontogenic changes typical of certain tuna species allow them to access additional sources of food during their growth [21], it is possible that the diversification of their diet with age could result in substantial modifications of their gut bacteriome.

The objective of this study was (i) to describe the composition of the gut bacteriome of the three most fished tuna species (skipjack-*Katsuwonus pelamis,* yellowfin-*Thunnus albacares* and bigeye-*Thunnus obesus*), and (ii) to investigate whether the fish size (as a proxy of their development stage) is a decisive factor in explaining the structure of this enteric community.

## 2. Materials and Methods

### 2.1. Sampling

The protocol consisted of collecting 18 individuals of each of the three tuna species targeted by the large-scale purse seine fleet. All individuals were caught between May and December 2019 (Figure 1) in the Eastern Atlantic Ocean (Gulf of Guinea and off the coast of Senegal) and sampled by the Exploited Tropical Pelagic Ecosystem Observatory (IRD, Ob7), as part of the multiannual European fishery data collection framework (DCF, financed by the European Maritime and Fisheries Fund, Article 77). Once the fish had been caught, they were stored onboard in chilled brine to lower their temperature to around −15 °C.

**Figure 1.** Fishing positions in the east Atlantic Ocean. Colored circles correspond to tuna caught from shoals near fish aggregating devices or from free swimming schools.

Each fish was weighed (whole weight, in kg) and lengthed (at fork, in cm). Size ratio (SR) between the largest and smallest individuals was calculated for each species, following the equation: SR = minimal size/maximal size. For each species, the individuals were grouped into three equivalent categories (in terms of number), named Small, Medium and Large, and corresponding, respectively, to the 6 smallest, medium and largest fish per species (Table 1). Considering the size at 50% sexual maturity of each species [22], individuals from the Small group can be considered as being mainly juvenile or sub-adult fish, and individuals from the Medium group as sub-adults or young adults, while the Large group for each species can be considered to be composed of adults. The complete set included male, female and immature fishes but no significant difference in size or weight were observed between these three categories ($p > 0.05$, Kruskal-Wallis test).

**Table 1.** Main morphometric and sexual traits of the fish sampled in the three size categories (small/medium/large). M, male; F, female; I, Immature.

| Size Class | SKIPJACK Size (cm) | SKIPJACK Weight (kg) | SKIPJACK Sex | YELLOWFIN Size (cm) | YELLOWFIN Weight (kg) | YELLOWFIN Sex | BIGEYE Size (cm) | BIGEYE Weight (kg) | BIGEYE Sex |
|---|---|---|---|---|---|---|---|---|---|
| Small | 30.5 | 0.6 | F | 66.1 | 5.8 | M | 71.4 | 8.0 | M |
| | 32.1 | 0.6 | I | 71.0 | 6.8 | F | 79.2 | 12.2 | F |
| | 34.8 | 0.7 | I | 75.3 | 8.7 | M | 84.4 | 15.9 | F |
| | 38.9 | 1.1 | I | 84.9 | 12.1 | F | 87.8 | 13.2 | M |
| | 40.2 | 1.4 | F | 87.8 | 13.3 | M | 94.2 | 19.8 | M |
| | 42.1 | 1.4 | I | 91.9 | 15.3 | M | 97.8 | 22.2 | F |
| Medium | 45.5 | 1.8 | M | 103.8 | 25.9 | I | 102.8 | 23.8 | F |
| | 46.7 | 2.1 | M | 109.3 | 28.3 | F | 109.5 | 29.9 | M |
| | 49.4 | 2.6 | F | 110.1 | 24.8 | F | 111.3 | 30.9 | F |
| | 51.0 | 2.7 | F | 116.8 | 32.3 | F | 115.7 | 36.6 | M |
| | 52.0 | 3.0 | F | 127.2 | 38.5 | F | 127.7 | 47.6 | F |
| | 55.5 | 3.9 | M | 131.3 | 50.4 | M | 132.5 | 51.7 | F |
| Large | 56.7 | 4.1 | F | 140.4 | 59.2 | M | 138.8 | 63.3 | M |
| | 59.2 | 4.7 | M | 145.2 | 58.9 | F | 142.2 | 68.5 | M |
| | 60.8 | 5.2 | M | 151.6 | 68.7 | M | 152.8 | 78.4 | F |
| | 63.5 | 7.7 | M | 158.8 | 81.9 | M | 155.0 | 90.0 | F |
| | 65.5 | 6.0 | F | 161.6 | 89.9 | M | 162.1 | 87.3 | F |
| | 67.5 | 6.0 | F | 164.3 | 71.0 | M | 166.8 | 99.9 | M |

### 2.2. Extraction of the Gut Bacteriome

After landing, the 54 frozen fish were transferred to the Laboratory of Microbiology of the Centre de Recherches Océanologiques (CRO) of Abidjan, where they were thawed and dissected. The time length between catch and dissection was approximately 40 days and it was comparable for each sampled fish. Briefly, the gastrointestinal tract was extracted from each individual and cut from below the stomach to 2 cm before the rectum (to avoid potential chilled brine intrusion), using sterile tools. Each gut was squeezed to expel the contents (from 3 to 15 mL) on a sterile surface, and the contents were homogenized before sampling [23] and conserved at −80 °C until the DNA extraction.

### 2.3. DNA Extraction, Amplification and Sequencing

Bacterial DNA was extracted from 250 ± 0.5 mg of gut ($n = 54$). All extractions were performed with the PowerSoil DNA Isolation Kit (Qiagen®, Hilden, Germany) following the manufacturer's instructions. DNA quality and quantity were assessed by spectrophotometry (NanoDrop®, Wilmington, DE, USA). The V3-V4 region of the 16S rDNA gene was amplified using universal bacterial primers modified for Illumina sequencing: 343F (5′- ACGGRAGGCAGCAG) [24] and 784R (5′- TACCAGGGTATCTAATCCT) [25]. The reaction mixture consisted of 12.5 µL of 2X Phusion Mix (New England Biolabs®, Ipswich, MA, USA), 1 µL of each primer at 10 µM (Eurofin®, Luxembourg), 10 ng of DNA template and enough molecular-grade $H_2O$ (Qiagen®) to reach a final volume of 25 µL. All samples were amplified in triplicate and pooled to avoid PCR bias in the taxonomic diversity of the community [26]. Successfully amplified samples ($n = 54$) were sequenced on the Illumina platform (Genoscreen®, Lille, France) using 2 × 250 bp MiSeq chemistry.

### 2.4. Treatment and Analysis of the Bacterial Sequences

In total 1,823,118 reads were obtained and were processed on RStudio (R v. 3.5.3) with the DADA2 pipeline (v. 1.10.1) [27], following the authors' tutorial (https://benjjneb.github.io/dada2/tutorial.html, accessed on 27 February 2019). The quality of the forward and reverse reads was assessed prior to the removing of adaptors and primers, based on their length. Reads were then filtered, trimmed and merged into 583,716 amplicon sequence variants (ASV), which have a greater resolution than the operational taxonomic unit (OTU) [27]. The chimeras were removed and the paired sequences were aligned with the SILVA 123 taxonomic database [28] to access their taxonomy. To compensate for varying sequencing efficiency, analyses were performed on a random subsample of 8979 sequences per sample, which corresponded to the sample with the smaller number of sequences after trimming and quality processing. Final taxonomic and ASVs tables were associated with the morphometric data using the *phyloseq* package [29]. Finally, relative abundances of ASVs in each sample were calculated by the *phyloseq* package and ASVs corresponding to archea, non-prokaryotes, chloroplasts and mitochondria were deleted.

To assess the alpha diversity, the Shannon diversity index was calculated by *phyloseq* for each sample. The composition of bacterial communities was represented at the order level, based on the relative abundance of ASVs in each sample and performed with *phyloseq* and *ggplot* packages. Finally, using the microbiome package [30], the *Core Microbiota*, defined here as all ASVs (relative abundance ≥ 1%) shared by at least 50% of the gut samples, was determined for the three size groups of the three tuna species.

### 2.5. Statistical Analysis

All the statistical analyses were carried out using RStudio (R v. 3.5.3). For each species, possible relationship between the alpha diversity index (Shannon) and the size were tested by linear regression, while Kruskal-Wallis and Wilcoxon non-parametric tests were used to observe the variation of the alpha diversity between the three species. Finally, the effects of the species then of the size on the bacterial composition was determined by one-factor PERMANOVA with 999 permutations of the Bray-Curtis matrix using the "adonis" function of the *vegan* package [31].

## 3. Results

### 3.1. Fish Morphometric Traits

For the three target species, the size ratio between the smallest and largest individuals was comparable, reaching 2.2, 2.5 and 2.3 for skipjack, yellowfin and bigeye tunas, respectively. Skipjack were significantly smaller (30.5 to 67.5 cm) than the yellowfin (66.1 to 164.3 cm) and the bigeye tuna (71.4 to 166.8 cm) ($p < 0.001$, Kruskal-Wallis test).

### 3.2. Alpha Diversity

The Shannon diversity index varied considerably for each of the three species of tuna (Figure 2A). On average, the index for the skipjack tuna was significantly lower than for the two other species (Kruskall-Wallis, $p < 0.05$). However, for all the three species, the Shannon index did not vary significantly along the size gradient ($R2_{Skipjack} = 0.13$, $p = 0.07$; $R2_{Yellowfin} = 0.11$, $p = 0.09$; $R2_{Bigeye} = 0.13$, $p = 0.07$) (Figure 2B).

### 3.3. Composition and Beta Diversity

Forty-five orders of bacteria belonging to 15 different classes were identified (Figure 3). *Actinobacteria*, *Alphaproteobacteria* and *Gammaproteobacteria* were the most represented classes, regardless of the species and size of tuna. Generally, there was a significant species-effect on the enteric bacteriome of tuna, which was particularly marked for the gut skipjack individuals, which was dominated by *Mycoplasmatales* (*Mollicutes*) (Figure 3).

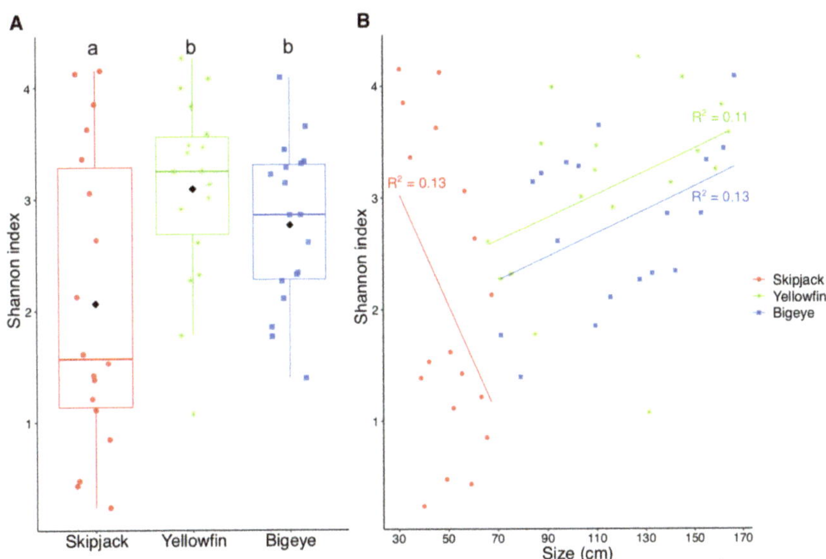


**Figure 2.** Representation of *alpha* taxonomic diversity (Shannon index) of enteric bacterial communities (ASVs) in the three tuna species (**A**), and according to the size of individuals (**B**). Boxplots represent the distribution of the alpha taxonomic diversity within each species. Different letters indicate significant differences (KW, $p < 0.05$) between groups within each square. * Significant correlation between the Shannon index and fish size (Pearson, $p < 0.05$).

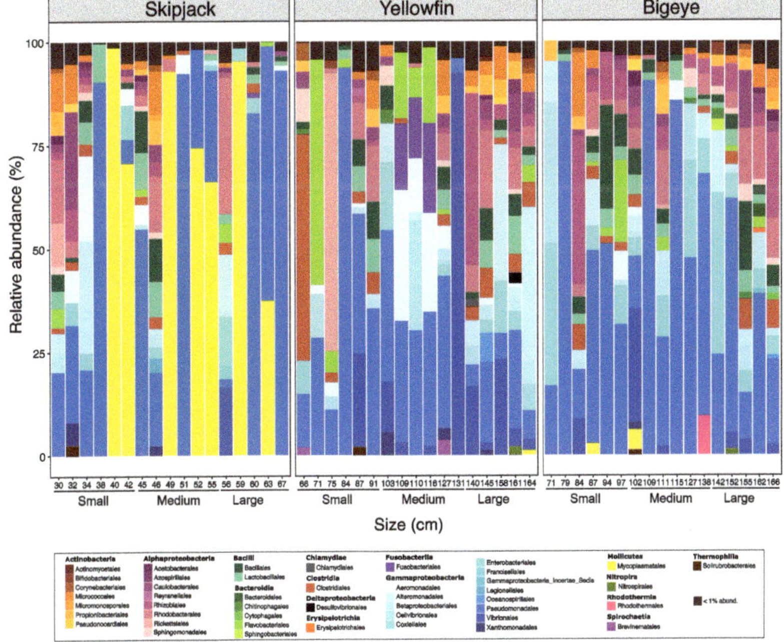


**Figure 3.** Relative abundances of the main bacterial classes in the gut of skipjack, yellowfin and bigeye tunas, in their size group (small/medium/large). Each bar corresponds to an individual fish. Bacterial classes showing a relative abundance lower than 1% were designated "<1% abund".

Regarding the variability in the gut microflora during fish growth, two different patterns were observed: while the composition of enteric bacterial communities was significantly affected by fish size for yellowfin and bigeyes tunas, that of skipjack remained relatively stable across the range of fish sizes (Table 2).

**Table 2.** Results of permutational ANOVAS (PERMANOVA, 999 permutations) performed on Bray-Curtis dissimilarity matrices to test the variation of bacterial community composition with respect to the size of the three tuna species. Bold values indicate a significant effect of the tested factor ($p < 0.05$).

| | SKIPJACK | | | | YELLOWFIN | | | | BIGEYE | | | |
|---|---|---|---|---|---|---|---|---|---|---|---|---|
| | df | Sum of Squares | F. Model | *p* Value | df | Sum of Squares | F. Model | *p* Value | df | Sum of Squares | F. Model | *p* Value |
| **SIZE** | 1 | 0.25 | 0.70 | 0.68 | 1 | 0.77 | 2.02 | **0.01** | 1 | 0.48 | 1.76 | **0.04** |
| **Residuals** | 16 | 5.7 | | | 16 | 6.11 | | | 16 | 4.48 | | |
| **Total** | 17 | 5.97 | | | 17 | 6.89 | | | 17 | 4.96 | | |

### 3.4. The Core Gut Microbiome

During their growth, each species hosted taxa that were common and specific to one or more size groups. Twelve genera of bacteria were found in the gut of all the three different tuna species (Table 3), some of them such as *Carnobacterium* sp., *Oceanisphaera* spp., *Pseudomonas* spp. and *Psychrobacter* spp. were even found in all the three size groups (small, medium and large). Each tuna species had also its own specific taxa, such as *Mycoplasma* sp. in skipjack, in all size categories, and to a lesser extent, *Corynebacterium* sp. and *Rahnella* spp. Yellowfin tuna hosted specific bacterial taxa mainly belonging to the *Alphaproteobacteria*, *Clostridia*, *Fusobacteriia* and *Gammaproteobacteria* classes, the latter being the most relevant, with taxa such as *Citenobacter* sp., *Aeromonas* sp., *Massilia* sp., and *Photobacterium* sp. Taxa such as *Microbacterium* sp., *Labrenzia* sp. *Vibrionimonas magnilacihabitans* and *Escherischia/Shigella* were specific to the bigeye's gut.

**Table 3.** Bacterial genera representative of the 'Core Microbiota' (determined with the microbiome package in R) in the gut of Skipjack, Yellowfin and Bigeye tunas, in the small (S), medium (M) and large (L) size categories. Taxa common to all the three tuna species are represented by grey squares. Red, green and blue squares correspond to unique taxa present in each species.

| | | SKIPJACK | | | YELLOWFIN | | | BIGEYE | | |
|---|---|---|---|---|---|---|---|---|---|---|
| Class | Genus Species | S | M | L | S | M | L | S | M | L |
| *Actinobacteria* | *Corynebacterium* sp. | | red | | | | | | | |
| | *Cutibacterium* sp. | | | | | | grey | | | grey |
| | ***Kocuria* sp.** | | grey | | grey | | | grey | | grey |
| | *Microbacterium* sp. | | | | | | | | blue | blue |
| *Alphaproteobacteria* | *Bradyrhizobium* sp. | | | | | | | blue | blue | blue |
| | ***Brevundimonas* sp.** | grey | | | grey | grey | grey | grey | grey | grey |
| | *Labrenzia* sp. | | | | | | | | blue | |
| | ***Novosphingobium* sp.** | grey | | | | grey | | | | grey |
| | ***Paracoccus* sp.** | grey | grey | | grey | | grey | grey | grey | grey |
| | *Ruegeria* sp. | | | | | | green | | | |
| *Bacilli* | *Brochothrix thermosphacta* | | | | green | | | | | |
| | ***Carnobacterium* sp.** | grey | grey | grey | grey | grey | grey | grey | grey | grey |
| | *Lactococcus* spp. | | | | | | green | | | |
| | ***Sporosarcina* spp.** | grey | grey | | grey | grey | grey | grey | grey | grey |
| | *Vagococcus salmoninarum* | | | | grey | | grey | grey | grey | grey |
| *Bacteroidia* | ***Ulvibacter* spp.** | | | grey | grey | grey | | | grey | |
| | *Vibrionimonas magnilacihabitans* | | | | | | | | blue | blue |

**Table 3.** *Cont.*

| | | SKIPJACK | | | YELLOWFIN | | | BIGEYE | | |
|---|---|---|---|---|---|---|---|---|---|---|
| **Class** | **Genus Species** | **S** | **M** | **L** | **S** | **M** | **L** | **S** | **M** | **L** |
| *Clostridia* | *Clostridium_sensu_stricto_7* spp. | | | | | | | | | |
| | *Gottschalkia* spp. | | | | | | | | | |
| | *Tissierella* spp. | | | | | | | | | |
| *Fusobacteriia* | *Psychrilyobacter* spp. | | | | | | | | | |
| *Gammaproteobacteria* | *Acinetobacter* spp. | | | | | | | | | |
| | *Acinetobacter haemolyticus* | | | | | | | | | |
| | *Aeromonas* sp. | | | | | | | | | |
| | *BD1-7_clade* spp. | | | | | | | | | |
| | ***Enhydrobacter aerosaccus*** | | | | | | | | | |
| | *Escherichia/Shigella* sp. | | | | | | | | | |
| | *Massilia* sp. | | | | | | | | | |
| | *Massilia timonae* | | | | | | | | | |
| | ***Oceanisphaera* spp.** | | | | | | | | | |
| | *Oceanisphaera ostreae* | | | | | | | | | |
| | *Photobacterium* spp. | | | | | | | | | |
| | *Photobacterium angustum* | | | | | | | | | |
| | *Photobacterium leiognathi* | | | | | | | | | |
| | ***Pseudomonas* spp.** | | | | | | | | | |
| | ***Psychrobacter* spp.** | | | | | | | | | |
| | *Psychrobacter fozii* | | | | | | | | | |
| | ***Psychrobacter maritimus*** | | | | | | | | | |
| | *Rahnella* spp. | | | | | | | | | |
| | ***Shewanella* sp.** | | | | | | | | | |
| *Mollicutes* | *Mycoplasma* sp. | | | | | | | | | |

Some taxa were characteristic of a size group, this being very marked for yellowfin which had the highest number of size-specific taxa. Taxa such as *Acinetobacter* sp. and *Aeromonas* sp. were specific to medium sized yellowfin and others such as *Photobacterium angustum* and *Photobacterium leiognathi*, both reported as histamine producing bacteria, were found only in large yellowfin (Table 3).

## 4. Discussion

In the three tuna species, the gut bacteriome was dominated by four main phyla: *Proteobacteria, Firmicutes, Bacteroidetes* and *Actinobactera*, which accounted for up to 95% of the bacteria identified in the gut. Such phyla are typically found in the intestinal microflora of marine and freshwater fish [6,32]. However, the diversity of the 16S rDNA gene sequences showed considerable differences between the enteric bacterial communities of the three species (Figures 2 and 3). Skipjack were distinguished from yellowfin and bigeye tunas by their particularly low species richness and a strong presence of *Firmicutes* (*Mollicutes, Mycoplasma* sp.). A low gut microbial diversity, associated with the dominance of *Mycoplasma* spp. was also reported for Atlantic salmon (*Salmo salar*) [17]. So far, within the marine environment, *Firmicutes* were thought to be the dominant phylum in herbivorous species [11,33], probably as they facilitate the digestion of cellulose. However, this metabolic function is not vital in tunas and Salmonids, which are carnivorous fish. By comparison with yellowfin and bigeye, skipjack tunas are also more subjected to risk of overheating as they grow mostly because of their lower capacity of thermoregulation [34]. The occurrence of temperature-induced changes in the gut microbiome is a well-known phenomenon in vertebrates and usually results in a disruption of the alpha-diversity towards a decrease in richness [35]. Identical microbiome responses to thermal shifts in phylogenetically distant animal taxa suggest the existence of a conserved mechanism, which could also apply in tuna. All together, these may explain the observed lowest alpha diversity in skipjack.

Although our results indicate a clear species effect on the enteric bacteriome of tuna (Figure 3), there is nevertheless a strong convergence in the structure of the gut microbiota between yellowfin and bigeye while skipjack (both belonging to the same genus *Thunnus*), whereas skipjack tuna (genus *Katsuwonus*) are in a separate branch). The composition of the gut bacteriome in tropical tuna could thus depend mainly on evolutive considerations [36]. Indeed, phylogenetically close fish usually host a similar bacterial flora [37]. However, the formation of the bacterial communities in the gastro-intestinal tract of fish is a complex process affected by other exogenic and endogenic factors such as the diet, the life style and the environment [6,37,38].

### 4.1. Microbial Changes during Fish Growth

For all the three species, the taxonomic structural diversity of enteric bacteria did not vary significantly with size (Figure 2). This is contradictory with recent reports showing a reduction of the alpha diversity with age, and therefore, with size, for Atlantic Salmon, olive flounders and zebra fish [17,18,39,40]. A reduction in diversity is usually associated with a diet that encouraged the growth of generalist bacteria or that included chemical compounds inhibiting certain more specialist bacteria [41]. Conversely, in terrestrial vertebrates (mammals, birds and reptiles) a positive correlation was observed between the enteric microbial diversity and the mass of the animal, independent of the age, phylogeny, diet or the structure of the digestive tract [42]. Overall, the changes in the structural diversity of enteric bacteria during bodily growth does not seem to follow a unidirectional pathway within the animal world and more studies are needed to identify the factors that govern this particular diversity.

The examination of the core microbiome demonstrated that changes in the proportions of the main taxa during growth were specific to each tuna species (Table 3). Although a core bacterial community was found across the three tuna species (comprising four major genera: *Carnobacterium, Oceanisphaera, Pseudomonas* and *Psychrobacter)*, other genera were specific to one or more size groups for each species. In particular, for yellowfin, *Acinetobacter* and *Aeromonas* were found in medium sized fish while *Cutibacterium, Lactococcus, Gottschalkia* and *Photobacterium* spp. were found only in large individuals. These taxa might, therefore, play a specific role in the late development stage of yellowfin.

### 4.2. Dietary Changes during Growth

The most striking result of our study was a drastic change in the composition of the gut microbiome of yellowfin and bigeye tuna during their growth, which was not observed for skipjack (Table 2, Figure 3). The fish size could have an effect on the selection of prey and some studies demonstrated that the proportion of fish in the tuna diet increases as the size of the tuna increases [43,44]. A vertical distribution of tuna species in the water column has been long reported, with the larger fish (bigeye and yellowfin) able to stay in deeper water than the smaller (skipjack), which gives them access to different types of prey [21]. Such modifications in the diet could be related to changes in nutritional needs depending on the development stage of the tuna. Indeed, ontogenic changes in yellowfin and bigeye tuna are generally observed when they reach about 45–50 cm, enabling them to dive into colder, deeper water [45], which would radically change their diet, and promote its diversification throughout growth. Quite the opposite, skipjack are physiologically enabled to reach these deeper waters with potential new preys and therefore remain in surface waters at all life stage, which could explain the homogeneity of their gut microbiota throughout growth. Conversely, although there was no yellowfin or bigeye smaller than 65 cm in our sampling, we suspect that the significant differences observed between the various growth stages was due to their ability to diversify their diet with age, going further and further to seek food.

### 4.3. Commensal and Potential Pathogenic Bacteria

In our study, the main bacterial genera forming the core microbiome of the tuna species included commensal and potential pathogenic taxa, some of them being common to all the three target species (Table 3). Of the commensal taxa, some *Carnobacterium* species, for example, are known to inhibit the action of certain fish pathogens [46,47]. A similar antagonistic activity of several *Pseudomonas* bacterial strains has also been reported [48]. Many *Kocuria* species are commensal bacteria found on mammals and have been isolated from the gut of rainbow trout (*Oncorhynchus mykiss*) and the Atlantic cod (*Ghadus morhua*) [49]. However, several species of *Carnobacterium* and *Kocuria* are also known to be pathogenic [49,50], like *Escherichia* and *Shigella,* which are enteric human pathogens able to establish in the gut of several fish in certain conditions. This has occurred, for example, in trout after consuming infected food and in Nile tilapia when the surrounding water had been contaminated [51,52]. Other *Sporocarcina* species are bacteria that spoil seafood and refrigerated meat [53]. Some species found in the tuna gut microbiome are also known to be histamine-producing bacteria (HPB) [54]. The histamine is produced by these bacteria from a precursor (histidine) by a bacterial enzyme (histidine decarboxylase) and is main global cause of food-poisoning from consuming fish [55]. HPB are often found in tuna when the cold chain is broken during landing, processing and handling fresh tuna [56,57]. In this study, two *Photobacterium* species (*Photobacterium angustum* and *Photobacterium leiognathi*) were found in the gut microbiome of large yellowfin. Other species such as *Photobacterium damselae, Photobacterium phosphoreum* and *Pseudomonas fluorescens* were found in some samples of all three tuna species. The relative abundances were low (<2%) and there was no clear relationship between the size of the tuna and the presence of these HPBs.

All these confirm that the fish gut typically hosts a complex and highly diversified bacterial community composed of a balance of commensal and pathogenic taxa, which contribute to the functioning of the gut and help to maintain the health of the host organism.

## 5. Conclusions

Our results revealed that the composition of the enteric microflora showed contrasting patterns between skipjack on one side, and yellowfin and bigeye tuna on the other side. Beside phylogeny, several other endogenic factors could explain the microbial differences and similarities between species, including the size which emerged as a major determinant of gut bacteriome in tuna. If significant changes in the intestinal microflora have been observed during the growth of yellowfin and bigeye tuna, the case of skipjack, by contrast, is interesting because of the relative stability of its microbiota and its unique composition. Overall, our study suggests that strong structural (and presumably functional) microbiological differences exist between species within the same family of fish, probably linked to their differential ability to grow in size, improve their mobility for foraging, and thus promote diet diversification.

**Author Contributions:** Conceptualization, Y.B.; E.G. and A.G.; methodology, E.G.; S.C.; L.D.; software, L.D. (Lucile Durand); E.G. and J.-C.A.; validation, Y.B.; formal analysis, L.D. (Lucile Durand); investigation, Y.B., E.R.d.; T.B.; E.G.; resources, A.G.; data curation, L.D.; writing—original draft preparation, Y.B. and L.D. (Laurent Dagorn); writing—review and editing, Y.B.; E.G.; T.B.; E.R.d.; L.D. (Laurent Dagorn), J.-C.A. and C.D. visualization, A.A.; supervision, Y.B.; E.G.; A.G. and C.D. project administration, Y.B.; funding acquisition, Y.B. All authors have read and agreed to the published version of the manuscript.

**Funding:** This research was funded by *THE MOME* Projet and Montpellier University of Excellence (MUSE), and the APC was funded by the *MOSANE* project (JEAI IRD). The data used here was collected within the framework of the Data Collection Framework program (Reg 2017/1004 and 2016/1251) co-financed by the IRD and Measure 77 of the European Maritime and Fisheries Fund (EMFF).

**Institutional Review Board Statement:** No applicable.

**Informed Consent Statement:** Not applicable.

**Data Availability Statement:** Not applicable.

**Acknowledgments:** We would like to thank the members of the IRD's Exploited Tropical Pelagic Ecosystems Observatory in Abidjan for their help in extracting the digestive tracts of tuna. We also grateful to the CRO staff for providing access to their laboratory and facilities, and to Monique Simier for her advices on the statistical analysis.

**Conflicts of Interest:** The authors declare no conflict of interest to publish the results.

## References

1. Stal, L.J.; Cretoiu, M.S. *The Marine Microbiome: An Untapped Source of Biodiversity and Biotechnological Potential;* Springer: Berlin/Heidelberg, Germany, 2016.
2. Apprill, A. Marine animal microbiomes: Toward understanding host–microbiome interactions in a changing ocean. *Front. Mar. Sci.* **2017**, *4*, 222. [CrossRef]
3. Trevathan-Tackett, S.M.; Sherman, C.D.H.; Huggett, M.J.; Campbell, A.H.; Laverock, B.; Hurtado-McCormick, V.; Macreadie, P.I. A horizon scan of priorities for coastal marine microbiome research. *Nat. Ecol. Evol.* **2019**, *3*, 1509–1520. [CrossRef]
4. Chiarello, M.; Auguet, J.C.; Graham, N.; Claverie, T.; Sucre, E.; Bouvier, C.; Rieuvilleneuve, F.; Bettarel, Y.; Villeger, S.; Bouvier, T. Exceptional but vulnerable microbial diversity in coral reef animal surface microbiomes. *Proc. R. Soc. B* **2020**, *287*, 1927. [CrossRef] [PubMed]
5. Tarnecki, A.M.; Burgos, F.A.; Ray, C.L.; Arias, C.R. Fish intestinal microbiome: Diversity and symbiosis unravelled by metagenomics. *J. Appl. Microbiol.* **2017**, *123*, 2–17. [CrossRef] [PubMed]
6. Egerton, S.; Culloty, S.; Whooley, J.; Stanton, C.; Ross, R.P. The gut microbiota of marine fish. *Front. Microbiol.* **2018**, *9*, 1–17. [CrossRef] [PubMed]
7. Butt, R.L.; Volkoff, H. Gut microbiota and energy homeostasis in fish. *Front. Endocrinol.* **2019**, *10*, 9. [CrossRef] [PubMed]
8. Gupta, V.K.; Minsuk, K.; Bakshi, U.; Cunningham, K.Y.; Davis III, J.M.D.; Lazaridis, K.N.; Nelson, H.; Chia, N.; Sung, J. A predictive index for health status using species-level gut microbiome profiling. *Nat. Commun.* **2020**, *11*, 4635. [CrossRef] [PubMed]
9. Hansen, G.H.; Olafsen, J.A. Bacterial interactions in early life stages of marine cold water fish. *Microb. Ecol.* **1999**, *38*, 1–26. [CrossRef]
10. Dhanasiri, A.K.S.; Brunvold, L.; Brinchmann, M.F.; Kornses, K.; Bergh, O.; Kiron, W. Changes in the intestinal microbiota of wild Atlantic cod *Gadus morhua* L. upon captive rearing. *Microb. Ecol.* **2011**, *1*, 20–30. [CrossRef]
11. Miyake, S.; Ngugo, D.K.; Stingl, U. Diet strongly influences the gut microbiota of surgeon fishes. *Mol. Ecol.* **2015**, *24*, 656–672. [CrossRef]
12. Clements, K.D.; Pasch, I.B.Y.; Moran, D.; Turner, S.J. Clostridia dominate 16S rRNA gene libraries prepared from the hindgut of temperate marine herbivorous fishes. *Mar. Biol.* **2007**, *150*, 1431–1440. [CrossRef]
13. Hovda, M.B.; Fontanillas, R.; McGurk, C.; Obach, A.; Rosnes, J.T. Seasonal variations in the intestinal microbiota of farmed Atlantic salmon (*Salmo salar* L.). *Aquac. Res.* **2012**, *43*, 154–159. [CrossRef]
14. Bano, N.; Smith, A.D.; Bennett, W.; Vasquez, L.; Hollibaugh, J.T. Dominance of *Mycoplasma* in the guts of the long jawed Mudsucker, *Gillichthys mirabilis*, from five California salt marshes. *Environ. Microbiol.* **2007**, *9*, 2636–2641. [CrossRef] [PubMed]
15. Talwar, C.; Nagar, S.; Lal, R.; Negi, R.K. Fish gut microbiome: Current approaches and future perspectives. *Indian J. Microbiol.* **2018**, *58*, 397–414. [CrossRef] [PubMed]
16. Perry, W.B.; Lindsay, E.; Payne, C.J.; Brodie, C.; Kazlauskaite, R. The role of the gut microbiome in sustainable teleost aquaculture. *Proc. R. Soc. B* **2020**, *287*, 20200184. [CrossRef] [PubMed]
17. Llewellyn, M.S.; McGinnity, P.; Dionne, M.; Letourneau, J.; Thonier, F.; Carvalho, G.R.; Creer, S.; Derome, N. The biogeography of the atlantic salmon (*Salmo salar*) gut microbiome. *ISME J.* **2016**, *10*, 1280–1284. [CrossRef] [PubMed]
18. Niu, K.M.; Lee, B.J.; Kothari, D.; Loo, W.D.; Hur, S.W.; Lim, S.G.; Kim, K.W.; Kim, N.N.; Kim, S.K. Dietary effect of low fish meal aquafeed on gut microbiota in olive flounder (*Paralichtys olivaceus*) at different growth stages. *MicrobiologyOpen* **2020**, *9*, e992. [CrossRef]
19. Zhao, R.; Symonds, J.E.; Walker, S.P.; Steiner, K.; Carter, C.G.; Bowman, J.P.; Nowak, B.F. Salinity and fish age affect the gut microbiota of farmed Chinook salmon (*Oncorhynchus tshawytscha*). *Aquaculture* **2020**, *528*, 735539. [CrossRef]
20. FAO. *The State of World Fisheries and Aquaculture 2018—Meeting the Sustainable Development Goals;* FAO: Rome, Italy, 2018.
21. Schaefer, K.M.; Fuller, D.W. Simultaneous behavior of skipjack (*Katsuwonus pelamis*), bigeye (*Thunnus obsesus*), and yellowfin (*T. albacares*) tunas, within large multi-species aggregations associated with drifting fish aggregating devices (FADs) in the equatorial eastern Pacific Ocean. *Mar. Biol.* **2013**, *160*, 3005–3014. [CrossRef]
22. Albaret, J.J. La reproduction de l'albacore *(Thunnus albacares)* dans le Golfe de Guinée. *Cah. ORSTOM. Série Océanographie* **1977**, *15*, 389–419.
23. Bettarel, Y.; Combe, M.; Adingra, A.; Ndiaye, A.; Bouvier, T.; Panfili, J.; Durand, J.D. Horde of phages in the fish gut of the tilapia *Sarotherodon Melanotheron*. *Sci. Rep.* **2018**, *8*, 11311. [CrossRef]

24. Klindworth, A.; Pruesse, E.; Schweer, T.; Peplies, J.; Quast, C.; Horn, M.; Glöckner, F.O. Evaluation of general 16S ribosomal RNA gene PCR primers for classical and next-generation sequencing-based diversity studies. *Nucleic Acids Res.* **2013**, *41*, 1. [CrossRef]
25. Andersson, A.F.; Lindberg, M.; Jakobsson, H.; Bäckhed, F.; Nyrén, P.; Engstrand, L. Comparative analysis of korean human gut microbiota by barcoded pyrosequencing. *PLoS ONE* **2008**, *3*, e2836. [CrossRef]
26. Perreault, N.N.; Andersen, D.T.; Pollard, W.H.; Greer, C.W.; Whyte, L.G. Characterization of the prokaryotic diversity in cold saline perennial springs of the Canadian high arctic. *Appl. Environ. Microbiol.* **2007**, *73*, 1532–1543. [CrossRef]
27. Callahan, B.J.; McMurdie, P.J.; Rosen, M.J.; Han, A.W.; Johnson, A.J.A.; Holmes, S.P. DADA2: High-resolution sample inference from Illumina amplicon data. *Nat. Met.* **2016**, *13*, 581–583. [CrossRef]
28. Quast, C.; Pruesse, E.; Yilmaz, P.; Gerken, J.; Schweer, T.; Yarza, P.; Peplies, J.; Glöckner, F.O. The SILVA ribosomal RNA gene database project: Improved data processing and web-based tools. *Nucleic Acid Res.* **2012**, *41*, D590–D596. [CrossRef] [PubMed]
29. McMurdie, P.J.; Holmes, S. phyloseq: An R package for reproducible interactive analysis and graphics of microbiome census data. *PLoS ONE* **2013**, *8*, e61271. [CrossRef] [PubMed]
30. Lahti, L.; Shetty, S.; Blake, T.; Salojarvi, J. Tools for Microbiome Analysis in R. Available online: https://microbiome.github.io/tutorials/ (accessed on 15 February 2019).
31. Dixon, P. VEGAN, a package of R functions for community ecology. *J. Veg. Sci.* **2003**, *14*, 927–930. [CrossRef]
32. Ghanbari, M.; Kneifel, W.; Domig, K.J. A new view of the fish gut microbiome: Advances from next-generation sequencing. *Aquaculture* **2015**, *448*, 464–475. [CrossRef]
33. Mouchet, M.A.; Bouvier, C.; Bouvier, T.; Troussellier, M.; Escalas, A.; Mouillot, D. Genetic difference but fonctional similarity among fish gut bacterial communities through molecular and biogeochemical fingerprints. *FEMS Microbiol. Ecol.* **2012**, *79*, 568–580. [CrossRef] [PubMed]
34. Kitchell, J.F.; Neill, W.H.; Dizon, A.E.; Magnusson, J.J. Bioenergetic spectra of skipjack and yellowfin tunas. In *The Physiologycal Ecology of Tunas*; Sharp, G.D., Dizon, A.E., Eds.; Academic Press: New York, NY, USA, 1978; pp. 357–368.
35. Sepulveda, J.; Moeller, A.H. The effects of temperature on animal gut microbiomes. *Front. Microbiol.* **2020**, *11*, 384. [CrossRef]
36. Sharpton, T.J. Role of the gut microbiome in vertebrate Evolution. *Msystems* **2018**, *3*, e00174-17. [CrossRef]
37. Sullam, K.E.; Essinger, S.D.; Lozupone, C.A.; O'Connor, M.P.; Rosen, G.L.; Knight, R.; Kilham, S.S.; Russell, J.A. Environmental and ecological factors that shape the gut bacterial communities of fish: A meta-analysis. *Mol. Ecol.* **2012**, *21*, 3363–3378. [CrossRef]
38. Givens, C.E.; Ransom, B.; Bano, N.; Hollibaugh, J.T. Comparison of the gut microbiomes of 12 bony fish and 3 shark species. *Mar. Ecol. Prog. Ser.* **2015**, *518*, 209–223. [CrossRef]
39. Burns, A.R.; Stephens, W.Z.; Stagaman, K.; Wong, S.; Rawls, J.F.; Guillemin, K.; Bohannan, B.J. Contribution of neutral processes to the assembly of gut microbial communities in the zebrafish over host development. *ISME J.* **2016**, *10*, 655–664. [CrossRef]
40. Heys, C.; Cheaib, B.; Busetti, A.; Kazlauskaite, R.; Sloan, W.T.; Ijaz, U.; Kaufmann, J.; McGinnity, P.; Llewellyn, M.S. Neutral processes dominate microbial community assembly in Atlantic salmon, *Salmo salar*. *Appl. Environ. Microbiol.* **2020**, *86*. [CrossRef] [PubMed]
41. Bolnick, D.I.; Snowberg, L.K.; Hirsch, P.E.; Lauber, C.L.; Org, E.; Parks, B.; Lusis, A.J.; Knight, R.; Caporaso, J.G.; Svanbäck, R. Individual diet has sex-dependent effects on vertebrate gut microbiota. *Nat. Commun.* **2014**, *5*, 4500. [CrossRef] [PubMed]
42. Godon, J.-J.; Arulazhagan, P.; Steyer, J.-P.; Hamelin, J. Vertebrate bacterial gut diversity: Size also matters. *BMC Ecol.* **2016**, *16*, 12. [CrossRef]
43. Dragovich, A. The Food of Bluefin Tuna (*Thunnus thynnus*) in the Western North Atlantic Ocean. *Trans. Am. Fish. Soc.* **1970**, *99*, 726–731. [CrossRef]
44. Dragovich, A.; Potthoff, T. Comparative study of food of skipjack and yellowfin tunas off the coast of west Africa. *Fish. Bull.* **1972**, *70*, 1087–1110.
45. Graham, J.B.; Dickson, K.A. Tuna comparative physiology. *Mar. Biol.* **2004**, *150*, 647–658. [CrossRef]
46. Ringø, E.; Schillinger, U.; Holzapfel, W. Antimicrobial activity of lactic acid bacteria isolated from aquatic animals and the use of lactic acid bacteria in aquaculture. In *Biology of Growing Animals*; Elsevier: Amsterdam, The Netherlands, 2005; Volume 2, pp. 418–453.
47. Nayak, S.K. Role of gastrointestinal microbiota in fish. *Aquac. Res.* **2020**, *41*, 1553–1573. [CrossRef]
48. Sivasubramanian, K.; Ravichandran, S.; Kavitha, R. Isolation and characterization of gut microbiota from some estuarine fishes. *Mar. Sci.* **2012**, *2*, 1–6. [CrossRef]
49. Pękala-Safińska, A. Contemporary threats of bacterial infections in freshwater fish. *J. Vet. Res.* **2018**, *62*, 261–267. [CrossRef] [PubMed]
50. Leisner, J.J.; Laursen, B.G.; Prévost, H.; Drider, D.; Dalgaard, P. *Carnobacterium*: Positive and negative effects in the environment and in foods. *FEMS Microbiol. Rev.* **2007**, *31*, 592–613. [CrossRef]
51. Del Rio-Rodriguez, R.E.; Inglis, V.; Millar, S.D. Survival of Escherichia coli in the intestine of fish. *Aquac. Res.* **1997**, *28*, 257–264. [CrossRef]
52. David, O.M.; Wandili, S.; Kakai, R.; Waindi, E.N. Isolation of *Salmonella* and *Shigella* from fish harvested from the Winam Gulf of Lake Victoria, Kenya. *J. Infect Dev. Ctries.* **2009**, *3*, 99–104. [CrossRef] [PubMed]
53. Tsuda, K.; Nagano, H.; Ando, A.; Shima, J.; Ogawa, J. Isolation and characterization of psychrotolerant endospore-forming *Sporosarcina* species associated with minced fish meat (surimi). *Int. J. Food Microbiol.* **2015**, *199*, 15–22. [CrossRef] [PubMed]

54. Bjornsdottir-Butler, K.; Green, D.P.; Bolton, G.E.; McClellan-Green, P.D. Control of histamine-producing bacteria and histamine formation in fish muscle by trisodium phosphate. *J. Food Sci.* **2015**, *80*, 1253–1258. [CrossRef]
55. Hungerford, J.M. Scombroid poisoning: A review. *Toxicon* **2010**, *56*, 231–243. [CrossRef] [PubMed]
56. Takahashi, H.; Ogai, M.; Miya, S.; Kuda, T.; Kimura, B. Effects of environmental factors on histamine production in the psychrophilic histamine-producing bacterium *Photobacterium iliopiscarium*. *Food Control.* **2015**, *52*, 39–42. [CrossRef]
57. Trevisani, M.; Cecchini, M.; Fedrizzi, G.; Corradini, A.; Mancusi, R.; Tothill, I.E. Biosensing the histamine producing potential of bacteria in tuna. *Front. Microbiol.* **2019**, *10*, 1844. [CrossRef] [PubMed]

MDPI

Article

# Commensal and Opportunistic Bacteria Present in the Microbiota in Atlantic Cod (*Gadus morhua*) Larvae Differentially Alter the Hosts' Innate Immune Responses

**Ragnhild Inderberg Vestrum [1,†], Torunn Forberg [1,†], Birgit Luef [1], Ingrid Bakke [1], Per Winge [2], Yngvar Olsen [2] and Olav Vadstein [1,*]**

[1] Department of Biotechnology and Food Science, Norwegian University of Science and Technology, 7491 Trondheim, Norway; ragnhild@morefish.no (R.I.V.); torfo@biomar.com (T.F.); b.luef@biology.leidenuniv.nl (B.L.); Ingrid.bakke@ntnu.no (I.B.)

[2] Department of Biology, Norwegian University of Science and Technology, 7491 Trondheim, Norway; per.winge@ntnu.no (P.W.); yngvar.olsen@ntnu.no (Y.O.)

* Correspondence: olav.vadstein@ntnu.no

† These authors contributed equally to this work.

**Citation:** Vestrum, R.I.; Forberg, T.; Luef, B.; Bakke, I.; Winge, P.; Olsen, Y.; Vadstein, O. Commensal and Opportunistic Bacteria Present in the Microbiota in Atlantic Cod (*Gadus morhua*) Larvae Differentially Alter the Hosts' Innate Immune Responses. *Microorganisms* **2022**, *10*, 24. https://doi.org/10.3390/microorganisms10010024

Academic Editor: Francesco Marotta

Received: 15 November 2021
Accepted: 21 December 2021
Published: 24 December 2021

**Publisher's Note:** MDPI stays neutral with regard to jurisdictional claims in published maps and institutional affiliations.



**Abstract:** The roles of host-associated bacteria have gained attention lately, and we now recognise that the microbiota is essential in processes such as digestion, development of the immune system and gut function. In this study, Atlantic cod larvae were reared under germ-free, gnotobiotic and conventional conditions. Water and fish microbiota were characterised by 16S rRNA gene analyses. The cod larvae's transcriptional responses to the different microbial conditions were analysed by a custom Agilent 44 k oligo microarray. Gut development was assessed by transmission electron microscopy (TEM). Water and fish microbiota differed significantly in the conventional treatment and were dominated by different fast-growing bacteria. Our study indicates that components of the innate immune system of cod larvae are downregulated by the presence of non-pathogenic bacteria, and thus may be turned on by default in the early larval stages. We see indications of decreased nutrient uptake in the absence of bacteria. The bacteria also influence the gut morphology, reflected in shorter microvilli with higher density in the conventional larvae than in the germ-free larvae. The fact that the microbiota alters innate immune responses and gut morphology demonstrates its important role in marine larval development.

**Keywords:** Atlantic cod; microbiota; innate immune system; germ-free; gnotobiotic

## 1. Introduction

The roles of the microbiota associated with vertebrate hosts, including fish, have received much attention over the last decade. Several studies have shown that the microbiota stimulates the immune system and functions as a barrier against potential pathogens [1–4], aids in epithelial development and maturation [4,5] and affects the digestion of nutrients [6,7]. There is a bias in the type of animals studied, and still relatively few studies are published on the function of microbiota in fish. Most of the bacteria associated with the fish are harmless or beneficial [8,9]. However, specific pathogens and opportunistic bacteria are also present [10,11], and bacteria present in the natural environment of the fish cause many of the infections that are associated with the mortality of marine fish larvae [12].

Germ-free animals have been popular tools used in studies of host–microbe interactions [13,14], and the use of gnotobiotic zebrafish is well-known [15,16]. Rawls et al. [16] observed that gut microbiota in zebrafish stimulated proliferation of intestinal epithelial cells, as previously seen for rodents [5,17]. They found 212 genes that were differentially regulated in germ-free fish compared to fish exposed to bacteria. Moreover, 59 of those gene-expression responses were observed in both mice and zebrafish. These genes are involved in epithelial proliferation, nutrient metabolism and innate immune responses [16].

Our group have developed a protocol and a cultivation system for germ-free and gnotobiotic Atlantic cod larvae [18]. Using this system, Forberg et al. [19,20] showed that bacteria regulate the transcription of genes involved in immune response and nutrient digestion in the cod larvae, and that the response differs depending on whether the bacteria are dead or alive.

However, at that time, molecular tools for Atlantic cod were poorly developed, and this allowed characterisation of only a limited part of the host transcriptome. Following the sequencing of the Atlantic cod genome [21], it was discovered that cod lack certain families of cell surface receptors, whereas others are expanded, compared with zebrafish and mice [21–24]. The functional consequences of this evolutionary divergence in immune response with regards to host–microbe interactions have, to our knowledge, not been studied. Thus, the knowledge about the early regulation of the immune system of cod and its responses to both pathogenic and non-pathogenic bacteria is still limited.

The aim of this study was to increase our knowledge about host responses induced by the early microbial colonisation of Atlantic cod larvae, focusing on the roles of microbes in regulation of the immune system and digestion. Newly hatched cod larvae were reared under three different conditions: germ-free, gnotobiotic (two probiotic candidate strains added) and conventional (uncontrolled microbial environments). The water- and fish-associated microbiota was characterised and analysed, and host responses were examined using a custom oligo microarray as well as transmission electron microscopy (TEM).

## 2. Materials and Methods

The experiment was carried out within the Norwegian animal welfare act guidelines, in accordance with the Animal Welfare Act of 20 December 1974, amended 19 June 2009, at a facility with permission to conduct experiments on fish (code 93) provided by the Norwegian Animal Research Authority (NARA). The experiment was approved by NARA.

### 2.1. Cod Larval Rearing

Atlantic cod eggs were delivered from Nofima Marin national breeding station (Havbruksstasjonen Tromsø, Norway). Upon arrival, the cod eggs (55–65-day degrees) were acclimatised in filtered (0.22 μm Micropore®) autoclaved (121 °C, 20 min) seawater (FASW) at 6 ± 1 °C, in the dark. Germ-free larvae were obtained according to the protocol of Forberg et al. [18] (information about germ-free verification in Text S1). Cod larvae (65 larvae in 2 L water) were reared under three different conditions: germ-free, conventional and gnotobiotic. For the gnotobiotic treatment, two different bacterial strains were added in equal amounts (final density of $10^6$ cells/mL) to the rearing bottles: *Microbacterium* ND 2–7 and *Vibrio* RD 5–30, both previously isolated from cod and identified as probiotic candidates (for details, see [25]) (information about live feed and bacterial cultures in Text S2). The rearing bottles representing the conventional condition were filled with microbially matured water, from a biofilter in a seawater lab-scale aquaculture system. After stocking, the temperature was increased by 1 °C/day until 12 °C was reached.

### 2.2. Sampling

Each treatment had 11 replicate bottles at trial start. Cod larvae were collected at 4 (only for DNA extraction), 8, 13 and 16 dph. Fish from one bottle were sampled at 4 dph, while 3 replicate bottles were sampled at 8, 13 and 16 dph. Water was sampled from one bottle at 1 and 4 dph, and three replicate bottles at 8, 13 and 16 dph. After sampling, the bottles were taken out of the experiment, thus reducing the number of replicate bottles with time. Larvae were sacrificed with an overdose of tricaine methanesulfonate (MS-222) prior to sampling, snap-frozen in liquid nitrogen and stored at −80 °C for further analyses. To investigate larval growth, 10 individual larvae from each sampled bottle were freeze-dried and weighed. More details regarding sampling procedures and DNA/RNA extraction are described in Supplementary Materials (Text S3).

### 2.3. Characterisation of Microbial Communities

The water and fish samples from the gnotobiotic treatment were analysed by DGGE. PCR products representing the V3 region of the 16 S rRNA gene were generated using a nested PCR protocol to avoid possible amplification of eukaryotic 18S rDNA [26]. The PCR was set up and analysed as described by Bakke et al. [27].

For in-depth analysis of the microbiota in the conventional treatment, Illumina MiSeq sequencing was performed based on total DNA extracted from water and larvae sampled at 1 (only water), 4, 8, 13 and 16 dph. Larval and water samples were prepared for Illumina MiSeq sequencing by amplification of the V4 region of the 16S rRNA gene, by using the following primers (bacteria-specific V4 primer, underlined and bold) including 5′ overhang, as suggested by Illumina:

515F F′ TCGTCGGCAGCGTCAGATGTGTATAAGAGACAGNNNN**GTGCCAGCM-GCCGCGGTAA** 3′ and

803 R 5′ GTCTCGTGGGCTCGGAGATGTGTATAAGAGACAGNNNN**CTACVVGGG-TATCTAAKCCBK** 3′.

The amplicon library preparation and processing of the Illumina sequencing data was performed as described by Vestrum et al. [28]. In short, for the first stage of amplification, the reactions were run for 38 cycles for water samples and 40 cycles for cod larval samples (98 °C 15 s, 55 °C 20 s, 72 °C 20 s), with 0.3 μM of each primer, 0.25 mM of each dNTP, 2 mM of $MgCl_2$, 12 μM of BSA, glycerol (10%), Phusion Hot Start II High-Fidelity DNA Polymerase and reaction buffer from Thermo Scientific in a total volume of 20 μL. All samples were normalised using the SequalPrepTM Normalisation Plate Kit (Invitrogen). A second PCR was performed to attach dual indices and Illumina sequencing adapters to the normalised v4 amplicons by using the Nextera XT Index Kit. The indexed PCR products were normalised as described above, pooled, and concentrated by using Amicon® Ultra-0.5 Centrifugal Filter Devices. The resulting amplicon library was sequenced on a MiSeq lane (Illumina, San Diego, CA, USA) with v4 reagents employing 260 bp paired-end reads at the Norwegian Sequencing Center at the University of Oslo, Norway. The Illumina sequencing data were processed with the high-performance USEARCH utility (version 11) (http://drive5.com/usearch/features.html (accessed on 22 March 2019). Taxonomy assignment was performed applying the Sintax script [29] with a confidence value threshold of 0.8 and the RDP reference dataset (version 16). OTUs of particular interest were further analysed with the RDP tools [30] Classifier and Sequence Match. OTUs representing algae, Archaea and Cyanobacteria/Chloroplast were removed from the OTU table. In addition, an OTU representing Propionibacterium acne, a well-known contaminant of DNA extraction kits [31], was removed. To remove biases due to variation in sequencing depth, analyses were performed on an OTU table that had been subsampled to 15500 sequencing reads for each sample. The subsampling threshold was chosen based on the sample with the lowest number of reads in order to keep all samples in the dataset and was performed to avoid bias due to differences in sequencing depth. The resulting Illumina sequencing data were deposited at the European Nucleotide Archive (accession numbers ERS8484975-ERS8484994).

### 2.4. Microarray Design, Hybridisation and Annotation

A custom, Agilent 44 k oligo microarray (A-MEXP-2226, ArrayExpress, EMBL-EBI) described by Kleppe et al. [32] was used and analysed as described by Vestrum et al. [33]. This microarray design is partly based on the Atlantic cod gene set described by Star et al. [21] as well as EST sequences from various cod tissues/developmental stages. The identified differentially regulated transcripts were used for biological term enrichment analysis and Gene Ontology term (GO term) annotation in DAVID (Database of Annotation, Visualisation and Integrated Discovery) [34,35] (using the official gene symbol for human homologues).

### 2.5. Electron Microscopy Procedures

For processing fish larvae for transmission electron microscopy (TEM), the protocol from Galloway et al. [36] was adopted. Shortly, germ-free and conventional reared fish larvae from 16 dph were fixed in a mixture of 2.5% paraformaldehyde, 2.5% glutardialdehyde, 0.5% sucrose and 0.11 M HEPES buffer (pH 7.4) and stored at 4 °C until further processing. Three larvae from each treatment were rinsed in 0.11 M HEPES buffer, post-fixed in 2% OsO4 in 1.5% potassium ferricyanide (final concentration), bulk contrasted in 1.5% uranyl acetate and dehydrated in ethanol before embedding in Epon. After polymerisation, 50–60 nm sections of the fish midgut were cut using a Leica UC6 Ultramicrotome. These ultrathin sections were collected on 200-mesh copper grids and contrasted with 4% uranyl acetate and 1% lead citrate. Sections were inspected with a FEI Company Tecnai 12 operated at 80 kV and imaged using a digital MORADA G3 CCD camera (EMSIS).

### 2.6. Intestinal Morphometry and Statistical Methods

Computerised morphometric measurements of microvilli lengths (from tip to base, l), diameter (2r) and abundance of microvilli ($\mu m^{-2}$) in the midgut of the fish were made using the image processing program iTEM (Olympus Soft Imaging Solutions GmbH). Microvilli parameters were measured according to the criteria of Brown [37]. Three fish larvae per treatment were investigated ($n = 3$). For length measurements, at least 65 microvilli per fish were analysed; in total, 318 microvilli within the germ-free and 294 microvilli within the conventional treatment were measured. For diameter measurements, at least 58 microvilli per fish were analysed; in total, 214 microvilli within the germ-free and 287 microvilli within the conventional treatment were measured. To quantify the abundance of microvilli, microvilli within a total area of around 400 $\mu m^2$ were counted. iTEM was used to adjust contrast in the images and to insert calibrated scale bars into images.

### 2.7. Statistical Analyses

Student's *t*-test (unpaired) was used to investigate significance in differences in Shannon indices, abundance of individual DGGE bands and larval growth measurements. Survival analysis was performed by the Kaplan–Meier method, and the Log-rank test was used for pairwise post hoc comparisons of survival across the groups. Ordination by Principal Coordinate Analysis (PCoA) based on Bray–Curtis similarities was used to visualize differences between sample groups, and one-way and two-way PERMANOVA based on Bray–Curtis similarities were used to test for statistically significant differences between sample groups. Similarity Percentage analysis (SIMPER) was used to identify OTUs responsible for differences (measured as Bray–Curtis similarities) between different sample groups. The multivariate analyses were performed using the program package PAST version 3.22 [38]. Venn diagrams were created using jvenn [39]. The Usearch commands Alpha_div and Sintax_summary were used to calculate alpha diversity indices and to generate taxa summary tables (at various taxonomic levels, as specified with the results), respectively.

Data from the intestinal morphometric study were statistically analysed with IBM SPSS Statistics (SPSS for Windows, version 26.0; SPSS Inc., Chicago, IL, USA). A Welch test was performed to investigate significant differences between the axenic- and conventional-treated fish regarding microvilli length, diameter and abundance of microvilli ($\mu m^{-2}$). Differences were considered statistically significant when $p \leq 0.05$.

## 3. Results

### 3.1. Larval Survival and Growth

Daily counts of dead larvae in the rearing bottles were used to calculate the percent of survival. Kaplan–Meier survival curves showed a clear separation of the conventional group vs. the germ-free and gnotobiotic group cumulative mortality, and this difference was highly significant (Log rank post hoc *p*-values 0.000032 and 0.000005 for the pairwise comparisons) (Supplementary Figure S1) (84.9%, 84.8% and 76.0% survival, respectively).

At 16 dph, the dry weight of germ-free larvae was significantly lower (average 75.5 µg) than the gnotobiotic and conventional larvae (average 96.9 and 110.1 µg, respectively) ($p = 0.017$ and 0.059, respectively) (Supplementary Figure S2).

### 3.2. Composition of Fish and Water Microbiota

The DGGE profiles for the samples from the gnotobiotic rearing bottles (Supplementary Figure S3), where only two bacterial strains were added, were consistently identical, except for the presence of one additional band in one fish sample at 8 dph. This suggests the presence of a contaminating bacterial strain. Even though bacteria were added to the same final cell density in the rearing water in the gnotobiotic treatment, *V. gallicus* clearly dominated both in water and fish samples. *Microbacterium* was detectable at low levels in water samples and present only in some fish samples.

Fish and water microbiota in the conventional treatment were characterised by amplicon sequencing. After quality trimming and chimera removal, 1,039,322 reads were obtained. Two water samples and two fish samples from 4 dph were removed due to low number of reads. The estimated total (Chao1) and observed number of OTUs for each sample (Figure 1) indicate a sequencing depth of on average 95% and 82% in fish and water samples, respectively. The observed richness was generally higher in water than in fish.

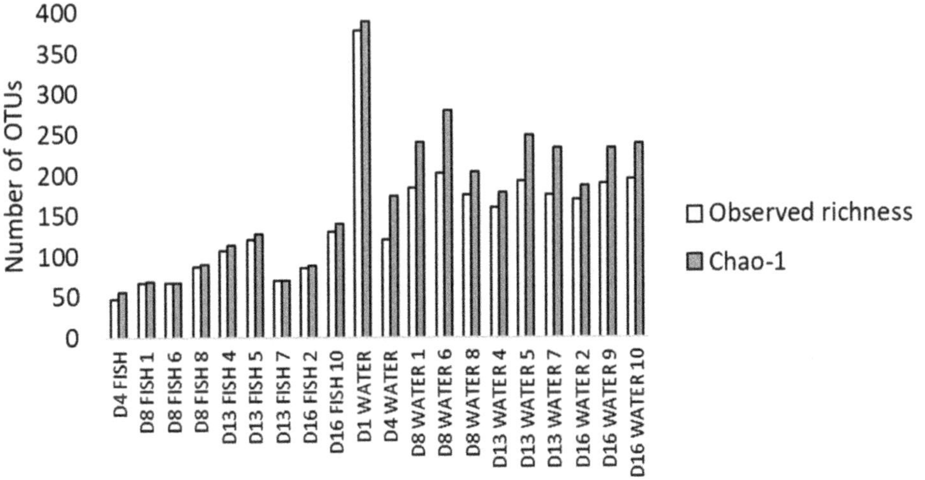


**Figure 1.** Observed richness (number of OTUs) and Chao1 values in fish and water microbiota at 1 (only water) to 16 dph (D1–D16). Numbers indicate the rearing bottle sampled. One fish sample consisted of 5 pooled larvae from one rearing bottle.

A PCoA ordination of the microbial communities in fish and water samples (Figure 2a) showed that the water and fish clustered separately. This was corroborated by Bray–Curtis similarities (Figure 2b). The PCoA plot also indicated that both the water and fish microbiota changed over time. There were significant differences between the water and fish microbiota both early (1/4–8 dph) and late (13–16 dph) in the experiment (one-way PERMANOVA, $p = 0.04$ and 0.01 for early and late, respectively) and also between water samples early and late in the experiment (one-way PERMANOVA, $p = 0.01$). There were no significant differences in the microbiota of fish samples early and late in the experiment.

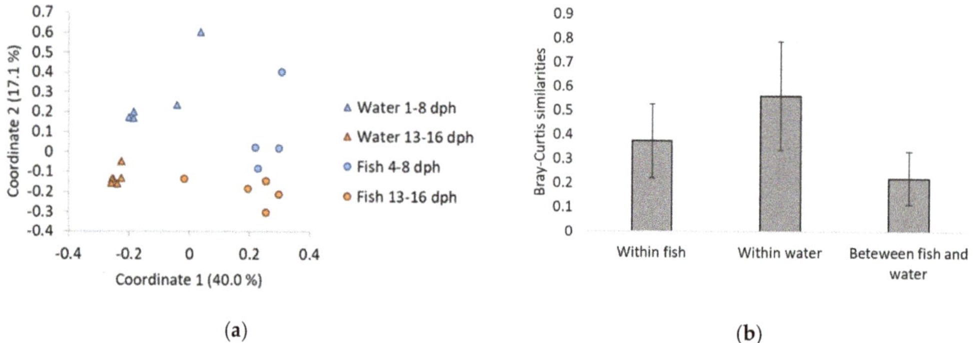


**Figure 2.** (**a**) PCoA ordination plot based on Bray-Curtis similarities for comparison of water and fish microbiota from samples taken early (1/4–8 dph) and late (13–16 dph) in the experiment. One fish sample consists of five pooled cod larvae from one rearing bottle. (**b**) Bray-Curtis similarities within fish and water microbiota, and between fish and water microbiota. Bars indicate standard deviation.

The results from PERMANOVA analysis are reflected in the taxonomic composition of the microbial communities at the order level (Figure 3). The relative abundance of Vibrionales was more than 14 times higher in fish (up to 44%) than in water ($\leq$3%). The relative abundances of Flavobacteriales in the water increased with time.

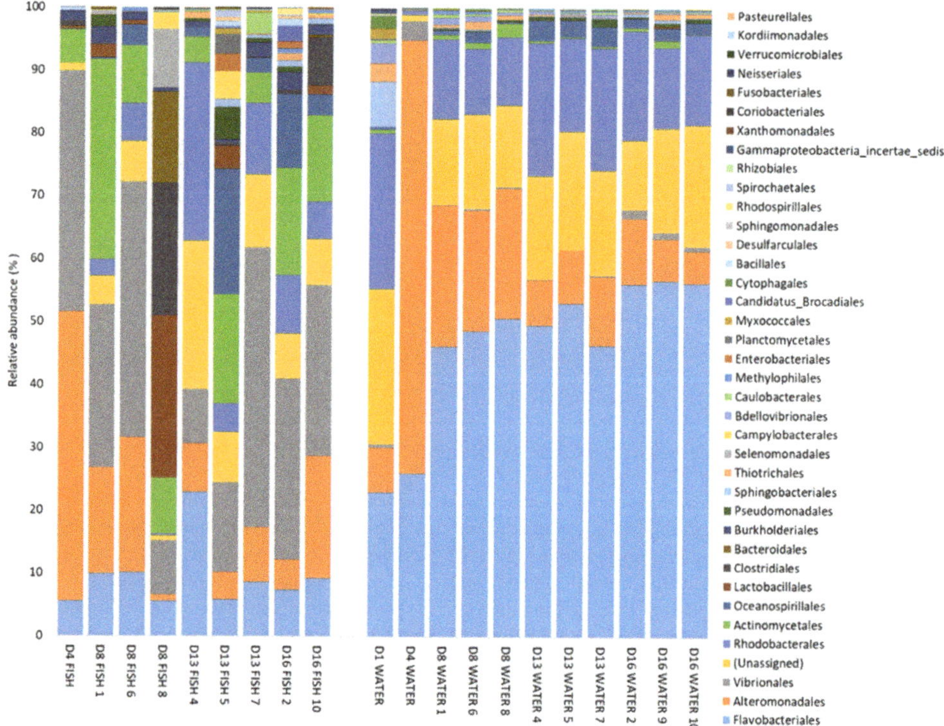


**Figure 3.** Taxa summary of bacterial orders detected at a relative abundance higher than 0.1% in at least one sample, at 1 to 16 dph (D1–D16). Numbers indicate the rearing bottle sampled. One fish sample consists of five pooled larvae from one rearing bottle.

OTU 1 (*Polaribacter*, Flavobacteriales order) and OTU 2 (*Vibrio*) were the most abundant OTUs in the dataset. Both OTUs were present in both fish and water samples, but OTU 1 was far more abundant in the water than in the fish (on average 31% and 4% of the reads, respectively). OTU 2 had a higher relative abundance in the fish than in the water (average 21% and 0.4%, respectively). The third most abundant OTU was OTU 4 (*Colwellia*), which was more abundant early in the experiment (1–8 dph), both for fish (average 9.2%) and water (average 14.2%) samples, than later in the experiment (13–16 dph) (average 1.1% and 2.8% in fish and water samples, respectively).

### 3.3. Gene Expression in Atlantic Cod Larvae

At 8 dph, no genes in the cod larvae were differentially expressed between the treatments. For 13 dph samples, there were still no differentially expressed genes between conventional and gnotobiotic larvae. However, the genes *G-protein-coupled receptor family C group 6 member A* (*gprc6a*) (involved in regulation of inflammation, metabolism and endocrine functions) and *rhamnose binding lectin* (*rbl*) (involved in innate immunity) were downregulated in both conventional and gnotobiotic larvae compared with germ-free larvae. The *zg16* and *zg16-like* genes (involved in innate immunity) were also downregulated in gnotobiotic larvae compared with germ-free larvae. Only one gene, *lect2*, was upregulated in conventional larvae compared with germ-free larvae.

For samples from 16 dph, 82 genes were downregulated and 97 were upregulated in conventional larvae compared with germ-free larvae. In gnotobiotic larvae, only 23 were downregulated and none were upregulated compared with germ-free larvae (Supplementary Tables S1–S3). Gnotobiotic and germ-free larvae generally showed similar expression profiles. Of the genes that were upregulated in conventional compared with germ-free larvae, 74% were also upregulated compared with gnotobiotic larvae (Supplementary Figure S4). Many of the genes that were upregulated in conventional compared with germ-free larvae are involved in innate immune responses and linked to signalling and glucose transport. Examples for immunity are, e.g., *interleukin 8* (*cxcl8*), *leukocyte cell-derived chemotaxin 2* (*lect2*) and *interleukin-1 receptor-activated kinase* (*irak1*), and for signalling/transport are, e.g., *solute carrier family 2 facilitated glucose transporter member 11-like* (*slc2a11*) and *solute carrier family 2 facilitated glucose transporter member 4-like* (*slc2a4*). Biological term enrichment analysis and Gene Ontology term (GO term) annotation in DAVID showed that 15 GO terms were enriched in conventional fish compared to germ-free fish (Figure 4). Several of the enriched GO terms, including the one with the highest number of genes ("regulation of nucleobase-containing metabolic process"), were related to growth and cell division. Other enriched GO terms were related to signalling and cell communication. Most of these GO terms were also enriched compared with gnotobiotic larvae (Supplementary Table S4). The KEGG pathway "bacterial invasion of epithelial cells" was enriched in conventional larvae compared with both gnotobiotic and germ-free fish.

However, 15 annotated genes had significantly lower expression in both conventional and gnotobiotic larvae compared with germ-free larvae. Interestingly, nine of these genes were involved in innate immune responses: *eosinophil peroxidase* (*epx*), *rbl*, *zymogen granule membrane protein 16* (*zgp16*), *myeloperoxidase precursor* (*mpo*), *Cytochrome b-245 heavy chain-like* (*cybb*), *immune-responsive gene 1 protein-like* (*Irg1*), *fish egg lectin* (*fel*), *N-acetylmuramoyl-L-alanine amidase-like* (*pglyrp1*) and *transmembrane protease serine 9-like* (*tmprss9*). Thus, the presence of bacteria, both as complex communities and simple gnotobiotic associations, downregulated some of the innate immune responses in the cod larvae.

Far more GO terms differed between germ-free and conventional larvae than between germ-free and gnotobiotic larvae (95 and 11, respectively). The most enriched GO terms in the germ-free larvae were "proteolysis", "negative regulation and regulation of metabolic processes", "signal transduction" and "adhesion and cell death" (Figure 5).

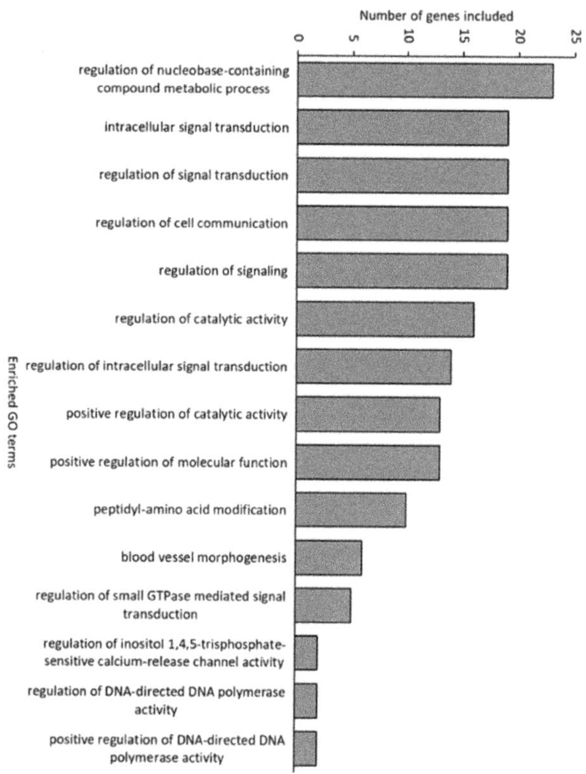


**Figure 4.** Significantly enriched Gene Ontology (GO) terms in conventional fish compared with germ-free fish.

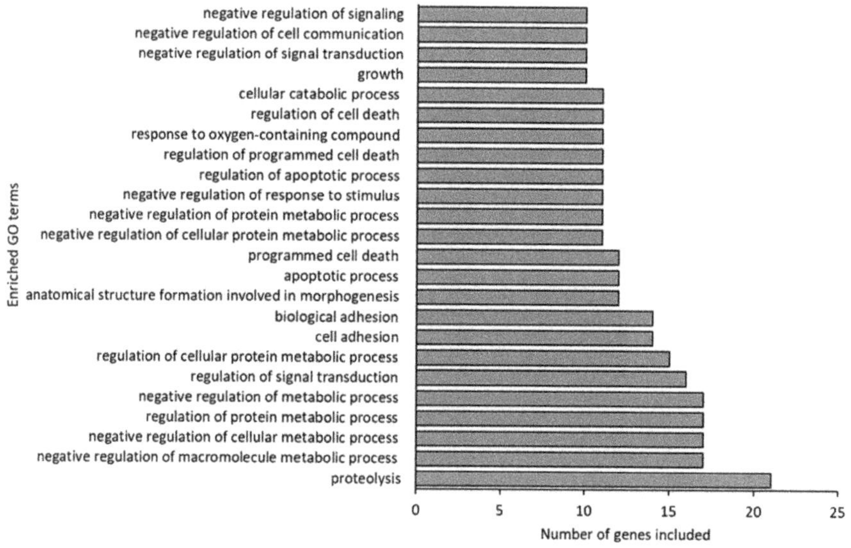


**Figure 5.** Enriched Gene Ontology (GO) terms (including 10 or more of the 52 input genes) in germ-free fish compared to conventional fish.

Of the 21 genes included in the GO term "proteolysis", 5 were involved in the KEGG pathway of "protein digestion and absorption". This indicates that a large fraction of the "proteolysis" GO term is related to the cod larvae's digestion.

### 3.4. Ultrastructure and Morphometric Analysis of the Intestinal Tissue

Comparison of the midgut ultrastructure, including tight junctions, microvilli disruption/damage, intercellular space and vacuoles, showed no significant differences between germ-free and conventional cod larvae. However, the mitochondria in germ-free cod larvae were distorted. The outer membrane showed discontinuities or was missing, and structures of the Christae were reduced or hardly visible (Figure 6). In contrast, the mitochondria in conventional cod larvae showed clear Christae and a clear double membrane.

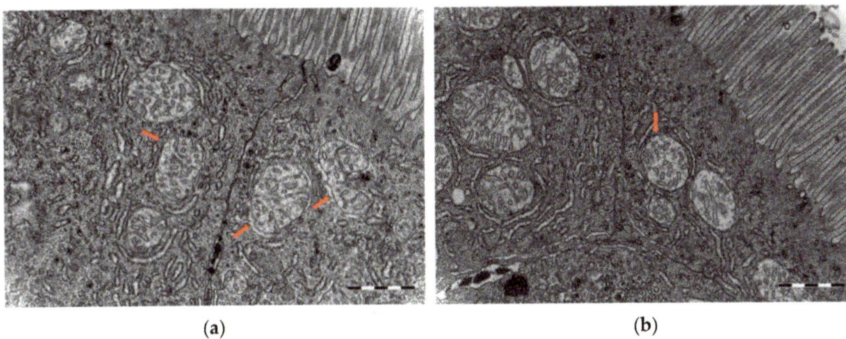

**Figure 6.** Transmission electron microscopy micrographs of mitochondria from the midgut of cod larvae at 16 dph reared under (**a**) germ-free and (**b**) conventional conditions. (**a**) Mitochondria in germ-free cod larvae were distorted, outer membrane showed discontinuities or was often missing and structures of Christae were reduced or hardly visible. (**b**) Mitochondria in conventional cod larvae showed clear Christae and a double membrane. Scale bars are 1 μm.

The microvillous brush borders in the midgut of cod larvae at 16 dph, reared under germ-free as well as conventional conditions, were well-defined and regular. Interestingly, germ-free cod larvae had significantly longer and significantly thicker microvilli than conventional cod larvae (for both analyses, Welch test, $p \leq 0.001$, Figure 7). Moreover, the abundance of microvilli ($\mu m^{-2}$) in the midgut was significantly lower in the germ-free larvae than in the conventional ones (Welch test, $p \leq 0.001$, insets in Figure 7).

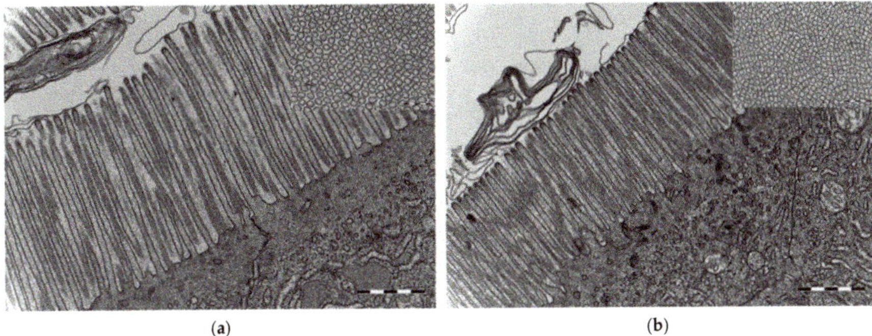

**Figure 7.** Transmission electron microscopy micrographs from the midgut of cod larvae at 16 dph reared under (**a**) germ-free and (**b**) conventional conditions. Microvilli in conventional cod larvae were shorter and thinner compared to the microvilli in germ-free cod larvae. Insets in (**a**,**b**): Transverse section of brush border—microvilli were closer to each other in the conventional than in the germ-free cod larvae. Scale bars are 1 μm and of insets 500 nm.

Microvilli were significantly closer to each other in the conventional than in the germ-free cod larvae. The morphometric measurements for the microvillous length, abundance and diameter are summarised in Table 1.

**Table 1.** Average length, diameter and abundance of microvilli in germ-free and conventional cod larvae at 16 dph. $n$ = 3 cod larvae per treatment were analysed.

| | **Treatment** | | | | | | | | |
|---|---|---|---|---|---|---|---|---|---|
| | **Microvillus Length (nm)** | | | **Microvillus Diameter (nm)** | | | **Abundance of Microvilli ($\mu m^{-2}$)** | | |
| | **Mean ± SE** | **Min** | **Max** | **Mean ± SE** | **Min** | **Max** | **Mean ± SE** | **Min** | **Max** |
| Germ-free | 2021.40 ± 31.62 | 1038.08 | 3183.59 | 107.01 ± 0.89 | 81.77 | 148.49 | 42.76 ± 1.03 | 25.39 | 68.43 |
| Conventional | 1703.04 ± 11.18 | 1211.07 | 2360.96 | 99.10 ± 0.53 | 77.51 | 123.92 | 54.14 ± 1.05 | 35.56 | 69.29 |

## 4. Discussion

In this study, gnotobiotic husbandry of Atlantic cod larvae was used as a tool to study host–microbe interactions. Cod larvae were reared under three different treatments: germ-free, gnotobiotic (*Vibrio gallicus* and *Microbacterium* added) and conventional. The only difference between the cod larvae in the conventional treatment and those in the gnotobiotic and germ-free treatments was that the microbial conditions in the conventional treatment were uncontrolled.

For the transcriptomic analysis, the major differences between the treatments were observed at 16 dph, and therefore the discussion is focused on this time point. At 16 dph, the gut of the larvae is larger and more developed, and thus the number of niches available for bacteria may be higher than at earlier life stages, and this may allow more bacteria to coexist through selection [40]. This was also reflected in our data, as the richness increased at this time point. Importantly, our analyses were based on pooled, homogenised whole fish, leading to conservative conclusions. Analyses at the organ level could possibly discover more differences in the gene expression patterns between the fish from the different treatments, also at earlier time points.

### 4.1. Microbial Environments

DGGE analyses of the microbiota from the gnotobiotic treatment showed that *V. gallicus* dominated both in water and fish samples, whereas *Microbacterium* was detectable at low levels in water samples but present in only some of the fish samples. This corroborated previous studies, where *V. gallicus* was found to adhere to and grow fast in mucus [25]. This could explain the higher abundance of *V. gallicus* compared to *Microbacterium* in the cod larvae in this experiment.

For the conventional treatment, we used 16S rDNA Illumina amplicon sequencing to characterise fish and water microbiota. The results show that the fish microbiota differed significantly from the water microbiota, corroborating earlier studies [33,41–43]. The water and larval microbiota were dominated by bacterial taxa considered to represent opportunistic, rapid-growing bacteria, such as Vibrionales, Actinomycetales, Alteromonadales and Flavobacteriales. This may be a result of r-selection in the water [44]. Pulses of organic matter originating from the addition of feed and fish defecation will create a high carrying capacity in the rearing bottles. When adding new water with lower carrying capacity to the rearing bottles, the microbe–microbe competition in the water is reduced, and this favoured growth of fast-growing, opportunistic species (r-strategists) [44]. Opportunistic bacteria could potentially be detrimental for the cod larvae, and this might be the reason for the lower survival observed in the conventional treatment than in the gnotobiotic and germ-free treatment. However, overall, the survival was good, and comparable to what is typically seen in first feeding experiments with cod larvae. In our dataset, 6 OTUs were classified as Vibrio. Using the Ribosomal database project (RDP) SeqMatch tool [30] to identify the most closely related type strains for each of them (Supplementary Table S5) showed that one OTU matched *V. campbellii* and one *V. anguillarum*. Both species are known fish pathogens [45].

Our results confirm that we had distinct microbial environments in the gnotobiotic and conventional treatments. The gnotobiotic treatment represents environmental conditions with non-detrimental host–microbe interactions. The conventional treatment, on the other hand, represents an environment characterised by the presence of opportunistic bacteria, which might have had detrimental effects on the fish. Thus, the conventional treatment does not represent a "natural" microbial environment, but rather a suboptimal microbial environment, including detrimental host–microbe interactions. This is reflected in the gene expression of the fish, and likely the reason why we see (1) increased expression of some genes related to inflammatory responses and oxidative stress, and (2) lower survival of conventionally reared fish than for the germ-free and gnotobiotic cod larvae.

### *4.2. Presence of Bacteria Downregulates Host Responses Related to Nutrient Utilisation and Innate Immune Responses*

Members of the gut microbiota in other species are well-known to aid in the digestion of, e.g., complex carbohydrates [46] and proteins [47]. Since the gut of cod larvae is functionally immature at hatching [48,49], the enzymatic activity of bacteria may aid in digestion of the live feed organisms. Even if survival was very high for the germ-free cod, and the intake of feed appeared similar in all treatments, they gained less weight than the conventional larvae. The transcriptional responses in the germ-free larvae support that they had difficulties in digesting feed, as the most enriched GO term in germ-free larvae compared with conventional larvae was proteolysis. Of the 21 genes included in the GO term "proteolysis", 5 were involved in the KEGG pathway of "protein digestion and absorption". This indicates that a large fraction of the "proteolysis" GO term is related to the host's digestion. Interestingly, germ-free rats also have higher activities of digestive enzymes, such as amylase and lipase, in their intestine than conventional rats [50]. Transcriptional responses related to fasting were downregulated in the conventional larvae compared with germ-free larvae. Host responses related to fasting were also found in the zebrafish study by Rawls et al. [16].

Similar to findings in germ-free rodents [51,52], the germ-free cod larvae showed significantly longer microvilli in the midgut than the conventional larvae (Figure 7, Table 1). This might be related to reduced renewal of the intestinal epithelium in these animals [52]. Similarly, albeit at a larger physiological scale, Willing et al. [53] found that germ-free pigs have longer villi and shorter crypts in their distal intestine, and that the shortening observed after colonisation was associated with increased cell turnover. They hypothesise that commensal bacteria contribute to enterocyte turnover through induction of inflammatory responses and cell apoptosis. In addition to the increased microvilli length, germ-free cod showed less microvilli per $\mu m^2$ in the midgut than conventional reared larvae. Microvilli not only increase the cellular surface area for absorption of nutrients, they also increase the number of digestive enzymes present on the cell surface. Thus, the lower density of microvilli in the germ-free larvae could explain the apparent reduced nutrient uptake in the larvae. The distorted mitochondria observed in the germ-free cod larvae (Figure 6) also support the hypothesis of a physiological starvation state in these larvae. Hailey et al. [54] demonstrated how lipids from the mitochondrial membrane are utilised in the biogenesis of autophagosomes under starvation conditions.

The gene expression analysis indicates that certain elements of the innate immune system of cod larvae are "turned on" in the larval stage of the fish but are subsequently regulated by host–microbiota interactions. Solbakken et al. [22] described how infection by a pathogenic bacterium, *Francisella noatunensis*, dampens the intracellular immune response to allow intracellular persistence of the pathogen. Our results indicate that generation of reactive oxygen species (ROS) was lower in conventional and gnotobiotic larvae than in germ-free larvae. ROS generation is tightly linked to mitochondrial metabolism, and thus could be elevated in the disintegrating mitochondria, as observed by electron microscopy analysis (Figure 6). ROS is also produced during recognition of non-self-substances and the immune response processes [55]. The main enriched GO terms in germ-free larvae were related to defence responses and responses to reactive oxygen species. This presumably higher ROS activity despite no bacteria present might be a consequence of the unique immune system of cod, or because phagocytes of germ-free larvae are activated by components of the live feed (such as algae). During infection by *F. noatunensis*, downregulation of ROS production in phagosomes is hypothesised to contribute to this pathogen's survival [22]. Another indication of high ROS generation in the germ-free cod was high expression of *immunoresponsive gene 1* (*irg1*). Irg1 controls macrophage function, by regulating metabolic pathways leading to increased mitochondrial ROS production that aids bacterial killing. In zebrafish, this gene was upregulated by bacterial infection [56], whereas in our experiment, the expression was lower in the fish exposed to bacteria.

Transcripts for key proteins involved in recognition of both peptidoglycans (PGLYRP) and lectins were also downregulated by the bacteria present in this study. Pglyrp proteins have been found to be expressed in zebrafish eggs, developing embryos and adult tissues that are in contact with the environment, and there are indications that they have an important role in the defence against bacteria in young fish [57,58]. Both rbl (Rhamnose-binding lectin) and fel (fish egg lectin) are known to enhance phagocytosis, and the expression of the genes is normally upregulated when the host is exposed to potentially harmful bacteria [59,60]. However, Thongda et al. [59] point to studies in catfish where *rbl* is highly upregulated by short-term fasting, indicating a link between the feeding status and the immune function. Thus, the downregulation of *rbl* in the gnotobiotic and conventional cod might be coupled to the observed transcriptional responses to fasting. Thus, enforcing the indications that the germ-free fish do not digest their food as well as the gnotobiotic and conventional fish.

It has previously been shown that the bacteria used in the gnotobiotic treatment are non-detrimental and may improve survival of cod larvae [25]. Thus, the downregulation of immune responses due to these bacteria may be a way of inducing tolerance to the colonising microbiota. However, similar downregulation observed in conventional cod larvae, where the microbiota was shown to have a slight but significantly negative effect on the survival, suggests that the immune system of the larvae is not capable of distinguishing friend from foe at this early life stage.

Several of the immune-related genes found to be downregulated in this experiment are involved in processes that have been reported to be upregulated by the presence of bacteria in gnotobiotic zebrafish and stickleback [16,61,62]. This illustrates how early host–microbiota responses differ across teleost species. In contrast to stickleback and zebrafish, Atlantic cod are heavily reliant on the innate immune response due to the loss of MHC-2, which is critical for initiation of antigen-specific immune response. In addition, cod toll-like receptor families (TLRs) have undergone genetic deletions and subsequent diversifications, possibly to compensate for the lack of the classical adaptive immunity [24,63]. This means that the type of receptors and downstream immune pathways will differ to some extent between cod and other vertebrates such as zebrafish and mice [64]. As an example, whereas LPS in high doses is lethal to mammals and zebrafish [61], Atlantic cod has a lower LPS response, with much higher LD50 values [64,65]. Analysis of the full genome revealed that Atlantic cod lacks TLR4, the mammalian LPS receptor that has a functional ortholog in zebrafish [21,66]. Despite the lack of a TLR4 ortholog, cod head kidney cells still reacted to LPS exposure by upregulation of certain immune and xenobiotic pathways [66].

### *4.3. Presence of Bateria Induces Host Responses Related to Inflammatory Responses and Signalling*

Even though several genes related to the innate immune system seem to be downregulated by the presence of bacteria, other innate immune system responses were induced by the presence of bacteria. *Cxcl8* (*interleukin 8*) and *lect2* (*leukocyte-cell derived chemotaxin 2*) (upregulated in both 13 and 16 dph larvae) were upregulated in both conventional and gnotobiotic larvae (*cxcl8* just below the cut-off value of log2 0.8-fold change in gnotobiotic larvae). These genes that encode chemokines that attract neutrophils by chemotactic activity are induced by inflammatory stimuli caused by, e.g., microbial stress, and have been identified in teleosts earlier [67–70]. Both genes were also upregulated by in vitro LPS stimulation of cod cell cultures [66]. In fish, *lect2* is assumed to have an important role in the inflammatory response, promoting phagocytic activity of macrophages [71,72]. Since *lect2* was upregulated at both 13 and 16 dph, we suspect that this gene may be an early and important bacterial response in cod.

The KEGG pathway "bacterial invasion of epithelial cells" and GO terms related to signalling and signal transduction were enriched in conventional larvae compared with both gnotobiotic and germ-free larvae. This seems plausible, as compared to a gnotobiotic community consisting of only two probiotic candidates, the microbiota in the conventional treatment could give a higher invasion pressure, more host–microbe interactions due to higher species richness and thus more signalling and host responses. For example, a gene linked to the innate immune system, *irak1* (*interleukin-1 receptor-associated kinase 1*), was upregulated in conventional larvae compared with germ-free larvae. This gene plays a critical role in initiating the innate immune response against pathogens and has been shown to be upregulated in fish after pathogen challenge [73]. This implies that *irak1* might participate in antibacterial immunity. In gnotobiotic fish, the expression of *irak1* was at the same level as in germ-free fish, indicating that the probiotic candidates used in the gnotobiotic treatment were detected as non-pathogens by the fish.

*Slc2a11* was the most upregulated gene (log2 fold change of 2.3) in conventional compared with germ-free larvae, and it was also upregulated in conventional compared with gnotobiotic larvae. This gene encodes a glucose transporter protein GLUT11, belonging to class II of these proteins [74]. In fish, glucose transport is important for several reasons: blood glucose levels change rapidly in response to environmental disturbances, increased plasma glucose levels may be an indicator of stress and glucose intolerance has been documented in fish [75]. The expression of another GLUT protein, GLUT4, belonging to class I of these proteins and encoded by the *slc2a4* gene, was also upregulated in conventional fish compared to the germ-free fish. This is the only insulin-sensitive member of class I, and it is expressed in insulin-sensitive tissues, such as heart, muscle and adipose tissue in cod [76]. These findings indicate that the microbial community in the conventional rearing bottles induced more stress on the fish than in the gnotobiotic treatment. This is in line with the fact that potential detrimental bacteria were found in the conventional rearing bottles, and that the bacterial strains added to the gnotobiotic flasks had been characterised as probiotic candidates in an earlier study [25].

## 5. Conclusions

To conclude, our results indicate that bacteria actively downregulate certain cod larvae immune responses, facilitating bacterial colonisation of mucosal surfaces. This concept of "downregulation" contrasts previous findings in zebrafish and stickleback and emphasizes the role of evolutionary history, highlighting the need to study host–microbe interactions in several teleost species. Similar to what has been shown in studies with other vertebrates, bacterial colonisation improved the nutritional state of the cod larvae, evident at both transcriptional and micromorphological levels and materialised as differences in growth rate. This study illustrates the dynamics between water- and host-associated microbiota and increases our insight into how Atlantic cod larvae respond physiologically and transcriptionally to bacterial colonisation.

**Supplementary Materials:** The following are available online at https://www.mdpi.com/article/10.3390/microorganisms10010024/s1, Figure S1: Kaplan–Meier survival curves, Figure S2: Dry weight of cod larvae, Figure S3: DGGE gel, Figure S4: Venn diagram, Table S1: List of down-regulated genes (cut-off value of log2 fold change > 0.8) in conventional cod larvae compared with germ-free cod larvae at 16 dph. Table S2: List of up-regulated genes (cut-off value of log2 fold change > 0.8) in conventional cod larvae compared with germ-free cod larvae at 16 dph. Table S3: List of down-regulated genes (cut-off value of log2 fold change > 0.8) in gnotobiotic cod larvae compared with germ-free cod larvae at 16 dph. Table S4: Enriched GO terms in conventional larvae compared with both germ-free and gnotobiotic cod larvae. Table S5: Taxonomy and most similar "type strains" for the six OTUs classified as Vibrio as inferred from the Ribosomal Database Project (RDP). Text S1: Verification of germ-free conditions, Text S2: Live feed and bacterial cultures, Text S3: Sampling procedures, Text S4: Statistical analyses.

**Author Contributions:** Conceptualisation, T.F. and O.V.; formal analysis, T.F., R.I.V., B.L. and P.W.; investigation, T.F., R.I.V. and B.L.; writing—original draft preparation, R.I.V., T.F. and B.L.; writing—review and editing, R.I.V., T.F., B.L., O.V., I.B., Y.O. and P.W.; visualisation, R.I.V.; supervision, O.V. and I.B.; project administration, O.V., T.F. and R.I.V.; funding acquisition, O.V., T.F. and B.L. All authors have read and agreed to the published version of the manuscript.

**Funding:** This study was performed with funding from the Faculty of Natural Sciences at the NTNU—Norwegian University of Science and Technology, the Norwegian Research Council (HAVBRUK, 233865), and a People Marie Curie Actions International Incoming Fellowship FP7-PEOPLE-2013-IIF (#625655).

**Institutional Review Board Statement:** The study was conducted according to the guidelines of the Declaration of Helsinki, and approved by the Ethics Committee of Norwegian Animal Facility Establishment 154 – NTNU – Sealab (protocol code 4174, 5 March 2012).

**Data Availability Statement:** Sequencing data are stored at European Nucleotide Archive, accession numbers ERS8484975-ERS8484994.

**Acknowledgments:** We thank Mari-Ann Østensen for invaluable help in conducting this experiment. Thanks to Tora Bardal and Rolf Erik Olsen for valuable training and input. The TEM work was performed at the Cellular and Molecular Imaging Core Facility (CMIC), Norwegian University of Science and Technology (NTNU). CMIC is funded by the Faculty of Medicine at NTNU and the Central Norway Regional Health Authority.

**Conflicts of Interest:** The authors declare no conflict of interest. The funders had no role in the design of the study; in the collection, analysis, or interpretation of data; in the writing of the manuscript, or in the decision to publish the results.

## References

1. Hiippala, K.; Jouhten, H.; Ronkainen, A.; Hartikainen, A.; Kainulainen, V.; Jalanka, J.; Satokari, R. The potential of gut commensals in reinforcing intestinal barrier function and alleviating inflammation. *Nutrients* **2018**, *10*, 988. [CrossRef] [PubMed]
2. Lazado, C.C.; Caipang, C.M.A. Probiotics–pathogen interactions elicit differential regulation of cutaneous immune responses in epidermal cells of Atlantic cod *Gadus morhua*. *Fish Shellfish Immunol.* **2014**, *36*, 113–119. [CrossRef] [PubMed]
3. Lazado, C.C.; Caipang, C.M.A.; Brinchmann, M.F.; Kiron, V. In vitro adherence of two candidate probiotics from Atlantic cod and their interference with the adhesion of two pathogenic bacteria. *Vet. Microbiol.* **2011**, *148*, 252–259. [CrossRef]
4. Kanther, M.; Rawls, J.F. Host–microbe interactions in the developing zebrafish. *Curr. Opin. Immunol.* **2010**, *22*, 10–19. [CrossRef]
5. Naito, T.; Mulet, C.; De Castro, C.; Molinaro, A.; Saffarian, A.; Nigro, G.; Bérard, M.; Clerc, M.; Pedersen, A.B.; Sansonetti, P.J.; et al. Lipopolysaccharide from crypt-specific core microbiota modulates the colonic epithelial proliferation-to-differentiation balance. *mBio* **2017**, *8*, e01680-17. [CrossRef] [PubMed]
6. Semova, I.; Carten, J.D.; Stombaugh, J.; Mackey, L.C.; Knight, R.; Farber, S.A.; Rawls, J.F. Microbiota Regulate Intestinal Absorption and Metabolism of Fatty Acids in the Zebrafish. *Cell Host Microbe* **2012**, *12*, 277–288. [CrossRef]
7. Sonnenburg, J.L.; Bäckhed, F. Diet–microbiota interactions as moderators of human metabolism. *Nature* **2016**, *535*, 56. [CrossRef] [PubMed]
8. Ringø, E.; Birkbeck, T. Intestinal microflora of fish larvae and fry. *Aquac. Res.* **1999**, *30*, 73–93. [CrossRef]
9. Vadstein, O.; Bergh, Ø.; Gatesoupe, F.-J.; Galindo-Villegas, J.; Mulero, V.; Picchietti, S.; Scapigliati, G.; Makridis, P.; Olsen, Y.; Dierckens, K.; et al. Microbiology and immunology of fish larvae. *Rev. Aquac.* **2013**, *5*, S1–S25. [CrossRef]
10. Olafsen, J.A. Interactions between fish larvae and bacteria in marine aquaculture. *Aquaculture* **2001**, *200*, 223–247. [CrossRef]

11. Vadstein, O.; Mo, T.A.; Bergh, Ø. Microbial Interactions, Prophylaxis and Diseases. In *Culture of Cold-Water Marine Fish*; Moksness, E., Kjørsvik, E., Olsen, Y., Eds.; Blackwell Publishing Ltd.: Hoboken, NJ, USA, 2003; pp. 28–72. [CrossRef]
12. Toranzo, A.E.; Magariños, B.; Romalde, J.L. A review of the main bacterial fish diseases in mariculture systems. *Aquaculture* **2005**, *246*, 37–61. [CrossRef]
13. Forberg, T.; Milligan-Myhre, K. Chapter 6—Gnotobiotic Fish as Models to Study Host–Microbe Interactions. In *Gnotobiotics*; Schoeb, T.R., Eaton, K.A., Eds.; Academic Press: Cambridge, MA, USA, 2017; pp. 369–383. [CrossRef]
14. Smith, K.; McCoy, K.D.; Macpherson, A.J. Use of axenic animals in studying the adaptation of mammals to their commensal intestinal microbiota. *Semin. Immunol.* **2007**, *19*, 59–69. [CrossRef] [PubMed]
15. Pham, L.N.; Kanther, M.; Semova, I.; Rawls, J.F. Methods for generating and colonizing gnotobiotic zebrafish. *Nat. Protoc.* **2008**, *3*, 1862. [CrossRef] [PubMed]
16. Rawls, J.F.; Samuel, B.S.; Gordon, J.I. Gnotobiotic zebrafish reveal evolutionarily conserved responses to the gut microbiota. *Proc. Natl. Acad. Sci. USA* **2004**, *101*, 4596–4601. [CrossRef]
17. Uribe, A.; Alam, M.; Midtvedt, T.; Smedfors, B.; Theodorsson, E. Endogenous Prostaglandins and Microflora Modulate DNA Synthesis and Neuroendocrine Peptides in the Rat Gastrointestinal Tract. *Scand. J. Gastroenterol.* **1997**, *32*, 691–699. [CrossRef]
18. Forberg, T.; Arukwe, A.; Vadstein, O. A protocol and cultivation system for gnotobiotic Atlantic cod larvae (*Gadus morhua* L.) as a tool to study host microbe interactions. *Aquaculture* **2011**, *315*, 222–227. [CrossRef]
19. Forberg, T.; Vestrum, R.I.; Arukwe, A.; Vadstein, O. Bacterial composition and activity determines host gene-expression responses in gnotobiotic Atlantic cod (*Gadus morhua*) larvae. *Vet. Microbiol.* **2012**, *157*, 420–427. [CrossRef]
20. Forberg, T.; Arukwe, A.; Vadstein, O. Two strategies to unravel gene expression responses of host–microbe interactions in cod (*Gadus morhua*) larvae. *Aquac. Res.* **2011**, *42*, 664–676. [CrossRef]
21. Star, B.; Nederbragt, A.J.; Jentoft, S.; Grimholt, U.; Malmstrom, M.; Gregers, T.F.; Rounge, T.B.; Paulsen, J.; Solbakken, M.H.; Sharma, A.; et al. The genome sequence of Atlantic cod reveals a unique immune system. *Nature* **2011**, *477*, 207–210. [CrossRef]
22. Solbakken, M.H.; Jentoft, S.; Reitan, T.; Mikkelsen, H.; Gregers, T.F.; Bakke, O.; Jakobsen, K.S.; Seppola, M. Disentangling the immune response and host-pathogen interactions in *Francisella noatunensis* infected Atlantic cod. *Comp. Biochem. Physiol. Part D Genom. Proteom.* **2019**, *30*, 333–346. [CrossRef]
23. Tørresen, O.K.; Rise, M.L.; Jin, X.; Star, B.; MacKenzie, S.; Jakobsen, K.S.; Nederbragt, A.J.; Jentoft, S. An improved version of the Atlantic cod genome and advancements in functional genomics: Implications for the future of cod farming. In *Genomics in Aquaculture*; Academic Press: Cambridge, MA, USA, 2016; pp. 45–72. [CrossRef]
24. Star, B.; Jentoft, S. Why does the immune system of Atlantic cod lack MHC II? *BioEssays* **2012**, *34*, 648–651. [CrossRef]
25. Fjellheim, A.J.; Klinkenberg, G.; Skjermo, J.; Aasen, I.M.; Vadstein, O. Selection of candidate probionts by two different screening strategies from Atlantic cod (*Gadus morhua* L.) larvae. *Vet. Microbiol.* **2010**, *144*, 153–159. [CrossRef] [PubMed]
26. Bakke, I.; De Schryver, P.; Boon, N.; Vadstein, O. PCR-based community structure studies of Bacteria associated with eukaryotic organisms: A simple PCR strategy to avoid co-amplification of eukaryotic DNA. *J. Microbiol. Methods* **2011**, *84*, 349–351. [CrossRef] [PubMed]
27. Bakke, I.; Skjermo, J.; Vo, T.A.; Vadstein, O. Live feed is not a major determinant of the microbiota associated with cod larvae (*Gadus morhua*). *Environ. Microbiol. Rep.* **2013**, *5*, 537–548. [CrossRef]
28. Vestrum, R.I.; Attramadal, K.J.K.; Vadstein, O.; Gundersen, M.S.; Bakke, I. Bacterial community assembly in Atlantic cod larvae (*Gadus morhua*): Contributions of ecological processes and metacommunity structure. *FEMS Microbiol. Ecol.* **2020**, *96*. [CrossRef]
29. Caporaso, J.G.; Lauber, C.L.; Walters, W.A.; Berg-Lyons, D.; Lozupone, C.A.; Turnbaugh, P.J.; Fierer, N.; Knight, R. Global patterns of 16S rRNA diversity at a depth of millions of sequences per sample. *Proc. Natl. Acad. Sci. USA* **2011**, *108*, 4516–4522. [CrossRef]
30. Wang, Q.; Garrity, G.M.; Tiedje, J.M.; Cole, J.R. Naïve Bayesian Classifier for Rapid Assignment of rRNA Sequences into the New Bacterial Taxonomy. *Appl. Environ. Microbiol.* **2007**, *73*, 5261–5267. [CrossRef]
31. Mollerup, S.; Friis-Nielsen, J.; Vinner, L.; Hansen, T.A.; Richter, S.R.; Fridholm, H.; Herrera, J.A.R.; Lund, O.; Brunak, S.; Izarzugaza, J.M.G.; et al. Propionibacterium acnes: Disease-Causing Agent or Common Contaminant? Detection in Diverse Patient Samples by Next-Generation Sequencing. *J. Clin. Microbiol.* **2016**, *54*, 980. [CrossRef]
32. Kleppe, L.; Edvardsen, R.B.; Furmanek, T.; Taranger, G.L.; Wargelius, A. Global transcriptome analysis identifies regulated transcripts and pathways activated during oogenesis and early embryogenesis in atlantic cod. *Mol. Reprod. Dev.* **2014**, *81*, 619–635. [CrossRef]
33. Vestrum, R.I.; Attramadal, K.J.K.; Winge, P.; Li, K.; Olsen, Y.; Bones, A.M.; Vadstein, O.; Bakke, I. Rearing water treatment induces microbial selection influencing the microbiota and pathogen associated transcripts of cod (*Gadus morhua*) larvae. *Front. Microbiol.* **2018**, *9*, 851. [CrossRef]
34. Huang, D.W.; Sherman, B.T.; Lempicki, R.A. Bioinformatics enrichment tools: Paths toward the comprehensive functional analysis of large gene lists. *Nucleic Acids Res.* **2008**, *37*, 1–13. [CrossRef]
35. Huang, D.W.; Sherman, B.T.; Zheng, X.; Yang, J.; Imamichi, T.; Stephens, R.; Lempicki, R.A. Extracting biological meaning from large gene lists with DAVID. *Curr. Protoc. Bioinform.* **2009**, *27*, 13.11.1–13.11.13. [CrossRef] [PubMed]
36. Galloway, T.F.; Kjørsvik, E.; Kryvi, H. Effect of temperature on viability and axial muscle development in embryos and yolk sac larvae of the Northeast Arctic cod (*Gadus morhua*). *Mar. Biol.* **1998**, *132*, 559–567. [CrossRef]
37. Brown, A.L., Jr. Microvilli of the human jejunal epithelial cell. *J. Cell Biol.* **1962**, *12*, 623–627. [CrossRef] [PubMed]

38. Hammer, Ø.; Harper, D.A.T.; Ryan, P.D. past: Paleontological statistics software package for education and data analysis. *Palaeontol. Electron.* **2001**, *4*, 9.
39. Bardou, P.; Mariette, J.; Escudié, F.; Djemiel, C.; Klopp, C. jvenn: An interactive Venn diagram viewer. *BMC Bioinform.* **2014**, *15*, 293. [CrossRef]
40. Nemergut, D.R.; Schmidt, S.K.; Fukami, T.; Neill, S.P.; Bilinski, T.M.; Stanish, L.F.; Knelman, J.E.; Darcy, J.L.; Lynch, R.C.; Wickey, P.; et al. Patterns and Processes of Microbial Community Assembly. *Microbiol. Mol. Biol. Rev.* **2013**, *77*, 342. [CrossRef]
41. Giatsis, C.; Sipkema, D.; Smidt, H.; Heilig, H.; Benvenuti, G.; Verreth, J.; Verdegem, M. The impact of rearing environment on the development of gut microbiota in tilapia larvae. *Sci. Rep.* **2015**, *5*, 18206. [CrossRef]
42. Califano, G.; Castanho, S.; Soares, F.; Ribeiro, L.; Cox, C.J.; Mata, L.; Costa, R. Molecular Taxonomic Profiling of Bacterial Communities in a Gilthead Seabream (*Sparus aurata*) Hatchery. *Front. Microbiol.* **2017**, *8*, 204. [CrossRef]
43. Giatsis, C.; Sipkema, D.; Smidt, H.; Verreth, J.; Verdegem, M. The colonization dynamics of the gut microbiota in tilapia larvae. *PLoS ONE* **2014**, *9*, e103641. [CrossRef]
44. Vadstein, O.; Attramadal, K.J.K.; Bakke, I.; Olsen, Y. K-Selection as Microbial Community Management Strategy: A Method for Improved Viability of Larvae in Aquaculture. *Front. Microbiol.* **2018**, *9*, 2730. [CrossRef]
45. Ina-Salwany, M.Y.; Al-saari, N.; Mohamad, A.; Mursidi, F.A.; Mohd-Aris, A.; Amal, M.N.A.; Kasai, H.; Mino, S.; Sawabe, T.; Zamri-Saad, M. Vibriosis in Fish: A Review on Disease Development and Prevention. *J. Aquat. Anim. Health* **2019**, *31*, 3–22. [CrossRef] [PubMed]
46. Sonnenburg, E.D.; Sonnenburg, J.L. Starving our Microbial Self: The Deleterious Consequences of a Diet Deficient in Microbiota-Accessible Carbohydrates. *Cell Metab.* **2014**, *20*, 779–786. [CrossRef] [PubMed]
47. Lazado, C.C.; Caipang, C.M.A.; Kiron, V. Enzymes from the gut bacteria of Atlantic cod, *Gadus morhua* and their influence on intestinal enzyme activity. *Aquac. Nutr.* **2012**, *18*, 423–431. [CrossRef]
48. Kjørsvik, E.; Pittman, K.; Pavlov, D. From fertilisation to the end of metamorphosis—Functional development. In *Culture of Cold-Water Marine Fish*; Moksness, E., Kjørsvik, E., Olsen, Y., Eds.; Blackwell Publishing: Hoboken, NJ, USA, 2004. [CrossRef]
49. Kjørsvik, E.; van der Meeren, T.; Kryvi, H.; Arnfinnson, J.; Kvenseth, P.G. Early development of the digestive tract of cod larvae, *Gadus morhua* L., during start-feeding and starvation. *J. Fish Biol.* **1991**, *38*, 1–15. [CrossRef]
50. Reddy, B.S.; Pleasants, J.R.; Wostmann, B.S. Pancreatic Enzymes in Germfree and Conventional Rats Fed Chemically Defined, Water-soluble Diet Free from Natural Substrates. *J. Nutr.* **1969**, *97*, 327–334. [CrossRef] [PubMed]
51. Hayes, C.L.; Dong, J.; Galipeau, H.J.; Jury, J.; McCarville, J.; Huang, X.; Wang, X.Y.; Naidoo, A.; Anbazhagan, A.N.; Libertucci, J.; et al. Commensal microbiota induces colonic barrier structure and functions that contribute to homeostasis. *Sci. Rep.* **2018**, *8*, 14184. [CrossRef] [PubMed]
52. Meslin, J.C.; Sacquet, E.; Delpal, S. Effects of microflora on the dimensions of enterocyte microvilli in the rat. *Reprod. Nutr. Développement* **1984**, *24*, 307–314. [CrossRef]
53. Willing, B.P.; Van Kessel, A.G. Enterocyte proliferation and apoptosis in the caudal small intestine is influenced by the composition of colonizing commensal bacteria in the neonatal gnotobiotic pig. *J. Anim. Sci.* **2007**, *85*, 3256–3266. [CrossRef] [PubMed]
54. Hailey, D.W.; Rambold, A.S.; Satpute-Krishnan, P.; Mitra, K.; Sougrat, R.; Kim, P.K.; Lippincott-Schwartz, J. Mitochondria Supply Membranes for Autophagosome Biogenesis during Starvation. *Cell* **2010**, *141*, 656–667. [CrossRef]
55. Kohchi, C.; Inagawa, H.; Nishizawa, T.; Soma, G.-I. ROS and innate immunity. *Anticancer Res.* **2009**, *29*, 817–821. [PubMed]
56. Hall, C.J.; Boyle, R.H.; Astin, J.W.; Flores, M.V.; Oehlers, S.H.; Sanderson, L.E.; Ellett, F.; Lieschke, G.J.; Crosier, K.E.; Crosier, P.S. *Immunoresponsive Gene 1* Augments Bactericidal Activity of Macrophage-Lineage Cells by Regulating β-Oxidation-Dependent Mitochondrial ROS Production. *Cell Metab.* **2013**, *18*, 265–278. [CrossRef] [PubMed]
57. Li, X.; Wang, S.; Qi, J.; Echtenkamp, S.F.; Chatterjee, R.; Wang, M.; Boons, G.J.; Dziarski, R.; Gupta, D. Zebrafish peptidoglycan recognition proteins are bactericidal amidases essential for defense against bacterial infections. *Immunity* **2007**, *27*, 518–529. [CrossRef]
58. Royet, J.; Gupta, D.; Dziarski, R. Peptidoglycan recognition proteins: Modulators of the microbiome and inflammation. *Nat. Rev. Immunol.* **2011**, *11*, 837–851. [CrossRef] [PubMed]
59. Thongda, W.; Li, C.; Luo, Y.; Beck, B.H.; Peatman, E. L-Rhamnose-binding lectins (RBLs) in channel catfish, *Ictalurus punctatus*: Characterization and expression profiling in mucosal tissues. *Dev. Comp. Immunol.* **2014**, *44*, 320–331. [CrossRef] [PubMed]
60. Vasta, G.R.; Nita-Lazar, M.; Giomarelli, B.; Ahmed, H.; Du, S.; Cammarata, M.; Parrinello, N.; Bianchet, M.A.; Amzel, L.M. Structural and functional diversity of the lectin repertoire in teleost fish: Relevance to innate and adaptive immunity. *Dev. Comp. Immunol.* **2011**, *35*, 1388–1399. [CrossRef]
61. Bates, J.M.; Akerlund, J.; Mittge, E.; Guillemin, K. Intestinal Alkaline Phosphatase Detoxifies Lipopolysaccharide and Prevents Inflammation in Zebrafish in Response to the Gut Microbiota. *Cell Host Microbe* **2007**, *2*, 371–382. [CrossRef]
62. Milligan-Myhre, K.; Small, C.M.; Mittge, E.K.; Agarwal, M.; Currey, M.; Cresko, W.A.; Guillemin, K. Innate immune responses to gut microbiota differ between oceanic and freshwater threespine stickleback populations. *Dis. Models Mech.* **2016**, *9*, 187–198. [CrossRef]
63. Solbakken, M.H.; Tørresen, O.K.; Nederbragt, A.J.; Seppola, M.; Gregers, T.F.; Jakobsen, K.S.; Jentoft, S. Evolutionary redesign of the Atlantic cod (*Gadus morhua* L.) Toll-like receptor repertoire by gene losses and expansions. *Sci. Rep.* **2016**, *6*, 25211. [CrossRef]

64. Seppola, M.; Mikkelsen, H.; Johansen, A.; Steiro, K.; Myrnes, B.; Nilsen, I.W. Ultrapure LPS induces inflammatory and antibacterial responses attenuated in vitro by exogenous sera in Atlantic cod and Atlantic salmon. *Fish Shellfish Immunol.* **2015**, *44*, 66–78. [CrossRef]
65. Bakkemo, K.R.; Mikkelsen, H.; Bordevik, M.; Torgersen, J.; Winther-Larsen, H.C.; Vanberg, C.; Olsen, R.; Johansen, L.H.; Seppola, M. Intracellular localisation and innate immune responses following Francisella noatunensis infection of Atlantic cod (*Gadus morhua*) macrophages. *Fish Shellfish Immunol.* **2011**, *31*, 993–1004. [CrossRef]
66. Holen, E.; Lie, K.K.; Araujo, P.; Olsvik, P.A. Pathogen recognition and mechanisms in Atlantic cod (*Gadus morhua*) head kidney cells: Bacteria (LPS) and virus (poly I:C) signals through different pathways and affect distinct genes. *Fish Shellfish Immunol.* **2012**, *33*, 267–276. [CrossRef] [PubMed]
67. Brugman, S.; Witte, M.; Scholman, R.C.; Klein, M.R.; Boes, M.; Nieuwenhuis, E.E.S. T Lymphocyte–Dependent and–Independent Regulation of *Cxcl8* Expression in Zebrafish Intestines. *J. Immunol.* **2014**, *192*, 484–491. [CrossRef]
68. Chai, Y.; Cong, B.; Yu, S.; Liu, Y.; Man, X.; Wang, L.; Zhu, Q. Effect of a LECT2 on the immune response of peritoneal lecukocytes against *Vibrio anguillarum* in roughskin sculpin. *Fish Shellfish Immunol.* **2018**, *74*, 620–626. [CrossRef] [PubMed]
69. Oehlers, S.H.B.; Flores, M.V.; Hall, C.J.; O'Toole, R.; Swift, S.; Crosier, K.E.; Crosier, P.S. Expression of zebrafish *cxcl8* (interleukin-8) and its receptors during development and in response to immune stimulation. *Dev. Comp. Immunol.* **2010**, *34*, 352–359. [CrossRef]
70. Van Der Aa, L.M.; Chadzinska, M.; Tijhaar, E.; Boudinot, P.; Verburg-van Kemenade, B.M.L. CXCL8 chemokines in teleost fish: Two lineages with distinct expression profiles during early phases of inflammation. *PLoS ONE* **2010**, *5*, e12384. [CrossRef] [PubMed]
71. Lin, B.; Chen, S.; Cao, Z.; Lin, Y.; Mo, D.; Zhang, H.; Gu, J.; Dong, M.; Liu, Z.; Xu, A. Acute phase response in zebrafish upon *Aeromonas salmonicida* and *Staphylococcus aureus* infection: Striking similarities and obvious differences with mammals. *Mol. Immunol.* **2007**, *44*, 295–301. [CrossRef] [PubMed]
72. Shi, X.; Zhang, Z.; Qu, M.; Ding, S.; Zheng, L. Genomic organization, promoter characterization and expression analysis of the leukocyte cell-derived chemotaxin-2 gene in *Epinephelus akaraa. Fish Shellfish Immunol.* **2012**, *32*, 1041–1050. [CrossRef]
73. Shan, S.J.; Liu, D.Z.; Wang, L.; Zhu, Y.Y.; Zhang, F.M.; Li, T.; An, L.G.; Yang, G.W. Identification and expression analysis of irak1 gene in common carp *Cyprinus carpio* L.: Indications for a role of antibacterial and antiviral immunity. *J. Fish Biol.* **2015**, *87*, 241–255. [CrossRef]
74. Scheepers, A.; Schmidt, S.; Manolescu, A.; Cheeseman, C.I.; Bell, A.; Zahn, C.; Joost, H.G.; Schürmann, A. Characterization of the human SLC2A11 (GLUT11) gene: Alternative promoter usage, function, expression, and subcellular distribution of three isoforms, and lack of mouse orthologue. *Mol. Membr. Biol.* **2005**, *22*, 339–351. [CrossRef]
75. Tseng, Y.C.; Hwang, P.P. Some insights into energy metabolism for osmoregulation in fish. *Comp. Biochem. Physiol. Part C Toxicol. Pharmacol.* **2008**, *148*, 419–429. [CrossRef] [PubMed]
76. Kamalam, B.S.; Medale, F.; Panserat, S. Utilisation of dietary carbohydrates in farmed fishes: New insights on influencing factors, biological limitations and future strategies. *Aquaculture* **2017**, *467*, 3–27. [CrossRef]

*Article*

# Microbial Shift in the Enteric Bacteriome of Coral Reef Fish Following Climate-Driven Regime Shifts

Marie-Charlotte Cheutin [1,*], Sébastien Villéger [1], Christina C. Hicks [2], James P. W. Robinson [2], Nicholas A. J. Graham [2], Clémence Marconnet [1], Claudia Ximena Ortiz Restrepo [1], Yvan Bettarel [1], Thierry Bouvier [1] and Jean-Christophe Auguet [1]

1 UMR MARBEC, Université de Montpellier, CNRS, Ifremer, IRD, 34095 Montpellier, France; sebastien.villeger@cnrs.fr (S.V.); marconnet.clemence@gmail.com (C.M.); claudiaximenaro@gmail.com (C.X.O.R.); yvan.bettarel@ird.fr (Y.B.); thierry.bouvier@cnrs.fr (T.B.); Jean-christophe.AUGUET@cnrs.fr (J.-C.A.)

2 Lancaster Environment Centre, Lancaster University, Lancaster LA1 4YQ, UK; christina.hicks@lancaster.ac.uk (C.C.H.); james.robinson@lancaster.ac.uk (J.P.W.R.); nick.graham@lancaster.ac.uk (N.A.J.G.)

* Correspondence: mccheutin@gmail.com

**Abstract:** Replacement of coral by macroalgae in post-disturbance reefs, also called a "coral-macroalgal regime shift", is increasing in response to climate-driven ocean warming. Such ecosystem change is known to impact planktonic and benthic reef microbial communities but few studies have examined the effect on animal microbiota. In order to understand the consequence of coral-macroalgal shifts on the coral reef fish enteric bacteriome, we used a metabarcoding approach to examine the gut bacteriomes of 99 individual fish representing 36 species collected on reefs of the Inner Seychelles islands that, following bleaching, had either recovered to coral domination, or shifted to macroalgae. While the coral-macroalgal shift did not influence the diversity, richness or variability of fish gut bacteriomes, we observed a significant effect on the composition ($R2 = 0.02$; $p = 0.001$), especially in herbivorous fishes ($R2 = 0.07$; $p = 0.001$). This change is accompanied by a significant increase in the proportion of fermentative bacteria (*Rikenella, Akkermensia, Desulfovibrio, Brachyspira*) and associated metabolisms (carbohydrates metabolism, DNA replication, and nitrogen metabolism) in relation to the strong turnover of *Scarinae* and *Siganidae* fishes. Predominance of fermentative metabolisms in fish found on macroalgal dominated reefs indicates that regime shifts not only affect the taxonomic composition of fish bacteriomes, but also have the potential to affect ecosystem functioning through microbial functions.

**Keywords:** coral-macroalgal shift; coral reef fish; enteric bacteriome; microbial functions; barcoding

**Citation:** Cheutin, M.-C.; Villéger, S.; Hicks, C.C.; Robinson, J.P.W.; Graham, N.A.J.; Marconnet, C.; Restrepo, C.X.O.; Bettarel, Y.; Bouvier, T.; Auguet, J.-C. Microbial Shift in the Enteric Bacteriome of Coral Reef Fish Following Climate-Driven Regime Shifts. *Microorganisms* **2021**, *9*, 1711. https://doi.org/10.3390/microorganisms9081711

Academic Editor: Ulrich (Uli) Stingl

Received: 13 July 2021
Accepted: 6 August 2021
Published: 11 August 2021

**Publisher's Note:** MDPI stays neutral with regard to jurisdictional claims in published maps and institutional affiliations.

## 1. Introduction

Coral reefs have increasingly been subject to critical disturbances leading to a decrease of coral cover [1], a loss of coral habitat biodiversity [2], and to a reduction in associated ecosystem services [3,4]. Among the multiple stressors driving reef ecosystem decline, sea surface warming is responsible for severe bleaching events worldwide and the subsequent mortality of corals. In addition to climatic anomalies, overexploitation of herbivore fishes and nutrient discharges derived from land run-off can reduce coral cover and enhance the proliferation of macroalgae [5,6]. Indeed "coral-macroalgae regime shifts" are frequent in post-disturbance reefs [7,8]. Increase in macroalgal cover affects the resilience of coral reefs by reducing the survival and growth of adult corals [9], and/or preventing the recruitment of juvenile corals [10]. Macroalgae also produce secondary metabolites that can induce the growth of pathogenic and fouling microorganisms, causing a physiological deterioration of the coral tissues [11] and a dysbiosis in their microbiome [12]. This shift is not only dramatic for coral fitness, but it also impacts the assemblage composition and trophic structure of the entire coral habitat [13,14] and endangers associated ecosystem services (i.e., protection of coastal communities against storms, provision of protein through reef fisheries, and generation of tourism related incomes) [3,15,16].

Among coral reef biota, fishes play a well-known central role in coral-macroalgae regime shifts since the loss of herbivorous fishes through overfishing is considered as one of the causes of dominance by macroalgae [5]. Changes in composition and abundance of fish assemblages related to coral-macroalgae regime shifts are well understood, leaving gaps in knowledge about the impact of macroalgal dominance on other ecological traits of fishes, such as their microbiota. The great diversity of coral reef fishes, with more than 6000 species described [17], combined with the high diversity of their biological traits, provide specific ecological niches both on their skin and within their bodies, which ultimately promote the development of taxonomically and functionally original microbial lineages compared to the surrounding environment [18,19]. In a recent study, Chiarello et al. (2020) [19] showed with a conservative estimation that coral reef animal microbiota may account for up to 2.5% of Earth's prokaryotic diversity, representing a hotspot of microbial diversity.

While our understanding of some components of fish microbiota such as viruses, archaea, and protists remain limited, their bacteriome has been more extensively studied in the recent years [18–22]. Most of these bacteria reside in the intestinal tract where they form complex communities and provide a range of essential functions linked to development, immunity, health, protection against pathogen invasion, and even influence behavior [23–25]. However, the most obvious and important role is the contribution of the fish bacteriome to the degradation and assimilation of large and complex molecules [20,26,27]. Evidence has accumulated that the gut bacteriome is not just a random set of microorganisms, but rather a highly variable community depending upon intrinsic fish factors such as diet or genetic background and extrinsic environmental conditions [18,21,22,28,29]. Nonetheless, our understanding of fish bacteriome variability is still scarce compared to terrestrial animals and is even more rare concerning coral reef fishes [30,31]. For example, environmental degradation or modification, such as urban sprawl and captivity, are known to have dramatic consequences on the enteric microbiome by altering the diet of wild animals, and thus impacts host fitness as observed in black howler [32,33] and other vertebrates [34–36]. Whether this impact also takes place in marine animals, and particularly the fish enteric microbiome, is poorly documented. The influence of coral bleaching induced regime shifts on coral reef fish bacteriomes remains unresolved. Furthermore, the loss of the most vulnerable fishes (i.e., corallivores) may induce an erosion of reef prokaryotic richness and their related functions [19]. Such events are clearly case studies to address fish gut microbiome responses and plasticity to environmental degradation. Moreover, understanding which bacterial lineages and associated functions are lost, and if there is compensation by other lineages, it is essential to better understand the consequences of bleaching-induced regime shifts on the functioning of coral reef ecosystems in general.

Coral reefs in Seychelles are located in the northern gyre of the western Indian Ocean (WIO), and are periodically subject to high marine heat waves with extreme SST associated with both El Niño and the Indian Ocean Dipole [37]. Severe bleaching and consequent regime shift events have occurred since the early 1980s [38] with two mass bleaching events in 1998 and 2016 that caused >90% and 70% of cover coral loss, respectively [37,39–41]. Following the 1998 coral bleaching event, Seychelles coral reefs underwent divergent trajectories, either recovering to a live coral condition or undergoing regime shifts to fleshy brown macroalgal dominance [6]. Here, we use these alternate Seychelles coral reef conditions to investigate the consequences of coral-macroalgal phase shift on the diversity and the structure of fish gut bacteriomes. First, we explored the diversity, richness, and composition associated with macroalgae and the enteric coral reef fish core bacteriomes. Second, we assessed the consequences of the regime shifts on the diversity, variability, and composition of the core bacteriomes at both taxonomic and functional levels.

## 2. Materials and Methods

### 2.1. Study Area and Sample Collection

Seven locations were sampled around Praslin and Mahe islands in January 2019, representing 4 recovering coral reefs (RCR), and 3 macroalgal dominated (mainly *Sargassum* and green turf algae) reefs (MSR) (Figure 1).

**Figure 1.** Sampling map of coral-dominated reefs (RCR1: RCR4 in blue) and macroalgae shifted reefs (MSR1: MSR3 in red) with their respective geomorphology (patch, carbonate or granitic). The pictures of coral-covered and macroalgae shifted reefs are represented respectively in blue squared (bottom right) and in red squared (top left). Photo credits: Nicholas A.J. Graham.

The recovering coral reefs had recovered their live coral following the 1998 coral bleaching event [6], but experienced 70% mortality in 2016, having a mean coral cover of 6% by 2017 [40]. Reef ecosystems of the Inner Seychelles support ecologically and phylogenetically diverse fish families. Species in the families *Siganidae*, *Lethrinidae*, *Lutjanidae*, *Acanthuridae*, *Scarinae*, *Mullidae*, *Labridae*, and *Haemulidae* together comprise >95% of total trap fishery catches [42]. Fish samples were collected using handlines and traps deployed from a small boat, using diverse baits (coconut, mackerel, seaweed). In order to take into account intraspecific variability of the gut bacteriome, up to 11 adult individuals of each species were sampled in each site. Immediately after capture, fishes were killed by cervical dislocation (following the European directive 2010/63/UE) and conserved on ice in coolers for dissection in the laboratory later the same day. The animal study was reviewed and approved by the Seychelles Fishing Authority (Memorandum of Understanding signed the 12 December 2018) and by the Lancaster University FST research Ethics review committee (approval number FST18132). At the laboratory (Seychelles Fishing Authority), fishes were placed in trays, washed with 70% ethanol, and the whole intestinal tract of each fish was extracted using sterile dissection tools following the protocol of Clements et al. (2007) [43] and Miyake et al. (2015) [20]. Briefly, we squeezed out the gut content (taking care to avoid contamination by gut wall cells) into a 2 mL sterile Eppendorf tube by rolling a sterile 1 mL micropipette on the intestinal tract starting from segments posterior to the stomach (spanning the midgut and hindgut) or from the 75% most distal part of the gut for fishes lacking stomachs. Gut contents were immediately flash frozen in liquid nitrogen and stored at −80 °C until ready for DNA extractions. A total of 99 fishes belonging to 36 species covering 19 genera and 9 families were sampled for their gut bacteriome (Table S1).

### 2.2. Fish Identification and Diet Type Definition

For all fishes, host taxonomic identification was performed using the reference book on reef fishes from the West Indian Sea [44]. Fish diet was described using categories as in Mouillot et al. (2014) [45], where Carnivores are separated into invertivores (MI) which mainly feed on mobile invertebrates (i.e., benthic species such as crustaceans) and piscivores (FC) (i.e., feeding on teleosts or cephalopods). Herbivores are divided into strict herbivores (H) eating fleshy macroalgae with browsing (*Siganidae*) and grazing (*Acanthuridae*) behaviors, and detritivores (HD) with scrappers (*Scarinae*), which bite dead pieces of coral and indirectly scrape away turf algae [46]. Finally, omnivorous fishes (OM) feed on both algae or cyanobacteria and small invertebrates (i.e., zooplankton such as copepods). We used a principal coordinates analysis (PCoA), based on Bray-Curtis dissimilarity, to illustrate their distribution through sampled sites (Figure S1). To assess the sources of variation (i.e., taxonomy and diet) in the Bray-Curtis matrix, we used a PERMANOVA analysis based on 1000 permutations [47] with the function adonis, in the vegan package [48].

### 2.3. DNA Extraction and 16S rDNA Gene Amplification

Total genomic DNA from 200 mg of homogenized intestinal contents and from swabs was extracted using the MagAttract PowerSoil® DNA kit according to the manufacturer instructions (MoBio Laboratories, Inc., Carlsbad, CA, USA) with automated processing and the liquid handling system KingFisher Flex™ (ThermoScientific®, Waltam, MA, USA). Nucleic acids were eluted in molecular water (Merck Millipore™, Burlington, MA, USA) and quantified on a NanoDrop 8000 ™ spectrophotometer (ThermoScientific®, Wilmington, MA, USA). The V4-V5 region of the 16S rDNA gene was targeted with the universal primers 515F-Y(5′-GTGYCAGCMGCCGCGGTAA-3′) and 926R (5′-CCGYCAATTYMTTTRAGT TT-3′) [49] coupled with platform specific Illumina adaptor sequences on the 5′ ends. Each 25 µL PCR reaction was prepared with 12.5 µL Taq Polymerase Phusion® High-Fidelity PCR Master Mix with GC Buffer (New England Biolabs®, Inc., Ipswich, MA, USA), 0.5 µL forward primer (10 µM), 0.5 µL reverse primer (10 µM), 1 µL template DNA, 0.75 µL DMSO, and 9.75 µL molecular water. PCR amplifications involved the following protocol: An initial 98 °C denaturing step for 30 s following by 35 cycles of amplification (10 s denaturation at 98 °C; 1 min at 60 °C annealing; 1.5 min extension at 72 °C), and a final extension of 10 min at 72 °C. Amplification and primer specificity were verified by electrophoresis on a 2.0% agarose gel for confirmation of ~450 bp amplicon size. All samples were amplified in triplicate and equally pooled for a final product of 50 µL. Extraction of blank samples used as DNA extraction controls were also performed. None of them were successfully amplified with the primers used in this study. Each amplicon pool was sequenced using the 2 × 250 bp Miseq chemistry on an Illumina MiSeq sequencing platform at the INRA GeT-PlaGE platform (Toulouse, France).

### 2.4. Sequence Processing

All analyses were carried out with R software 3.6.2 (https://www.r-project.org/, accessed on 9 August 2021) [50] and are available on GitHub: https://github.com/mccheutin/Seychelles.git, accessed on 9 August 2021.

Sequence reads were processed using the DADA2 pipeline (v.1.12.1) in R [51], following the pipeline's tutorial (https://benjjneb.github.io/dada2/tutorial.html, accessed on 9 August 2021). Briefly, sequences were trimmed and filtered based on read quality profiles (maxN = 0; maxEE = (2, 2); truncQ = 2; and truncLen = (240, 240)), error correct, dereplicated and amplicon sequence variants (ASVs) were inferred [52]. Forward and reverse ASVs were merged and pooled in a count table where chimera were identified and removed. Taxonomy assignment was performed using the SILVA reference database (release 132) [53]. The ASVs count table, their taxonomy, and their sequences were organized in a phyloseq object using the phyloseq package (v.1.28.0) [54], on R. ASVs assigned to the kingdom Eukarya, Archaea, and to chloroplast, were removed before computing any further analysis. Bacterial genera known as potential kit contaminants were also removed from our datasets using the list described in Salter et al. (2014) [55]. Overall, 40 genera corresponding to 12% of the total reads were removed (Supplementary File 1). Our final dataset consisted of 1,042,080 sequences belonging to 5129 ASVs.

### 2.5. *Defining the Core Bacteriome of Reef Organisms*

As observed in many animal microbiomes [18,56], ASVs may span a range from permanent to transient inhabitants. Closely associated ASVs should be more considered when thinking about holobiont ecology [57] since these core taxa may have evolved in close association with their hosts for a long time period [58–60]. Here, core bacteriomes were independently identified by examining the species abundance distribution (SAD), patterns of each ASV, and by partitioning the SAD into core and satellite ASVs [61] for the gut (Figure S2A) and for the macroalgae (Figure S2B). For this purpose, the index of dispersion for each ASV was calculated as the ratio of the variance to the mean abundance (VMR) multiplied by the occurrence. This index was used to model whether lineages follow a Poisson distribution (i.e., stochastic distribution), falling between the 2.5% and 97.5% confidence interval of the $\chi 2$ distribution [62]. Index values less than 1 mean that the ASV is under-dispersed compared to the Poisson distribution, so that it spreads uniformly and can be considered as a core ASV. Index values higher than 1 mean that the ASV is over-dispersed, i.e., the ASV is clustered and corresponds to a satellite ASV. Fish and macroalgae core bacteriomes consisted of 531,930 sequences (254 ASVs) and 109,550 sequences (310 ASVs), respectively. All analyses detailed below were performed on the core microbiome.

### 2.6. *Inference of ASVs Habitat Preference*

We used a BLASTn approach on the nr/nt database and the habitat-associated metadata to the closest ASV match to infer the habitat preference of the 254 ASVs constituting the enteric core bacteriome of reef fishes [22,28]. Only blast results with an identity >95% and a sequence coverage >95% were kept. Information concerning the isolation source contained in the GenBank fields "isolation source", "host", and "title" of each closest blast were extracted using a dedicated python script and parsed into "Animal", "Environment" (i.e., free living bacteria associated with sediment, soil or water), and "Other" habitat categories. For ASVs associated with animals, we further categorized the isolation sources into specific hosts (i.e., fish, marine invertebrates, terrestrial vertebrates, and unknown animals) and organ (i.e., gut, tissue, and other organs) categories (Supplementary File 2). In order to associate these habitat preferences to the phylogenetic affiliation of each ASV, core bacteriome ASVs were aligned against the silva.nr_v132 reference database using mothur v.1.35.1 ([63]; https://mothur.org/, accessed on 9 August 2021) before being imported into the ARB software ([64]; http://www.arb-home.de/, accessed on 9 August 2021) and loaded with the SILVA (v.138) reference database [53]. A base frequency filter was applied to exclude highly variable positions before adding sequences to the maximum parsimony backbone tree using the parsimony quick add marked tool implemented in ARB. The tree and the associated categories were drawn and visualized using the interactive Tree of Life (iTOL) web server ([65]; https://itol.embl.de/, accessed on 9 August 2021).

### 2.7. *Computation of Alpha and Beta-Diversity of Bacteriomes*

In order to correct for the uneven sequencing depth among samples, 1041 sequences were randomly sub-sampled within each sample using the "rarefy_even_depth" function from the phyloseq R-package v.1.28 [54] (Figure S3A). Good's coverage estimator [66] was

99.9 $\pm$ 0.1 indicating that the coverage was still excellent after rarefaction. Taxonomic diversity of each microbial community (fish gut or macroalgae swab) was measured using the richness (number of ASV) and the Shannon's index H, computed on ASV relative abundance, and later exponentially transformed to express it as effective number of species (ENS) [67]. Taxonomic dissimilarities between pairs of bacteriome samples were assessed using the Bray-Curtis dissimilarity computed on relative abundances of ASV.

### *2.8. Functional Diversity Predictions of Bacteriome*

Using the 16S rRNA gene information, predictions of metabolic functions for Bacteria were performed using Tax4Fun2 v.1.1.5 [68] with a clustering threshold set at 99%, following the tutorial of the algorithm (https://github.com/bwemheu/Tax4Fun2, accessed on 9 August 2021). In order to account for all ASVs, the predicting functional profiles were then proceeded using the minimum blast identity to reference at 78%. Among the 7279 KOs predicted by Tax4Fun2, about 23% are involved in at least two different metabolic pathways (until 15 for some KOs) and 33% are unknown or hypothetical proteins. These KOs are thus not indicators of a particular function and are a source of an additional and false functional redundancy, hardly ever taken into account in the literature. To avoid this bias, we created a new functional table containing 3261 unique KOs, involved in only one metabolic pathway.

### *2.9. Statistical Tests*

First, the gut and macroalgal bacteriomes were compared in richness and in composition by measuring the alpha and beta-diversity. In the same way, to understand the influence of the reef condition on the bacteriomes, we compared the same measures between reef conditions for both macroalgae and fishes. Since the effect of reef condition could have been masked by the effect of diet or phylogeny, we removed this by analyzing the dataset at different community levels (i.e., inside trophic guilds, family and species level). Only levels with at least a triplicate per reef condition were tested. Comparison of alpha diversity indices (richness and entropy) was achieved using a Kruskal-Wallis test (999 permutations) in the vegan R-package followed by a post-hoc Dunn test (999 perm, $p$-value corrected by Bonferroni's method) in order to identify which group means differed. To determine beta-diversity changes, significant sources of variation in bacteriome Bray-Curtis dissimilarity matrices were assessed using permutational analysis of variance (PERMANOVA) with the adonis function from the vegan package.

ASV biomarkers of bacteriomes of macroalgae, carnivorous, and herbivorous fish were identified using the LEfSe algorithm [69]. The first analysis step was a non-parametric Kruskal-Wallis (KW) sum-rank test allowing the detection of taxa with significant differential abundance. Biological consistency was subsequently investigated using a pairwise Wilcoxon test. Finally, linear discriminant analysis (LDA) was used to estimate the effect size of each differentially abundant taxon. Alpha values of 0.05 were used for KW and Wilcoxon tests and a threshold of 3 was used for logarithmic LDA scores. The same analysis was used to identify functional biomarkers (i.e., KO) of the *Scarinae* and *Siganidae* bacteriomes.

## 3. Results

### *3.1. Sampling Size and Composition of Fish Catch between Reef Conditions*

The fish species distribution and sample size were highly variable among reefs. The sampling size of caught fish species was higher in recovering coral reefs (RCR), with 27 species sampled compared to 17 species in macroalgae shifted reefs (MSR). Fish community composition and fish diet behavior differed significantly between RCR and MSR (Figure S1A,B) with a higher abundance of the scrapers *Scarinae* in RCR while grazers *Siganidae* are more abundant in MSR (Table S1), conforming with the underwater visual census (UVC) data [42]. In contrast, carnivorous species, overall represented by the *Lethrinidae* and *Lutjanidae* families, were distributed in both reefs with 11 *Lethrinidae* in RCR and 15 in

MSR and six *Lutjanidae* in RCR and eight in MSR. Only four species have been sampled in triplicate in both RCR and MSR (i.e., *Scarus ghobban*, *Lethrinus mahsena*, *Lethrinus enigmaticus*, and *Aprion virescens*) (Table S1).

### 3.2. Composition and Diversity of the Fish Core Gut Bacteriome

A total of 254 bacterial ASVs representing 63% of the total reads formed the core bacteriome of the 99 fish gut samples (Supplementary File 2). This core bacteriome was dominated by the *Proteobacteria* (dominated by the order *Vibrionales*) and the *Firmicutes* phyla (mainly constituted by the order *Clostridiales*) that represented collectively more than 67% of the sequences (Figure S4). Other less abundant phyla such as the *Bacteroidetes* (8%), *Fusobacteria* (8%), *Spirochaetes* (5%), *Planctomycetes* (3%), *Cyanobacteria* (3%), *Verrucomicrobia* (3%), and *Tenericutes* (2%) constituted the rest of the fish core bacteriome.

BLASTn analysis revealed that 70% (178) of the bacterial ASVs were closely related to sequences previously retrieved from animal microbiomes (Figure 2). In addition, 45% (115 ASV) belonged to the *Akkermansiaceae*, *Desulfovibrionaceae*, *Vibrionaceae*, *Rikenellaceae*, *Fusobacteriaceae*, and *Lachnospiraceae* families, and matched preferentially sequences previously reported in the intestinal tract of fish from the *Siganidae*, *Acanthuridae*, and *Scarinae* families (Figure 2, Supplementary File 2), indicating a certain degree of conservation for a significant part of the coral fish gut bacteriome.

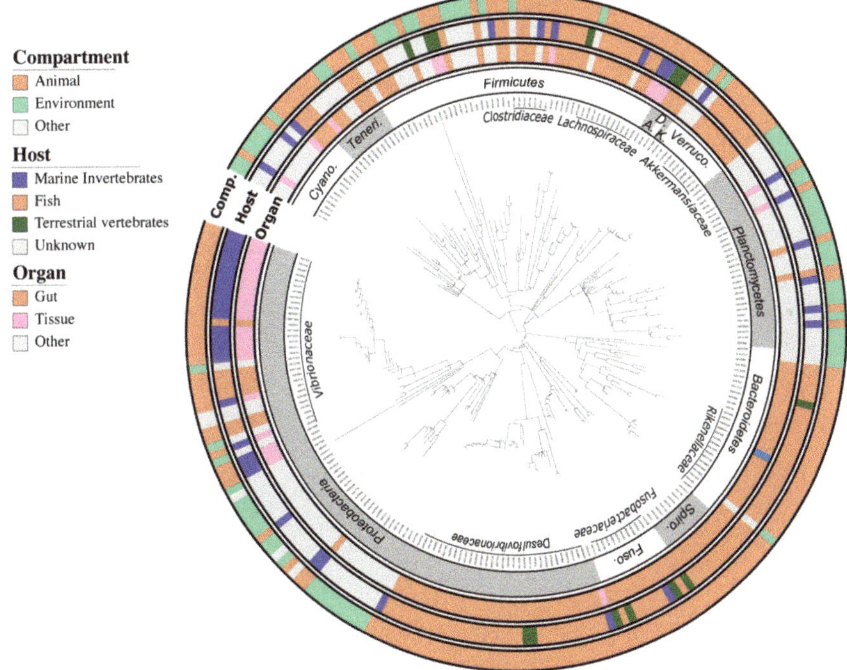


**Figure 2.** Maximum parsimony phylogenetic tree of the 254 ASVs from the fish gut core bacteriome. The 16S rRNA sequences were inserted into the original SILVA (release 138.1) tree using parsimony criteria with the Bacteria filter excluding highly variable positions. The inner ring represents the order level nomenclature following the taxonomy provided by default in the SILVA bacterial tree. The three outer rings depict the habitat preferences of each ASV described here as three categories (i.e., habitat, specific host, and organ) clustered from the environmental information associated with each closest blast. The tree was drawn using the web-based interface interactive tree of life (iTOL). Abbreviations: Cyano. = *Cyanobacteria*; Teneri. = *Tenericutes*; A. = *Actinobacteria*; D. = *Deferribacteres*; K. = *Kiritimatiellaeota*; Verruco. = *Verrucomicrobia*; Spiro. = *Spirochaetes*; Fuso. = *Fusobacteria*.

In addition to these fish gut specialists, 17% (44 ASV) of the core ASVs, mainly affiliated to the *Vibrionaceae*, *Pirellulaceae*, *Lachnospiraceae*, and *Endozoicomonadaceae* families, were best related to sequences associated with other marine animal bacteriomes, such as corals or sponges, indicating that another significant part of the fish gut bacteriome maybe symbiotic generalists distributed among other marine organisms. The composition of fish core gut bacteriomes differed significantly (PERMANOVA $p = 0.001$; $R2 = 0.07$) from macroalgae bacteriomes (Figure S5A) which were dominated by bacteria from the *Proteobacteria* (56%), *Bacteroidetes* (25%), and *Cyanobacteria* (10%) phyla. Both richness and diversity of fish core gut bacteriomes were half that of macroalgae bacteriomes (Figure S5B). Herbivore bacteriome shared 2.5 times more ASVs with the macroalgal bacteriome than the carnivore one (21 vs. 8) (Figure S6A), mainly belonging to the Orders *Bacteroidales* (i.e., *Rickenella*), and *Clostridiales* (i.e., *Lachnoclostridium*). Fish gut bacteriomes were also more variable in their composition than macroalgae bacteriomes as indicated by a significantly higher dispersion (Figure S5B).

### 3.3. *Alteration of the Coral Reef Significantly Disrupts Herbivore but Not Carnivore Bacteriomes*

The reef condition explained a small but significant amount of the variability in bacteriome community composition among all fishes (Table 1; Figure 3A).

**Table 1.** Results of PERMANOVA on the 29 core bacteriomes of macroalgae and 99 enteric core bacteriomes of reef fish ($n$ = sampling size). For a relevant sampling size (If not "–"), diet, taxonomy and the reef condition (RCR vs. MSR) were tested (999 perms). Signif. codes for $p$-value: *** $\leq 0.001$; ** $\leq 0.01$; * $\leq 0.05$ or not (NS).

| | Diet | Family | Genus | Species | Reef Condition |
|---|---|---|---|---|---|
| Algae ($n$ = 29) | – | – | – | **R2 = 0.13 (***)** | **R2 = 0.07 (**)** |
| Fish ($n$ = 99) | **R2 = 0.04 (***)** | **R2 = 0.16 (***)** | **R2 = 0.27 (***)** | **R2 = 0.46 (***)** | **R2 = 0.02 (***)** |
| Herbivores ($n$ = 44) | – | **R2 = 0.14 (***)** | **R2 = 0.24 (***)** | **R2 = 0.43 (***)** | **R2 = 0.07 (***)** |
| *Scarinae* ($n$ = 22) | – | – | **R2 = 0.14 (**)** | – | **R2 = 0.09 (**)** |
| *S.ghobban* ($n$ = 7) | – | – | – | – | R2 = 0.28 (NS.) |
| *Siganidae* ($n$ = 17) | – | – | – | **R2 = 0.20 (**)** | – |
| Carnivores ($n$ = 53) | – | **R2 = 0.10 (**)** | **R2 = 0.24 (**)** | **R2 = 0.44 (*)** | R2 = 0.02 (NS.) |
| *Lutjanidae* ($n$ = 14) | – | – | R2 = 0.08 (NS.) | R2 = 0.34 (NS.) | R2 = 0.10 (NS.) |
| *A.virescens* ($n$ = 7) | – | – | – | – | R2 = 0.17 (NS.) |
| *Lethrinidae* ($n$ = 26) | – | – | – | R2 = 0.21 (NS.) | R2 = 0.05 (NS.) |
| *L.mahsena* ($n$ = 10) | – | – | – | – | R2 = 0.12 (NS.) |
| *L.enigmaticus* ($n$ = 6) | – | – | – | – | R2 = 0.17 (NS.) |

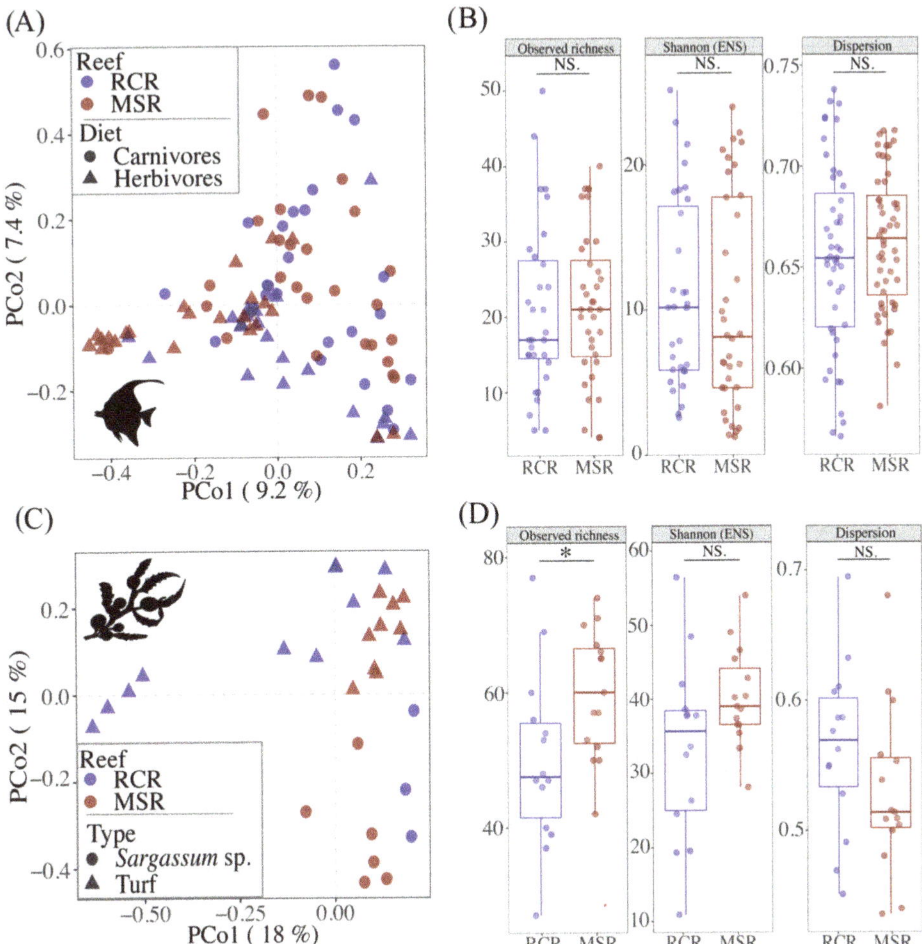


**Figure 3.** Comparison of the alpha and beta diversity of fish gut and macroalgae bacteriomes in function of the condition of the reef (i.e., coral covered vs. macroalgae shifted). (**A**,**C**) Principal coordinates analysis (PCoA) plots illustrating Bray-Curtis distances between pairs of bacteriome samples. Bacteriomes are colored according to the reef condition, while the shape represents (**A**) fish diet or (**C**) macroalgae type. (**B**,**D**) Boxplots representing the alpha diversity, expressed as the observed richness and the Shannon's index H-exponentially transformed in effective number of species (ENS), and the dispersion (distance to the centroid for each sample type grouping) calculated for each bacteriome sample. Horizontal brackets indicate pairs which differ significantly: *** $\leq 0.001$; ** $\leq 0.01$; * $\leq 0.05$) or not (NS) with a Wilcoxon test.

However, the reef condition neither appeared as a significant driver of variability, nor of bacteriome diversity between fish individuals (Figure 3B). Similarly, the ordination of macroalgae bacteriomes in a PCoA showed a clear separation between CCR and RCR (Figure 3C) which explained 7% of the variance in the community composition for macroalgae bacteriomes (Table 1). In addition, richness of macroalgae bacteriomes were 80% higher in MSR (Figure 3D). For fishes, diet was one of the main drivers of gut bacteriome composition as indicated by a PERMANOVA analysis (Table 1, Figure 3A). The gut bacteriome of herbivores (i.e., grazers, scrapers, browsers, and the two omnivorous *Cantherines pardalis*) was characterized by the enrichment of 12 biomarkers, genera belonging mainly to the *Desulfovibrionales*, *Bacteroidales*, and *Fusobacteriales*, while eight genera belonging mainly to the *Clostridiales* and *Vibrionales* appeared as biomarkers for carnivores (i.e., invertivores and piscivores) (Figure S6B).

The reef condition significantly affected the composition of gut bacteriome of herbivorous fishes (R2 = 0.07; $p$ = 0.001, Figure 4, Table 1).

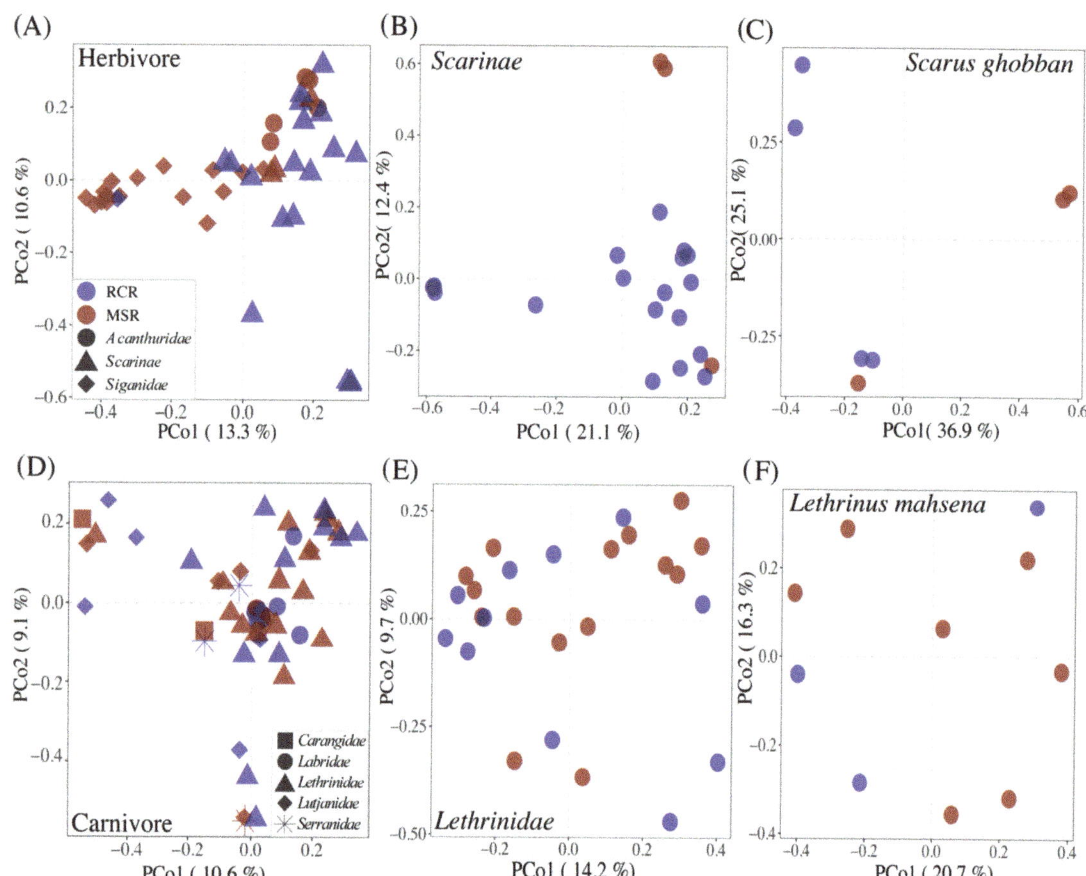


**Figure 4.** Principal coordinates analysis (PCoA) plots illustrating Bray-Curtis distances between pairs of bacteriome for (**A**) herbivorous fishes, (**B**) *Scarinae,* (**C**) *S.ghobban,* (**D**) carnivorous fishes, (**E**) *Lethrinidae,* and (**F**) *L. mahsena.* Bacteriomes are colored according to the reef condition: RCR in blue and MSR in red. See Figure S6 for the *Lutjanidae, A. virescens,* and *L. enigmaticus* results.

However, part of this effect may be driven by the strong fish species turnover among herbivores between coral and macroalgae dominated reefs. Indeed, 16 out of 17 of *Siganidae* and *Acanthuridae* fishes were distributed in MSR, while 19 out of 22 *Scarinae* fishes (mainly represented by *Scarus ghobban*) were present in RCR (Figure 4A). Fish phylogeny was a strong and significant determinant of bacteriome composition at the family (R2 = 0.14; $p$ = 0.001), genus (R2 = 0.24; $p$ = 0.001), and species (R2 = 0.43; $p$ = 0.001) levels. In order to exclude this effect, we analyzed the herbivore dataset at the family and species levels (for *Scarinae* and *Scarus ghobban*, the only herbivores distributed in both reef conditions). In this way, we corroborated the fact that gut bacteriome composition did differ as a result of reef condition (Figure 4B,C). This effect was marginal at the species level probably due to the low number of samples (R2 = 0.28; $p$ = 0.078, N = 7). Differences in the composition of herbivore core bacteriomes among reef conditions was driven by changes in the relative abundance of biomarkers of the *Scarinae* and *Siganidae* families (Figure 5).

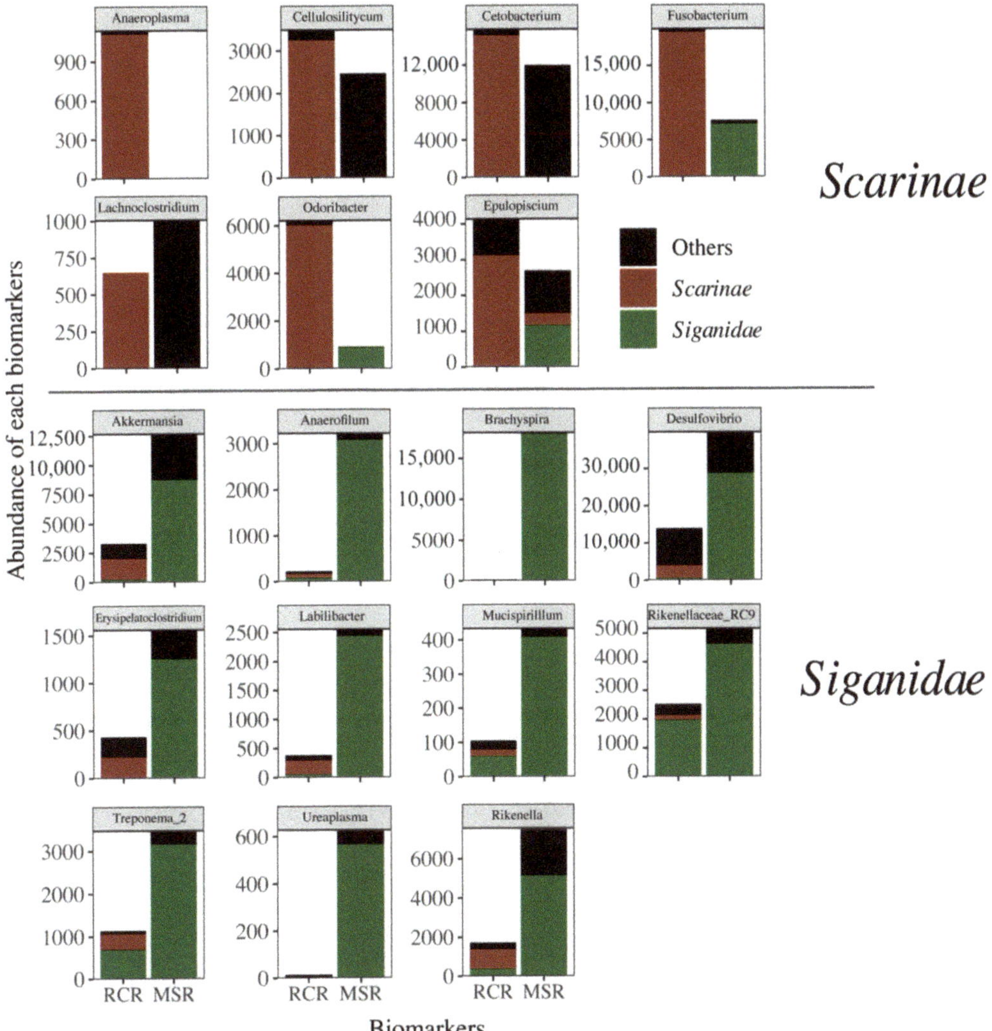


**Figure 5.** Abundance of each biomarker at Genus level related to *Scarinae* and *Siganidae* (delineated using a LEFSE approach) in RCR and MSR. Contribution of *Scarinae* (red), *Siganidae* (green), and other fish families (black) is indicated on each biomarker.

Bacteriome abundance in MSR was lower for six of the seven *Scarinae* biomarkers and one (i.e., *Anaeroplasma*) was totally absent. *Fusobacterium* and *Odoribacter* biomarkers were only present in *Siganidae* in MSR (Figure 5). These biomarkers accounted for 0.8% on average of the *Scarinae* bacteriomes and 0.4% of the whole dataset. The decrease in *Scarinae* biomarkers was paralleled by a significant decrease in the abundance of 207 specific KOs (Kegg Orthologs) mainly involved in host lipid (i.e., fatty acids, butanoate, propanoate, and glycerophospholipid metabolisms) and glucose homeostasis (Figure 6).

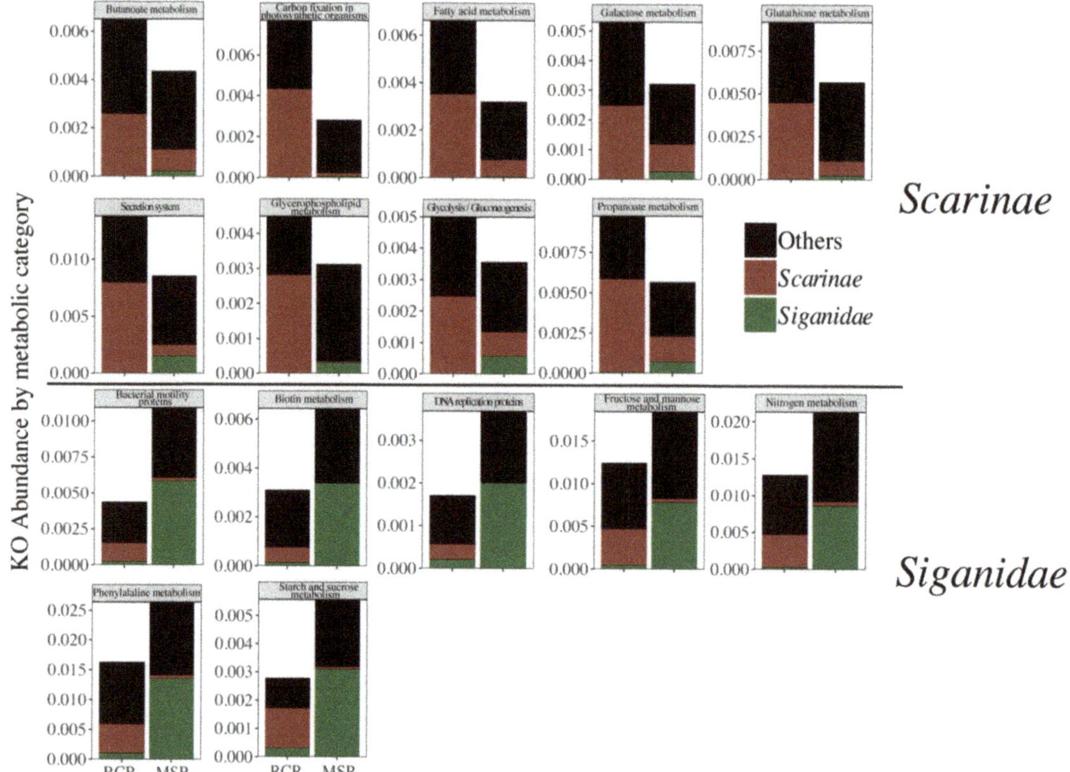


**Figure 6.** Abundance of each Kegg Ortholog (KO) merged by metabolic pathway, related to enteric bacteriomes of Scarinae (red), Siganidae (green), and other fish families (black) (delineated using a LEfSe approach) in RCR and MSR.

In contrast, *Siganidae* biomarkers, all efficient anaerobes fermenters of plant and algal polysaccharides [30], showed a significant increase in MSR and one new appeared (i.e., *Brachyspira*), accounting for 5.9% on average of the *Siganidae* bacteriomes and 2.0% of the whole dataset. This increase came also with an increase in KOs notably involved in carbohydrates metabolism (starch, sucrose, fructose, and mannose), DNA replication, and nitrogen metabolism suggesting higher rates of fermentation and a stimulation of bacterial growth (Figure 6). Reef condition neither appeared as a significant driver of herbivore bacteriome variability, nor of bacteriome diversity (Figure S7).

Contrary to herbivores, we did not detect a significant effect of reef condition on any of the bacteriome diversity facets (i.e., alpha diversity, beta diversity, and variability) of carnivorous fishes (Table 1, Figure 4, Figure S8). The fish family was the only driver of difference in microbiome composition (R2 = 0.10; $p$ = 0.007).

## 4. Discussion

Macroalgal shifted reefs (MSR) are often considered degraded systems in which drastic changes to biotic communities occur, particularly reef fishes [70,71]. So far, the "microbial phase shift" [72] consecutive to a macroalgae regime shift has been studied only in free living microbial communities [73–76] and primary producer microbiomes [11,77,78]. Here, we pinpointed for the first time the influence of such a shift on the gut bacteriomes of Seychelles reef fishes.

The observations from Robinson et al. (2019) [42] indicated that biodiversity losses were more severe in shifted-reefs resulting in novel fish compositions. This conformed with the different fish functions (e.g., browsing and grazing activities) found in MSR compared to recovering coral reefs (RCR) (Figure S1). Alterations to habitat directly affect coral-dependent fish species [79] such as coral dwellers [80,81] and corallivores [82], and promote the replacement of these highly specialized species by opportunistic species that live in areas of low relief and rubble [70,83–85]. In agreement, fish communities from MSR were characterized by a depletion in *Scarinae*, which are scavengers feeding the epilithic layers present on corals [46], and the dominance of browsers and grazers of the *Siganidae* and *Acanthuridae* families [20] (Figure S1, Table S1). By conditioning the availability of their nutritional resources, regime shifts influenced the occurrence of these two herbivorous fish families (Table S1). Among opportunistic species, invertivorous fishes are believed to benefit from a carbon flow cascade in which the important release of dissolved organic material in algae-dominated reefs stimulates microbial production ultimately fueling benthic invertebrate biomass [86,87]. In this study, invertivores, essentially represented by fishes from the *Lutjanidae* and *Lethrinidae* families, were however uniformly distributed among RCR and MSR.

Several lines of evidence indicate that microorganisms play an active role in the transition from coral dominance to fleshy algae through the DDAM positive feedback loop (dissolved organic carbon, disease, algae, microorganism) [88,89]. In this mechanism, exudation of labile organic matter by turf and macroalgae promotes an increase in microbial abundance and activity, as well as a change in the composition towards copiotrophic and potentially pathogenic microbial taxa, ultimately causing a physiological deterioration of the coral tissues [11] and a dysbiosis in their microbiome [90]. Except for a recent study [91], disruption of the planktonic microbial composition [73,75,76,92,93] and coral microbiomes [11,12,77,78,90] is a recurrent pattern in MSR. Accordingly, we observed here a significant difference in the composition of macroalgae bacteriomes between MSR and RCR accompanied with an increase in bacterial richness in MSR (Figure 3C,D). Macroalgae bacteriomes in MSR were enriched in *Alphaproteobacteria* (*Ahrensia* sp. and *Albimonas* sp.) and *Gammaproteobacteria* (*Leucothrix* sp.). Enrichment in *Gammaproteobacteria* and particularly from the *Leucothrix* genera, which contains filamentous species known to provoke massive invertebrate egg and larvae mortalities [94], agrees with the DDAM model predicting that a proliferation of macroalgae leads to an increase in copiotrophic microorganisms with the potential to create disease. Altogether, these results indicate a microbialization [75] of the MSR studied here, although we did not assess microbial abundance in our sampling.

We showed that 45% of the ASVs composing the core fish gut bacteriomes corresponded to fish specialists, mainly belonging to the *Desulfovibrionaceae*, *Vibrionaceae*, *Akkermansiaceae*, *Fusobacteriaceae*, and *Lachnospiraceae* families, often retrieved in studies investigating the gut microbiome of coral fishes [21,22,28,30,95,96]. In addition, a significant part of core ASVs were symbiotic generalists shared among marine organisms indicating a potentially important connectivity of fish gut bacteriomes with their surrounding habitat and animal-associated microbial communities, through feeding activity and defecation. This suggests that perturbations of their habitat microbiome related to macroalgal regime shifts could translocate to their own microbiome. Indeed, although fish diet and taxonomy were major determinants of fish gut bacteriome composition, this latter differed significantly between RCR and MSR (Table 1, Figure 4). Shifts in the fish gut microbiome may reflect changes in diet in degraded habitats. While this has never been observed before in coral reef ecosystems, in disturbed continental areas where their nutritional resources were modified or even absent, the composition of black howler monkey enteric microbiomes responded to habitat perturbations [32,97]. Since macroalgae regime shifts represent an important modification of their main nutritional resources, we hypothesized a strong effect on herbivorous fish gut bacteriomes. In agreement, the reef condition explained a significant amount of the variance for herbivorous fish, while we failed to detect any significant effect for carnivorous fishes. One explanation may be related to the fact that carnivorous fishes seem to have a larger dietary niche width than obligate herbivores [98] that would allow them to forage in adjacent healthy areas of the reef [99]. Our sampling strategy did not allow us to detect a significant effect of coral-macroalgal shift at the intra-species level. To overcome this limitation related to the high intra-specific variability observed in fish gut bacteriomes, future investigations should focus on species present in both MSR and CCR and with a significant increase in the number of individuals (more than 10) in each condition and species.

Rather than a dysbiosis, the significant response of herbivorous gut bacteriome composition to the condition of the reef reflected the loss or gain of specific bacterial taxa associated with the strong turnover of their hosts, particularly *Scarinae* and *Siganidae* fishes, between RCR and MSR (Figure 6). This result indicates a certain degree of conservation for a significant part of the coral reef fish gut bacteriome, but also agrees with a recent study showing that loss of the most vulnerable reef animals, and among them fishes, due to reef degradation would induce a significant loss of the reef prokaryotic richness [19]. While we did not observe an erosion of bacteriome diversity in MSR, nor an increase of bacteriome variability among individuals expected under the Anna Karenina principle [100], we did record a significant reduction or loss of *Scarinae* biomarkers and associated functional genes towards the prevalence of bacterial fermenters associated with *Siganidae*. In addition, we also observed a concomitant increase in abundance of KOs involved in carbohydrate metabolism (starch, sucrose, fructose, and mannose), DNA replication, and nitrogen metabolism, suggesting higher rates of fermentation and a stimulation of bacterial growth in MSR. Seaweeds such as *Sargassum* and turf algae are rich in sulfated polysaccharides and high carbohydrate food is well known to promote rates of gastrointestinal fermentation [101]. ASVs constituting *Siganidae* biomarkers were closely related to sequences previously retrieved from *Siganus canaliculatus* (Supplementary File 2). Indeed, bacteria from the genera *Desulfovibrio* (sulfate reducing bacteria), *Rickenella*, *Brachyspira* (anaerobic fermentative bacteria), and *Akkermansia* (mucin degrading bacteria) were found to be part of the core bacteriome of *Siganidae* [102,103], accounting for 5.9% on average of their bacteriomes and 2.0% of the whole dataset. These taxa may be of importance for host digestive function in MSR, in particular for the fermentation of sulfated algal polysaccharides. For example, members of the *Rikenella* genus are known to degrade celluloses into short chain fatty acids (SCFA) available for the host through microbial fermentation [30,104]. The prevalence of these fermentative bacteria is in line with the high fermentation rates observed within herbivorous fish hindguts [105], particularly in *Siganidae* [106] and further suggest a well-suited adaptation of *Siganidae* bacteriomes to the consumption of algae. We acknowledge that these predicted functions based on barcoding data should be corroborated by future transcriptional or proteomic studies that could address the consequence of coral-macroalgal shift on the fermentative activity of microbes associated with reef fish. Nonetheless, the predominance of fermentative metabolisms in MSR indicated that regime shifts not only affect the taxonomic composition of fish bacteriomes, but has the potential to also affect ecosystem functioning through microbial functions.

## 5. Conclusions

Identifying the mechanisms and consequences of bleaching-induced benthic regime shifts for reef microbiota is vital for understanding the resilience of these habitats to changing ocean conditions. Here, we showed that a "microbial phase shift" occurred following a macroalgae regime shift, which was translocated to the gut bacteriome of herbivore reef fishes affecting their composition and potentially their functional role in the reef ecosystem. This response reflected the loss or gain of specific bacterial taxa associated with the strong turnover of their hosts between RCR and MSR. A pattern that maybe reflects a long-term effect of regime shifts. The consequences of increasing recurrence of "coral-macroalgae regime shifts" on reef animal microbiota and reef functioning is an emerging field of reef ecology. Further work should investigate the repercussions of microbiota dysbiosis consecutive to habitat degradation impacts on both host fitness and ecosystem functioning.

**Supplementary Materials:** The following are available online at https://www.mdpi.com/article/10.3390/microorganisms9081711/s1. Figure S1: PCoA illustrating similarity of fish catch composition and diets; Figure S2: Species abundance distribution (SAD) pattern of bacterial ASVs; Figure S3: Rarefaction curves for each fish gut and macroalgae bacteriomes; Figure S4: Treemaps of the constitutive phyla and their representative families; Figure S5: Beta and alpha diversity of fish gut and macroalgae bacteriomes; Figure S6: Venn diagram and polar histogram; Figure S7: Boxplots representing the alpha diversity; Figure S8: PCoA illustrating Bray-Curtis distances between pairs of bacteriome samples from the *Lutjanidae*, *A. virescens*, and *L. enigmaticus*; Table S1: Inventory of collected species; File S1: List of extraction kit contaminants; File S2: Blast results and the taxonomy.

**Author Contributions:** Conceptualization, M.-C.C., S.V., C.C.H., J.P.W.R., N.A.J.G., T.B. and J.-C.A.; data curation, J.P.W.R.; formal analysis, M.-C.C., C.M., J.-C.A.; funding acquisition, C.C.H., N.A.J.G. and J.-C.A.; investigation, M.-C.C., S.V., T.B. and J.-C.A.; methodology, M.-C.C. and J.-C.A.; project administration, S.V., T.B., J.P.W.R., N.A.J.G. and J.-C.A.; supervision, S.V., T.B. and J.-C.A.; writing—original draft, M.-C.C. and J.-C.A.; writing—review and editing, M.-C.C., S.V., C.C.H., J.P.W.R., N.A.J.G., C.X.O.R., Y.B. and T.B. All authors have read and agreed to the published version of the manuscript.

**Funding:** This project was funded by a grant from the CNRS INSU EC2CO "Microbien" program project BIOMEVO, a grant from the CNRS international emerging actions (IEA) program project FISHμBIOM, the Royal Society, and an ERC grant (759457) to Lancaster University. Marie Charlotte Cheutin was supported by a Ph.D grant from the French Ministry of Higher Education and Research.

**Institutional Review Board Statement:** The study was approved by Lancaster University FST research Ethics review committee (approval number FST18132) and according to the guidelines of the MoU with the Seychelles Fishing Authority (signed the 12 December 2018). An agreement with Seychelles' government over the transport of biological materials was signed with the Lancaster University (signed the 21 December 2019) and issued by UK DEFRA under the project ITMIP19.0048.

**Data Availability Statement:** Raw Illumina Miseq sequence data for each sample obtained in this study were deposited in the NCBI Sequence Read Archive (SRA) under BioProject accession no. PRJNA674042.

**Acknowledgments:** We thank the Seychelles Fishing Authority research team for assistance in the field and with sample preparation (Andrew Souffre, Rodney Melanie, Achille Pascal, Marie-Corinne Balett, Stephanie Marie, Clara Belmont) and Nathalie Bodin for assistance with study design and logistics. We are grateful to three trap fishers who provided boat time and fisheries expertise, and assisted with sample collection. We are grateful to the genotoul sequencing platform in Toulouse (Occitanie, France), for providing the MiSeq Illumina sequencing.

**Conflicts of Interest:** The authors declare that the research was conducted in the absence of any commercial or financial relationships that could be construed as a potential conflict of interest.

## References

1. Hughes, T.P. Climate Change, Human Impacts, and the Resilience of Coral Reefs. *Science* **2003**, *301*, 929–933. [CrossRef]
2. Jackson, J.B.C. Ecological extinction and evolution in the brave new ocean. *Proc. Natl. Acad. Sci. USA* **2008**, *105*, 11458–11465. [CrossRef] [PubMed]
3. Woodhead, A.J.; Hicks, C.C.; Norström, A.V.; Williams, G.J.; Graham, N.A.J. Coral reef ecosystem services in the Anthropocene. *Funct. Ecol.* **2019**, *33*, 1023–1034. [CrossRef]
4. McWilliam, M.; Pratchett, M.S.; Hoogenboom, M.O.; Hughes, T.P. Deficits in functional trait diversity following recovery on coral reefs. *Proc. R. Soc. B Biol. Sci.* **2020**, *287*, 20192628. [CrossRef]
5. Mumby, P.J. The impact of exploiting grazers (Scaridae) on the dynamics of Caribbean coral reefs. *Ecol. Appl.* **2006**, *16*, 747–769. [CrossRef]
6. Graham, N.A.J.; Jennings, S.; MacNeil, M.A.; Mouillot, D.; Wilson, S.K. Predicting climate-driven regime shifts versus rebound potential in coral reefs. *Nature* **2015**, *518*, 94–97. [CrossRef]
7. Norström, A.; Nyström, M.; Lokrantz, J.; Folke, C. Alternative states on coral reefs: Beyond coral–macroalgal phase shifts. *Mar. Ecol. Prog. Ser.* **2009**, *376*, 295–306. [CrossRef]
8. Cheal, A.J.; MacNeil, M.A.; Cripps, E.; Emslie, M.J.; Jonker, M.; Schaffelke, B.; Sweatman, H. Coral-macroalgal phase shifts or reef resilience: Links with diversity and functional roles of herbivorous fishes on the Great Barrier Reef. *Coral Reefs* **2010**, *29*, 1005–1015. [CrossRef]

9. McCook, L.; Jompa, J.; Diaz-Pulido, G. Competition between corals and algae on coral reefs: A review of evidence and mechanisms. *Coral Reefs* **2001**, *19*, 400–417. [CrossRef]
10. Hughes, T.P.; Rodrigues, M.J.; Bellwood, D.R.; Ceccarelli, D.; Hoegh-Guldberg, O.; McCook, L.; Moltschaniwskyj, N.; Pratchett, M.S.; Steneck, R.S.; Willis, B. Phase Shifts, Herbivory, and the Resilience of Coral Reefs to Climate Change. *Curr. Biol.* **2007**, *17*, 360–365. [CrossRef]
11. Morrow, K.M.; Bromhall, K.; Motti, C.A.; Munn, C.B.; Bourne, D.G. Allelochemicals Produced by Brown Macroalgae of the Lobophora Genus Are Active against Coral Larvae and Associated Bacteria, Supporting Pathogenic Shifts to Vibrio Dominance. *Appl. Environ. Microbiol.* **2017**, *83*. [CrossRef] [PubMed]
12. Morrow, K.M.; Liles, M.R.; Paul, V.J.; Moss, A.G.; Chadwick, N.E. Bacterial shifts associated with coral-macroalgal competition in the Caribbean Sea. *Mar. Ecol. Prog. Ser.* **2013**, *488*, 103–117. [CrossRef]
13. Burkepile, D.E.; Hay, M.E. Herbivore vs. nutrient control of marine primary producers: Context-dependent effects. *Ecology* **2006**, *87*, 3128–3139. [CrossRef]
14. Nyström, M.; Graham, N.A.J.; Lokrantz, J.; Norström, A.V. Capturing the cornerstones of coral reef resilience: Linking theory to practice. *Coral Reefs* **2008**, *27*, 795–809. [CrossRef]
15. Hughes, T.P.; Barnes, M.L.; Bellwood, D.R.; Cinner, J.E.; Cumming, G.S.; Jackson, J.B.C.; Kleypas, J.; van de Leemput, I.A.; Lough, J.M.; Morrison, T.H.; et al. Coral reefs in the Anthropocene. *Nature* **2017**, *546*, 82–90. [CrossRef]
16. Bellwood, D.R.; Pratchett, M.S.; Morrison, T.H.; Gurney, G.G.; Hughes, T.P.; Álvarez-Romero, J.G.; Day, J.C.; Grantham, R.; Grech, A.; Hoey, A.S.; et al. Coral reef conservation in the Anthropocene: Confronting spatial mismatches and prioritizing functions. *Biol. Conserv.* **2019**, *236*, 604–615. [CrossRef]
17. Bellwood, D.R.; Hughes, T.P.; Folke, C.; Nyström, M. Confronting the coral reef crisis. *Nature* **2004**, *429*, 827–833. [CrossRef] [PubMed]
18. Chiarello, M.; Auguet, J.C.; Bettarel, Y.; Bouvier, C.; Claverie, T.; Graham, N.A.J.; Rieuvilleneuve, F.; Sucré, E.; Bouvier, T.; Villéger, S. Skin microbiome of coral reef fish is highly variable and driven by host phylogeny and diet. *Microbiome* **2018**, *6*, 1–14. [CrossRef]
19. Chiarello, M.; Auguet, J.-C.; Graham, N.A.J.; Claverie, T.; Sucré, E.; Bouvier, C.; Rieuvilleneuve, F.; Restrepo-Ortiz, C.X.; Bettarel, Y.; Villéger, S.; et al. Exceptional but vulnerable microbial diversity in coral reef animal surface microbiomes. *Proc. R. Soc. B Biol. Sci.* **2020**, *287*, 20200642. [CrossRef] [PubMed]
20. Miyake, S.; Ngugi, D.K.; Stingl, U. Diet strongly influences the gut microbiota of surgeonfishes. *Mol. Ecol.* **2015**, *24*, 656–672. [CrossRef] [PubMed]
21. Pratte, Z.A.; Besson, M.; Hollman, R.D.; Stewarta, F.J. The gills of reef fish support a distinct microbiome influenced by hostspecific factors. *Appl. Environ. Microbiol.* **2018**, *84*, 1–15. [CrossRef]
22. Scott, J.J.; Adam, T.C.; Duran, A.; Burkepile, D.E.; Rasher, D.B. Intestinal microbes: An axis of functional diversity among large marine consumers. *Proc. R. Soc. B Biol. Sci.* **2020**, *287*, 20192367. [CrossRef]
23. Fischbach, M.A.; Segre, J.A. Signaling in Host-Associated Microbial Communities. *Cell* **2016**, *164*, 1288–1300. [CrossRef]
24. Li, X.; Zhou, L.; Yu, Y.; Ni, J.; Xu, W.; Yan, Q. Composition of Gut Microbiota in the Gibel Carp (Carassius auratus gibelio) Varies with Host Development. *Microb. Ecol.* **2017**, *74*, 239–249. [CrossRef]
25. Piazzon, M.C.; Calduch-Giner, J.A.; Fouz, B.; Estensoro, I.; Simó-Mirabet, P.; Puyalto, M.; Karalazos, V.; Palenzuela, O.; Sitjà-Bobadilla, A.; Pérez-Sánchez, J. Under control: How a dietary additive can restore the gut microbiome and proteomic profile, and improve disease resilience in a marine teleostean fish fed vegetable diets. *Microbiome* **2017**, *5*, 164. [CrossRef] [PubMed]
26. Ngugi, D.K.; Miyake, S.; Cahill, M.; Vinu, M.; Hackmann, T.J.; Blom, J.; Tietbohl, M.D.; Berumen, M.L.; Stingl, U. Genomic diversification of giant enteric symbionts reflects host dietary lifestyles. *Proc. Natl. Acad. Sci. USA* **2017**, *114*, E7592–E7601. [CrossRef]
27. Butt, R.L.; Volkoff, H. Gut microbiota and energy homeostasis in fish. *Front. Endocrinol. (Lausanne)* **2019**, *10*, 6–8. [CrossRef] [PubMed]
28. Sullam, K.E.; Essinger, S.D.; Lozupone, C.A.; O'Connor, M.P.; Rosen, G.L.; Knight, R.; Kilham, S.S.; Russell, J.A. Environmental and ecological factors that shape the gut bacterial communities of fish: A meta-analysis. *Mol. Ecol.* **2012**, *21*, 3363–3378. [CrossRef] [PubMed]
29. de Bruijn, I.; Liu, Y.; Wiegertjes, G.F.; Raaijmakers, J.M. Exploring fish microbial communities to mitigate emerging diseases in aquaculture. *FEMS Microbiol. Ecol.* **2018**, *94*, 1–12. [CrossRef]
30. Clements, K.D.; Angert, E.R.; Montgomery, W.L.; Choat, J.H. Intestinal microbiota in fishes: What's known and what's not. *Mol. Ecol.* **2014**, *23*, 1891–1898. [CrossRef] [PubMed]
31. Neave, M.J.; Apprill, A.; Aeby, G.; Miyake, S.; Voolstra, C.R. *Microbial Communities of Red Sea Coral Reefs*; Springer International Publishing: Cham, Switzerland, 2019; pp. 53–68. [CrossRef]
32. Amato, K.R.; Yeoman, C.J.; Kent, A.; Righini, N.; Carbonero, F.; Estrada, A.; Rex Gaskins, H.; Stumpf, R.M.; Yildirim, S.; Torralba, M.; et al. Habitat degradation impacts black howler monkey (*Alouatta pigra*) gastrointestinal microbiomes. *ISME J.* **2013**, *7*, 1344–1353. [CrossRef]
33. Kohl, K.D.; Amaya, J.; Passement, C.A.; Dearing, M.D.; Mccue, M.D. Unique and shared responses of the gut microbiota to prolonged fasting: A comparative study across five classes of vertebrate hosts. *FEMS Microbiol. Ecol.* **2014**, *90*, 883–894. [CrossRef] [PubMed]

34. Borbón-García, A.; Reyes, A.; Vives-Flórez, M.; Caballero, S. Captivity shapes the gut microbiota of Andean bears: Insights into health surveillance. *Front. Microbiol.* **2017**, *8*, 1–12. [CrossRef] [PubMed]
35. Clayton, J.B.; Vangay, P.; Huang, H.; Ward, T.; Hillmann, B.M.; Al-Ghalith, G.A.; Travis, D.A.; Long, H.T.; Van Tuan, B.; Van Minh, V.; et al. Captivity humanizes the primate microbiome. *Proc. Natl. Acad. Sci. USA* **2016**, *113*, 10376–10381. [CrossRef]
36. Schmidt, E.; Mykytczuk, N.; Schulte-Hostedde, A.I. Effects of the captive and wild environment on diversity of the gut microbiome of deer mice (*Peromyscus maniculatus*). *ISME J.* **2019**, *13*, 1293–1305. [CrossRef]
37. Gudka, M.; Obura, D.; Mbugua, J.; Ahamada, S.; Kloiber, U.; Holter, T. Participatory reporting of the 2016 bleaching event in the Western Indian Ocean. *Coral Reefs* **2020**, *39*. [CrossRef]
38. Faure, G.; Guillaume, M.; Payri, C.; Thomassin, B.; Vanpraet, M.; Vasseur, P. Massive bleaching and death of corals in the Mayotte Reef ecosystem (SW Indian-Ocean). *Comptes Rendus l'académie des Sci. Série III-Sciences la vie-Life Sci.* **1984**, *299*, 637–642.
39. Wilkinson, C.; Lindén, O.; Cesar, H.; Hodgson, G.; Rubens, J.; Strong, A.E. Ecological and socioeconomic impacts of 1998 coral mortality in the Indian Ocean: An ENSO impact and a warning of future change? *Ambio* **1999**, *28*, 188–196.
40. Wilson, S.K.; Robinson, J.P.W.; Chong-Seng, K.; Robinson, J.; Graham, N.A.J. Boom and bust of keystone structure on coral reefs. *Coral Reefs* **2019**, *38*, 625–635. [CrossRef]
41. Graham, N.A.J.; Wilson, S.K.; Jennings, S.; Polunin, N.V.C.; Bijoux, J.P.; Robinson, J. Dynamic fragility of oceanic coral reef ecosystems. *Proc. Natl. Acad. Sci. USA* **2006**, *103*, 8425–8429. [CrossRef] [PubMed]
42. Robinson, J.P.W.; Wilson, S.K.; Jennings, S.; Graham, N.A.J. Thermal stress induces persistently altered coral reef fish assemblages. *Glob. Chang. Biol.* **2019**, *25*, 2739–2750. [CrossRef]
43. Clements, K.D.; Pasch, I.B.Y.; Moran, D.; Turner, S.J. Clostridia dominate 16S rRNA gene libraries prepared from the hindgut of temperate marine herbivorous fishes. *Mar. Biol.* **2007**, *150*, 1431–1440. [CrossRef]
44. Taquet, M.; Delinguer, A. *Poissons de l'Océan Indien et de la Mer Rouge*; Quae: Versailles, 2007; ISBN 978-2-7592-0045-0.
45. Mouillot, D.; Villéger, S.; Parravicini, V.; Kulbicki, M.; Arias-González, J.E.; Bender, M.; Chabanet, P.; Floeter, S.R.; Friedlander, A.; Vigliola, L.; et al. Functional over-redundancy and high functional vulnerability in global fish faunas on tropical reefs. *Proc. Natl. Acad. Sci. USA* **2014**, *111*, 13757–13762. [CrossRef]
46. Clements, K.D.; German, D.P.; Piché, J.; Tribollet, A.; Choat, J.H. Integrating ecological roles and trophic diversification on coral reefs: Multiple lines of evidence identify parrotfishes as microphages. *Biol. J. Linn. Soc.* **2017**, *120*, 729–751. [CrossRef]
47. McArdle, B.H.; Anderson, M.J. Fitting multivariate models to community data: A comment on distance-based redundancy analysis. *Ecology* **2001**, *82*, 290–297. [CrossRef]
48. Oksanen, J.; Blanchet, F.G.; Michael, F.; Kindt, R.; Legendre, P.; McGlinn, D.; Minchin, P.R.; O'Hara, R.B.; Simpson, G.L.; Peter, S.; et al. Vegan: Community Ecology Package. 2019. Available online: inecol.mx (accessed on 9 August 2021).
49. Parada, A.E.; Needham, D.M.; Fuhrman, J.A. Every base matters: Assessing small subunit rRNA primers for marine microbiomes with mock communities, time series and global field samples. *Environ. Microbiol.* **2016**, *18*, 1403–1414. [CrossRef] [PubMed]
50. R Core Team R. A language and environment for statistical computing. 2020. Available online: r-project.org (accessed on 9 August 2021).
51. Callahan, B.J.; McMurdie, P.J.; Rosen, M.J.; Han, A.W.; Johnson, A.J.A.; Holmes, S.P. DADA2: High-resolution sample inference from Illumina amplicon data. *Nat. Methods* **2016**, *13*, 581–583. [CrossRef]
52. Glassman, S.I.; Martiny, J.B.H. Broadscale Ecological Patterns Are Robust to Use of Exact. *mSphere* **2018**, *3*, e00148-18. [CrossRef] [PubMed]
53. Quast, C.; Pruesse, E.; Yilmaz, P.; Gerken, J.; Schweer, T.; Yarza, P.; Peplies, J.; Glöckner, F.O. The SILVA ribosomal RNA gene database project: Improved data processing and web-based tools. *Nucleic Acids Res.* **2013**, *41*, 590–596. [CrossRef]
54. McMurdie, P.J.; Holmes, S. Phyloseq: An R Package for Reproducible Interactive Analysis and Graphics of Microbiome Census Data. *PLoS ONE* **2013**, *8*, e61217. [CrossRef] [PubMed]
55. Salter, S.J.; Cox, M.J.; Turek, E.M.; Calus, S.T.; Cookson, W.O.; Moffatt, M.F.; Turner, P.; Parkhill, J.; Loman, N.J.; Walker, A.W. Reagent and laboratory contamination can critically impact sequence-based microbiome analyses. *BMC Biol.* **2014**, *12*, 87. [CrossRef]
56. Cariveau, D.P.; Elijah Powell, J.; Koch, H.; Winfree, R.; Moran, N.A. Variation in gut microbial communities and its association with pathogen infection in wild bumble bees (Bombus). *ISME J.* **2014**, *8*, 2369–2379. [CrossRef]
57. Madhusoodanan, J. News Feature: Do hosts and their microbes evolve as a unit? *Proc. Natl. Acad. Sci. USA* **2019**, *116*, 14391–14394. [CrossRef]
58. Berg, R. The indigenous gastrointestinal microflora. *Trends Microbiol.* **1996**, *4*, 430–435. [CrossRef]
59. Martinson, V.G.; Danforth, B.N.; Minckley, R.L.; Rueppell, O.; Tingek, S.; Moran, N.A. A simple and distinctive microbiota associated with honey bees and bumble bees. *Mol. Ecol.* **2011**, *20*, 619–628. [CrossRef]
60. Koch, H.; Abrol, D.P.; Li, J.; Schmid-Hempel, P. Diversity and evolutionary patterns of bacterial gut associates of corbiculate bees. *Mol. Ecol.* **2013**, *22*, 2028–2044. [CrossRef]
61. Magurran, A.E.; Henderson, P.A. Explaining the excess of rare species in natural species abundance distributions. *Nature* **2003**, *422*, 714–716. [CrossRef] [PubMed]
62. Krebs, C.J. *Ecological Methodology*; Pearson; Addison-Wesley Educational Publishers: Menlo Park, 1999; ISBN 9780321021731.

63. Schloss, P.D.; Westcott, S.L.; Ryabin, T.; Hall, J.R.; Hartmann, M.; Hollister, E.B.; Lesniewski, R.A.; Oakley, B.B.; Parks, D.H.; Robinson, C.J.; et al. Introducing mothur: Open-source, platform-independent, community-supported software for describing and comparing microbial communities. *Appl. Environ. Microbiol.* **2009**, *75*, 7537–7541. [CrossRef] [PubMed]
64. Ludwig, W.; Strunk, O.; Westram, R.; Richter, L.; Meier, H.; Yadhukumar, A.; Buchner, A.; Lai, T.; Steppi, S.; Jacob, G.; et al. ARB: A software environment for sequence data. *Nucleic Acids Res.* **2004**, *32*, 1363–1371. [CrossRef]
65. Letunic, I.; Bork, P. Interactive tree of life (iTOL) v3: An online tool for the display and annotation of phylogenetic and other trees. *Nucleic Acids Res.* **2016**, *44*, W242–W245. [CrossRef] [PubMed]
66. Chao, A.; Lee, S.-M.; Chen, T.-C. A generalized Good's nonparametric coverage estimator. *Chin. J. Math.* **1988**, *16*, 189–199.
67. Jost, L. Entropy and diversity. *Oikos* **2006**, *113*, 363–375. [CrossRef]
68. Wemheuer, F.; Taylor, J.A.; Daniel, R.; Johnston, E.; Meinicke, P.; Thomas, T.; Wemheuer, B. Tax4Fun2: Prediction of habitat-specific functional profiles and functional redundancy based on 16S rRNA gene sequences. *Environ. Microbiome* **2020**, *15*, 11. [CrossRef]
69. Segata, N.; Izard, J.; Waldron, L.; Gevers, D.; Miropolsky, L.; Garrett, W.S.; Huttenhower, C. Metagenomic biomarker discovery and explanation. *Genome Biol.* **2011**, *12*, R60. [CrossRef]
70. Chong-Seng, K.M.; Mannering, T.D.; Pratchett, M.S.; Bellwood, D.R.; Graham, N.A.J. The influence of coral reef benthic condition on associated fish assemblages. *PLoS ONE* **2012**, *7*, e42167. [CrossRef]
71. Graham, N.A.J.; Robinson, J.P.W.; Smith, S.E.; Govinden, R.; Gendron, G.; Wilson, S.K. Changing role of coral reef marine reserves in a warming climate. *Nat. Commun.* **2020**, *11*, 2000. [CrossRef] [PubMed]
72. Silveira, C.B.; Cavalcanti, G.S.; Walter, J.M.; Silva-Lima, A.W.; Dinsdale, E.A.; Bourne, D.G.; Thompson, C.C.; Thompson, F.L. Microbial processes driving coral reef organic carbon flow. *FEMS Microbiol. Rev.* **2017**, *41*, 575–595. [CrossRef]
73. McDole, T.; Nulton, J.; Barott, K.L.; Felts, B.; Hand, C.; Hatay, M.; Lee, H.; Nadon, M.O.; Nosrat, B.; Salamon, P.; et al. Assessing Coral Reefs on a Pacific-Wide Scale Using the Microbialization Score. *PLoS ONE* **2012**, *7*, e43233. [CrossRef]
74. Meirelles, P.M.; Amado-Filho, G.M.; Pereira-Filho, G.H.; Pinheiro, H.T.; De Moura, R.L.; Joyeux, J.C.; Mazzei, E.F.; Bastos, A.C.; Edwards, R.A.; Dinsdale, E.; et al. Baseline assessment of mesophotic reefs of the Vitória-Trindade Seamount Chain based on water quality, microbial diversity, benthic cover and fish biomass data. *PLoS ONE* **2015**, *10*, e0130084. [CrossRef]
75. Haas, A.F.; Fairoz, M.F.M.; Kelly, L.W.; Nelson, C.E.; Dinsdale, E.A.; Edwards, R.A.; Giles, S.; Hatay, M.; Hisakawa, N.; Knowles, B.; et al. Global microbialization of coral reefs. *Nat. Microbiol.* **2016**, *1*, 16042. [CrossRef]
76. Roach, T.N.F.; Abieri, M.L.; George, E.E.; Knowles, B.; Naliboff, D.S.; Smurthwaite, C.A.; Kelly, L.W.; Haas, A.F.; Rohwer, F.L. Microbial bioenergetics of coral-algal interactions. *PeerJ* **2017**, *2017*, 1–19. [CrossRef] [PubMed]
77. Haas, A.F.; Nelson, C.E.; Kelly, L.W.; Carlson, C.A.; Rohwer, F.; Leichter, J.J.; Wyatt, A.; Smith, J.E. Effects of coral reef benthic primary producers on dissolved organic carbon and microbial activity. *PLoS ONE* **2011**, *6*, e27973. [CrossRef] [PubMed]
78. Vega Thurber, R.; Burkepile, D.E.; Correa, A.M.S.; Thurber, A.R.; Shantz, A.A.; Welsh, R.; Pritchard, C.; Rosales, S. Macroalgae Decrease Growth and Alter Microbial Community Structure of the Reef-Building Coral, Porites astreoides. *PLoS ONE* **2012**, *7*, e44246. [CrossRef]
79. Munday, P.L.; Jones, G.P.; Pratchett, M.S.; Williams, A.J. Climate change and the future for coral reef fishes. *Fish Fish.* **2008**, *9*, 261–285. [CrossRef]
80. Munday, P.; Jones, G.; Caley, M. Habitat specialisation and the distribution and abundance of coral-dwelling gobies. *Mar. Ecol. Prog. Ser.* **1997**, *152*, 227–239. [CrossRef]
81. Gardiner, N.; Jones, G. Habitat specialisation and overlap in a guild of coral reef cardinalfishes (Apogonidae). *Mar. Ecol. Prog. Ser.* **2005**, *305*, 163–175. [CrossRef]
82. Pratchett, M.S. Dietary overlap among coral-feeding butterflyfishes (Chaetodontidae) at Lizard Island, northern Great Barrier Reef. *Mar. Biol.* **2005**, *148*, 373–382. [CrossRef]
83. Bellwood, D.R.; Hughes, T.P.; Hoey, A.S. Sleeping Functional Group Drives Coral-Reef Recovery. *Curr. Biol.* **2006**, *16*, 2434–2439. [CrossRef] [PubMed]
84. Feary, D.; Almany, G.; Jones, G.; McCormick, M. Coral degradation and the structure of tropical reef fish communities. *Mar. Ecol. Prog. Ser.* **2007**, *333*, 243–248. [CrossRef]
85. Silveira, C.B.; Silva-Lima, A.W.; Francini-Filho, R.B.; Marques, J.S.M.; Almeida, M.G.; Thompson, C.C.; Rezende, C.E.; Paranhos, R.; Moura, R.L.; Salomon, P.S.; et al. Microbial and sponge loops modify fish production in phase-shifting coral reefs. *Environ. Microbiol.* **2015**, *17*, 3832–3846. [CrossRef] [PubMed]
86. Alexander, B.E.; Liebrand, K.; Osinga, R.; van der Geest, H.G.; Admiraal, W.; Cleutjens, J.P.M.; Schutte, B.; Verheyen, F.; Ribes, M.; van Loon, E.; et al. Cell Turnover and Detritus Production in Marine Sponges from Tropical and Temperate Benthic Ecosystems. *PLoS ONE* **2014**, *9*, e109486. [CrossRef] [PubMed]
87. Mueller, B.; de Goeij, J.M.; Vermeij, M.J.A.; Mulders, Y.; van der Ent, E.; Ribes, M.; van Duyl, F.C. Natural Diet of Coral-Excavating Sponges Consists Mainly of Dissolved Organic Carbon (DOC). *PLoS ONE* **2014**, *9*, e90152. [CrossRef] [PubMed]
88. Smith, J.E.; Shaw, M.; Edwards, R.A.; Obura, D.; Pantos, O.; Sala, E.; Sandin, S.A.; Smriga, S.; Hatay, M.; Rohwer, F.L. Indirect effects of algae on coral: Algae-mediated, microbe-induced coral mortality. *Ecol. Lett.* **2006**, *9*, 835–845. [CrossRef]
89. Barott, K.L.; Rohwer, F.L. Unseen players shape benthic competition on coral reefs. *Trends Microbiol.* **2012**, *20*, 621–628. [CrossRef]
90. Morrow, K.M.; Ritson-Williams, R.; Ross, C.; Liles, M.R.; Paul, V.J. Macroalgal Extracts Induce Bacterial Assemblage Shifts and Sublethal Tissue Stress in Caribbean Corals. *PLoS ONE* **2012**, *7*, e44859. [CrossRef] [PubMed]

91. Beatty, D.; Clements, C.; Stewart, F.; Hay, M. Intergenerational effects of macroalgae on a reef coral: Major declines in larval survival but subtle changes in microbiomes. *Mar. Ecol. Prog. Ser.* **2018**, *589*, 97–114. [CrossRef]
92. Nelson, C.E.; Goldberg, S.J.; Wegley Kelly, L.; Haas, A.F.; Smith, J.E.; Rohwer, F.; Carlson, C.A. Coral and macroalgal exudates vary in neutral sugar composition and differentially enrich reef bacterioplankton lineages. *ISME J.* **2013**, *7*, 962–979. [CrossRef]
93. Meirelles, P.M.; Soares, A.C.; Oliveira, L.; Leomil, L.; Appolinario, L.R.; Francini-Filho, R.B.; De Moura, R.L.; De Barros Almeida, R.T.; Salomon, P.S.; Amado-Filho, G.M.; et al. Metagenomics of coral reefs under phase shift and high hydrodynamics. *Front. Microbiol.* **2018**, *9*, 1–13. [CrossRef] [PubMed]
94. Aiken, D.E.; Waddy, S.L. Aquaculture. In *Biology of the Lobster*; Elsevier: Amsterdam, The Netherlands, 1995; pp. 153–175.
95. Ghanbari, M.; Kneifel, W.; Domig, K.J. A new view of the fish gut microbiome: Advances from next-generation sequencing. *Aquaculture* **2015**, *448*, 464–475. [CrossRef]
96. Egerton, S.; Culloty, S.; Whooley, J.; Stanton, C.; Ross, R.P. The gut microbiota of marine fish. *Front. Microbiol.* **2018**, *9*, 873. [CrossRef]
97. Barelli, C.; Albanese, D.; Donati, C.; Pindo, M.; Dallago, C.; Rovero, F.; Cavalieri, D.; Michael Tuohy, K.; Christine Hauffe, H.; De Filippo, C. Habitat fragmentation is associated to gut microbiota diversity of an endangered primate: Implications for conservation. *Sci. Rep.* **2015**, *5*, 14862. [CrossRef]
98. Hayden, B.; Palomares, M.L.D.; Smith, B.E.; Poelen, J.H. Biological and environmental drivers of trophic ecology in marine fishes-a global perspective. *Sci. Rep.* **2019**, *9*, 11415. [CrossRef] [PubMed]
99. Green, A.L.; Maypa, A.P.; Almany, G.R.; Rhodes, K.L.; Weeks, R.; Abesamis, R.A.; Gleason, M.G.; Mumby, P.J.; White, A.T. Larval dispersal and movement patterns of coral reef fishes, and implications for marine reserve network design. *Biol. Rev.* **2015**, *90*, 1215–1247. [CrossRef]
100. Zaneveld, J.R.; McMinds, R.; Thurber, R.V. Stress and stability: Applying the Anna Karenina principle to animal microbiomes. *Nat. Microbiol.* **2017**, *2*, 17121. [CrossRef] [PubMed]
101. Van Soest, J.P. *Nutritional Ecology of the Ruminant*; Cornell University Press: Ithaca, NY, USA, 1994.
102. Jones, J.; DiBattista, J.D.; Stat, M.; Bunce, M.; Boyce, M.C.; Fairclough, D.V.; Travers, M.J.; Huggett, M.J. The microbiome of the gastrointestinal tract of a range-shifting marine herbivorous fish. *Front. Microbiol.* **2018**, *9*, 2000. [CrossRef] [PubMed]
103. Zhang, X.; Wu, H.; Li, Z.; Li, Y.; Wang, S.; Zhu, D.; Wen, X.; Li, S. Effects of dietary supplementation of Ulva pertusa and non-starch polysaccharide enzymes on gut microbiota of Siganus canaliculatus. *J. Oceanol. Limnol.* **2018**, *36*, 438–449. [CrossRef]
104. Miller, D.A.; Suen, G.; Bruce, D.; Copeland, A.; Cheng, J.-F.; Detter, C.; Goodwin, L.A.; Han, C.S.; Hauser, L.J.; Land, M.L.; et al. Complete Genome Sequence of the Cellulose-Degrading Bacterium Cellulosilyticum lentocellum. *J. Bacteriol.* **2011**, *193*, 2357–2358. [CrossRef]
105. Mountfort, D.O.; Campbell, J.; Clements, K.D. Hindgut Fermentation in Three Species of Marine Herbivorous Fish. *Appl. Environ. Microbiol.* **2002**, *68*, 1374–1380. [CrossRef]
106. Clements, K.D.; Choat, J.H. Fermentation in Tropical Marine Herbivorous Fishes. *Physiol. Zool.* **1995**, *68*, 355–378. [CrossRef]

Article

# Is the Intestinal Bacterial Community in the Australian Rabbitfish *Siganus fuscescens* Influenced by Seaweed Supplementation or Geography?

**Valentin Thépot [1,*], Joel Slinger [2,3], Michael A. Rimmer [1], Nicholas A. Paul [1] and Alexandra H. Campbell [4]**

1 School of Science, Technology and Engineering, University of the Sunshine Coast, Maroochydore DC, QLD 4558, Australia; mrimmer@usc.edu.au (M.A.R.); npaul@usc.edu.au (N.A.P.)
2 CSIRO Agriculture and Food, Bribie Island Research Centre, Woorim, QLD 4507, Australia; joel.slinger@csiro.au
3 Institute of Marine and Antarctic Studies, University of Tasmania, Launceston, TAS 7250, Australia
4 School of Health and Behavioural Sciences, University of the Sunshine Coast, Maroochydore DC, QLD 4558, Australia; acampbe1@usc.edu.au
* Correspondence: valentin.thepot@research.usc.edu.au

**Abstract:** We recently demonstrated that dietary supplementation with seaweed leads to dramatic improvements in immune responses in *S. fuscescens*, a candidate species for aquaculture development in Asia. Here, to assess whether the immunostimulatory effect was facilitated by changes to the gut microbiome, we investigated the effects of those same seaweed species and four commercial feed supplements currently used in aquaculture on the bacterial communities in the hindgut of the fish. Since we found no correlations between the relative abundance of any particular taxa and the fish enhanced innate immune responses, we hypothesised that *S. fuscescens* might have a core microbiome that is robust to dietary manipulation. Two recently published studies describing the bacteria within the hindgut of *S. fuscescens* provided an opportunity to test this hypothesis and to compare our samples to those from geographically distinct populations. We found that, although hindgut bacterial communities were clearly and significantly distinguishable between studies and populations, a substantial proportion (55 of 174 taxa) were consistently detected across all populations. Our data suggest that the importance of gut microbiota to animal health and the extent to which they can be influenced by dietary manipulations might be species-specific or related to an animals' trophic level.

**Keywords:** functional ingredient; immunity; core microbiome; macroalga and rabbitfish

**Citation:** Thépot, V.; Slinger, J.; Rimmer, M.A.; Paul, N.A.; Campbell, A.H. Is the Intestinal Bacterial Community in the Australian Rabbitfish *Siganus fuscescens* Influenced by Seaweed Supplementation or Geography?. *Microorganisms* **2022**, *10*, 497. https://doi.org/10.3390/microorganisms10030497

Academic Editor: Konstantinos Ar. Kormas

Received: 13 January 2022
Accepted: 17 February 2022
Published: 23 February 2022



## 1. Introduction

Changes in the communities of microbiota within the gastro-intestinal tracts of fish (their 'gut microbiomes') have recently been linked to impacts on fish health and condition, including their metabolism, overall size, and immune responses [1]. This emerging understanding has great potential to facilitate the development of sustainable aquaculture industries because harnessing the positive effects of gut microbiomes on fish health could reduce the reliance of the industry on antibiotics and other chemotherapies [2]. However, we still know very little about the structure or function of gut microbiomes in most farmed fish species, or how they can be enhanced to improve yields in aquaculture [3]. Indeed, understanding the role of microbiomes in the health and resilience of marine and aquatic animals was recently highlighted as a key knowledge gap in the field of marine microbiome research [4].

Aquaculture recently replaced wild fisheries as the main source of seafood globally [5] and its importance in the provision of protein is likely to increase [5,6]. Two of the greatest threats to the sustainability of aquaculture are (1) its reliance on fish meal and fish oil from increasingly depleted wild fisheries [7]; and (2) disease [8]. The use of plant-based feed alternatives has been proposed as a potential solution to the unsustainable use of wild fish products [9], however, these novel ingredients, which fish rarely encounter naturally, can create novel challenges, such as stunted growth, increased mortality, and gut inflammation, especially in highly valuable carnivorous species [9,10].

Diet can strongly influence the structure of gut microbiomes in some fish [11,12] and dietary supplementation has been suggested as a tool to improve disease resistance in aquaculture [13]. However, dietary supplementation with plant materials from terrestrial environments can have negative outcomes, including reduced diversity in the microbial communities that persist within fish gastrointestinal (GI) tracts and associated negative health outcomes for farmed fish [13,14].

Our recent review revealed the exciting potential of seaweeds (marine 'plants') as immunostimulants for farmed fish, and also highlighted major gaps in our understanding, including the potential mode/s of action of successful immunostimulants (i.e., direct or mediated by the hosts' microbiome; [15]). Indeed, the effects of most of the dietary supplements currently in use or of interest, on the microbiomes in the GI tracts of commercially important fish, are completely unknown, especially for lower-trophic level (i.e., herbivorous) fish.

We recently demonstrated that dietary supplementation with several species of seaweed caused significant stimulation of parts of the innate immune response of the mottled rabbitfish *Siganus fuscescens* [16], however, the mode/s of action of this immunostimulatory effect remain unknown. Here, we explore whether observed changes in the innate immune responses of experimental *S. fuscescens* were correlated with shifts in the GI microbiomes of those fish and thus, whether a mode of action of dietary seaweed immunostimulation may have been microbially-mediated. Seaweeds used in this trial included members of the red, green, and brown taxonomic groups and species that produce a broad range of bioactive, natural compounds (e.g., bromoform in *Asparagopsis taxiformis*, caulerpin in *Caulerpa taxifolia*; and terpenoids in *Sargassum* sp. [17–19]). Compounds from these species all have antimicrobial activity [19–21] demonstrated in laboratory assays. We hypothesised that changes in innate immune responses in *S. fuscescens* resulting from seaweed supplementation would be correlated with changes in gut microbiomes and provide evidence that a potential microbially-mediated mode of action of seaweed immunostimulants. This study targeted the hindgut bacterial community based on literature reporting that the hindgut is the part of the GI tract that contains a higher proportion of resident rather than transient microbiota [22], suggesting that this region may be more reflective of the 'true' fish GI microbiome.

Rabbitfish of the Siganidae family are marine herbivores presently receiving increased attention due to their attractive attributes for aquaculture [23–26] and their range-shifting ability and associated indirect impacts of warming waters on temperate ecosystems ("Tropicalization") [27]. The mottled rabbitfish (*Siganus fuscescens*) was the focus of our study because of its candidature for aquaculture development in Asia [24] and we aimed to provide baseline information about the geographical and temporal variation of the gut microbiome of *S. fuscescens* to support the development of a sustainable farming industry for this species.

When comparing hindgut microbial communities in fish fed different diets during our experiments, we became interested in the possible existence of a core microbiome in this species. 'Core microbiomes' are variably defined in the literature (e.g., based on 50%, 90% or 100% prevalence cut-offs [28]) and remain a subject of active debate and research [4]. Since we know so little about natural variation in the intestinal microbiomes of farmed fish in general and herbivorous marine fish in particular, and because of our emerging interest in a possible core microbiome in this species, we also compared our data to those from two recently published studies that characterised hindgut bacterial communities from conspecific populations of *S. fuscescens*. Our aims here were two-fold: firstly, to provide some ecological context for our results and, secondly, to provide baseline data on the microbiome of Siganids to help facilitate the aquaculture development of this fish species.

## 2. Material and Methods

Mottled rabbitfish (*Siganus fuscescens*: ranging from 15 cm/70 g to 21 cm/189 g) were captured between February and March 2018 using a drag net (15 m long by 2.1 m deep with a 2.5 cm mesh size) on rocky reefs at Moffat Beach, Queensland Australia (26°47′21.7″ S 153°08′36.0″ E; Figure 1). This collection was carried out under a "General fisheries permit" (# 195305) issued by the Queensland Department of Agriculture and Fisheries (Fisheries Act 1994). Feeding trials were conducted at the Bribie Island Research Centre (BIRC) on Bribie Island, Queensland, Australia (27°03′15.9″ S 153°11′42.9″ E). After collection, fish were transferred to BIRC in an oxygenated 500 L tank. The newly captured fish were treated with hydrogen peroxide (200 mg/L for 30 min) to rid them of potential external pathogens and parasites as per BRIC biosecurity requirements. Although it was not the aim of this experiment, it is possible that this treatment might have had effects on the GI microbial communities of the fish. However, since all fish were exposed to the same hydrogen peroxide treatment, we assumed that the treatment effect was even and thus did not affect the current feeding trial. Following this, the fish were transferred to three 1000 L fibreglass tanks where they were acclimatised and fed the control (unsupplemented 'Native' pellets from Ridley Aquafeeds Ltd.) diet for at least two weeks. The Native diet has been formulated for Australian native carnivorous freshwater fish species and was chosen based on its low protein (38% protein, 10% fat content, and 15 MJ/kg gross energy) compared to other commercially available diets. All activities were approved by the animal ethics committee of the University of the Sunshine Coast (ANS1751).

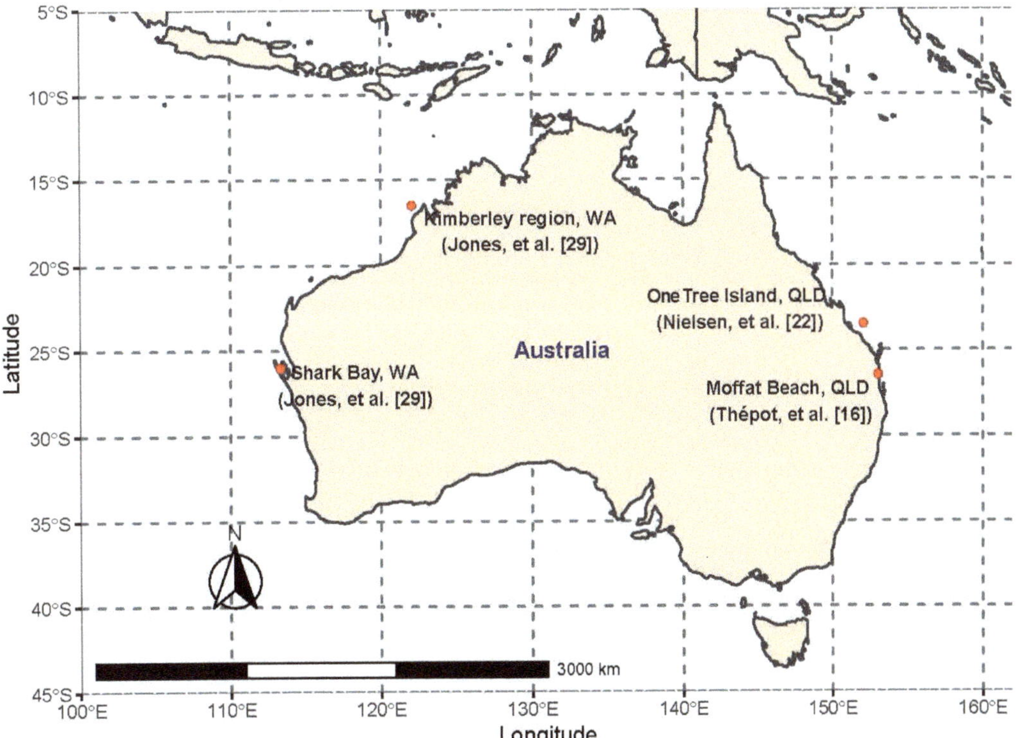


**Figure 1.** Map of mainland Australia including the four *Siganus fuscescens* sampling locations from the three studies; Kimberley region (pink) and Shark Bay (dark pink) from Jones, et al. [29], One Tree Island (blue) from Nielsen, et al. [22] and Moffat Beach (black) from Thépot, et al. [16].

## 2.1. Seaweed and Experimental Diets

We aimed to screen multiple species of taxonomically and chemically diverse seaweeds for their effects on the hindgut microbiomes of *S. fuscescens*. Eleven species of seaweed (5 red, 3 brown, and 3 green species) were evaluated as functional ingredients in feeding trials with *S. fuscescens* (Supplementary Table S1, hereafter referred to by genus). Four commercially available 'aquafeed' supplements were also evaluated: (i) Hilyses ® (MarSyt Inc., Elizabethtown, PA, USA), a hydrolysed yeast culture derived from the sugarcane fermentation process (and a source of β-glucans), (ii) sodium alginate, the anionic polysaccharide extracted from brown seaweeds, (iii) the cyanobacteria spirulina (high strength organic spirulina, Swiss Wellness Pty Ltd., Collingwood, VIC, Australia) and (iv) cracked and window refractance encysted (>95%) dried biomass of the microalga *Haematococcus pluvialis*, which is a source of astaxanthin. Together there was a total of 15 supplement treatments in the trial. The proximate composition of each supplement was determined following the recommended methods of the Association of Official Analytical Chemists [30], with the protein estimation using a factor of 5 to multiply the seaweed nitrogen content as recommended by Angell, et al. [31]. The source of each species, their morphological and chemical features of interest of the supplements from the four groups (red, green, and brown seaweed and aquafeed supplements) are described in Supplementary Table S1.

For the preparation of the seaweed-supplemented diets, fresh seaweeds were rinsed with saltwater (34.5 ppt) to remove sand and biological contaminants. They were then spun in a washing machine (Fisher and Paykel 5.5 kg Quick Smart, East Tamaki, New Zealand) on spin cycle (1000 rpm) for 5 min to remove excess water, frozen at −80 °C, and then lyophilised in a freeze dryer (Thermo Savant model MODULYOD-230, Waltham, MA, USA) for 3 days at approximately −44 °C and 206 mbar. Once dried, each seaweed species was vacuum-sealed in individual bags with silica desiccant and stored at −20 °C until used. The 'control' (unsupplemented) diets for experimental *S. fuscescens* were produced using the commercial aquafeed 'Native' (Ridley Aquafeeds Ltd., Brisbane, QLD, Australia). The pellets (1.5 kg in total) were powdered then added to a blender (Hobart A120, Silverwater, NSW, Australia) with deionised water (30% weight/weight) and combined for approximately 10 min at low speed (agitator rpm of 104) using a dough hook to produce a stiff dough. The dough was extruded through a 4 mm die onto trays which were then placed in a fan-forced oven overnight at 50 °C. Once dried, the feed was packaged in airtight bags and stored at 4 °C until required. All 15 experimental diets (supplemented with seaweed or aquafeed supplement) were made in the same manner but received supplements at 3% dietary inclusion which were powdered and sieved through a 300 μm mesh prior to the addition of water during the blending step. The use of an unsupplemented, control diet is standard practice in aquaculture to test the effect that specific ingredients may have on fish [14,15,32–34]. The use of wild fish as control would be inadequate because wild fish are not exposed to the same conditions as fish in aquaculture (e.g., artificial diet and captivity) thus one could not conclude if differences between wild fish control and fish fed treatment diets would be a link to (1) the treatment ingredient, (2) the artificial diet used to convey that ingredient or (3) the captivity effect (e.g., exposure to artificial light and filtered seawater).

## 2.2. Experimental Design

This study used three replicate plastic tanks (55 L) for each of the 16 dietary treatments (including the control; $n = 3$) to have a total of 48 tanks, all of which were then stocked with three fish each (144 fish in total). Due to variation in sizes, fish ($N = 144$) were sorted into two size classes: 'small' (ranging from 15 cm/70.5 g to 18 cm/112.1 g) or 'large' (ranging from 18 cm/112.4 g to 21 cm/189.2 g). The 144 fish were randomly allocated into 48 tanks so that each replicate tank contained 3 fish with at least one small and one large fish. The exact mass and length of each fish were recorded and used as a covariate in analyses assessing the influence of fish diet on microbial community diversity. As processing limitations were forecasted for the end of the trial, the fish were stocked in a staggered manner with one tank per treatment stocked each day over three days to allow for the sampling of one tank per treatment each day over three days at the end of the screening trial.

Fish were fed one of 16 different diets, comprising of 15 experimental diets and a control diet. Each treatment consisted of three replicate tanks and 9 fish (3 fish per replicate tank). Therefore 'Tank' was a random factor nested within the fixed factor of 'Diet'. To enable staggered sampling at the end of the experiment and ensure that all fish were exposed to the treatment diets for the same time period (two weeks), one out of three tanks from each dietary treatment was stocked with fish each day, over three days. The *Ulva* dietary treatment included only 2 replicate tanks after the loss of one tank due to water and air supply issues.

Fish were fed by hand at 3% body weight twice a day (10:00 and 15:00) for a period of 14 days. The reason for the trial lasting 14 days is based on our previous review [15], which revealed that trials where fish were fed seaweed as a functional ingredient lasted on average 14 days to conduct blood immunochemistry analysis for innate immune responses. No differences in feed consumption between tanks or treatments were observed as fish in all tanks consumed the total of both morning and afternoon feed allocations in each tank (visual inspection during handfeeding). During the trial, the water temperature was maintained at 27 °C, and the pH was within the range of 7.9 and 8.1. The system was operated as flow-through, with fresh seawater (34–35 ppt) pumped from approximately 300 m off the beach adjacent to the research station then through a series of 16 spin disk filters (40 μm) and 10 multimedia filters (~10–15 μm), after which it received ozone treatment from two 100 $gO_3/h$ generator units (WEDECO OCS-GSO30, Herford, Germany). The ozone-treated seawater was then pumped via ultraviolet filters, providing 80 $mJ/cm^2$, to two 4 m × 2.2 m granular activated carbon vessels for a contact time of >9 min to remove unwanted by-products from the ozone treatment. Finally, the seawater was pumped to a header tank, which fed directly into a pipe system delivering treated seawater to this experiment. The system was maintained in a temperature and light controlled room kept at 24–26 °C and on a 24L:0D dim central light regime.

### 2.3. Sample Collection and Preparation

After the feeding trial (14 days), the fish were subjected to a 24 h fasting period. The fish were euthanized in 10 ppt Aqui-S® (Lower Hutt, New Zealand), then the entire digestive tract from each fish was aseptically excised and placed in a Falcon tube (50 mL) before being snap-frozen and stored at −80 °C until further processing could occur.

### 2.4. Innate Immune Variables Measured

The full methodological details of samples obtained for analysis of innate immune responses to the different seaweed diets are described in Thépot, et al. [16]. Briefly, we obtained blood samples to assess cellular innate immune responses, which included the phagocytic activity/index and respiratory burst activity and we also obtained serum samples to assess humoral innate immune parameters including lysozyme activity and haemolytic activity.

### 2.5. DNA Extraction

To compare the hindgut microbiomes of fish fed different experimental diets, DNA was isolated from the hindgut and digesta of one randomly selected small and large fish from each tank (except for the *Ulva fasciata* treatment which only had 2 tanks). After the samples thawed, 0.25 g (approximately 0.5 cm length) of hindgut containing digesta was sampled. Our rationale for choosing to sample the hindgut with digesta was based on results published by Nielsen, et al. [22] and Jones, et al. [29], which suggested that this part of the microbiome was more representative of the host's GI microbiome rather than the more transient and food associated microbiome of the midgut. We defined the section of the distal intestine starting 1 cm internally to the anal pore as "hindgut", referred to as such hereafter. Digesta containing hindgut samples were placed directly into the isolation buffer in PowerBead tubes from the PowerSoil DNA isolation kit (Mo Bio, San Diego, CA, USA). Microbial DNA was isolated from the hindgut samples following the manufacturer's instructions and then stored at −20 °C.

### 2.6. Sixteen S rRNA Gene Sequencing and Bioinformatics

From the isolated DNA, the 16S rRNA gene was amplified using PCR following previously published methods [35–37]. Briefly, the hypervariable region V3–V4 was targeted using the primers 341F (5′-CCTAYGGGRBGCASCAG-3′) and 806R (5′-GGACTACNNGGGTATCTAAT-3′) at the Australian Genome Research Facility (AGRF, Melbourne, VIC, Australia), who then sequenced the amplicons on a MiSeq platform (2 × 300 bp; MCS v3.1.0.13, San Diego, CA, USA), and the resulting reads were analysed with Illumina bcl2fastq pipeline v2.20.0.422 (San Diego, CA, USA). Demultiplexed paired-end reads were assembled by aligning the forward and reverse reads using Quantitative Insights into Microbial Ecology QIIME2 v2018.8; (available at http://qiime.org/, accessed on 10 January 2021) [38]. To ensure that comparisons were made from sequences assigned in the same hypervariable region (V4) of the comparison studies (below), the raw data from the current study was trimmed using the cutadapt package [39], using the 515F (5′-GTGCCAGCMGCCGCGGTAA-3′) and 806R (5′-GGACTACHVGGGTWTCTAAT-3′) primers as per Yu, et al. [40]. Trimmed sequences were processed and denoised using the DADA2 package v1.16.0 (available at https://www.bioconductor.org/packages/release/bioc/html/dada2.html, accessed on 10 January 2021) [41] and QIIME2 (v2018.8) software, with amplicon sequence variants (ASVs) tables constructed and aligned against the Silva 16S rRNA 99% reference database (release v132) (available at https://www.arb-silva.de/, accessed on 10 January 2021) [42]. Due to practical and budget restraints, the DNA samples were sequenced in two separate runs on the same machine at the same facility (AGRF). Bioinformatical and statistical steps were included to ensure comparability between the two sequencing runs (see below). Raw sequences have been deposited in the National Center for Biotechnology Information (NCBI) sequence read archive (SRA) under the bioproject number PRJNA649307.

Approximately 95.1% (398,112) of total reads were quality filtered and retained through this process. Subsequent quality filtering included the removal of singletons, chimeric sequences, mitochondrial DNA, and unassigned or eukaryotic ASVs. This resulted in a total of 1250 ASVs from 48 samples. Rarefaction to 6290 counts was performed to account for uneven sequencing depth among samples ( Supplementary Figure S1). This resulted in the removal of one replicate from the *Laurencia* treatment (4_1_s; 874 counts) and the removal of 52 ASVs no longer present after rarefaction, leaving a total of 1,198 ASVs and 47 samples.

### 2.7. Comparisons with Previously Published Data on the Immune Response of the Same Individuals of S. fuscescens

The recently published paper [16] characterised the immune response of rabbitfish from a large experiment ($n$ = 9 fish, 3 fish per tank). Here a subset of these fish per tank ($n$ = 3) were randomly selected and are then related back to the immune data of those individual fish. These were included in MDS and PERMANOVA as per the Data Analysis (below).

### 2.8. Comparisons with Previously Published Sequences of the Hindgut Microbiota from Wild Populations of S. fuscescens

Two recently published papers [22,29] also characterised microbial communities in the hindgut of wild-caught *S. fuscescens*. With the permission of those authors and the provision of raw sequence data from those papers, we compared the microbiomes of our captive fish (fed experimental diets) to the results obtained from fish caught from wild populations on the east and western coastlines of Australia, respectively. Nielsen, et al. [22] characterised GI microbiomes in wild populations of *S. fuscescens* captured nearby One Tree Island (23°30′27.0″ S, 152°05′30.5″ E) in the tropical Great Barrier Reef (GBR), whereas Jones, et al. [29] sampled wild fish from two populations in Western Australia (WA), including the subtropical Shark Bay (26°01′47.28″ S, 113°33′12.49″ E) and the tropical Kimberley region (16°51′14.57″ S, 122°10′39.45″ E).

Raw sequence data were retrieved from the NCBI Short Read Archive (SRA; Jones, et al. [29]; accession number PRJNA356981 and Nielsen, et al. [22]; accession number PRJNA396430) using the SRA Toolkit software(v2.10.9, available at https://github.com/ncbi/sra-tools, accessed on 10 January 2021) and processed as demultiplexed fastq files. Raw data from both comparison studies (sequenced in the V4 region) were also processed using the cutadapt package [39] to remove respective primer sequences. From this point, the same bioinformatic pipeline as detailed above was used, with identical denoising, filtering and taxonomic reference database (Silva 16S rRNA gene 99% reference database, release v132) applied. Subsequent quality filtering included the removal of singletons, chimeric sequences, mitochondrial DNA, and unassigned or eukaryotic ASVs. Approximately 95.1% (713,284) of total reads were quality filtered and retained through this process. Subsequent quality filtering included the removal of singletons, chimeric sequences, mitochondrial DNA, and unassigned or eukaryotic ASVs. This resulted in 3160 ASVs from 86 samples. Rarefaction to 6290 counts was performed to account for uneven sequencing depth among studies and samples. This resulted in the removal of the same replicate (from the *Laurencia* treatment; (4_1_s; 874 counts) and the removal of 76 ASVs no longer present after rarefaction, leaving a total of 3084 ASVs and 85 samples (Supplementary Figure S2).

### 2.9. Data Analysis and Statistics

After processing, data were imported into R v3.6.3 (available at https://www.r-project.org/, accessed on 9 October 2020) [43] using the package phyloseq [44] for statistical analysis and visualisations. The effects of the different diets and the overall relationship between 5 innate immune parameters (lysozyme activity, phagocytic activity, and index, haemolytic activity: ACH50 and respiratory burst activity), 6 health indicators (erythrocytes: RBC, leukocytes: WBC, mean corpuscular volume: MCV and hepatosomatic index: HSI), fish weight and the most abundant 17 ASVs (representing >1% relative abundance) in the hindgut of *S. fuscescens* fed the different diets was explored in a non-metric multidimensional scaling (NMDS) using Euclidian distance and compared between treatments using PERMANOVA. Differences between means are considered significant at $p < 0.05$. Alpha diversity of microbial communities, Observed ASVs, and Shannon-Weaver index (hereafter "Shannon index"), were compared among fish fed different diets and later, between different studies, using Kruskal-Wallis tests. For the rest of the analyses, to allow the comparison of both sequencing runs on a shared number of ASVs, the rarefied ASVs were agglomerated at the genus level. Venn diagrams were used to show the number of shared ASVs among samples and studies and were constructed using the Limma package [45]. Beta diversity was visualised using non-metric multidimensional scaling (NMDS) ordinations and Bray-Curtis and unweighted UniFrac community dissimilarity indices and compared between treatments and fish length as a covariate using PERMANOVA [46]. ASV level differences in relative abundance between each treatment and the control were evaluated using multiple one-way ANOVAs, with square root transformed data to meet the assumptions of homogeneity of variance and improve normality.

In order to compare our fish as one population ("Sunshine Coast") to the other 3 wild populations of *S. fuscescens*, the 47 fish fed the different treatment diets in our trial were combined under the "Sunshine Coast" population. The four geographically distinct rabbitfish populations were analysed using pairwise comparisons of changes in the relative abundances of raw, un-rarefied data using Wald tests in the DESeq2 function [47] where the $p$-values were adjusted using the Benjamini and Hochberg method. ASV level differences between each population (Shark Bay, Kimberley, GBR, and Sunshine Coast) were evaluated using multiple one-way ANOVAs, with square root transformed data to meet the assumptions of homogeneity of variance and improve normality. Significant ANOVA ($p < 0.05$) results were followed by a Tukey's HSD post hoc test. Additionally, the package *microbiome* [48] was used as in previous studies [49,50], to identify ASVs that were part of a core microbiome in fish from the four geographic populations. In the literature, core microbiomes are variably defined, but the most common definitions we found were that microbial taxa are considered part of a 'core microbiome' when they are present in 50%, 90%, or 100%, (e.g., [28]), of sampled individuals. We, therefore, applied all three prevalence thresholds to assess the possibility of a core microbiome in the hindgut of this species across studies.

## 3. Results

### 3.1. Relationship between Innate Immune Response and Microbiome Taxonomic Composition

Despite clear and often dramatic influences of seaweed diets on several innate immune parameters, there were no overall, community-level differences in the hindgut bacterial communities between fish fed the different treatment diets or any relationships between microbiota and the innate immune parameters we measured (PERMANOVA: $F = 0.39$, $p = 0.846$; Figure 2). However, 13 ASVs correlated with other measurements, including fish haemolytic activity, respiratory burst activity, and haematocrit (Figure 2B, Supplementary Table S2). The respiratory burst activity of the fish was not positively correlated with any ASV but it was negatively correlated to ASV7160 (unidentified *Firmicutes*; Figure 2A,B and Supplementary Table S2). Similarly, haemolytic activity (ACH50) was negatively correlated with ASV2081 (*Arcobacter* sp.), which was a highly abundant taxon in the hindgut of the fish fed the control diet (which also had low ACH50; Figure 3A and Supplementary Table S3). The relationship between ASV2081 and fish ACH50 is unclear as fish fed diets supplemented with *Asparagopsis* had significantly higher ACH50 than the other fish [16], although there were no differences in the relative abundance of this ASV between diet types. Rather, seven treatments had higher relative abundance while eight treatments had a lower relative abundance of ASV2081 compared to the fish fed *Asparagopsis* (Supplementary Table S3).

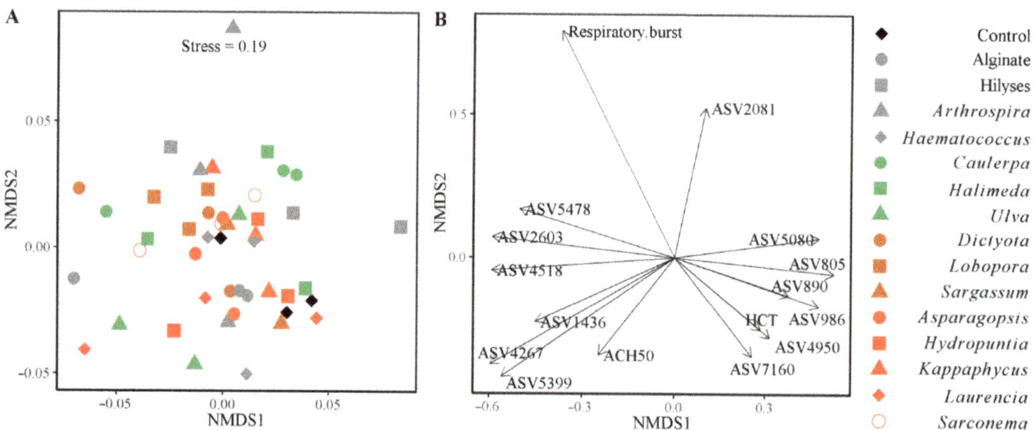


**Figure 2.** (**A**) Non-metric multidimensional scaling (NMDS) plot (Euclidean distance) of the 5 innate immune parameters (lysozyme activity, phagocytic activity and index, haemolytic activity: ACH50 and respiratory burst activity), the 6 health indicators (erythrocytes: RBC, leukocytes: WBC, mean corpuscular volume: MCV and hepatosomatic index: HSI), fish weight and most abundant 17 ASVs (representing >1% relative abundance) in the hindgut of *S. fuscescens* fed the different dietary treatments with the individual fish and (**B**) plot of the original variables (innate immune response and health indicators) loaded as vectors in NMDS space (with loading >0.7; $p < 0.05$). The different colours represent the fish fed the different groups of seaweed (green, brown, and red), the positive controls (grey), and those fed the unsupplemented control diet (black) ($n = 3$ data points per treatment refer to 3 replicate tanks comprised of one sub-sampled fish [small or large]).

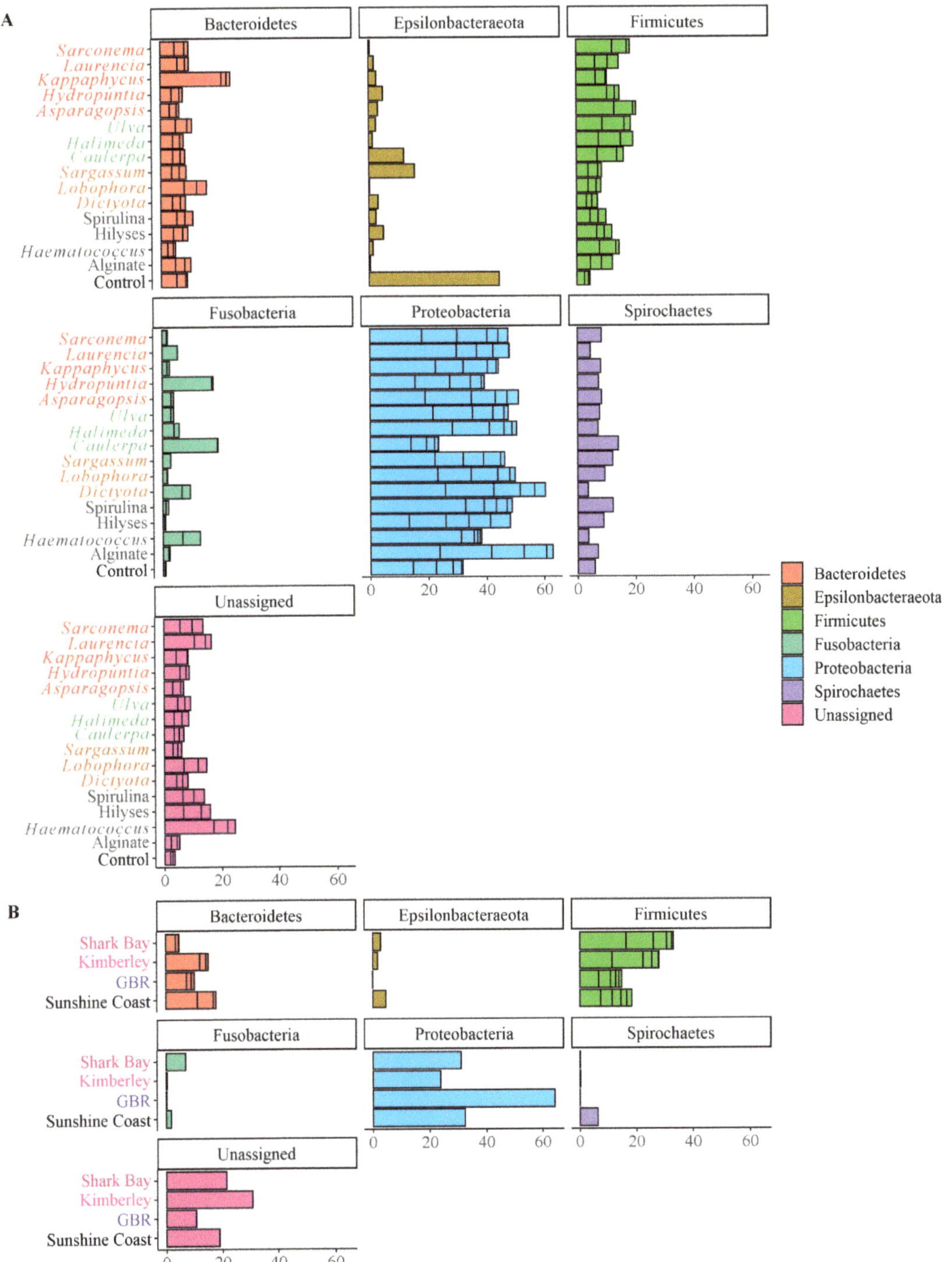


**Figure 3.** Mean relative abundance of phyla in the hindgut of the fish fed the different treatments and from the different geographical populations of *S. fuscescens*. Phyla contributing to >1% abundance to

the microbial communities of the hindgut of *S. fuscescens* (**A**) fed the 16 diets of the current study and (**B**) of the fish from the current study and those from the three wild populations of this fish. On y-axes, red text indicates that diets were supplemented with a species of red seaweed, green text indicates green seaweed, and brown text indicates supplementation with brown seaweed. Aquafeed supplements are indicated in light grey with the control in black. The fish from Eastern Australia (GBR) are in blue and those from Western Australia (Shark Bay and Kimberley) are in pink.

### 3.2. Bacterial Community Diversity

In total, we recovered 1198 ASVs after rarefaction from the hindgut of *Siganus fuscescens* (*N* = 47) used in our experiment. To allow the comparison of both sequencing runs from this trial at the ASV level, the rarefied ASVs abundance agglomerated at the genus level. This left 113 assigned genera in total from all 5 treatment groups (red, green, brown seaweed, and aquafeed supplements and control). Out of these 113 taxa, the hindgut of the fish fed the control and supplemented diets shared 63 taxa (Figure 4), with further overlaps with and between the seaweed groups (Figure 4). Hindguts of fish fed control diets had fewer taxa (69 taxa) compared to fish fed supplemented diets, which were all similar with 98 taxa in fish fed red seaweeds, 95 taxa in fish fed green seaweeds, 93 taxa in fish fed 'aquafeed' supplements and 87 taxa in fish fed brown seaweeds (Figure 4).

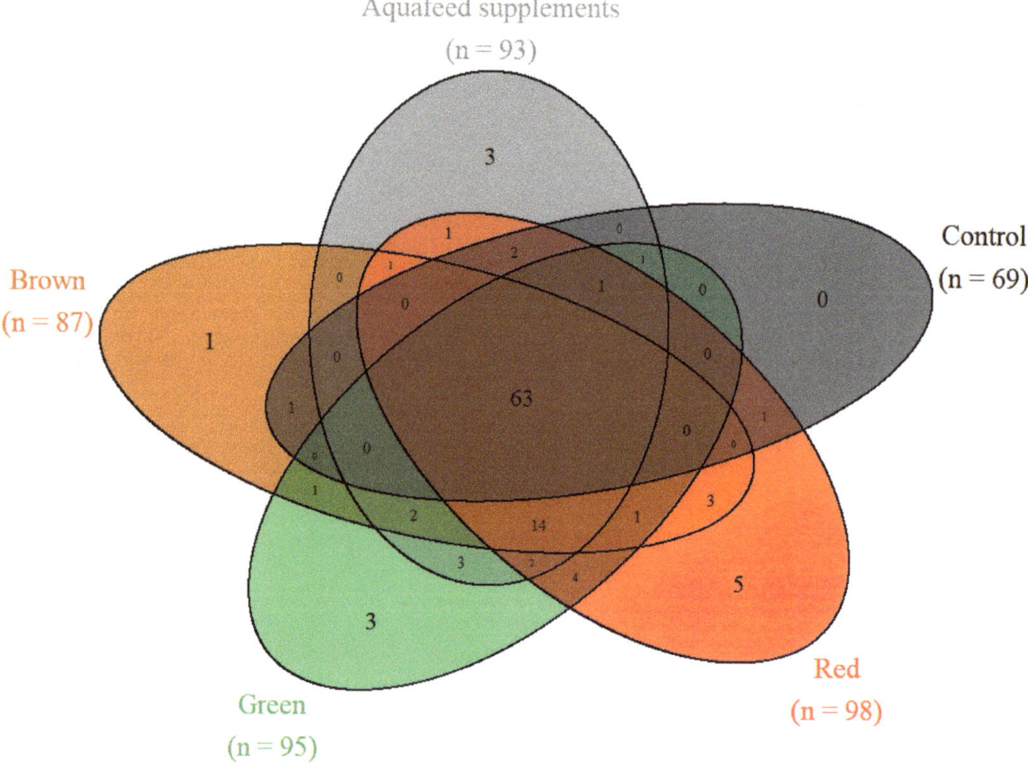


**Figure 4.** Venn diagram of the shared ASVs between the hindgut of the fish fed the different diets based on their functional groups. Shared and unique ASVs in the hindgut of the mottled rabbitfish (*S. fuscescens*) fed the control diet or diets supplemented with reds (*N* = 15), greens (*N* = 9) or browns (*N* = 9) seaweeds or aquafeed supplements (existing industry dietary additives; *N* = 12).

Although there appear to have been some dissimilarities, there were no statistically significant differences in alpha or beta diversity indices between treatments ($p > 0.05$; Figure 5A,B) or between treatment groups ($p > 0.05$; Figure 5C,D).

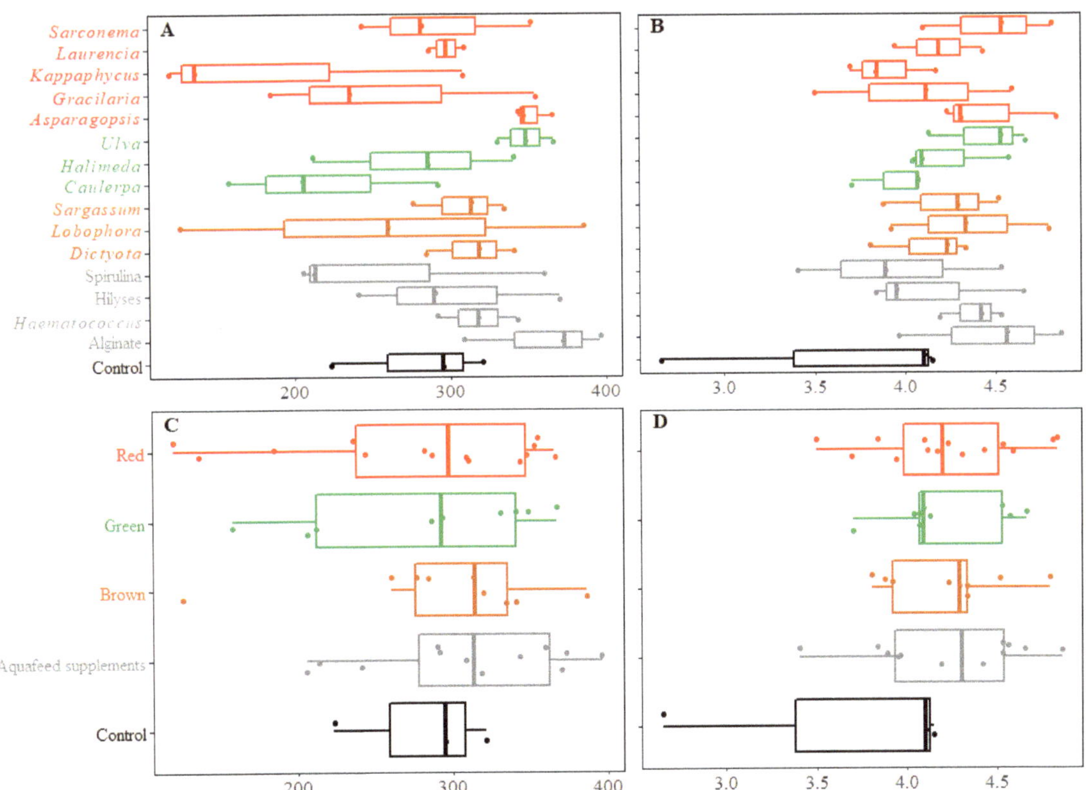


**Figure 5.** Alpha diversity indices in relation to dietary treatments. Alpha diversity analysis using species richness (Observed ASVs; (**A**,**C**)) and species diversity (Shannon index; (**B**,**D**)) for all treatments (**A**,**B**) and for the different functional groups of seaweeds used in feeding trials (**C**,**D**).

The beta diversity results (PERMANOVA based on Bray-Curtis and unweighted UniFrac measures; $p = 0.99$ and $p = 0.209$ respectively) did not show clear differences between the composition of hindgut microbial communities in fish fed different diets compared to the control fish (Figure 6). There was also no effect of fish size (weight or length) on the microbiome when each size variable was added (independently) as a covariate in separate PERMANOVAs (weight: $p = 0.104$ and length: $p = 0.15$ for unweighted UniFrac measure).

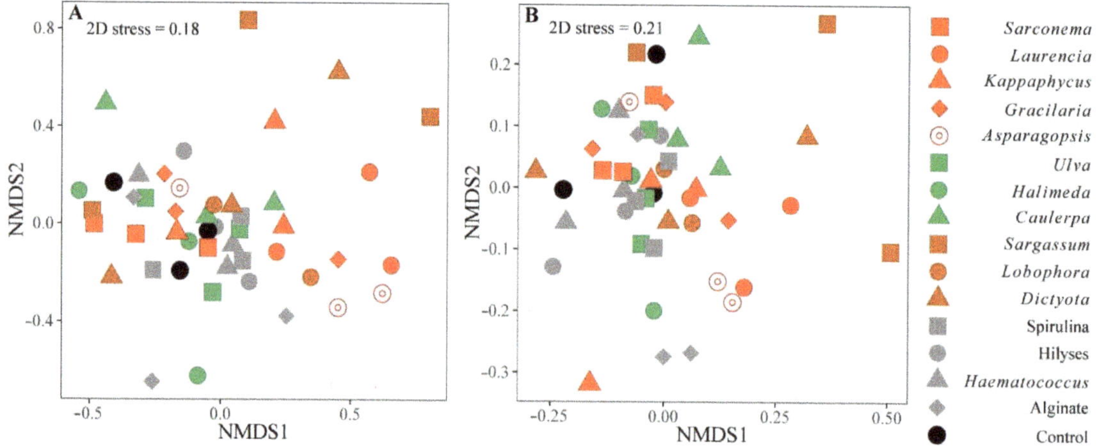


**Figure 6.** Beta diversity indices in relation to the dietary treatments using NMDS on rarefied ASVs abundance using Bray-Curtis (**A**) and unweighted UniFrac (**B**) dissimilarities between the genus-subset hindgut bacterial communities of *S. fuscescens* fed the supplemented or control diets. Symbol colours correspond to diet treatment type, including brown seaweed (brown symbols), red seaweed (red symbols), green seaweed (green symbols), Aquafeed supplements (grey symbols), and control diets (black symbols).

*3.3. Microbiome Taxonomic Composition*

Of the 113 taxa detected post rarefaction and agglomeration to the genus level, only 17 represented more than 1% of the total abundance. Of those 17 taxa, eight belonged to the phylum *Firmicutes*, three belonged to the *Proteobacteria*, and another two each to *Bacteroidetes* and *Fusobacteria*. The most abundant phylum (*Proteobacteria*) was represented by just three taxa and accounted for an average 44.0% ± 1.42% relative abundance across our samples (Mean ± SE). The most abundant taxon was the genus *Desulfovibrio* and represented between 11% (sodium alginate fed fish) and 34% (*Dictyota* fed fish; Figure 3A). The second most abundant phylum was the *Firmicutes* with 20.4% ± 1.4% abundance. *Firmicutes* was the only phylum that differed significantly between the different diets in our screening trial (ANOVA, $F = 3.07$, $p = 0.003$), with the lowest relative abundance observed in the hindgut of fish fed the control diet (9.3% ± 4.0%) compared to an average of 21.1% ± 1.4% for all other treatments (Figure 3A). Fish fed the *Haematococcus* sp. and *Halimeda* sp. diets had the highest relative abundance (28.8% ± 4.2% and 28.5% ± 9.9% respectively) of *Firmicutes* and the average value for the seaweed supplements was 20.5% ± 1.7%. Conversely, fish on the control diet seemed to possess a higher proportion of bacteria in their hindgut belonging to the phylum *Epsilonbacteraeota* (Figure 3A) although this consistent observation could not be resolved statistically.

At the family level, there were no significant differences between fish from the different diet treatments except for the relative abundance of *Ruminococcaceae* (3.5% ± 1.6%; $F = 3.08$, $p = 0.004$) which was lower in fish fed control diets compared to those fed supplemented diets, with the highest relative abundance for that taxon observed in fish fed the calcified green seaweed *Halimeda* sp. (13.0% ± 6.3%; ANOVA, $F = 3.07$, $p = 0.003$). The relative abundance of bacteria from the *Arcobacteraceae* family also appeared to be higher in hindguts of fish fed control diets (Figure 3A and Supplementary Table S3). However, despite the magnitude of these differences, they were not significant when compared to supplemented diets overall or individually.

Although, most of the bacterial genera in our samples were unidentified (53%), we did find some differences between the communities in the hindgut of fish fed any of the supplemented diets compared to those fed the control diet at the genus level. For example, although the relative abundance of *Fusobacterium* spp. was low and variable across all fish, including those fed with supplemented diets (average 1.6% ± 0.5%), it had extremely low abundance (0.3% ± 0.2%) in the hindgut of fish fed the control diet (ANOVA, $F = 2.135$, $p = 0.036$; Supplementary Table S3).

### 3.4. Comparison with Wild Populations

To provide ecological context for our results and additional information for the development of bespoke aquafeed for this fish species, we compared our microbiome data, obtained from fish collected on the subtropical Sunshine Coast, Australia, in 2018, to those obtained from conspecific hindgut microbiomes from individuals in populations located ~4000 km west off the Western Australian (WA) coastline (Shark Bay and the Kimberley Coast) and ~350 km north on the Great Barrier Reef (GBR; One Tree Island) during 2015 and 2016. In total, we recovered 3084 ASVs after rarefaction (6290 sequence depth; Supplementary Figure S2) from hindguts of *S. fuscescens* from the three combined studies (85 samples in total; with $n = 47$ fish from our study, $n = 16$ from Nielsen, et al. [22] and $n = 22$ (made of fish from Kimberley $n = 16$, and from Shark Bay $n = 6$) from Jones, et al. [29].

To compare the three studies, taxa agglomeration was performed at the genus level. This led to a comparable list of 174 taxa in total, including 134 taxa that were present in our study on the Sunshine Coast, 110 taxa that were detected in fish from the GBR; Nielsen, et al. [22], 110 taxa that were detected in Shark Bay and 130 present in the Kimberley (WA) fish; Jones, et al. [29] (Figure 7). We found 55 ASVs that were common in hindguts of mottled rabbitfish from all three populations and 35 ASVs that were present in 50% of all samples (Figure 7 and Supplementary Table S4). When the prevalence threshold was increased, only 9 ASVs were found to be shared between 90% of the individual fish and only 6 ASVs were present in 100% of individuals sampled across all three studies (Supplementary Tables S5 and S6). Fish from the Sunshine Coast had the highest number of unique taxa (i.e., those that we did not detect from other populations) followed by those from the GBR (Figure 7). Fish from the two populations in WA shared more taxa than any other two *S. fuscescens* population. Fish from the Sunshine Coast shared more taxa with the GBR fish than either of the two WA populations (Figure 7). There was a marginally non-significant difference in alpha diversity between the four populations in terms of the number of observed ASVs: $F = 7.30$, $p = 0.06$) and the Shannon index: $F = 85.81$, $p = 0.12$) of *S. fuscescens* (Figure 8).

The analyses of beta diversity revealed strong differences between the four populations of *S. fuscescens* (PERMANOVAs based on Bray-Curtis and unweighted UniFrac measures; $F = 13.39$, $p = 0.001$ and $F = 16.34$, $p = 0.001$ respectively; Figure 9). Omitting the data from our trial (Sunshine Coast), the hindgut microbiome of the other three *S. fuscescens* populations (Shark Bay, Kimberley, and GBR) were also significantly different from each other (PERMANOVAs, $F = 14.87$, $p = 0.001$, and $F = 8.97$, $p = 0.001$ for Bray-Curtis and unweighted UniFrac respectively; Figure 9A,B). The beta diversity of the two wild populations from Western Australia (Shark Bay and Kimberley) differed significantly based on Bray-Curtis (PERMANOVAs, $F = 2.66$, $p = 0.009$) but not on UniFrac (PERMANOVAs, $F = 1.14$, $p = 0.29$).

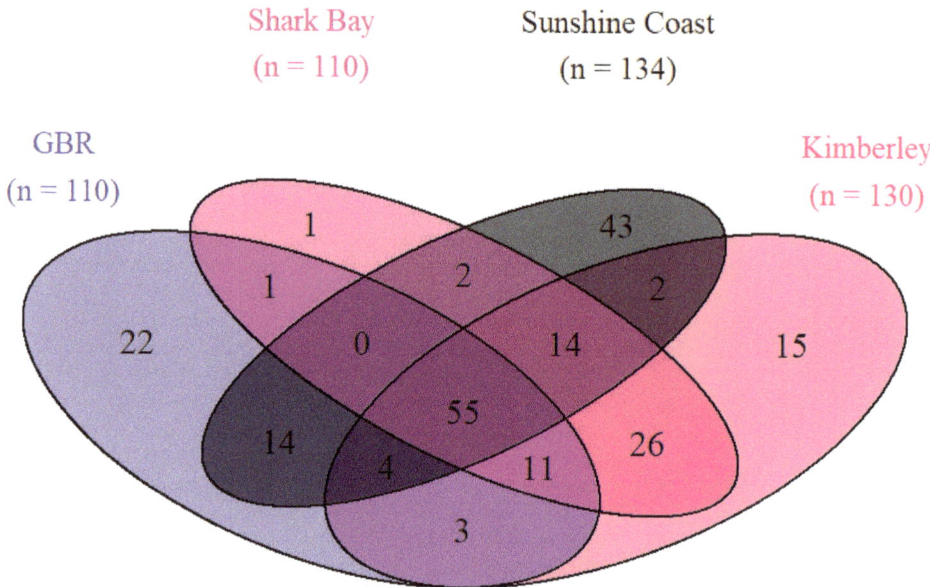


**Figure 7.** Venn diagram of the shared ASVs between the hindgut of the fish from the different geographical populations. Shared and unique ASVs in the hindgut of the mottled rabbitfish (*Siganus fuscescens*) from the current study (Sunshine Coast: $N = 47$) and the three wild *S. fuscescens* populations (GBR: $N = 16$; Kimberley: $N = 16$ and Shark Bay: $N = 6$).

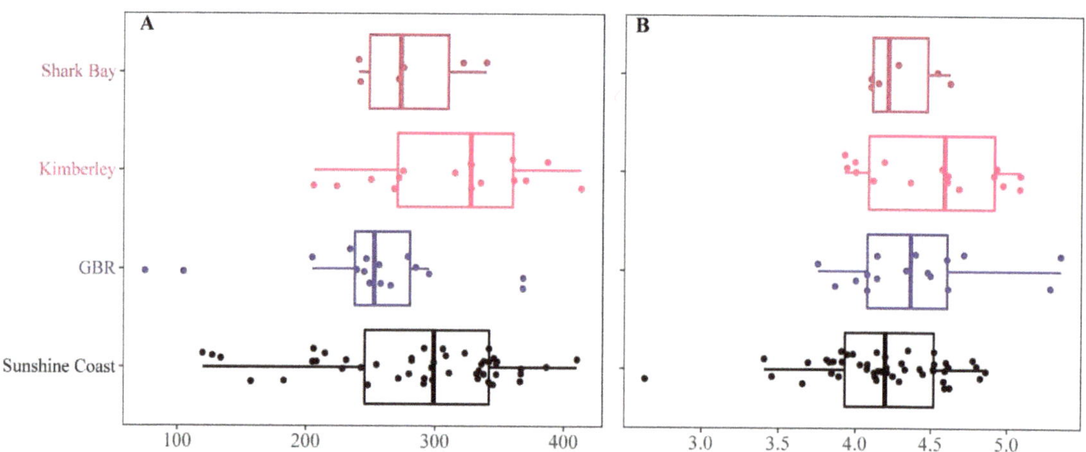


**Figure 8.** Alpha diversity indices in relation to the fish's geographical population. Alpha diversity analysis using species richness. Observed ASVs; (**A**) and species diversity. Shannon index; (**B**) from the four *Siganus fuscescens* populations.

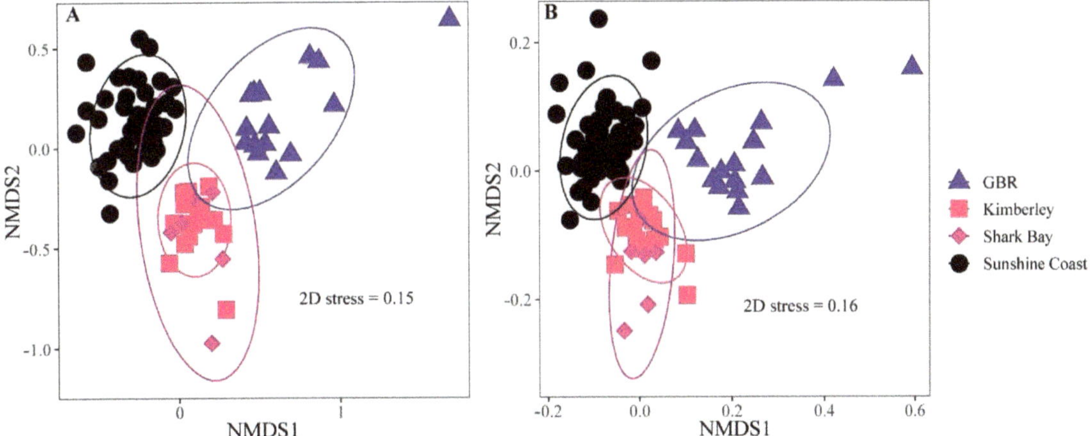


**Figure 9.** Beta diversity indices in relation to the fish's geographical population. NMDS based on Bray-Curtis (**A**) and unweighted UniFrac (**B**) similarities of the rarefied ASV abundance in the hindgut of *S. fuscescens* individuals collected by Nielsen et al., (GBR) Jones et al., (Kimberley and Shark Bay) and the current study (Sunshine Coast). The ellipses represent the 95% confidence interval.

### *3.5. Microbiome Taxonomic Composition of the Three Studies*

The three studies were clearly distinguishable from each other with respect to the relative abundance of many differentially abundant ASVs (Figure 3, Supplementary Figure S3 and Supplementary Tables S2 and S3). Fish from the GBR appeared to have the highest relative abundance of *Proteobacteria* (52.6% ± 2.9%) compared to all the other fish including those in our study (30.9% ± 2.2%) and the fish from WA (26.5% ± 1.9% and 20.6 % ± 2.0% for Shark Bay and the Kimberley fish respectively), however, this was not resolved statistically (ANOVA, $F = 1.45$, $p = 0.236$; Figure 3B). The GBR fish were the only ones without ASVs from the *Spirochaetes* (ANOVA, $F = 69.08$, $p < 0.001$), and they also had the lowest relative abundance of *Fusobacteria* (0.2% ± 0.0%, ANOVA, $F = 6.86$, $p < 0.001$) and *Epsilonbacteraeota* (0.06% ± 0.02%, ANOVA, $F = 3.34$, $p = 0.023$), which across all the other fish represented an average of 3.8% relative abundance (Figure 3B). The four most abundant ASVs in the GBR fish represented 75.2% ± 2.2% of the relative abundance compared to 71.9% ± 2.2% and 61.3% ± 1.9% in Western Australia and the Sunshine Coast fish, respectively (Figure 3B and Supplementary Table S3).

Fish from Western Australia had more similar hindgut microbiomes to the fish from our feeding trial compared to those from the GBR. For example, the relative abundance of *Fusobacteria* in Shark Bay fish was comparable to those in our study, with (13.6% ± 3.5% and 10.6 % ± 0.7% respectively, Tukey 'Sunshine Coast vs Shark Bay' adjusted $p = 0.778$) but it was significantly lower in those from the Kimberley (9.6% ± 1.2%, Tukey 'Shark Bay vs. Kimberley' adjusted $p = 0.049$) and the GBR (6.1% ± 0.6%, Tukey 'Shark Bay vs. GBR adjusted $p = 0.018$; Figures 3B and 8). Furthermore, the *Bacteroidetes*, which represented 15.0% ± 2.9% of the community in the hindgut of fish from the Kimberley and 16.2% ± 0.8% from fish in our study (Tukey 'Sunshine Coast vs Kimberley' adjusted $p = 0.757$), significantly higher than for GBR fish (Tukey 'Sunshine Coast vs GBR' adjusted $p = 0.0.24$).

One key difference between our study and the others was the absence or very low relative abundance of *Spirochaetes* (0–0.17% ± 0.0%; ANOVA, $F = 69.08$, $p < 0.001$) in the GBR and WA fish, compared to our samples which had a relative abundance of 5.7% ± 0.5% (Tukey 'Sunshine Coast vs GBR', 'Sunshine Coast vs Shark Bay', 'Sunshine Coast vs Kimberley' adjusted $p < 0.001$; Figure 3B). Furthermore, 65 ASVs significantly differed (Wald tests, adjusted $p < 0.05$) in abundance between the hindgut microbiome of one or more of the four geographically distinct populations of rabbitfish from the 3 studies (Supplementary Figure S3 and Table S5). Compared to the GBR and WA, our fish tended to have increased relative abundances of ASVs representing >1% abundance with assigned genera. These ASVs included *Treponema* spp., *Romboutsia* spp., *Turicibacter* spp., and *Ruminococcaceae* UCG-014, which all consistently and significantly represented greater proportions of the microbial communities in the hindguts of our fish compared to the other populations (Supplementary Figure S3 and Table S5).

There were some exceptions to this pattern. For example, abundances of *Akkermansia* spp. and *Tyzzerella* spp. were significantly lower in our fish than in the other studies (ANOVA, $F = 69.08$, $p < 0.001$; Supplementary Figure S3 and Table S5). The fish from both sites in WA also had significantly higher relative abundances of *Rikenella* spp. and *Sedimentibacter* spp. than our fish, while the GBR population had higher relative abundances of *Terrisporobacter* spp. and *Staphylococcus* spp. than any of the other geographical locations (Supplementary Figure S3 and Table S5). The most similar populations were those from the two sites in WA (only 12 significantly different ASVs between those; Supplementary Table S5). Our fish appeared to be most similar to those from Shark Bay in WA, which had only 27 ASVs with significantly different relative abundances compared to the other two populations which had more than 53 (Supplementary Figure S3B and Table S5). The two populations with the highest amount of ASVs which significantly differed between their hindgut were the fish from the GBR and those from our study with 65 different ASVs (Supplementary Figure S3F and Table S5).

Despite those differences, the core microbiome analyses revealed that out of the 55 shared assigned taxa between the three studies, 35 were present in 50% of all the fish sampled (43 out 86 fish; Supplementary Table S4). Only seven out of the 35 assigned taxa could be assigned to genera and these included *Fusobacterium* sp., *Romboutsia* sp., *Treponema* sp., *Arcobacter* sp., *Alistipes* sp., *Odoribacter* sp., and *Brenakia* sp. (Supplementary Table S4). Finally, 13 taxa represented between 66% and 85% of the total relative abundance in the hindgut of *S. fuscescens* regardless of its geographical population. When the prevalence threshold was increased to 90%, *Alistipes* sp. was the only taxon out of 9 ASVs that were identified at the genus level, while the six taxa that were present in all individual fish were unidentified at the genus level (Supplementary Tables S6 and S7).

## 4. Discussion

Supplementation of diets with seaweeds and commercially available aquafeed supplements had only subtle effects on the diversity and composition of hindgut microbial communities in the rabbitfish *Siganus fuscescens*. Therefore, none of the strong, seaweed-derived immune responses reported previously [16] showed any correlations to changes in microbiome diversity or composition. The only exception was the bacterial genus *Fusobacterium*, which was enhanced in the hindgut of fish fed diets supplemented with seaweed or other functional ingredients. This result is surprising and diverges from an emerging understanding of links between gut microbiomes and health and immunity in animals. In our study, hindgut microbiomes remained remarkably consistent between treatments including the control fish suggesting that (i) the dietary supplementations (3% dietary inclusion) which led to profound immune responses in the fish were insufficient to elicit a strong change in the fish GI microbiomes, (ii) the immune response we observed in experimental fish was unlikely to have been microbially mediated, (iii) there is potential for the existence of a stable, core GI microbiome in *S. fuscescens*, for which we conducted a further investigation with published data from two other studies [22,29].

Despite some clear and expected differences between the GI microbiomes in fish from our study and the two other studies, 55 out of 174 assigned taxa were shared, and 13 of those represented between 66% and 85% of the total relative abundance in all fish. These observations from geographically and temporally distinct populations—including our fish which were all fed experimental diets based upon commercial fish pellets developed for carnivorous fish (with and without seaweed and other supplements)—provide compelling, initial evidence for a possible core microbiome in this opportunistic omnivorous subtropical fish species. The hindgut microbiome of *S. fuscescens* appears surprisingly stable, despite experimental manipulations of diet at levels that are known to be able to fundamentally change the outcomes of production and other fish traits [13,51].

### 4.1. Lack of Correlation between Microbiomes and Innate Immune Responses

The number of studies exploring the effect of seaweed dietary supplements on both the immune and gut microbiome of fish is limited [15]. However, there is evidence of shifts in bacterial communities associated with fish gastrointestinal tracts after seaweed treatment (Thepot et al., 2021b). These shifts tended to be associated with improved immune responses, including up-regulation of immune-related genes [32,52]. Several studies also reported that fish fed seaweed had reduced levels of potentially pathogenic bacteria including *Aeromonas hydrophila* [53,54] and improvements in humoral immune defences including lysozyme activity and respiratory burst activity [54]. However, in our trial, those seaweed species that induced strong immune responses in *S. fuscescens* (e.g., *Asparagopsis taxiformis* and *Dictyota intermedia* [16]) were not correlated to any changes in hindgut microbial composition from the same experiment.

These observations suggest that the effects of seaweeds, especially *Asparagopsis taxiformis* on the immune responses in *S. fuscescens*, were direct and not mediated by microbiomes in the hindgut of the fish. However, it is possible that the seaweed dietary supplements had effects on the microbiomes of the fish outside of their hindgut, including the skin and gills, which have previously been reported to be locations where the microbial community can be influenced by diet [55]. Furthermore, we preferred to measure in situ immune responses through various immunochemistry tests rather than their gene expression as a proxy for immune stimulation because the relationship between mRNA transcripts and protein abundance is often quite low (~30–40%) [56]. It is possible that the immune related genes of the fish fed the seaweed diets might have correlated with the observed changes in the fish hindgut microbial communities as per previous studies [32,52]. Future studies investigating the mechanisms involved in fish immunostimulation should include measurements of both the fish immunochemistry, their GI microbiome, and the relevant immune-related gene expression.

### 4.2. Effects of Diet on the Hindgut Microbiome of S. fuscescens

The subtle effects of the dietary supplements in this trial are surprising considering the potent immunostimulatory effects some of the supplements had and their diverse natural product composition [16]. Some previous studies have observed dramatic effects of experimental diets on the intestinal microbiome of farmed fish with comparable inclusion rates and experimental designs (i.e., *Sparus aurata* and *Seriola lalandi* [57–59]), whereas others found that the hindgut microbiome of cultured fish was relatively stable and did not appear to show much overall change to dietary manipulation (i.e., *Oncorhynchus mykiss* and *Siganus canaliculatus* [26,33,60]). The studies that detected strong changes typically supplemented fish diets with probiotics or other functional feeds for 4–8-week experiments, longer than our experiments but with a similar dietary inclusion ratio. On the other hand, Wong, et al. [60] fed rainbow trout experimental diets including grains for a period of 10 months and observed only subtle changes in the fish intestinal microbial communities. Lyons, et al. [33] supplemented the diet of rainbow trout for 15 weeks but in this case with a microalgal meal at a level of 5% and found that whilst addition increased diversity, the overall structure of microbiomes in the hindgut of the fish were not significantly altered. Similarly, Zhang, et al. [26] found that supplementing *Siganus canaliculatus* (a color morph of *S. fuscescens* [61]) with 10% of the green seaweed *Ulva pertusa* for a period of 8 weeks did not significantly alter the microbial diversity in the intestinal communities in the fish. They concluded that a strong core microbiome existed, comprised of 86 operational taxonomic units, which were shared across their fish regardless of the dietary treatment [26]. Out of the

86 OTUs reported in Zhang, et al. [26], three, namely *Arcobacter* sp., *Fusobacterium* sp., and *Treponema* sp. were also identified as core members of *S. fuscescens* hindgut microbiome.

Although seaweed is a highly diverse group (>10,000 species) that produce a wide range of secondary metabolites with bioactive properties and are increasingly used as animal supplements, there is a gap in the literature regarding their potential as a dietary supplement to shape the intestinal microbiome of animals including fish [62,63]. Ours was the first study to test so many seaweed species in one experiment, including several known for their bioactive natural products (e.g., *Asparagopsis taxiformis*, *Caulerpa taxifolia*, and *Sargassum* spp.). Indeed, there was some evidence for the effects of these chemically 'rich' seaweed supplements. For example, fish fed diets supplemented with the green alga *Caulerpa taxifolia* had significantly higher levels of *Fusobacterium* spp. and similarly enhanced *Cetobacterium* spp. and *Treponema* spp. *C. taxifolia* produces many interesting bioactive compounds [21] and its presence on reefs can completely alter sediment microbiomes through chemical modifications of the substrate (see [64] and references therein). This seaweed is typically avoided by native herbivorous fish (e.g., *Girella tricuspidata*) and invertebrate grazers in Australia [65] and can be toxic to invertebrates forced to consume it in feeding trials [65,66]. The red seaweed *A. taxiformis* was also expected to significantly modulate the GI microbiome of the rabbitfish fed that supplement due to its production and storage of potent antimicrobial bioactives [20] and its fast modulatory effect on the rumen microbes of ruminants [67,68]. However, the abundance of some ASVs was only slightly enhanced in fish from that treatment (e.g., *Romboutsia* sp.) and these changes were not statistically significant.

### 4.3. Comparing Separate Studies of the Hindgut Microbiome of *S. fuscescens*

Not surprisingly, the geographically distinct populations of *S. fuscescens* that were sampled by different teams at different times from different places had significantly different hindgut microbiomes. However, there was still substantial overlap, with almost 18% of bacterial taxa present across all populations. Interestingly the hindgut microbiome of our fish appeared to have more in common with that of the fish from Western Australia (~4000 km away) than with the geographically closer population of *S. fuscescens* from the GBR (~380 km away), which seemed more distinct than other populations. Overall, the hindgut microbiome composition of the fish from our study (Sunshine Coast) was most similar to those from Shark Bay in Western Australia, which is at a similar latitude.

Potential explanations for these groupings are that all of the seaweed genera fed to our fish have tropical, subtropical, or temperate distributions and are common on the eastern coast of Australia, with many also occurring on the west [69]. It is, therefore, possible that some of the similarities between populations were the result of similar native diets that include these seaweed species. Another explanation for the similarities observed between our fish, and those from Shark Bay more specifically, could be the similar abiotic and biotic factors, given that the Kimberley and GBR sites are both tropical and the Sunshine Coast and Shark Bay sites are both sub-tropical locations. Furthermore, the fish from our screening trial were collected near shore (<1 km away), as were the fish from Shark Bay, whereas the fish from the Kimberley site was approximately 25 km from the coast and finally the most distinct hindgut microbiome was found in the fish from One Tree Island on the GBR which is about 70 km from the coast. The impact of rivers, agriculture, and other human or land-associated impacts may be more important in nearshore areas, which could explain some of the differences observed here. This hypothesis is further supported by the fact that the diet of the fish in our trial (97% commercial pellet designed for carnivorous fish with 3% seaweed inclusion) would be drastically different from that of the wild fish populations which predominantly would feed on seaweed [22,29]. Similarly, low spatial and temporal variation was observed in the gut microbiomes of larvae from another rabbitfish species, *Siganus guttatus*, across three sites separated by up to 390 km across a three-year sampling program [70]. This observation of stability in the microbiome led the authors to propose that *Siganus* spp. have a core microbiome in their GI tracts [22,26,34,70].

### 4.4. Does *Siganus Fuscescens* Have a Core Microbiome?

Despite experimental manipulations of diets with taxonomically and chemically diverse seaweeds and samples originating from populations in locations separated by up to 4000 km around the Australian coast (the Sunshine Coast in southeast Queensland; present study, Shark Bay and the Kimberley site in Western Australia; Jones, et al. [29], and the Great Barrier Reef in North East Queensland; Nielsen, et al. [22]), more than 50% of all mottled rabbitfish included shared nearly one-third of their hindgut microbiota, with between 6 and 35 taxa belonging to a potential core microbiome in this species, depending upon which threshold is used [28]. Definitions of a 'core microbiome' are still the topic of debate and disagreement in the literature [71], however, the identification of a core microbiome was recently highlighted as one of the first steps required to link microbial community structure and diversity to its function and, importantly, the role it plays for its host [4]. Our observations provide further evidence for the potential existence of a core microbiome in this species and support previous suggestions that *Siganus* species may have a core microbiome that is robust to dietary manipulations and large geographical distances as per other *Siganus* spp. [22,26,34,70]. By helping fish maintain homeostasis in new and changing environments, the existence of a core microbiome could confer performance advantages to fish in aquaculture settings and could also be a mechanism for their success as tropical invaders into temperate waters. However, further, more targeted work is needed to confirm whether *S. fuscescens* does indeed have a core microbiome and importantly, the functional roles of any core microbial taxa throughout the life of *Siganus* fish in the wild and on farms.

## 5. Conclusions

Immunostimulatory effects of dietary supplementation with seaweeds in the mottled rabbitfish *Siganus fuscescens* appear not to be microbially mediated. Rather, fish had remarkably stable hindgut microbiomes that were only subtly influenced by dietary manipulation with diverse seaweeds (including several with highly bioactive natural products) and commercial products. These results are contrary to emerging studies from other fish species and animals-including humans-and suggest that the effects of diet and functional feeds on animal gut microbiomes and resulting health may be species-specific and influenced by trophic levels. Our observations provide some preliminary evidence that a conserved core microbiome may exist within the hindgut of this fish species and provide other baseline data about temporal and spatial variation in the hindgut bacterial communities within this candidate aquaculture species, which may support the sustainable development of this industry.

**Supplementary Materials:** The following supporting information can be downloaded at: https://www.mdpi.com/article/10.3390/microorganisms10030497/s1, Figure S1: Rarefaction curves for the current study, Figure S2: Rarefaction curves for the current study and the two wild *S. fuscescens* studies, Figure S3: Differentially abundant ASVs between hindgut of the fish from the four geographical populations, Table S1: Physical and chemical attributes of the different supplements used in the current study, Table S2: Vector loading for all measured variables and their respective permutation-based $p$-values, Table S3: List of all ASVs present in >1% relative abundance in the hindgut of the mottled rabbitfish samples in the current study and the two wild sampled fish from Western and Eastern Australia, Table S4: Taxonomy and details of the 35 ASVs identified as core microbiome member in *S. fuscescens*, Table S5: List of the differentially abundant ASVs that significantly ($P$-*adjust* < 0.05) differed in the 6 pairwise comparisons of the 4 geographically distinct *S. fuscescens* populations, Table S6: Taxonomy and details of the 6 ASVs identified as core microbiome members in *S. fuscescens* using a prevalence threshold of 100% and Table S7: Taxonomy and details of the 9 ASVs identified as core microbiome members in S. fuscescens using a prevalence threshold of 90%.

**Author Contributions:** Conceptualization, V.T. and N.A.P.; Data curation, J.S. and A.H.C.; Formal analysis, V.T.; Funding acquisition, N.A.P.; Investigation, V.T. and A.H.C.; Methodology, V.T., J.S., M.A.R., N.A.P. and A.H.C.; Project administration, A.H.C.; Software, J.S.; Supervision, M.A.R. and N.A.P.; Validation, M.A.R.; Visualization, V.T. and J.S.; Writing–original draft, V.T.; Writing–review

& editing, V.T., J.S., M.A.R., N.A.P. and A.H.C. All authors have read and agreed to the published version of the manuscript.

**Funding:** This work was part of a PhD program funded by the University of the Sunshine Coast (USC) and supported by the Australian Centre for International Agricultural Research (ACIAR) under the project: Accelerating the development of finfish mariculture in Cambodia through south-south research cooperation with Indonesia (FIS/2016/130). Additional funds (Student Award) were received from the Crawford Fund which supported additional hindgut microbial analyses.

**Institutional Review Board Statement:** The study was conducted according to the guidelines of the Declaration of Helsinki, and approved by the Animal Ethics Committee of the University of the Sunshine Coast (Animal Ethics approval ANS1751 granted on the 22/12/17).

**Informed Consent Statement:** Not applicable.

**Data Availability Statement:** Raw sequences have been deposited in the National Center for Biotechnology Information (NCBI) sequence read archive (SRA) under the bioproject number PRJNA649307. Raw sequence data from the previously published studies were retrieved from the NCBI SRA Jones, et al. [29]; accession number PRJNA356981 and Nielsen, et al. [22]; accession number PRJNA396430) using the SRA Toolkit software.

**Acknowledgments:** We thank ACIAR for their support of this research, as well as the Crawford Fund for supporting analysis of the hindgut microbial diversity. We also thank Pacific Biotechnologies Pty Ltd. for generously providing the astaxanthin rich *Haematococcus pluvialis* biomass used in this trial. Finally, we thank Ian Tuart, Amalya Valle and Christina Bezuidenhout for their help with the sampling of the fish. The views in this work are the sole responsibility of the authors and do not necessarily reflect the views of USC, ACIAR or the Crawford Fund.

**Conflicts of Interest:** The authors declare no conflict of interest. USC, ACIAR and the Crawford Fund had no role in the design of the study; in the collection, analysis, or interpretation of data; in the writing of the manuscript, or in the decision to publish the results.

## References

1. Yukgehnaish, K.; Kumar, P.; Sivachandran, P.; Marimuthu, K.; Arshad, A.; Paray, B.A.; Arockiaraj, J. Gut microbiota metagenomics in aquaculture: Factors influencing gut microbiome and its physiological role in fish. *Rev. Aquac.* **2020**, *12*, 1903–1927. [CrossRef]
2. Haygood, A.M.; Jha, R. Strategies to modulate the intestinal microbiota of Tilapia (Oreochromis sp.) in aquaculture: A review. *Rev. Aquac.* **2018**, *10*, 320–333. [CrossRef]
3. Infante-Villamil, S.; Huerlimann, R.; Jerry, D.R. Microbiome diversity and dysbiosis in aquaculture. *Rev. Aquac.* **2021**, *13*, 1077–1096. [CrossRef]
4. Trevathan-Tackett, S.M.; Sherman, C.D.H.; Huggett, M.J.; Campbell, A.H.; Laverock, B.; Hurtado-McCormick, V.; Seymour, J.R.; Firl, A.; Messer, L.F.; Ainsworth, T.D.; et al. A horizon scan of priorities for coastal marine microbiome research. *Nat. Ecol. Evol.* **2019**, *3*, 1509–1520. [CrossRef] [PubMed]
5. Froehlich, H.E.; Jacobsen, N.S.; Essington, T.E.; Clavelle, T.; Halpern, B.S. Avoiding the ecological limits of forage fish for fed aquaculture. *Nat. Sustain.* **2018**, *1*, 298–303. [CrossRef]
6. Béné, C.; Barange, M.; Subasinghe, R.; Pinstrup-Andersen, P.; Merino, G.; Hemre, G.I.; Williams, M. Feeding 9 billion by 2050—Putting fish back on the menu. *Food Secur.* **2015**, *7*, 261–274. [CrossRef]
7. Cottrell, R.S.; Blanchard, J.L.; Halpern, B.S.; Metian, M.; Froehlich, H.E. Global adoption of novel aquaculture feeds could substantially reduce forage fish demand by 2030. *Nat. Food* **2020**, *1*, 301–308. [CrossRef]
8. Stentiford, G.D.; Sritunyalucksana, K.; Flegel, T.W.; Williams, B.A.P.; Withyachumnarnkul, B.; Itsathitphaisarn, O.; Bass, D. New Paradigms to Help Solve the Global Aquaculture Disease Crisis. *PLoS Pathog.* **2017**, *13*, e1006160. [CrossRef]
9. Hua, K.; Cobcroft, J.M.; Cole, A.; Condon, K.; Jerry, D.R.; Mangott, A.; Praeger, C.; Vucko, M.J.; Zeng, C.; Zenger, K.; et al. The future of aquatic protein: Implications for protein sources in aquaculture diets. *One Earth* **2019**, *1*, 316–329. [CrossRef]
10. Glendcross, B.D.; Booth, M.; Allan, G.L. A feed is only as good as its ingredients—A review of ingredient evaluation strategies for aquaculture feeds. *Aquac. Nutr.* **2007**, *13*, 17–34. [CrossRef]
11. Sylvain, F.-É.; Holland, A.; Bouslama, S.; Audet-Gilbert, É.; Lavoie, C.; Val Adalberto, L.; Derome, N.; McBain Andrew, J. Fish Skin and Gut Microbiomes Show Contrasting Signatures of Host Species and Habitat. *J. Appl. Environ. Microbiol.* **2020**, *86*, e00789-20. [CrossRef]
12. Miyake, S.; Ngugi, D.K.; Stingl, U. Diet strongly influences the gut microbiota of surgeonfishes. *J. Mol. Ecol.* **2015**, *24*, 656–672. [CrossRef] [PubMed]

13. Piazzon, M.C.; Calduch-Giner, J.A.; Fouz, B.; Estensoro, I.; Simó-Mirabet, P.; Puyalto, M.; Karalazos, V.; Palenzuela, O.; Sitjà-Bobadilla, A.; Pérez-Sánchez, J. Under control: How a dietary additive can restore the gut microbiome and proteomic profile, and improve disease resilience in a marine teleostean fish fed vegetable diets. *Microbiome* **2017**, *5*, 164. [CrossRef] [PubMed]
14. Michl, S.C.; Ratten, J.-M.; Beyer, M.; Hasler, M.; LaRoche, J.; Schulz, C. The malleable gut microbiome of juvenile rainbow trout (*Oncorhynchus mykiss*): Diet-dependent shifts of bacterial community structures. *PLoS ONE* **2017**, *12*, e0177735. [CrossRef] [PubMed]
15. Thépot, V.; Campbell, A.H.; Rimmer, M.A.; Paull, N.A. Meta-analysis of the use of seaweeds and their extracts as immunostimulants for fish: A systematic review. *Rev. Aquac.* **2021**, *13*, 907–933. [CrossRef]
16. Thépot, V.; Campbell, A.H.; Paul, N.A.; Rimmer, M.A. Seaweed dietary supplements enhance the innate immune response of the mottled rabbitfish, *Siganus fuscescens*. *Fish Shellfish Immunol.* **2021**, *113*, 176–184. [CrossRef]
17. Machado, L.; Magnusson, M.; Paul, N.A.; Kinley, R.; de Nys, R.; Tomkins, N. Identification of bioactives from the red seaweed *Asparagopsis taxiformis* that promote antimethanogenic activity in vitro. *J. Appl. Phycol.* **2016**, *28*, 3117–3126. [CrossRef]
18. Lemée, R.; Pesando, D.; Durand-Clément, M.; Dubreuil, A.; Meinesz, A.; Guerriero, A.; Pietra, F. Preliminary survey of toxicity of the green alga *Caulerpa taxifolia* introduced into the Mediterranean. *J. Appl. Phycol.* **1993**, *5*, 485–493. [CrossRef]
19. Yende, S.; Harle, U.; Chaugule, B. Therapeutic potential and health benefits of *Sargassum* species. *Pharmacogn. Rev.* **2014**, *8*, 1–7. [CrossRef]
20. Paul, N.A.; de Nys, R.; Steinberg, P. Chemical defence against bacteria in the red alga *Asparagopsis armata*: Linking structure with function. *Mar. Ecol. Prog. Ser.* **2006**, *306*, 87–101. [CrossRef]
21. Stabili, L.; Fraschetti, S.; Acquaviva, M.I.; Cavallo, R.A.; De Pascali, S.A.; Fanizzi, F.P.; Gerardi, C.; Narracci, M.; Rizzo, L. The potential exploitation of the Mediterranean invasive alga *Caulerpa cylindracea*: Can the invasion be transformed into a gain? *Mar. Drugs* **2016**, *14*, 210. [CrossRef] [PubMed]
22. Nielsen, S.; Walburn, J.W.; Vergés, A.; Thomas, T.; Egan, S. Microbiome patterns across the gastrointestinal tract of the rabbitfish *Siganus fuscescens*. *PeerJ* **2017**, *2017*, e3317. [CrossRef] [PubMed]
23. Hu, Z.; Xu, J.; Chai, X.; Shi, J.; Wu, Z. Preliminary study on monoculture and polyculture modes for *Siganus fuscescens* in sea net cage. *Fish. Mod.* **2008**, *35*, 26–28.
24. Li, Y.; Zhang, Q.; Liu, Y. Rabbitfish–an Emerging Herbivorous Marine Aquaculture Species. *Aquac. China Success Stories Mod. Trends Hum. Reprod. Physiol.* **2018**, 329–334.
25. Osako, K.; Saito, H.; Kuwahara, K.; Okamoto, A. Year-round high arachidonic acid levels in herbivorous rabbit fish *Siganus fuscescens* tissues. *Lipids* **2006**, *41*, 473–489. [CrossRef]
26. Zhang, X.; Wu, H.; Li, Z.; Li, Y.; Wang, S.; Zhu, D.; Wen, X.; Li, S. Effects of dietary supplementation of *Ulva pertusa* and non-starch polysaccharide enzymes on gut microbiota of *Siganus canaliculatus*. *J. Oceanol. Limnol.* **2018**, *36*, 438–449. [CrossRef]
27. Vergés, A.; Doropoulos, C.; Malcolm, H.A.; Skye, M.; Garcia-Pizá, M.; Marzinelli, E.M.; Campbell, A.H.; Ballesteros, E.; Hoey, A.S.; Vila-Concejo, A.; et al. Long-term empirical evidence of ocean warming leading to tropicalization of fish communities, increased herbivory, and loss of kelp. *Proc. Natl. Acad. Sci. USA* **2016**, *113*, 13791–13796. [CrossRef]
28. Huse, S.M.; Ye, Y.; Zhou, Y.; Fodor, A.A. A Core Human Microbiome as Viewed through 16S rRNA Sequence Clusters. *PLoS ONE* **2012**, *7*, e34242. [CrossRef]
29. Hall, M.B. Determination of starch, including maltooligosaccharides, in animal feeds: Comparison of methods and a method recommended for AOAC collaborative study. *J. AOAC Int.* **2008**, *92*, 42–49. [CrossRef]
30. Angell, A.; Mata, L.; de Nys, R.; Paul, N. The protein content of seaweeds: A universal nitrogen-to-protein conversion factor of five. *J. Appl. Phycol.* **2015**, *28*, 511–524. [CrossRef]
31. Cordero, H.; Guardiola, F.A.; Tapia-Paniagua, S.T.; Cuesta, A.; Meseguer, J.; Balebona, M.C.; Moriñigo, M.Á.; Esteban, M.Á. Modulation of immunity and gut microbiota after dietary administration of alginate encapsulated Shewanella putrefaciens Pdp11 to gilthead seabream (*Sparus aurata* L.). *Fish Shellfish Immunol.* **2015**, *45*, 608–618. [CrossRef] [PubMed]
32. Lyons, P.P.; Turnbull, J.F.; Dawson, K.A.; Crumlish, M. Effects of low-level dietary microalgae supplementation on the distal intestinal microbiome of farmed rainbow trout *Oncorhynchus mykiss* (Walbaum). *Aquaculture* **2017**, *48*, 2438–2452. [CrossRef]
33. Hoseinifar, S.H.; Ringø, E.; Shenavar Masouleh, A.; Esteban, M.Á. Probiotic, prebiotic and synbiotic supplements in sturgeon aquaculture: A review. *Rev. Aquac.* **2016**, *8*, 89–102. [CrossRef]
34. Jones, J.; DiBattista, J.D.; Stat, M.; Bunce, M.; Boyce, M.C.; Fairclough, D.V.; Travers, M.J.; Huggett, M.J. The microbiome of the gastrointestinal tract of a range-shifting marine herbivorous fish. *Front. Microbiol.* **2018**, *9*, 2000. [CrossRef] [PubMed]
35. Zhang, J.; Kobert, K.; Flouri, T.; Stamatakis, A. PEAR: A fast and accurate Illumina Paired-End reAd mergeR. *Bioinformatics* **2014**, *30*, 614–620. [CrossRef]
36. Caporaso, J.G.; Kuczynski, J.; Stombaugh, J.; Bittinger, K.; Bushman, F.D.; Costello, E.K.; Fierer, N.; Pena, A.G.; Goodrich, J.K.; Gordon, J.I. QIIME allows analysis of high-throughput community sequencing data. *Nat. Methods.* **2010**, *7*, 335–336. [CrossRef]
37. DeSantis, T.Z.; Brodie, E.L.; Moberg, J.P.; Zubieta, I.X.; Piceno, Y.M.; Andersen, G.L. High-density universal 16S rRNA microarray analysis reveals broader diversity than typical clone library when sampling the environment. *Microb. Ecol.* **2007**, *53*, 371–383. [CrossRef]
38. Bolyen, E.; Rideout, J.R.; Dillon, M.R.; Bokulich, N.A.; Abnet, C.; Al-Ghalith, G.A.; Alexander, H.; Alm, E.J.; Arumugam, M.; Asnicar, F. QIIME 2: Reproducible, interactive, scalable, and extensible microbiome data science. *Nat. Biotechnol.* **2019**, *37*, 852–857. [CrossRef]

39. Martin, M. Cutadapt removes adapter sequences from high-throughput sequencing reads. *EMBnet. J.* **2011**, *17*, 10–12. [CrossRef]
40. Yu, Y.; Lee, C.; Kim, J.; Hwang, S. Group-specific primer and probe sets to detect methanogenic communities using quantitative real-time polymerase chain reaction. *Biotechnol. Bioenng.* **2005**, *89*, 670–679. [CrossRef]
41. Callahan, B.J.; McMurdie, P.J.; Rosen, M.J.; Han, A.W.; Johnson, A.J.A.; Holmes, S.P. DADA2: High-resolution sample inference from Illumina amplicon data. *Nat. Methods* **2016**, *13*, 581–583. [CrossRef] [PubMed]
42. Quast, C.; Pruesse, E.; Yilmaz, P.; Gerken, J.; Schweer, T.; Yarza, P.; Peplies, J.; Glöckner, F.O. The SILVA ribosomal RNA gene database project: Improved data processing and web-based tools. *Nucleic Acids Res.* **2012**, *41*, 590–596. [CrossRef] [PubMed]
43. Core-Team, R. *R: A Language and Environment for Statistical Computing*; European Environment Agency: Copenhagen, Denmark, 2013.
44. McMurdie, P.J.; Holmes, S. Phyloseq: An R package for reproducible interactive analysis and graphics of microbiome census data. *PLoS ONE* **2013**, *8*, e61217. [CrossRef] [PubMed]
45. Ritchie, M.E.; Phipson, B.; Wu, D.; Hu, Y.; Law, C.W.; Shi, W.; Smyth, G.K. limma powers differential expression analyses for RNA-sequencing and microarray studies. *Nucleic Acids Res.* **2015**, *43*, e47. [CrossRef] [PubMed]
46. Dixon, P. VEGAN, a package of R functions for community ecology. *J. Veg. Sci.* **2003**, *14*, 927–930. [CrossRef]
47. Love, M.I.; Huber, W.; Anders, S. Moderated estimation of fold change and dispersion for RNA-seq data with DESeq2. *Genome Biol.* **2014**, *15*, 550. [CrossRef]
48. Lahti, L.; Shetty, S.; Blake, T.; Salojarvi, J. Tools for microbiome analysis in R, version 1.1.10013. 2017. Available online: http://microbiome.github.com/microbiome (accessed on 10 January 2021).
49. Meriggi, N.; Di Paola, M.; Vitali, F.; Rivero, D.; Cappa, F.; Turillazzi, F.; Gori, A.; Dapporto, L.; Beani, L.; Turillazzi, S. *Saccharomyces cerevisiae* induces immune enhancing and shapes gut microbiota in social wasps. *Front. Microbiol.* **2019**, *10*, 2320. [CrossRef]
50. Turgay, E.; Steinum, T.M.; Eryalçın, K.M.; Yardımcı, R.E.; Karataş, S. The influence of diet on the microbiota of live-feed rotifers (*Brachionus plicatilis*) used in commercial fish larviculture. *FEMS Microbiol. Lett.* **2020**, *367*, fnaa020. [CrossRef]
51. Rimoldi, S.; Torrecillas, S.; Montero, D.; Gini, E.; Makol, A.; Valdenegro, V.V.; Izquierdo, M.; Terova, G. Assessment of dietary supplementation with galactomannan oligosaccharides and phytogenics on gut microbiota of European sea bass (*Dicentrarchus Labrax*) fed low fishmeal and fish oil based diet. *PLoS ONE* **2020**, *15*, e0231494. [CrossRef]
52. Vidal, S.; Tapia-Paniagua, S.T.; Moriñigo, J.M.; Lobo, C.; García de la Banda, I.; Balebona, M.D.C.; Moriñigo, M.Á. Effects on intestinal microbiota and immune genes of *Solea senegalensis* after suspension of the administration of *Shewanella putrefaciens* Pdp11. *Fish. Shellfish Immunol.* **2016**, *58*, 274–283. [CrossRef]
53. Sutili, F.J.; Kreutz, L.C.; Flores, F.C.; da Silva, C.d.B.; Kirsten, K.S.; Voloski, A.P.d.S.; Frandoloso, R.; Pinheiro, C.G.; Heinzmann, B.M.; Baldisserotto, B. Effect of dietary supplementation with citral-loaded nanostructured systems on innate immune responses and gut microbiota of silver catfish (*Rhamdia quelen*). *J. Funct. Foods* **2019**, *60*, 103454. [CrossRef]
54. Asaduzzaman, M.; Iehata, S.; Moudud Islam, M.; Kader, M.A.; Ambok Bolong, A.-M.; Ikeda, D.; Kinoshita, S. Sodium alginate supplementation modulates gut microbiota, health parameters, growth performance and growth-related gene expression in Malaysian Mahseer *Tor tambroides*. *Aquac. Nutr.* **2019**, *25*, 1300–1317. [CrossRef]
55. Chiarello, M.; Auguet, J.-C.; Bettarel, Y.; Bouvier, C.; Claverie, T.; Graham, N.A.J.; Rieuvilleneuve, F.; Sucré, E.; Bouvier, T.; Villéger, S. Skin microbiome of coral reef fish is highly variable and driven by host phylogeny and diet. *Microbiome* **2018**, *6*, 147. [CrossRef]
56. Vogel, C.; Marcotte, E.M. Insights into the regulation of protein abundance from proteomic and transcriptomic analyses. *Nat. Rev. Genet.* **2012**, *13*, 227–232. [CrossRef]
57. Kormas, K.A.; Meziti, A.; Mente, E.; Frentzos, A. Dietary differences are reflected on the gut prokaryotic community structure of wild and commercially reared sea bream (*Sparus aurata*). *Microbiol. Open* **2014**, *3*, 718–728. [CrossRef]
58. Ramírez, C.; Romero, J. The microbiome of *Seriola lalandi* of wild and aquaculture origin reveals differences in composition and potential function. *Front. Microbiol.* **2017**, *8*, 1844. [CrossRef]
59. Wilkes Walburn, J.; Wemheuer, B.; Thomas, T.; Copeland, E.; O'Connor, W.; Booth, M.; Fielder, S.; Egan, S. Diet and diet-associated bacteria shape early microbiome development in Yellowtail Kingfish (*Seriola lalandi*). *Microb. Biotechnol.* **2019**, *12*, 275–288. [CrossRef] [PubMed]
60. Wong, S.; Waldrop, T.; Summerfelt, S.; Davidson, J.; Barrows, F.; Kenney, P.B.; Welch, T.; Wiens, G.D.; Snekvik, K.; Rawls, J.F.; et al. Aquacultured rainbow trout (*Oncorhynchus mykiss*) possess a large core intestinal microbiota that is resistant to variation in diet and rearing density. *J. Appl Environ. Microbiol* **2013**, *79*, 4974–4984. [CrossRef] [PubMed]
61. Hsu, T.H.; Adiputra, Y.T.; Burridge, C.P.; Gwo, J.C. Two spinefoot colour morphs: Mottled spinefoot *Siganus fuscescens* and white-spotted spinefoot *Siganus canaliculatus* are synonyms. *J. Fish. Biol.* **2011**, *79*, 1350–1355. [CrossRef]
62. Gupta, S.; Abu-Ghannam, N. Bioactive potential and possible health effects of edible brown seaweeds. *Trends Food Sci. Technol.* **2011**, *22*, 315–326. [CrossRef]
63. Pereira, R.C.; Costa-Lotufo, L.V. Bioprospecting for bioactives from seaweeds: Potential, obstacles and alternatives. *Rev. Bras. Farm.* **2012**, *22*, 894–905. [CrossRef]
64. Gribben, P.E.; Thomas, T.; Pusceddu, A.; Bonechi, L.; Bianchelli, S.; Buschi, E.; Nielsen, S.; Ravaglioli, C.; Bulleri, F. Below-ground processes control the success of an invasive seaweed. *J. Ecol.* **2018**, *106*, 2082–2095. [CrossRef]
65. Gollan, J.R.; Wright, J.T. Limited grazing pressure by native herbivores on the invasive seaweed *Caulerpa taxifolia* in a temperate Australian estuary. *Mar. Freshw. Res.* **2006**, *57*, 685–694. [CrossRef]

66. Boudouresque, C.F.; Lemée, R.; Mari, X.; Meinesz, A. The invasive alga *Caulerpa taxifolia* is not a suitable diet for the sea urchin *Paracentrotus lividus*. *Aquat. Bot.* **1996**, *53*, 245–250. [CrossRef]
67. Li, X.; Norman, H.C.; Kinley, R.D.; Laurence, M.; Wilmot, M.; Bender, H.; de Nys, R.; Tomkins, N. *Asparagopsis taxiformis* decreases enteric methane production from sheep. *Anim. Prod. Sci.* **2018**, *58*, 681–688. [CrossRef]
68. Roque, B.M.; Brooke, C.G.; Ladau, J.; Polley, T.; Marsh, L.J.; Najafi, N.; Pandey, P.; Singh, L.; Kinley, R.; Salwen, J.K. Effect of the macroalgae *Asparagopsis taxiformis* on methane production and rumen microbiome assemblage. *Anim. Microbiome* **2019**, *1*, 3. [CrossRef]
69. Huisman, J.M. *Marine Plants of Australia*; University of Western Australia Press: Perth, Australia, 2019.
70. Le, D.; Nguyen, P.; Nguyen, D.; Dierckens, K.; Boon, N.; Lacoere, T.; Kerckhof, F.-M.; De Vrieze, J.; Vadstein, O.; Bossier, P. Gut microbiota of migrating wild rabbit fish (*Siganus guttatus*) larvae have low spatial and temporal variability. *Microb. Ecol.* **2020**, *79*, 539–551. [CrossRef]
71. Berg, G.; Rybakova, D.; Fischer, D.; Cernava, T.; Vergès, M.-C.C.; Charles, T.; Chen, X.; Cocolin, L.; Eversole, K.; Corral, G.H.; et al. Microbiome definition re-visited: Old concepts and new challenges. *Microbiome* **2020**, *8*, 103. [CrossRef]

Article

# Isolation and Identification of Bacteria with Surface and Antibacterial Activity from the Gut of Mediterranean Grey Mullets

**Rosanna Floris [1,*], Gabriele Sanna [1], Laura Mura [1], Myriam Fiori [1], Jacopo Culurgioni [1], Riccardo Diciotti [1], Carmen Rizzo [2], Angelina Lo Giudice [3], Pasqualina Laganà [4] and Nicola Fois [1]**

[1] AGRIS-Sardegna, Agricultural Research Agency of Sardinia, Bonassai, 07100 Sassari, Italy; gabsanna@agrisricerca.it (G.S.); lmura@agrisricerca.it (L.M.); mfiori@agrisricerca.it (M.F.); jculurgioni@agrisricerca.it (J.C.); rdiciotti@agrisricerca.it (R.D.); nfois@agrisricerca.it (N.F.)

[2] Stazione Zoologica Anton Dohrn-Ecosustainable Marine Biotechnology Department, Sicily Marine Centre, Villa Pace, Contrada Porticatello 29, 98167 Messina, Italy; carmen.rizzo@szn.it

[3] Institute of Polar Sciences, National Research Council (ISP-CNR), 98122 Messina, Italy; angelina.logiudice@cnr.it

[4] Department of Biomedical and Dental Sciences and Morphofunctional Imaging, University of Messina, Torre Biologica 3p, AOU 'G. Martino, Via C. Valeria, s.n.c., 98125 Messina, Italy; plagana@unime.it

* Correspondence: rfloris@agrisricerca.it; Tel.: +39-079-284-2331

**Citation:** Floris, R.; Sanna, G.; Mura, L.; Fiori, M.; Culurgioni, J.; Diciotti, R.; Rizzo, C.; Lo Giudice, A.; Laganà, P.; Fois, N. Isolation and Identification of Bacteria with Surface and Antibacterial Activity from the Gut of Mediterranean Grey Mullets. *Microorganisms* **2021**, 9, 2555. https://doi.org/10.3390/microorganisms9122555

Academic Editor: Konstantinos Ar. Kormas

Received: 9 November 2021
Accepted: 7 December 2021
Published: 10 December 2021

**Publisher's Note:** MDPI stays neutral with regard to jurisdictional claims in published maps and institutional affiliations.

**Abstract:** Fish gut represents a peculiar ecological niche where bacteria can transit and reside to play vital roles by producing bio-compounds with nutritional, immunomodulatory and other functions. This complex microbial ecosystem reflects several factors (environment, feeding regimen, fish species, etc.). The objective of the present study was the identification of intestinal microbial strains able to produce molecules called biosurfactants (BSs), which were tested for surface and antibacterial activity in order to select a group of probiotic bacteria for aquaculture use. Forty-two bacterial isolates from the digestive tracts of twenty Mediterranean grey mullets were screened for testing emulsifying (E-24), surface and antibiotic activities. Fifty percent of bacteria, ascribed to *Pseudomonas aeruginosa*, *Pseudomonas* sp., *P. putida* and *P. anguilliseptica*, *P. stutzeri*, *P. protegens* and *Enterobacter ludwigii* were found to be surfactant producers. Of the tested strains, 26.6% exhibited an antibacterial activity against *Staphylococcus aureus* (10.0 ± 0.0–14.5 ± 0.7 mm inhibition zone), and among them, 23.3% of isolates also showed inhibitory activity vs. *Proteus mirabilis* (10.0 ± 0.0–18.5 ± 0.7 mm inhibition zone) and 6.6% vs. *Klebsiella pneumoniae* (11.5 ± 0.7–17.5 ± 0.7 mm inhibition zone). According to preliminary chemical analysis, the bioactive compounds are suggested to be ascribed to the class of glycolipids. This works indicated that fish gut is a source of bioactive compounds which deserves to be explored.

**Keywords:** intestinal microflora; fish gut; biosurfactants; grey mullets; natural antibiotics

## 1. Introduction

The study of biodiversity for exploring new biological sources is considered a suitable approach in the *bioprospecting* field for the discovery of new bioactive molecules in nature [1]. Marine environments represent some of the most interesting places for the isolation of new metabolites, due to their unique and variable physical surroundings which induce the producer microorganisms to develop metabolic and physiological capabilities for adapting to diverse habitats covering a wide range of thermal, pressure, salinity, pH and nutrient conditions [2–4]. Indeed, the identification and production of broad-spectrum activity microbial compounds have been obtained from different aquatic ecosystems and matrices [5–7]. However, to date, most of the marine microbial world is still scarcely explored, and the interest in novel compounds remains the main driver of different research projects.

Fish are generally valued for their qualities as goods, in the form of food protein, fishmeal, fish oil, for aquaculture and in the pharmaceutical industry for the production of medicine [3]. In particular, fish gut was found to be a peculiar ecological niche where bacteria transit and reside to play different vital roles (protective barriers against pathogens, promotion of fish immunity, fish nutrition, and so on) [8]. To date, different studies have described fish intestinal microbiota as a reflection of the environment and a variety of other factors (genotype, physiological status, fish behaviour, feeding habit); however, these works concerned the composition of the microbial community, the isolation, identification of microorganisms and the possible use of single bacterial culture or consortia of strains to promote fish growth and health [8–13]. Recently, the importance of marine fish gut microbiota has been highlighted in terms of the production of biofilm barriers formed by extracellular polymeric substances able to protect the host organism and protect it against pathogens [14]. Previous studies on the production of metabolites with surface activities, called biosurfactants (BSs), from fish intestinal microbial content, have highlighted the importance of the intestinal tract of fish in the search for new molecules [6]. The BS-producing microorganisms are ubiquitous, inhabiting both water (seawater, freshwater and ground water) and land (soil, sediment and sludge), as well as environments characterized by extreme conditions of pH, temperature or salinity (e.g., hyper saline sites and oil reservoirs) [15]. BSs have been commonly studied for their bioremediation properties and for antibacterial, antifungal and antiviral activities [16], and they have been found in polluted environments [17] as well as in different biological matrices [18]. Recently, the immunomodulatory role of microbial BSs has been highlighted by Giri et al. [19], and the antibacterial activity of fish gut, associated with bacteriocinogenic bacteria, was detected by Mukherjee et al. [20]. Nevertheless, the gastrointestinal microflora of fish remains a little-explored subject of basic and applied research.

Mugilidae, commonly known as grey mullet, comprise the highest number of fish species, and are among the most ubiquitous teleost families in the coastal waters of the world [21]. Grey mullets have been described as mud-eaters, as well as detritus, deposit and interface feeders; their gastrointestinal tracts, 1.5–4.6 times longer than the total body length, are arranged in several convolutions and wrapped in a peritoneal connective tissue which acts as a fat storage [22]. For these reasons, these fish constitute an interesting source of bacteria with peculiar biochemical characteristics due to the relative importance of detritus and algae in the diet. As regards Mediterranean Mugilidae, they are the main representative fish species in Sardinian estuaries and lagoons. They are appreciated in the food market, and *Mugil cephalus* roe, called "bottarga", represent an added value product and a highly prized delicacy in the southern Mediterranean [23]. Different studies have focused on the genetics [24], biogeography, distribution [25] and potential probiotics of these fish [26–28], but to the best of our knowledge, no works on the production of bioactive compounds from the intestinal microflora of the Mugilidae have been published. The aim of this study was to identify and investigate BS microbial producers isolated from the gut of different Mediterranean mullets, *Mugil cephalus*, *Chelon ramada*, *Chelon labrosus* and *Chelon saliens*, in order to assess them as a source of natural added-value bioactive compounds to be selected for aquaculture field and bioremediation.

## 2. Materials and Methods

### 2.1. Study Area, Sampling and Microbiological Analysis

Twenty wild-caught mullets, *Mugil cephalus* ($n = 4$), *Chelon ramada* ($n = 6$), *Chelon labrosus* ($n = 8$) and *Chelon saliens* ($n = 2$), destined for the local food market, were captured by professional fisheries on 27 September 2018 (autumn) and 10 February 2019 (winter). The study area was Santa Giusta lagoon, an 8 $km^2$ area with a mean depth of approximately 1 m (Sardinia, Italy; coordinates: Lat 39°52′N, Long 8°35′E). Water temperature (Temp: 12.5–28.0 °C), salinity (Sal: 28.0–44.0 ppm) and dissolved oxygen (DO: 7.0–11.5 mg $L^{-1}$) were measured in situ using a YSI 6600 v2 (YSI Inc., Yellow Springs, OH, USA) multi-parameter probe. This lagoon is peculiar for its recurring ecological instability due to different anthropogenic impacts since 2000 [29]. The fish (average size and weight, 30.1 ± 5.8 cm and 277.4 ± 183 g, respectively) were transported inside a refrigerated bag to the Bonassai laboratory within 6–8 h, weighed using a scale (d = 0.01 g) and measured in length (d = 0,1 cm). The entire intestine (mean weight 14.0 ± 6.0 g) was aseptically removed from each fish, diluted (10% *w*/*v*) in saline solution (0.9% NaCl) and homogenized in plastic bags by a Stomacher® 400 (FermionX Ltd., Worthing, UK) at room temperature. Samples were made by mixing the guts of two individuals of each species in order to obtain five samples for each date, up to a total of ten samples. Serial dilutions of the homogenate were prepared, and 100 μL of each dilution were spread on Marine agar (MA, Himedia, Mumbay, India) plates in duplicate and incubated at 30 °C for 48 h, for the enumeration of heterotrophic marine bacteria.

Bacterial colonies were randomly isolated and streaked onto fresh medium four times to obtain pure cultures. The purified isolates were stored at −80 °C in a 15% (*v*/*v*) glycerol-Nutrient broth (NB, Conda Pronadisa, Madrid, Spain) solution. The strains were assayed for the BS production, as follows.

### 2.2. Screening of Bacteria for Biosurfactant Production

#### 2.2.1. Detection of the BSs in the Culture Broth

Each isolate was tested for BS production using a battery of screening tests, performed in three independent experiments as replicates. A loopful of each strain from well-grown MA plates was used to inoculate the bacterial cultures in 100 mL sterile Erlenmeyer flasks containing 50 mL of Bushnell–Haas medium (BH, Himedia) supplemented with sunflower oil (2% *v*/*v*). The cultures were incubated at 25–28 °C under shaking for 48–96 h, and growth was monitored after 24, 48, 72 and 96 h by measuring the optical density at 600 nm ($OD_{600\ nm}$) with a spectrophotometer (Cary 1E UV-Visible Spectrophotometer, Varian Instruments, Sugarland, TX, USA); pH values were also registered. Bacterial cultures were screened for BS production using standard screening tests, performed as described in the following sections. Sodium dodecyl sulphate (SDS) and Tween 80 were used as positive controls, whereas distilled water and BH medium plus sunflower oil were used as negative controls.

Bacterial pure cultures were tested during the late stationary growth phase by means of the emulsification index (E-24) and the drop-collapse assay.

#### 2.2.2. Emulsification Index (E-24)

The emulsifying capacity was evaluated using the emulsification index detection (E-24) according to [30]. An equal volume of kerosene and culture broth (2 mL) was vortexed at high speed for 2 min in test tubes and allowed to stand for 24 h. The E-24 index is given as the height of the emulsified layer (cm) divided by the total height of the liquid column (cm) and expressed as a percentage.

#### 2.2.3. Drop-Collapse Assay

The tests were performed using the polystyrene lids of a 96-microtiter 12.7-by 8.5-cm BRANDplates® (Greiner Bio-One GmbH, Frickenhausen, Germany) according to [31], with some modifications. A 1.8 μL aliquot of diesel oil was added to each lid's well and equilibrated for 24 h at room temperature. A 5 μL amount of the culture broth was then added to the surface of the oil previously placed in the centre of each well. The shape of the drop on the oil surface was inspected after 1 min. BS-producing cultures, which collapsed, giving flat drops, were scored as positive (+), while those which gave rounded drops and remained beaded were scored as negative (–).

#### 2.2.4. Surface Tension Measurement

All isolates that proved to be positive in the previous screenings were tested with the Wilhelmy Plate method. The surface tension was determined on the cell-free supernatant of the bacterial cultures after centrifugation at 10,000 rpm for 20 min at 4 °C. Supernatants were stored at −20 °C and successively measured with a digital tensiometer using the

Wilhelmy Plate method according to [18]. A surface tension lower than 40 mN·m$^{-1}$ was considered as an index of BS production.

### 2.3. Biosurfactant Extraction and Thin Layer Chromatography (TLC)

The BSs-producing bacterial cultures were centrifuged at 10,000 rpm for 20 min and the supernatants characterized by thin-layer chromatography (TLC) according to [32]. The BSs were extracted from 5 mL of supernatants. The pH of the supernatants was adjusted to 2.0 with 1 N HCl and left at 4 °C overnight [33]. The extraction of the BSs was carried out twice, adding an equal volume of chloroform:methanol (2:1, *v*/*v*). The mixture was vigorously shaken for 1 min and allowed to stand until the phase separation. The organic layer (lower phase) was retained and concentrated under vacuum using a rotary evaporator at 40 °C. Successively, the extracts were weighted to acquire the amount of the crude yield, re-suspended in 200 µL chloroform:methanol (2:1, *v*/*v*) mixture and analysed by thin-layer chromatography (TLC) on silica gel plate (© Millipore Corporation, Burlington, Massachusetts, USA). The solvent system used was chloroform:methanol:acetic acid:water (65:15:1:1, *v*/*v*/*v*/*v*), and for detecting the less polar compounds, hexane:ether:acetic acid (70:30:2, *v*/*v*/*v*) was used. The TLC run lasted for approximately 90 min. The TLC plates were stained in two different solutions. The sugar moieties were identified by staining the plates with anisaldehyde (Sigma-Aldrich, Burlington, MA, USA):glacial acetic acid:sulphuric acid:ethanol (0.15:1:2:37, *w*/*v*/*v*/*v*), while, the fatty acid moieties were stained with an ammonium molybdate (Sigma-Aldrich, Burlington, MA, USA):cerium sulphate (Sigma-Aldrich, Burlington, MA, USA):sulphuric acid:water (3:0.5:8.5:23.45, *w*/*w*/*v*/*v*) solution. The colour of the spot on the plate was developed by heating inside an oven at 150 °C. The TLC patterns of the extracts were compared with those of three different standards for the identification of the BSs: Sophorolipids (S) (Sopholiance, Reims, France), a Trehalose Lipid Tetraester (crude extract) (T) (Karlsruhe Institute of Technology, Karlsruhe, Germany), Phospholipid mixture for HPLC (Supelco, Bellefonte, Pennsylvania) and a mix of Rhamnolipids R-95 (R) (Sigma-Aldrich, Burlington, MA, USA). Each retardation factor (Rf) was calculated by dividing the distance of the considered TLC fraction run in the TLC plate from the origin by the distance of the solvent from the same origin.

### 2.4. Bacterial Identification

Bacterial isolates were identified by partial 16S rRNA gene sequencing. Bacterial cell preparation for DNA extraction were performed using a Qiagen kit (DNeasy® Blood & Tissue Kit, Hilden, Germany). Universal primers designed to amplify approximately 1300 bp of *Escherichia coli* 16S rRNA gene were used [34]. The sequences were: forward primer 63f (50 -CAG GCC TAA CAC ATG CAA GTC-30) and reverse primer 1387r (50 -GGG CGG WGT GTA CAA GGC-30). PCR mixture contained from 50 to 100 ng DNA template, 1 µL of each primer (50 pmol µL$^{-1}$) (Sigma Genosys, The Woodlands, Texas, USA) and 20 µL of ready-to-use PCR master mix containing *Taq* polymerase (MegaMix 2MM-5, µ Microzone Limited, Stourbridge, UK), to give a total reaction of 25 µL. The PCR conditions were: 30 cycles of denaturation at 94 °C for 1 min, annealing at 58 °C for 1 min and elongation at 72 °C for 2 min, with a final elongation at 72 °C for 10 min. Purification of the amplicons for sequence study was carried out as described in the Qiagen (QIAquick® PCR Purification Kit, Hilden, Germany) protocol. Partial sequences were determined by BMR Genomics s.r.l (Padova, Italy). The sequencing results were submitted for homology searches by BLAST (Basic Logical Alignment Search Tool) [35] after unreliable sequences at the 3′ and 5′ ends were removed using the software Chromas, version 1.43 (Griffin University, Brisbane, Qld, Australia). The NCBI GenBank nucleotide database (National Center for Biotechnology Information, http://www.ncbi.nml.nih.gov accessed on October 2021) was used for sequence pairing. The identities were determined on the highest score basis. Nucleotide sequences were deposited in the NCBI GenBank database under the accession numbers MW369461- MW369487 [11] and OK342256-OK342267 (This study). A phylogenetic tree was reconstructed using the 16S rRNA gene sequences obtained in the study and the reference strains ENA AF094713 *P. aeruginosa* ATCC 10145$^T$, ENA X60410 *Aeromonas media* ATCC 339007$^T$ and ENA AJ853891 *Enterobacter ludwigii* DSMZ 16688$^T$. As an out-group, the strain NR074804 *Cellvibrio japonicus* strain Ueda 107 was utilized for the analysis [36]. A Clustal W Multiple alignment was obtained by MEGA X [37]. The final dataset was included using 820 bp positions. The best fit DNA evolution model selected on the IQTREE webserver [38] was K2P G4 [39]. A phylogenetic tree was inferred by the Maximum Likelihood algorithm [40,41] in IQTREE with default parameters. The bootstrap test (1000 replicates) [42] was used to evaluate the robustness of the tree topology.

### 2.5. *BS Antibacterial Activity*

#### 2.5.1. Bacterial Pathogens

The antibacterial activity was tested against the following bacterial pathogens: *Pseudomonas aeruginosa* H1628, *Staphylococcus aureus* H1670, *Klebsiella pneumoniae* H1637, *Proteus mirabilis* H1643 and *Aeromonas hydrophila* H1563. The strains were previously isolated from human clinical specimens and identified to the species level by API 20 E, API 20 NE and API STAPH profiles (bioMérieux, Marcy l'Etoile, France). Lab strains are maintained at −20 °C in Tryptone Soy Broth (TSB, Difco) supplemented with 15% glycerol.

#### 2.5.2. Antibacterial Activity

The inhibitory activity was tested on cell-free supernatants (CFSs) and crude extracts (CEs) using the standard disk diffusion method (DDM) (Kirby Bauer test), as accepted by the National Committee for Clinical Laboratory Standards (NCCLS 2000). Details are described below.

- Cell-Free Supernatants

CFSs were obtained through centrifugation at 10,000 rpm at 4 °C for 20 min of cell culture aliquots, and filter-sterilized on nitrocellulose membranes (pore diameter 0.22 µm). Each CFS was ten-fold concentrated prior to testing using a concentrator (Concentrator 5301, Eppendorf AG, Hamburg, Germany), as described by [43]. Bacterial pathogens were suspended in 3 mL of a saline solution (NaCl 0.9%, *w*/*v*) in order to achieve a turbidity of McFarland 0.5 standard (containing around $1.5 \times 10^8$ cells/mL), and the suspensions were spread-plated on plates of TSA supplemented with 1% (*w*/*v*) NaCl (TSA1), in triplicates. Aliquots (60 µL) of each CFS were used to soak sterile cellulose discs (6 mm diameter), which were laid on the medium surface previously inoculated with pathogenic strains. Distilled water (20 µL) was used to soak sterile disks as a negative control, while commercially available disks (6 mm in diameter, Oxoid) containing chloramphenicol (30 µg), amoxicillin (30 µg) and gentamycin CN30 (30 µg) were used as a positive control. The plates were incubated for 24 h at 37 °C.

- Crude Extracts

The extraction of CE was performed on 50 mL aliquots of each CFS (obtained as described in the section above). Firstly, CFSs were acidified with phosphoric acid (85%, *v*/*v*), and bioactive molecules were extracted twice in ethyl acetate (cell-free supernatant: ethyl acetate ratio 1:1.25, *v*/*v*). Ethyl acetate was totally evaporated at room temperature and extracts were collected [7]. Based on the total amount, each CE was dissolved in a proper volume of ethyl acetate in order to obtain 6 mg of extract in a final volume of 20 µL. After complete solvent evaporation, the disks were placed onto TSA1 plates inoculated with the target pathogens. Disks soaked with ethyl acetate and submitted to evaporation were used as negative controls, while positive controls were performed as for CFSs tests. Plates were incubated overnight at 37 °C. The diameter of complete inhibition zones was measured, and means and standard deviations ($n$ = 3) were calculated. The results were codified as weak activity for inhibition zone lower than 8 mm [44] European pharmacopoeia.

## 3. Results

### 3.1. Enumeration of Bacteria and Colony Isolation

Bacterial counts on MA medium showed values of heterotrophic marine bacteria from $10 \times 10^3$ to $10 \times 10^4$ colony forming units (CFU) in autumn and from $12 \times 10^4$ to $40 \times 10^4$ CFU in winter. Forty-two bacterial colonies were isolated from different mullet species (strains 1–26 from fish sampled in autumn and strains 28–56 in winter).

### 3.2. Screening of Bacteria for BS Production

In this study, the intestinal bacterial strains isolated from mullet grey fish showed a diversity in the bioactive performances. Thirty-three out of 42 strains were able to utilize sunflower oil for their growth as the sole energy and carbon source in BH medium at 25–28 °C after 72 h. The application of different screening methods allowed the selection of a "group" of intestinal strains as surfactant producers in three independent experiments. The drop-collapse method was used as a first screening test for identifying the "bioactive" microbes. Table 1 shows the results of all the used screening tests. By means of the drop-collapse assay, a surface activity of different intensity was detected: eight *Pseudomonas aeruginosa* (i.e., strains 1, 3, 5, 6, 9, 12, 13 and 15) gave a strong positive score (+++), two *Pseudomonas aeruginosa* (i.e., strains 8 and 26) were scored as pretty good surface active cultures (++), eight intestinal strains were ascribed to *Pseudomonas* spp. (i.e., strains 10, 19, 22, 25, 41, 45, 51 and 56) and one isolate, *Enterobacter* sp. (strain 28), presented a discrete activity (+), while seven *Pseudomonas* spp. isolates (i.e., strains 17, 18, 20, 23, 24, 47 and 55), five strains ascribed to *Aeromonas* spp. (i.e., strains 11, 30, 35, 37 and 40) and two unidentified cultures showed a weak or absent activity, scored as (weak) or (-) (Table 1). The emulsifying cultures showed stable and compact emulsions with kerosene at the end of the exponential and/or during the stationary growth phase (OD 600 nm = 2.0–3.0) and remained stable over 1–2 months without any significant change in the index values. The values reached by means of the emulsification index (E-24), were from 0 to 70% after 72 h of incubation (Table 1). Generally, the greatest E-24(%) values were observed in the strains which had given the strongest positive score by the drop-collapse test, except for *P. aeruginosa* (strain 8), which showed a good activity by the drop-collapse assay but a null value of E-24(%), while strain 12 showed a weak E-24(%) and a high score by the drop-collapse test.

**Table 1.** BS-producing bacteria from grey mullets' guts: bacterial affiliations (similarity 99–100%), performed tests: (mean ± SD) and TLC results. Highest E24 values are highlighted in bold.

| Strain | Fish Species | Bacterial Affiliation | GeneBank Accession Number | Drop Collapse | E-24 (%) | Surface TensionmN·m$^{-1}$ | BS Type |
|---|---|---|---|---|---|---|---|
| 1 | CR | *Pseudomonas aeruginosa* | MW369461 | +++ | **70.5 ± 9.1** | 36.5 ± 0.1 | Rhamnolipid |
| 3 | CR | *Pseudomonas aeruginosa* | OK342256 | +++ | **68.0 ± 12.7** | 37.1 ± 0.1 | Rhamnolipid |
| 5 | CR | *Pseudomonas aeruginosa* | OK342257 | +++ | **77.0 ± 0.0** | 36.9 ± 0.4 | Rhamnolipid |
| 6 | CR | *Pseudomonas aeruginosa* | MW369462 | +++ | **56.4 ± 0.0** | 37.1 ± 0.1 | Less polar compound |
| 8 | CR | *Pseudomonas aeruginosa* | OK342258 | ++ | 0.0 ± 0.0 | 37.1 ± 0.1 | nd |
| 9 | CR | *Pseudomonas aeruginosa* | OK342259 | +++ | **57.7 ± 1.8** | 37.2 ± 0.3 | nd |
| 10 | CR | *Pseudomonas alcaligenes* | MW369463 | + | **50.0 ± 1.8** | 37.2 ± 0.3 | nd |

**Table 1.** *Cont.*

| Strain | Fish Species | Bacterial Affiliation | GeneBank Accession Number | Drop Collapse | E-24 (%) | Surface TensionmN·m$^{-1}$ | BS Type |
|---|---|---|---|---|---|---|---|
| 11 | CR | *Aeromonas caviae* | MW369464 | - | 0.0 ± 0.0 | 43.0 ± 0.1 | nd |
| 12 | CR | - | - | +++ | 15.4 ± 21.8 | 36.9 ± 0.1 | nd |
| 13 | CR | *Pseudomonas aeruginosa* | MW369465 | +++ | **51.3 ± 3.6** | 36.9 ± 0.1 | Rhamnolipid |
| 15 | CR | *Pseudomonas aeruginosa* | MW369466 | +++ | **59.0 ± 3.6** | 36.6 ± 0.6 | Rhamnolipid |
| 16 | CR | - | - | weak | 0.0 ± 0.0 | 35.35 ± 0.6 | nd |
| 17 | CR | *Pseudomonas mendocina* | MW369467 | - | 20.5 ± 0 | nd | nd |
| 18 | CR | *Pseudomonas putida* | OK342260 | weak | **33.3 ± 3.6** | 36.1 ± 0.1 | Less polar compound |
| 19 | MC | *Pseudomonas* sp. | OK342261 | + | **28.2 ± 3.6** | 35.2 ± 0.0 | Less polar compound |
| 20 | MC | *Pseudomonas alcaliphila* | MW369468 | weak | 0.0 ± 0.0 | 35.0 ± 0.4 | nd |
| 21 | MC | - | - | weak | **25.6 ± 14.5** | nd | nd |
| 22 | MC | *Pseudomonas* sp. | OK342262 | + | **25.6 ± 0.0** | 35.1 ± 0.2 | Less polar compound |
| 23 | MC | *Pseudomonas* sp. | OK342263 | weak | 0.0 ± 0.0 | 36.5 ± 0.1 | nd |
| 24 | MC | *Pseudomonas khazarica* | MW369469 | weak | 0.0 ± 0.0 | nd | nd |
| 25 | MC | *Pseudomonas* sp. | OK342264 | + | 0.0 ± 0.0 | 35.3 ± 0.1 | Less polar compounds |
| 26 | MC | *Pseudomonas aeruginosa* | MW369470 | ++ | **33.3 ± 0.0** | 37.6 ± 0.3 | Less polar compound |
| 28 | CS | *Enterobacter ludwigii* | MW369471 | + | 0.0 ± 0.0 | 37.9 ± 0.1 | nd |
| 30 | CS | *Aeromonas media* | MW369472 | weak | 0.0 ± 0.0 | 39.4 ± 0.9 | nd |
| 35 | CS | *Aeromonas taiwanensis* | MW369473 | weak | 0.0 ± 0.0 | 43.2 ± 0.1 | nd |
| 37 | CL | *Aeromonas media* | MW369474 | - | 0.0 ± 0.0 | 35.9 ± 0.1 | nd |
| 40 | CL | *Aeromonas media* | MW369476 | - | 0.0 ± 0.0 | 46.1 ± 0.3 | nd |
| 41 | CL | *Pseudomonas anguilliseptica* | MW369477 | + | **32.1 ± 5.4** | 35.2 ± 0.6 | nd |
| 45 | CL | *Pseudomonas stutzeri* | OK342265 | + | 0.0 ± 0.0 | 36.3 ± 0.1 | nd |
| 47 | CL | *Pseudomonas protegens* | MW369478 | weak | 0.0 ± 0.0 | 40.5 ± 0.4 | nd |
| 51 | CL | *Pseudomonas protegens* | OK342266 | + | 0.0 ± 0.0 | 35.5 ± 0.1 | nd |
| 55 | CL | *Pseudomonas protegens* | MW369480 | - | 0.0 ± 0.0 | 37.7 ± 0.1 | nd |
| 56 | CL | *Pseudomonas* sp. | OK342267 | + | 0.0 ± 0.0 | 39.9 ± 0.1 | nd |

nd: not detected; CR: Chelon ramada, MC: Mugil cephalus; CS: Chelon saliens; CL: Chelon labrosus.

On the other hand, the results of surface tension, measured by the Wilhelmy Plate method, showed a lower variability between strains, detecting values from 35 to 46 (mN·m$^{-1}$) (Table 1). Figure 1 shows the emulsification indexes E-24(%) and the surface tension activity of the detected "bioactive" intestinal bacteria.

The most interesting strains are represented by eight isolates (strain 1, 3, 5, 6, 9, 10, 13, 15), showing an emulsification index E-24(%) from 50.0 to 77.0% and a surface tension from 36.5 to 37.2 (mN·m$^{-1}$).

### 3.3. Bacterial Identification

Thirty strains, which showed bioactivity, were identified by 16S rRNA gene partial sequencing. Table 1 shows fish origin, phylogenetic affiliation and accession number of the studied intestinal strains. The gut microbiota of the mullets was ascribed to 14 different species: *Pseudomonas aeruginosa* (9 strains), *Pseudomonas* sp. (5 strains), *A. media* (3 strains), *P. protegens* (3 strains), *P. alcaligenes* (1 strain), *P. mendocina* (1 strain), *P. putida* (1 strain), *P. alcaliphila* (1 strain), *P. khazarica* (1 strain), *P. anguilliseptica* (1 strain), *P. stutzeri* (1 strain), *Aeromonas caviae* (1 strain), *A. taiwanensis* (1 strain), *Enterobacter ludwigii* (1 strain). Figure 2 shows the phylogenetic tree reconstructed using the 16S rRNA gene sequences. Different groups and subgroups were obtained with respect to the corresponding reference species type strains and among themselves. The most heterogeneous group was represented by *Pseudomonas* spp., while *Aeromonas* spp. and *Enterobacter* sp. formed well distinguished clusters. The outgroup strain NR074804 *Cellvibrio japonicus* strain Ueda 107 was separated from all the others.

### 3.4. BSs Extracts and Thin Layer Chromatography (TLC)

Figure 3 shows the BSs yield extracts of representative intestinal bacteria. The BS producers which gave the highest yield extracts (values from 6–6.42 g $L^{-1}$) were strains 6, 13, 15.

The chromatographic analyses (TLC) of the BSs extracted from intestinal bacterial supernatants showed two types of glycolipid compounds (Figure 4). The TLC silica gel glass plates stained by anisaldehyde (carbohydrates) (Figure 4a,c) and cerium sulphate (lipids) (Figure 4b,d) indicate a group of specific TLC fractions (retardation factor Rf = 0.42), which presumably represents the di-rhamnolipid structures, while a group of other fractions (Rf = 0.75) detects the mono-rhamnolipid molecules (Figure 4a,b). These defined compounds were characterized by the same Rf values as the rhamnolipid standard and were exhibited by bacterial strains 1, 3, 5 and 13 (Figure 4a,b). Another type of molecule was also detected by the TLC analyses but not separated using the first solvent system indicated above (strains 19 and 26, Figure 4a,b). These less polar compounds from strains 6, 18, 19, 22, 25 and 26 were separated using a less polar solvent system, as described above, and gave different profiles of the TLC fractions (from the bottom, Rf = 0.20, 0.30, 0.60) (Figure 4c,d).

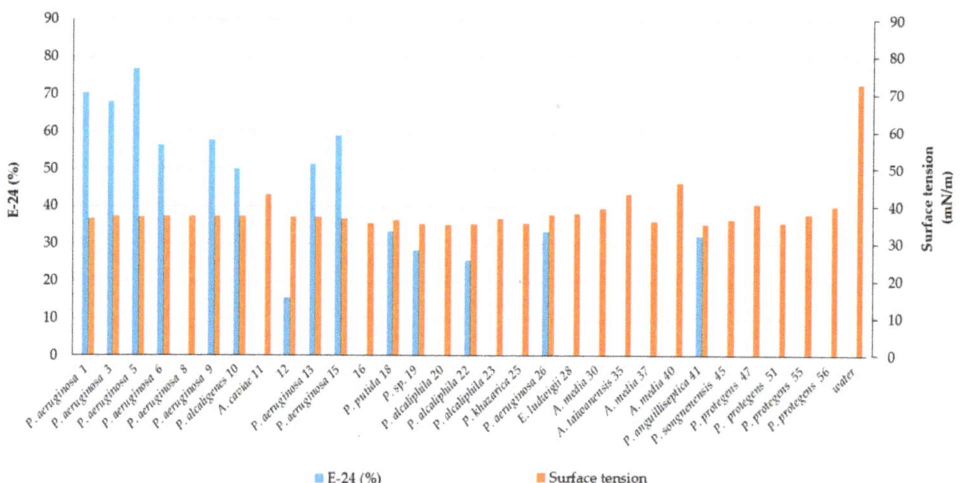


**Figure 1.** Emulsification index E-24(%) and the surface tension activity (mN·$m^{-1}$) of intestinal bacterial cultures from mullet species.

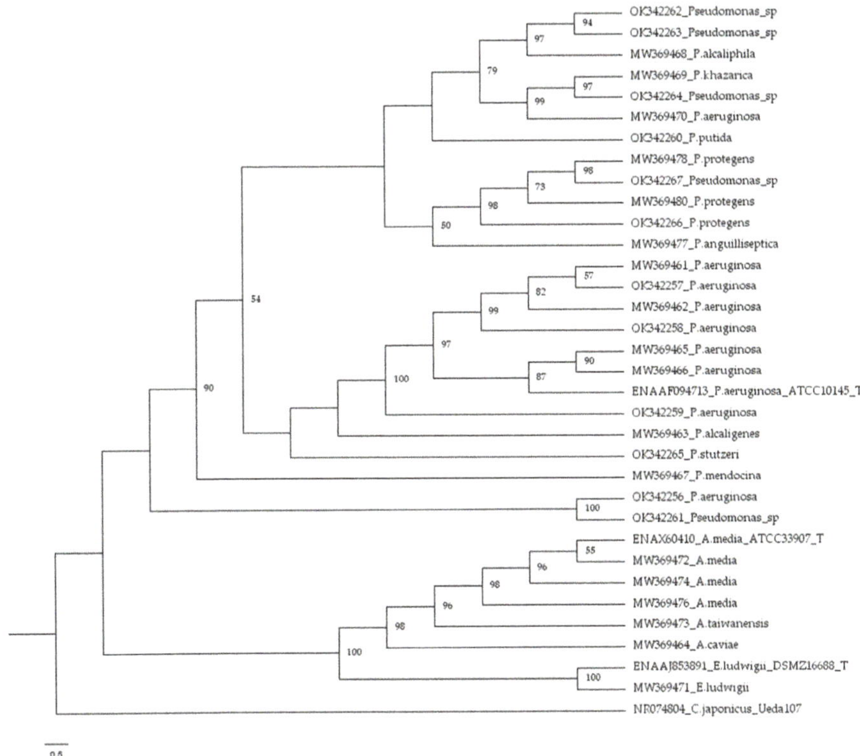


**Figure 2.** Phylogenetic tree based on 16S rRNA gene sequences comparison between the intestinal strains and reference collection strains. NR074804 *Cellvibrio japonicus* strain Ueda 107 was used as the outgroup strain. Each node indicates the percentage of the obtained bootstrap values higher than 50% of 1000 replicates. The scale bar indicates sequence divergence.

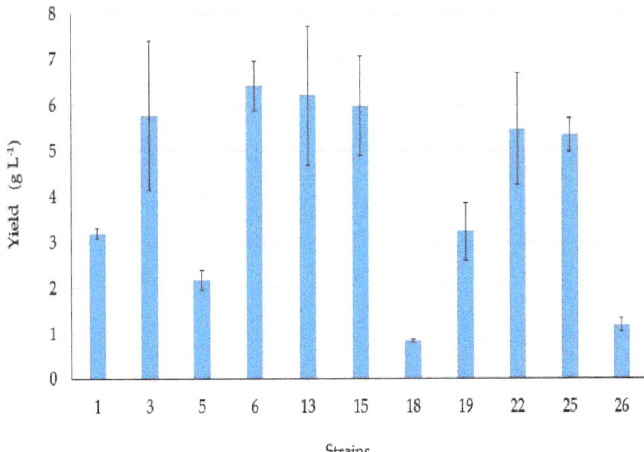


**Figure 3.** Yield of BS extracts (g $L^{-1}$) from intestinal bacterial cell-free supernatants. Error bars indicate standard error (SE).

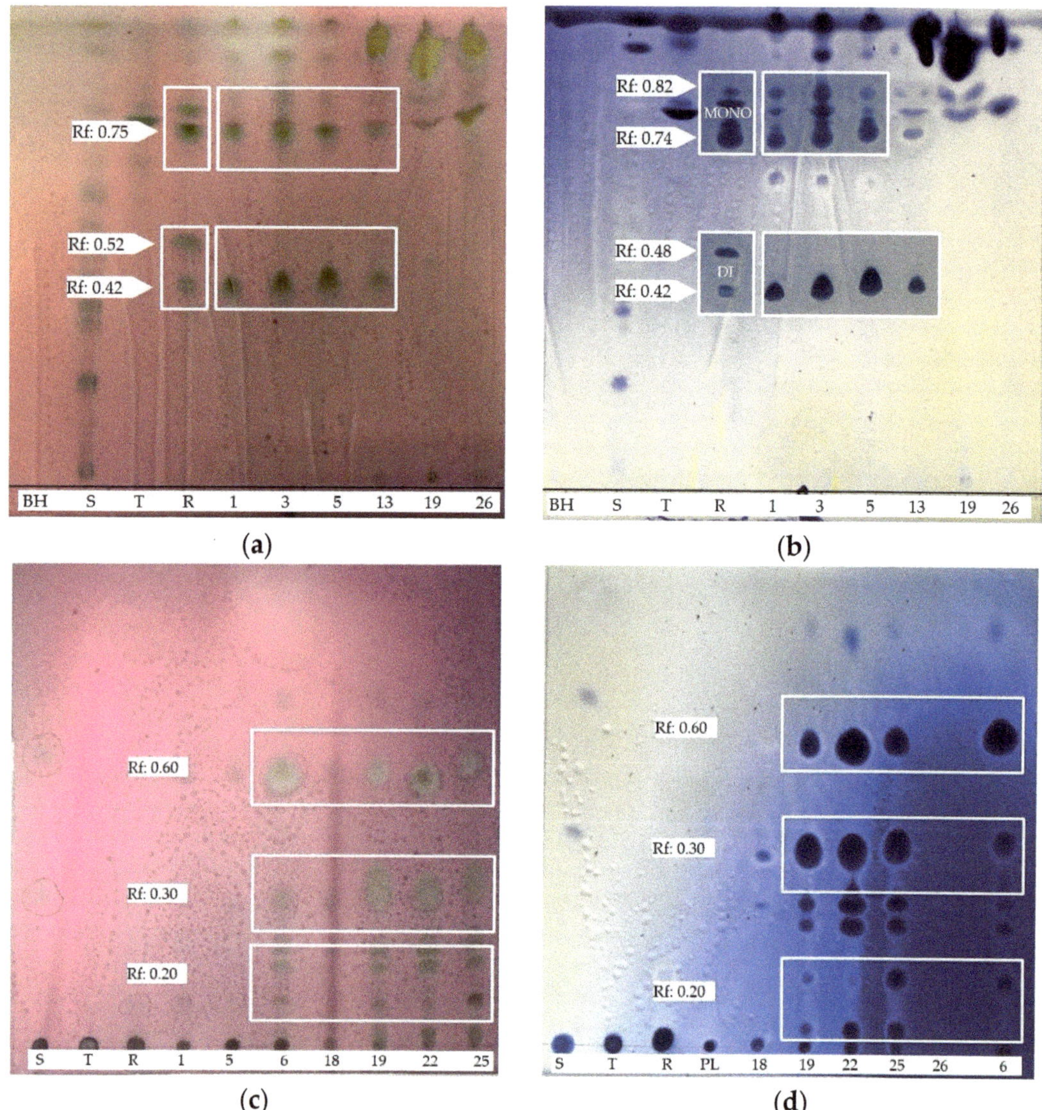


**Figure 4.** Examples of TLC plates of intestinal BS extracts stained for detecting sugars (**a**) and (**c**) and lipids (**b**) and (**d**). BH: Bushnell–Haas broth; S: sophorolipids; T: trealose lipids; R: rhamnolipids; PL: phospholipids; (**a**) and (**b**) = solvent system: chloroform: acetic acid:methanol:water (65:15:1:1); (**c**) and (**d**) = solvent system: chloroform:exane:ether:acetic acid (70:30:2).

### 3.5. Antibacterial Activities

CFSs exhibited inhibitory activity against the target strains *S. aureus* H1610 and *P. mirabilis* H1643, while no inhibition was evidenced against the target strains *P. aeruginosa* H1328, *K. pneumoniae* H1637 and *A. hydrophila* H1563 (Table 2).

Of the tested CFSs, 26.6% exhibited antibacterial activity, with halos $\geq$10 mm compared to the target strain *S. aureus* H1610 (Figure 5a). Specifically, CFSs 1 and 56 (from *Pseudomonas aeruginosa* and *Pseudomonas* sp., respectively) recorded inhibitory halos of 13.5 $\pm$ 0.7 mm, while CFS 15 (from *Pseudomonas aeruginosa*) showed the highest inhibitory activity of 14 ± 0.0 mm. CFS 6 (*Pseudomonas aeruginosa*) showed a weak activity, indicated as a positive response. Of the CFSs, 50% resulted as active against the target strain *P. mirabilis* H1643, with seven CFSs exhibiting inhibition activity for halos ≥10 mm (*Pseudomonas aeruginosa* 1 and 3, *Pseudomonas* sp. 22, *Pseudomonas aeruginosa* 26, *Aeromonas media* 37 and CFSs from strains 12 and 16), four CFSs (CFSs from *Pseudomonas aeruginosa* 13, *Enterobacter ludwigii* 28, *Pseudomonas protegens* 47 and *Pseudomonas* sp. 56), with only weak activity showing inhibition halos ≤10 mm (8 ± 00 mm, 8 ± 00 mm, 9 ± 00 mm, 7 ± 00 mm, respectively). In this case, the highest inhibitory activity was shown by the CFS of *Pseudomonas aeruginosa* 3, with an inhibition halo of 18.5 ± 0.7 mm (Table 2 and Figure 5b).

**Table 2.** Antibacterial activity of supernatants and crude extracts (mm) in agar diffusion assay against bacterial pathogens. Values are expressed as the mean ± standard deviation of three replicates. Highest values are highlighted in bold.

| | **Cell-Free Supernatants (CFSs) and Crude Extracts (CEs) (mm)** | | | | | | | |
|---|---|---|---|---|---|---|---|---|
| **Test** | ***S. aureus*** **H1610** | | ***P. mirabilis*** **H1643** | | ***K. pneumoniae*** **H1637** | | ***A. hydrophila*** **H1563** | |
| | **CFSs** | **CEs** | **CFSs** | **CEs** | **CFSs** | **CEs** | **CFSs** | **CEs** |
| *Pseudomonas aeruginosa* 1 | **13.5 ± 0.7** | 5.5 ± 0.7 | **15 ± 0.0** | - | - | **17.5 ± 0.7** | - | - |
| *Pseudomonas aeruginosa* 3 | **10.5 ± 2.1** | 5.5 ± 0.7 | **18.5 ± 0.7** | - | - | 8.0 ± 0.0 | - | - |
| *Pseudomonas aeruginosa* 5 | - | 5.5 ± 0.7 | - | **15.5 ± 0.7** | - | 6.0 ± 0.0 | - | - |
| *Pseudomonas aeruginosa* 6 | + | 8.5 ± 0.7 | - | **12.5 ± 0.7** | - | **11.5 ± 0.7** | - | - |
| *Pseudomonas aeruginosa* 8 | - | 6.5 ± 0.7 | + | - | - | 7.0 ± 0.0 | - | 6.5 ± 0.7 |
| *Pseudomonas aeruginosa* 9 | - | **14.5 ± 0.7** | - | - | - | **12.0 ± 0.0** | - | - |
| *Pseudomonas alcaligenes* 10 | **12 ± 0.0** | 7.5 ± 0.7 | - | **16.0 ± 0.0** | - | + | - | - |
| *Aeromonas caviae* 11 | - | 8.5 ± 0.7 | - | - | - | + | - | - |
| Unidentified 12 | 9 ± 1.4 | 7.5 ± 0.7 | **17.0 ± 1.4** | - | - | 7.5 ± 0.7 | - | - |
| *Pseudomonas aeruginosa* 13 | **12.5 ± 0.7** | **12.5 ± 0.7** | 7.0 ± 0.0 | - | - | **12.5 ± 0.7** | - | 7.0 ± 0.0 |
| *Pseudomonas aeruginosa* 15 | **14 ± 0.0** | **12.5 ± 0.7** | - | **14.0 ± 0.0** | - | **12.5 ± 0.7** | - | - |
| Unidentified 16 | - | - | **10.0 ± 0.0** | - | - | - | - | - |
| *Pseudomonas putida* 18 | - | - | + | - | - | - | - | - |
| *Pseudomonas* sp. 19 | - | 7.5 ± 0.7 | - | - | - | 8.0 ± 0.0 | - | + |
| *Pseudomonas alcaliphila* 20 | - | - | - | - | - | - | - | - |
| *Pseudomonas* sp. 22 | - | 7.5 ± 0.7 | **13.0 ± 0.0** | - | - | 5.5 ± 0.7 | - | - |
| *Pseudomonas* sp. 23 | - | + | + | - | - | - | - | - |
| *Pseudomonas* sp. 25 | 9 ± 0.0 | + | - | **12.0 ± 0.0** | - | 6.5 ± 0.7 | - | - |
| *Pseudomonas aeruginosa* 26 | - | - | **10 ± 0.0** | - | - | - | - | - |
| *Enterococcus ludwigii* 28 | - | 5.5 ± 0.7 | 9.0 ± 0.0 | - | + | + | - | - |
| *Aeromonas media* 30 | - | 5.5 ± 0.7 | - | - | - | 6.0 ± 0.0 | - | - |
| *Aeromonas taiwanensis* 35 | - | 5.5 ± 0.7 | - | - | - | 6.0 ± 0.0 | - | - |
| *Pseudomonas protegens* 37 | - | - | **10.0 ± 0.0** | - | - | - | - | - |
| *Aeromonas media* 40 | - | 7.5 ± 0.7 | - | **15.5 ± 0.7** | - | + | - | + |
| *Pseudomonas anguilliseptica* 41 | - | + | - | - | - | - | - | 8.0 ± 0.0 |
| *Pseudomonas stutzeri* 45 | - | - | + | - | - | - | - | - |
| *Pseudomonas protegens* 47 | - | 9.5 ± 0.7 | 8.0 ± 0.0 | - | - | - | - | 6.5 ± 0.7 |
| *Pseudomonas protegens* 51 | - | 7.0 ± 0.7 | - | - | - | 7.5 ± 0.7 | - | - |
| *Pseudomonas protegens* 55 | **10 ± 0.0** | 9.5 ± 0.7 | - | - | - | - | - | - |
| *Pseudomonas* sp. 56 | **13.5 ± 0.7** | 5.5 ± 0.7 | 8.0 ± 0.0 | - | - | + | - | - |
| Negative control | 0.0 ± 0.0 | | 0.0 ± 0.0 | | 0.0 ± 0.0 | | 0.0 ± 0.0 | |
| Chloramphenicol | 21 ± 0.0 | | - | | + | | 30.0 ± 0.0 | |
| Gentamycin CN30 | - | | 14 | | 8.0 ± 0.0 | | 18.0 ± 0.0 | |
| Amoxycillin | - | | - | | - | | - | |

The CEs evidenced antibacterial activity against more target strains, namely *S. aureus* H1610, *P. mirabilis* H1643, *K. pneumoniae* H1637 and *A. hydrophila* H1563, while no activity was recorded against *P. aeruginosa* H1628. Of the CEs, 76.7% and 66.7% were active against *S. aureus* H1610 and *K. pneumoniae* H1637, respectively, while 20% of the CEs showed inhibitory activity against *P. mirabilis* H1643 and *A. hydrophila* H1563 (Figure 5). Specifically, three CEs (*Pseudomonas aeruginosa* 9, 13 and 15) exhibited antibacterial activity, showing halos ≥10 mm against *S. aureus* H1610, with the highest inhibition for 9 (14.5 ± 0.7 mm) (Table 2). The rest of the tested CEs exhibited antibacterial activity, with halos ranging from 5.5 ± 0.7 to 9.5 ± 0.7 mm, and two weak responses have been recorded for CEs from *Pseudomonas* sp. 25 and *Pseudomonas anguilliseptica* 41 (Table 2 and Figure 4a). Six CEs, showing inhibitory activity against the target strain *P. mirabilis* H1643, evidenced inhibitory ≥10.00 ± 0.0 mm, ranging from 12.0 ± 0.7 mm (CEs *Pseudomonas aeruginosa* 6 and *Pseudomonas* sp. 25) to 16.0 ± 0.0 (CE *Pseudomonas alcaligenes* 10) (Table 2 and Figure 5b). Five CEs resulted active against *K. pneumoniae* H1637 with halos ≥10.0 mm, and four weak responses have been evidenced by CEs from *Pseudomonas alcaligenes* 10, *Aeromonas caviae* 11, *Enterobacter ludwigii* 28, *Aeromonas media* 40 and *Pseudomonas* sp. 56. The highest activity was obtained by CE of *Pseudomonas aeruginosa* 1 (17.5 ± 0.7 mm) (Table 2 and Figure 4c).

Finally, the antibacterial activity against *A. hydrophila* was exhibited in all cases, with halos ≤10.0 mm, with two weak responses (CEs from *Aeromonas media* 40 and *Pseudomonas* sp. 19) and four inhibitory halos ranging from 6.5 ± 0.7 to 8.0 ± 0.0 mm (Table 2 and Figure 5d).

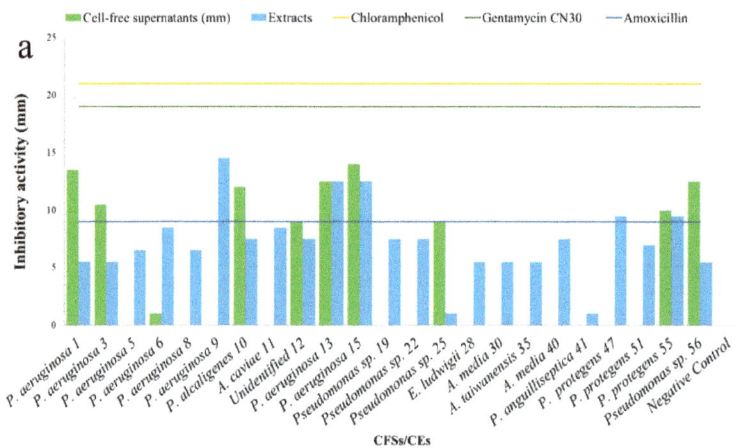


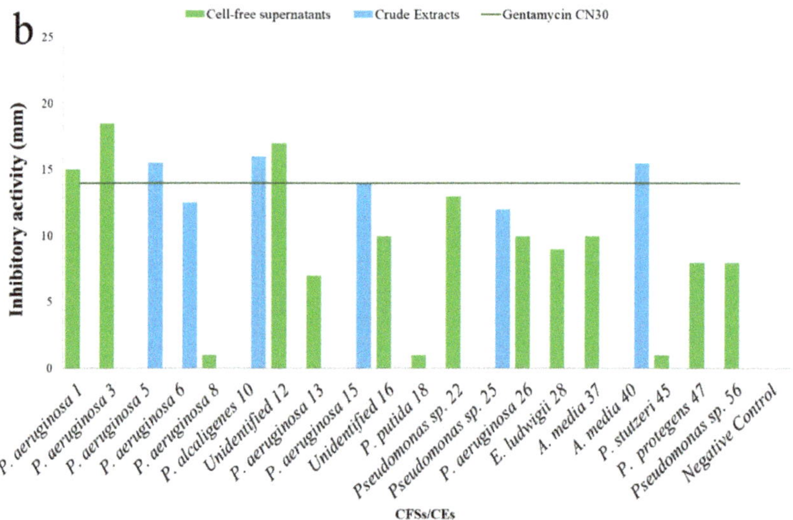


**Figure 5.** *Cont.*

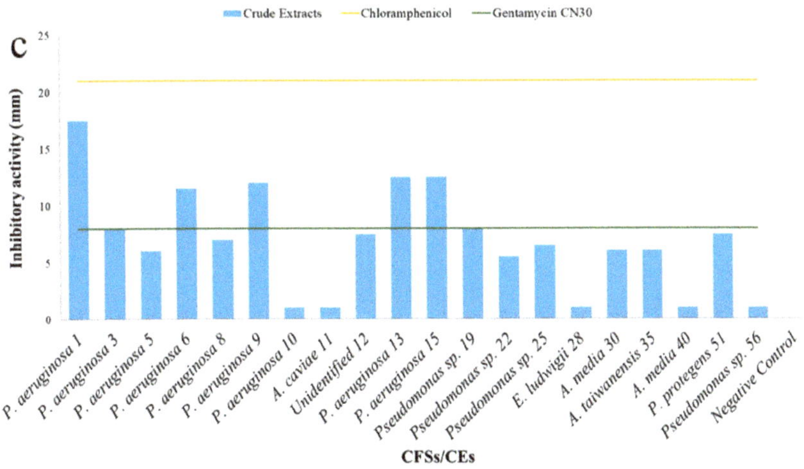


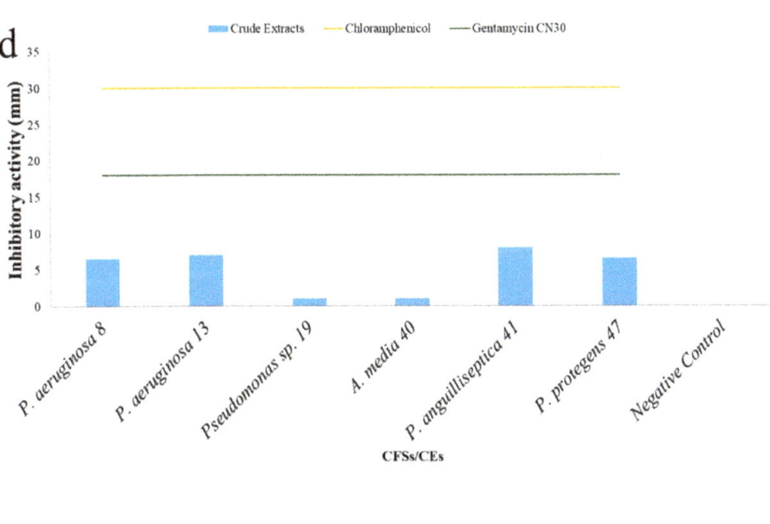


Cell-free supernatants Crude extracts

Chloramphenicol Gentamycin CN30 Amoxicillin

**Figure 5.** Inhibitory activity exhibited against the target *S. aureus* H1610 (**a**), *P. mirabilis* H1643 (**b**), *K. pneumoniae* H1637 (**c**) and *A. hydrophila* H1563 (**d**) by concentrated supernatants (CFSs green) and crude extracts (CEs blue) obtained by bacterial isolates.

## 4. Discussion

Fish gut is a place where several metabolic activities take place and where bacteria can enter, reside or transit; for this reason, it deserves to be explored as a marine source of new added-value compounds. As an example, bacterial extracellular polymeric substances seem to play a pivotal role in the formation of complex biofilm architecture in marine fish gut, as demonstrated for the luminous bacteria isolated from the gastrointestinal tract of White Sea fish [14]. In the present study, bacterial strains from the intestinal tract of different mullet species from a brackish peculiar environment were assayed for the production of secondary metabolites such as biosurfactants with surface and antimicrobial compounds. The battery of tests applied led to the selection of 50% of strains able to produce molecules with a different spectrum of emulsifying and surface activities. However, culture conditions play a crucial role in the growth of a strain and in its production of a particular metabolite, as reported by many authors who tried to discover the optimum culture conditions and suitable hydrocarbon source to achieve the maximum yield of these compounds [32,45–47]. The study concerns bacterial isolates from different species of wild mullets sampled in two different seasons (autumn and winter). Noteworthy, the analyses of cell-free supernatants obtained from bacterial cultures isolated from fish in autumn provided evidence of their good emulsifying properties and their significant reduction in surface tension, while for most of the strains isolated from the gut of fish captured in winter, only a discrete surface tension activity was scored in their supernatants. The different behaviours of the bacterial culture bioactivities are interesting and can be ascribed to the environmental conditions, which varied a great deal in the lagoon during the two sampling periods. Indeed, the aquatic environment of capture is known for its recurring ecological instability due to different anthropogenic impacts [29]. These findings strengthen the idea of the relevant role of factors, such as temperature and salinity, on the variability of fish intestinal bacteria other than nutritional factors, host, fish habit and metabolic activity [11]. This is quite realistic because previous microbiological studies have indicated that the aqueous habitat influences fish microbial gut flora [10,11]. Moreover, the results reported in this work (surface tension values from $35.05 \pm 0.4$ mN·m$^{-1}$ to $43.01 \pm 1.2$ mN·m$^{-1}$) seem similar to those registered in other studies on *Pseudomonas* spp. (29–50 mN·m$^{-1}$) [48–51]. In accordance with the present results, these authors stated that the production of the identified BSs (rhamnolipids) is probably connected to external conditions such as nutrient limitation or other environmental factors, thus playing a crucial role in modulating bacterial behaviour over microbial community life and environmental changes. However, the BS producer performances, observed in this study, were different from those found in other microbiological studies on the intestinal tract of *Sparus aurata* from different aquatic environments, during the winter season, where lower (E-24) values and stronger surface activities were registered for the majority of the tested intestinal strains [6]. Throughout this study, the interfacial activity and the emulsification capacity do not always correlate, and this is in line with what has been highlighted by several authors [6,15,18]. Overall, it is interesting to observe that the best performers for BS production mainly belonged to *Pseudomonas* spp. which represent ubiquitous bacteria in nature and were already found to be part of grey mullet and other fish species cultivable microflora [6,10,11,52]. The phylogenetic tree obtained using the 16S ribosomal RNA gene sequences indicated the presence of a heterogenous cluster of *Pseudomonas* spp. in the mullet intestinal bacterial flora, forming interesting groups and subgroups of strains that deserve to be further investigated for taxonomic purposes by other housekeeping genes [36]. Moreover, in this work, the preliminary chemical structure of BSs from a group of representative strains was analysed. On the basis of our results, it is suggested that these are ascribed to two classes of glycolipid molecules: rhamnolipids and less polar compounds. The glycolipid biosurfactants have recently gained special attention for their eco-friendly nature and high efficiency in biodegradation as well as other special activities such as pesticidal, antifungal and antibacterial activities [53]. Indeed, as reported in other studies, *Pseudomonas* genus is able to synthesize the BSs of a diverse chemical nature, and the more widely studied ones are low molecular weight compounds called rhamnolipids [48,54]. Bacterial rhamnolipids biosynthesis was first elucidated in the Gram-negative opportunistic pathogen *P. aeruginosa*, which can synthesize a range of rhamnolipid congeners (approximately 60), di-rhamnolipids (the most abundant) and mono rhamnolipids. They present low toxicity and high biodegradability and are naturally produced at different concentrations by other *Pseudomonas* spp., such as *P. fluorescens*, *P. chlororaphis*, *P. putida* and *P. mendocina* [51,55,56], though their level of production is low compared to *P. aeruginosa* strains [57]. However, *Pseudomonas* strains isolated from the gut of gilthead seabream from different Sardinian aquatic environments also resulted in the production of glycolipid compounds, although showing slight differences in the TLC profiles, with respect to this study [6]. Furthermore, it is important to highlight that there is a great variability in the bioactive performances of this type of molecules, and there are marked differences between biosurfactants and bioemulsifiers. Although both BS types can efficiently emulsify two immiscible liquids, bioemulsifiers are said to possess only emulsifying activity and not surface activity [58]. In any case, from a practical point of view, bacteria producing surface-active compounds, such as rhamnolipids, are thought to solubilise insoluble substrates such as hydrocarbons and to promote the uptake and the biodegradation of poorly soluble substrates, enhancing their bioavailability and subsequent metabolism [55]. Noteworthy, these natural compounds also act as immune modulators, virulence factors and antimicrobial agents and are involved in surface mobility and bacterial biofilm development [55]. Our experiments also referred to the antibacterial activity, another aspect of BS-producing strains which has important practical implications. In this study, CFSs and CEs showed different inhibitory activity toward target strains. Indeed, while CFSs resulted active only against two targets, the extracts resulted active, to a different extent, against all target strains. Interestingly, in some cases, the inhibitory activities achieved values similar or equal to those of the positive control. This is the case for many CFSs and CEs, which showed inhibition halos equal to or higher than those obtained using Gentamycin CN30 against the target *P. mirabilis* H1643. This is also true for the inhibition exhibited by both CFSs and CEs against the target *K. pneumonia* H1637. In particular, four *Pseudomonas* isolates (strains 1,6,13,15) from *Chelon ramada* were found to be the most effective, showing E-24 > 50%, a vigorous collapse in the drop collapse-test and antibacterial activities against diverse Gram-negative and Gram-positive pathogens, except for *P. aeruginosa* H1628. As reported in previous similar studies, extracts and supernatants showed differentiated activity [6]. Thus, the search for new molecules with antibacterial function finds pivotal potential application in the aquaculture field, which is seriously threatened by the spread of infectious diseases and, more seriously, by antibiotic-resistant bacteria caused by the excessive use of drugs [59]. Aquaculture production is projected to rise from 40 million tonnes by 2008 to 82 tonnes in 2050. Moreover, by 2030, farm-raised fish would account for nearly two-thirds of the world's seafood intake, according to estimates by the United Nations Food & Agriculture Organization (FAO, 2010) [60]. The excessive use of antibiotics and the problems linked to this in aquaculture have been questioned by [61], who claim that the improved rearing methods may lead to the existence of antibiotic residues in seafood, with the consequences of destructing the immune system of the host. Recently, new scientific strategies are moving towards the supplementation of fish diets with several additives (i.e., probiotics, prebiotics, immunostimulants, vaccines) which could improve animal survival and wellness [62]. In this panorama, the development of a sustainable aquaculture industry is challenged by the limited availability of natural resources as well as the impact of the industry on the environment [63].

For all these reasons, the identification of bacteria able to produce natural compounds as suitable alternative to common antibiotics that reduce intestinal pathogens in animals and humans is of great importance and can increase the amount of information on the possible influence of intestinal bacteria on the health/well-being of fish. Consequently, the attention of fish farming practices should focus on this topic because their aim is to produce in large quantities while respecting the environment and animal welfare, in accordance with strict European rules on microbiological criteria for food market safety (Reg. ECN°2073/2005). Besides, the obtained results have important economic implications for an aquacultured species such as *Mugil cephalus*, which is highly appreciated for its eggs, processed to obtain seafood which is known by different names, such as Avgotaracho (Greece), Karasumi (Japan) or Bottarga (Italy), depending on the geographical production area [23,64]. Finally, the presence of bacteria producing substances with surfactant activity deriving from a brackish transitional environment deserves attention because these compounds could also have applications in bioremediation and represent an important biotechnological potential that can be furtherly investigated.

## 5. Conclusions

In conclusion, the present research has led to the selection of bacterial strains with interesting biotechnologically traits from the gut of grey mullets and has confirmed that intestinal microbiota is a promising source of new and biologically active pharmaceutical agents to control fish health and to preserve the environment. Additionally, the study of BS-producing bacteria associated with fish intestine is of relevance for our understanding of their ecological role in the symbiotic and antagonist interaction with the host and between themselves and for understanding whether the production of bioactive compounds might represent a biological strategy for protecting fish against gut and liver inflammations, as an immune response and for survival with respect to the surrounding environment.

**Author Contributions:** Conceptualization, R.F., G.S., C.R. and N.F.; methodology, R.F., G.S., L.M., M.F., J.C., R.D., C.R. and P.L.; software, R.F., G.S., C.R. and L.M.; validation, all the authors.; formal analysis, R.F., G.S. and C.R.; investigation, R.F., G.S. and C.R.; resources, N.F.; data curation, R.F., G.S. and C.R.; writing—original draft preparation, R.F., G.S. and C.R.; writing—review and editing, all the authors; visualization, all the authors; supervision, R.F., G.S., C.R. and A.L.G.; project administration, N.F.; funding acquisition, N.F. All authors have read and agreed to the published version of the manuscript.

**Funding:** This research was funded by Sardinian Region contribution: L.R. n. 6/2012 art-3, comma34. Interventi regionali per l'attuazione della strategia comunitaria in agricoltura.

**Institutional Review Board Statement:** Not applicable.

**Informed Consent Statement:** Not applicable.

**Acknowledgments:** The authors acknowledge the Cooperativa Pescatori "Santa Giusta" for providing fish samples, and Johannes Kugler for providing TLC standards.

**Conflicts of Interest:** The authors declare no conflict of interest.

## References

1. De Pascale, D.; De Santi, C.; Fu, J.; Landfald, B. The microbial diversity of Polar environments is a fertile ground for bioprospecting. *Mar. Genom.* **2012**, *8*, 15–22. [CrossRef] [PubMed]
2. Bălașa, A.F.; Chircov, C.; Grumezescu, A.M. Marine Biocompounds for Neuroprotection—A Review. *Mar. Drugs* **2020**, *18*, 290. [CrossRef]
3. Caruso, G.; Floris, R.; Serangeli, C.; Di Paola, L. Fishery Wastes as a Yet Undiscovered Treasure from the Sea: Biomolecules Sources, Extraction Methods and Valorization. *Mar. Drugs* **2020**, *18*, 622. [CrossRef] [PubMed]
4. Floris, R.; Rizzo, C.; Giudice, A.L. Biosurfactants from Marine Microorganisms. In *Metabolomics—New Insights into Biology and Medicine*; IntechOpen: London, UK, 2020.
5. Dusane, D.H.; Matkar, P.; Venugopalan, V.P.; Kumar, A.R.; Zinjarde, S.S. Cross-Species Induction of Antimicrobial Compounds, Biosurfactants and Quorum-Sensing Inhibitors in Tropical Marine Epibiotic Bacteria by Pathogens and Biofouling Microorganisms. *Curr. Microbiol.* **2011**, *62*, 974–980. [CrossRef]
6. Floris, R.; Scanu, G.; Fois, N.; Rizzo, C.; Malavenda, R.; Spanò, N.; Giudice, A.L. Intestinal bacterial flora of Mediterranean gilthead sea bream (*Sparus aurata* Linnaeus) as a novel source of natural surface active compounds. *Aquac. Res.* **2018**, *49*, 1262–1273. [CrossRef]
7. Rizzo, C.; Gugliandolo, C.; Giudice, A.L. Exploring Mediterranean and Arctic Environments as a Novel Source of Bacteria Producing Antibacterial Compounds to be Applied in Aquaculture. *Appl. Sci.* **2020**, *10*, 4006. [CrossRef]

8. Wang, A.R.; Ran, C.; Ringø, E.; Zhou, Z.G. Progress in fish gastrointestinal microbiota research. *Rev. Aquac.* **2018**, *10*, 626–640. [CrossRef]
9. Carnevali, O.; de Vivo, L.; Sulpizio, R.; Gioacchini, G.; Olivotto, I.; Silvi, S.; Cresci, A. Growth improvement by probiotic in European sea bass juveniles (*Dicentrarchus labrax*, L.), with particular attention to IGF-1, myostatin and cortisol gene expression. *Aquaculture* **2006**, *258*, 430–438. [CrossRef]
10. Floris, R.; Manca, S.; Fois, N. Microbial ecology of intestinal tract of gilthead sea bream (Sparus aurata Linnaeus, 1758) from two coastal lagoons of Sardinia (Italy). *Trans. Water Bullet* **2013**, *7*, 4–12. [CrossRef]
11. Floris, R.; Sanna, G.; Satta, C.; Piga, C.; Sanna, F.; Logliè, A.; Fois, N. Intestinal Microbial Ecology and Fillet Metal Chemistry of Wild Grey Mullets Reflect the Variability of the Aquatic Environment in a Western Mediterranean Coastal Lagoon (Santa Giusta, Sardinia, Italy). *Water* **2021**, *13*, 879. [CrossRef]
12. Li, J.; Ni, J.; Wang, C.; Li, X.; Wu, S.; Zhang, T.; Yu, Y.; Yan, Q. Comparative study on gastrointestinal microbiota of eight fish species with different feeding habits. *J. Appl. Microbiol.* **2014**, *117*, 1750–1760. [CrossRef]
13. Yukgehnaish, K.; Kumar, P.; Sivachandran, P.; Marimuthu, K.; Arshad, A.; Paray, B.A.; Arockiaraj, J. Gut microbiota metagenomics in aquaculture: Factors influencing gut microbiome and its physiological role in fish. *Rev. Aquac.* **2020**, *12*, 1903–1927. [CrossRef]
14. Burtseva, O.; Baulina, O.; Zaytseva, A.; Fedorenko, T.; Chekanov, K.; Lobakova, E. In vitro Biofilm Formation by Bioluminescent Bacteria Isolated from the Marine Fish Gut. *Microb. Ecol.* **2021**, *81*, 932–940. [CrossRef] [PubMed]
15. Malavenda, R.; Rizzo, C.; Michaud, L.; Gerçe, B.; Bruni, V.; Syldatk, C.; Hausmann, R.; Lo Giudice, A. Biosurfactant production by Arctic and Antarctic bacteria growing on hydrocarbons. *Pol. Biol.* **2015**, *38*, 1565–1574. [CrossRef]
16. Satpute, S.K.; Banat, I.; Dhakephalkar, P.K.; Banpurkar, A.G.; Chopade, B.A. Biosurfactants, bioemulsifiers and exopolysaccharides from marine microorganisms. *Biotechnol. Adv.* **2010**, *28*, 436–450. [CrossRef]
17. Panjiar, N.; Sachan, S.G.; Sachan, A. Screening of bioemulsifier-producing micro-organisms isolated from oil-contaminated sites. *Ann. Microbiol.* **2015**, *65*, 753–764. [CrossRef]
18. Rizzo, C.; Michaud, L.; Hörmann, B.; Gerçe, B.; Syldatk, C.; Hausmann, R.; De Domenico, E.; Giudice, A.L. Bacteria associated with sabellids (Polychaeta: Annelida) as a novel source of surface active compounds. *Mar. Pollut. Bull.* **2013**, *70*, 125–133. [CrossRef]
19. Giri, S.S.; Kim, H.J.; Kim, S.; Kim, S.; Kwon, J.; Bin Lee, S.; Park, S.C. Immunomodulatory Role of Microbial Surfactants, with Special Emphasis on Fish. *Int. J. Mol. Sci.* **2020**, *21*, 7004. [CrossRef]
20. Mukherjee, A.; Banerjee, G.; Mukherjee, P.; Ray, A.K.; Chandra, G.; Ghosh, K. Antibacterial substances produced by pathogen inhibitory gut bacteria in Labeo rohita: Physicochemical characterization, purification and identification through MALDI-TOF mass spectrometry. *Microb. Pathog.* **2019**, *130*, 146–155. [CrossRef]
21. Crosetti, D. *Biology, Ecology and Culture of Grey Mullet (Mugilidae)*; Version Date 20151214; CRC Press Taylor & Francis Group: Boca Raton, FL, USA, 2016; pp. 42–127.
22. Thomson, J.M. The Mugilidae of the world. *Mem. Queens. Mus.* **1997**, *41*, 457–562.
23. Caredda, M.; Addis, M.; Pes, M.; Fois, N.; Sanna, G.; Piredda, G.; Sanna, G. Physicochemical, colorimetric, rheological parameters and chemometric discrimination of the origin of Mugil cephalus' roes during the manu-facturing process of Bottarga. *Food Res. Int.* **2018**, *108*, 128–135. [CrossRef] [PubMed]
24. Durand, J.-D.; Blel, H.; Shen, K.; Koutrakis, E.; Guinand, B. Population genetic structure of Mugil cephalus in the Mediterranean and Black Seas: A single mitochondrial clade and many nuclear barriers. *Mar. Ecol. Prog. Ser.* **2013**, *474*, 243–261. [CrossRef]
25. Turan, C.M.; Gürlek, D.; Ergüden, D.; Yağlıoğlu, D.; Öztürk, B. Systematic status of nine mullet species (Mugilidae) in the Mediterranean Sea. *Turk. J. Fish. Aquat. Sci.* **2011**, *11*, 315–321. [CrossRef]
26. Kumar, R.S.; Kanmani, P.; Yuvaraj, N.; Paari, K.; Pattukumar, V.; Arul, V. Purification and characterization of enterocin MC13 produced by a potential aquaculture probiont Enterococcus faeciumMC13 isolated from the gut of Mugil cephalus. *Can. J. Microbiol.* **2011**, *57*, 993–1001. [CrossRef]
27. Lin, Y.-H.; Chen, Y.-S.; Wu, H.-C.; Pan, S.-F.; Yu, B.; Chiang, C.-M.; Chiu, C.-M.; Yanagida, F. Screening and characterization of LAB-produced bacteriocin-like substances from the intestine of grey mullet (*Mugil cephalus* L.) as potential biocontrol agents in aquaculture. *J. Appl. Microbiol.* **2013**, *114*, 299–307. [CrossRef]
28. Cui, J.; Xiao, M.; Liu, M.; Wang, Z.; Liu, F.; Guo, L.; Meng, H.; Zhang, H.; Yang, J.; Deng, D.; et al. Coupling metagenomics with cultivation to select host-specific probiotic micro-organisms for subtropical aquaculture. *J. Appl. Microbiol.* **2017**, *123*, 1274–1285. [CrossRef]
29. Sechi, N.; Fiocca, F.; Sannio, A.; Luglié, A. Santa Giusta Lagoon (Sardinia): Phytoplankton and nutrients before and after waste water diversion. *J. Limnol.* **2001**, *60*, 194–200. [CrossRef]
30. Techaoei, S.; Leelapornpisid, P.; Santiarwarn, D.; Lumyong, S. Preliminary screenings of biosurfac-tants-producing microorganisms isolated from hot spring and garages in northern Thailand. *Curr. Appl. Sci. Technol.* **2007**, *7*, 38–43.
31. Bodour, A.A.; Drees, K.P.; Maier, R.M. Distribution of Biosurfactant-Producing Bacteria in Undisturbed and Contaminated Arid Southwestern Soils. *Appl. Environ. Microbiol.* **2003**, *69*, 3280–3287. [CrossRef]
32. Samadi, N.; Fazel, M.R.; Abadian, N.; Akhavan, A.; Tahzibi, A.; Jamalifar, H. Biosurfactant Production by the Strain Isolated from Contaminated Soil. *J. Biol. Sci.* **2007**, *7*, 1266–1269. [CrossRef]
33. Jamal, P.; Wan Nawawi, W.M.F.; Alam, M.Z. Optimum medium components for biosurfactant production by Klebsiella pneumoniae WMF02 utilizing sludge palm oil as a substrate. *Austral. J. Basic Appl. Sci.* **2012**, *6*, 100–108.

34. Marchesi, J.R.; Sato, T.; Weightman, A.J.; Martin, T.A.; Fry, J.C.; Hiom, S.J.; Wade, W.G. Design and evaluation of useful bacterium-specific PCR primers that amplify genes coding for bacterial 16S rRNA. *Appl. Environ. Microbiol.* **1998**, *64*, 795–799. [CrossRef]
35. Altschul, S.F.; Gish, W.; Miller, W.; Myers, E.W.; Lipman, D.J. Basic local alignment search tool. *J. Mol. Biol.* **1990**, *215*, 403–410. [CrossRef]
36. Gomila, M.; Penã, A.; Mulet, M.; Lalucat, J.; Garcia-Valdes, E. Phylogenomics and systematics in Pseudomonas. *Front. Microbiol.* **2015**, *6*, 214. [CrossRef]
37. Kumar, S.; Stecher, G.; Li, M.; Knyaz, C.; Tamura, K. MEGA X: Molecular Evolutionary Genetics Analysis across Computing Platforms. *Mol. Biol. Evol.* **2018**, *35*, 1547–1549. [CrossRef] [PubMed]
38. Nguyen, L.-T.; Schmidt, H.A.; Von Haeseler, A.; Minh, B.Q. IQ-TREE: A Fast and Effective Stochastic Algorithm for Estimating Maximum-Likelihood Phylogenies. *Mol. Biol. Evol.* **2015**, *32*, 268–274. [CrossRef]
39. Strimmer, K.; von Haeseler, A. Likelihood-mapping: A simple method to visualize phylogenetic content of a sequence alignment. *Proc. Natl. Acad. Sci. USA* **1997**, *94*, 6815–6819. [CrossRef]
40. Felsenstein, J. Evolutionary trees from DNA sequences: A maximum likelihood approach. *J. Mol. Evol.* **1981**, *17*, 368–376. [CrossRef] [PubMed]
41. Kalyaanamoorthy, S.; Quang Minh, B.; Wong, T.K.F.; von Haeseler, A.; Jermiin, L.S. ModelFinder: Fast model selection for accurate phylogenetic estimates. *Nat. Meth.* **2017**, *14*, 587–589. [CrossRef] [PubMed]
42. Felsenstein, J. Confidence Limits on Phylogenies: An Approach Using the Bootstrap. *Evolution* **1985**, *39*, 783–791. [CrossRef]
43. Simonetta, A.C.; de Velasco, L.G.M.; Frisón, L.N. Antibacterial activity of enterococci strains against Vibrio cholerae. *Lett. Appl. Microbiol.* **1997**, *24*, 139–143. [CrossRef] [PubMed]
44. Council of Europe. *Pharmacopoeia*, 3rd ed.; Council of Europe: Strasbourg, France, 1997; p. 121.
45. Twigg, M.S.; Baccile, N.; Banat, I.M.; Déziel, E.; Marchant, R.; Roelants, S.; Van Bogaert, I.N.A. Microbial biosurfactant research: Time to improve the rigour in the reporting of synthesis, functional characterization and process development. *Microb. Biotechnol.* **2021**, *14*, 147–170. [CrossRef]
46. Rizzo, C.; Michaud, L.; Syldatk, C.; Hausmann, R.; De Domenico, E.; Giudice, A.L. Influence of salinity and temperature on the activity of biosurfactants by polychaete-associated isolates. *Environ. Sci. Pollut. Res.* **2013**, *21*, 2988–3004. [CrossRef]
47. Rizzo, C.; Syldatk, C.; Hausmann, R.; Gerçe, B.; Longo, C.; Papale, M.; Conte, A.; De Domenico, E.; Luigi Michaud, L.; Lo Giudice, A. The demosponge *Halichondria* (Halichondria) panicea (Pallas, 1766) as a novel source of bio-surfactant-producing bacteria. *J. Basic. Microbiol.* **2018**, *58*, 532–542. [CrossRef]
48. Tuleva, B.K.; Ivanov, G.R.; Christova, N.E. Biosurfactant Production By A New Pseudomonas Putida Strain. *Z. Für Nat. C* **2002**, *57*, 356–360. [CrossRef] [PubMed]
49. Rahman, P.; Gakpe, E. Production, Characterisation and Applications of Biosurfactants-Review. *Biotechnology* **2008**, *7*, 360–370. [CrossRef]
50. Meliani, A.; Bensoltane, A. The ability of some Pseudomonas strains to produce biosurfactants. *Poult. Fish. Wildl. Sci.* **2014**, *2*, 112. [CrossRef]
51. Twigg, M.S.; Tripathi, L.; Zompra, A.; Salek, K.; Irorere, V.; Gutierrez, T.; Spyroulias, G.A.; Marchant, R.; Banat, I.M. Identification and characterisation of short chain rhamnolipid production in a previously uninvestigated, non-pathogenic marine pseudomonad. *Appl. Microbiol. Biotechnol.* **2018**, *102*, 8537–8549. [CrossRef]
52. Nikouli, E.; Meziti, A.; Antonopoulou, E.; Mente, E.; Kormas, K.A. Gut Bacterial Communities in Geographically Distant Populations of Farmed Sea Bream (*Sparus aurata*) and Sea Bass (*Dicentrarchus labrax*). *Microorganisms* **2018**, *6*, 92. [CrossRef]
53. Shu, Q.; Lou, H.; Wei, T.; Liu, X.; Chen, Q. Contributions of Glycolipid Biosurfactants and Glycolipid-Modified Materials to Antimicrobial Strategy: A Review. *Pharmaceutics* **2021**, *13*, 227. [CrossRef] [PubMed]
54. Chrzanowski, L.; Lawniczak, L.; Czaczyk, K. Why do microorganisms produce rhamnolipids? *World J. Microbiol. Biotechnol.* **2011**, *28*, 401–419. [CrossRef]
55. Abdel-Mawgoud, A.M.; Lépine, F.; Déziel, E. Rhamnolipids: Diversity of structures, microbial origins and roles. *Appl. Microbiol. Biotechnol.* **2010**, *86*, 1323–1336. [CrossRef]
56. Soberón-Chávez, G.; Hausmann, R.; Déziel, E. Editorial: Biosurfactants: New Insights in Their Biosynthesis, Production and Applications. *Front. Bioeng. Biotechnol.* **2021**, *9*, 769899. [CrossRef] [PubMed]
57. Soberón-Chávez, G.; González-Valdez, A.; Soto-Aceves, M.P.; Cocotl-Yañez, M. Rhamnolipids produced by Pseudomonas: From molecular genetics to the market. *Microb. Biotechnol.* **2021**, *14*, 136–146. [CrossRef]
58. Uzoigwe, C.; Burgess, J.G.; Ennis, C.J.; Rahman, P.K.S.M. Bioemulsifiers are not biosurfactants and require different screening approaches. *Front. Microbiol.* **2015**, *6*, 245. [CrossRef] [PubMed]
59. Okocha, R.C.; Olatoye, I.O.; Adedeji, O.B. Food safety impacts of antimicrobial use and their residues in aquaculture. *Public Health Rev.* **2018**, *39*, 21. [CrossRef]
60. Food and Agriculture Organization. *The State of World Fisheries and Aquaculture 2010*; Food and Agriculture Organization: Rome, Italy, 2010; p. 179. [CrossRef]

61. Ramírez, C.; Gutiérrez, M.S.; Venegas, L.; Sapag, C.; Araya, C.; Caruffo, M.; López, P.; Reyes-Jara, A.; Toro, M.; González-Rocha, G.; et al. Microbiota composition and susceptibility to florfenicol and oxytetracycline of bacterial isolates from mussels (*Mytilus* spp.) reared on different years and distance from salmon farms. *Environ. Res.* **2021**, *204*, 112068. [CrossRef] [PubMed]
62. Dawood, M.A.; Koshio, S.; Esteban, M. Ángeles Beneficial roles of feed additives as immunostimulants in aquaculture: A review. *Rev. Aquac.* **2018**, *10*, 950–974. [CrossRef]
63. Costa-Pierce, B.A.; Bartley, D.M.; Hasan, M.; Yusoff, F.; Kaushik, S.J.; Rana, K.; Lemos, D.; Bueno, P.; Ya-kupitiyage, A. 2012. Responsible use of resources for sustainable aquaculture. In *Farming the Waters for People and Food, Proceedings of the Global Conference on Aquaculture, Phuket, Thailand, 22–25 September 2010*; Subasinghe, R.P., Arthur, J.R., Bartley, D.M., De Silva, S.S., Halwart, M., Hishamunda, N., Mohan, C.V., Sorgeloos, P., Eds.; FAO: Rome, Italy; NACA: Bangkok, Thailand, 2010; pp. 113–147.
64. Bledsoe, G.; Bledsoe, C.; Rasco, B. Caviars and Fish Roe Products. *Crit. Rev. Food Sci. Nutr.* **2003**, *43*, 317–356. [CrossRef]

Article

# *Lacticaseibacillus casei* ATCC 393 Cannot Colonize the Gastrointestinal Tract of Crucian Carp

**Hongyu Zhang [1,2], Xiyan Mu [1,2], Hongwei Wang [2], Haibo Wang [1,2], Hui Wang [1,2], Yingren Li [1,2], Yingchun Mu [2,3,*], Jinlong Song [2,3,*] and Lei Xia [2,*]**

[1] Fishery Resource and Environment Research Center, Chinese Academy of Fishery Sciences, Beijing 100141, China; zhanghy@cafs.ac.cn (H.Z.); muxiyan@cafs.ac.cn (X.M.); xuehu1110@aliyun.com (H.W.); wanghui@cafs.ac.cn (H.W.); liyr@cafs.ac.cn (Y.L.)

[2] Chinese Academy of Fishery Sciences, Beijing 100141, China; wanghongwei@cafs.ac.cn

[3] Key Laboratory of Control of Quality and Safety for Aquatic Products (Ministry of Agriculture and Rural Affairs), Chinese Academy of Fishery Sciences, Beijing 100141, China

* Correspondence: muyc@cafs.ac.cn (Y.M.); songjl@cafs.ac.cn (J.S.); xialei666@126.com (L.X.)

**Abstract:** Lactic acid bacteria (LAB) are commonly applied to fish as a means of growth promotion and disease prevention. However, evidence regarding whether LAB colonize the gastrointestinal (GI) tract of fish remains sparse and controversial. Here, we investigated whether *Lacticaseibacillus casei* ATCC 393 (Lc) can colonize the GI tract of crucian carp. Sterile feed irradiated with $^{60}$Co was used to eliminate the influence of microbes, and 100% rearing water was renewed at 5-day intervals to reduce the fecal–oral circulation of microbes. The experiment lasted 47 days and was divided into three stages: the baseline period (21 days), the administration period (7 days: day −6 to 0) and the post-administration period (day 1 to 19). Control groups were fed a sterile basal diet during the whole experimental period, whereas treatment groups were fed with a mixed diet containing Lc ($1 \times 10^7$ cfu/g) and spore of *Geobacillus stearothermophilus* (Gs, $1 \times 10^7$ cfu/g) during the administration period and a sterile basal diet during the baseline and post-administration periods. An improved and highly sensitive selective culture method (SCM) was employed in combination with a transit marker (a Gs spore) to monitor the elimination of Lc in the GI tract. The results showed that Lc (<2 cfu/gastrointestine) could not be detected in any of the fish sampled from the treatment group 7 days after the cessation of the mixed diet, whereas Gs could still be detected in seven out of nine fish at day 11 and could not be detected at all at day 15. Therefore, the elimination speed of Lc was faster than that of the transit marker. Furthermore, high-throughput sequencing analysis combined with SCM was used to reconfirm the elimination kinetics of Lc in the GI tract. The results show that the Lc in the crucian carp GI tract, despite being retained at low relative abundance from day 7 (0.11% ± 0.03%) to 21, was not viable. The experiments indicate that Lc ATCC 393 cannot colonize the GI tract of crucian carp, and the improved selective culture in combination with a transit marker represents a good method for studying LAB colonization of fish.

**Keywords:** *Lacticaseibacillus casei*; colonization; crucian carp; gastrointestinal tract; $^{60}$Co irradiation sterilization; transit marker; *Geobacillus stearothermophilus*; high-throughput sequencing

**Citation:** Zhang, H.; Mu, X.; Wang, H.; Wang, H.; Wang, H.; Li, Y.; Mu, Y.; Song, J.; Xia, L. *Lacticaseibacillus casei* ATCC 393 Cannot Colonize the Gastrointestinal Tract of Crucian Carp. *Microorganisms* **2021**, *9*, 2547. https://doi.org/10.3390/microorganisms9122547

Academic Editors: Konstantinos Ar. Kormas and James S. Maki

Received: 30 September 2021
Accepted: 3 December 2021
Published: 9 December 2021

## 1. Introduction

Given the restrictions and prohibitions regarding the use of chemicals and antibiotics, there is an increasing demand for safe, cost-effective, and environmentally friendly feed supplements that possess exceptional benefits for farmed fish such as phytogenics, prebiotics and probiotics [1]. One of therapeutic benefits of probiotics are that they can colonize or temporally colonize gastrointestinal (GI) tract and thereby modulate the intestinal microbiota via competitive adherence and exclusion, resulting in the production of beneficial substances for the host [2,3]. Colonization is one of the most important characteristics when evaluating the application of probiotics in animal rearing. LAB are one of the most widely used and studied bacteria in aquaculture, but their colonization in the intestinal tract of fish remains highly debated. Tian et al. [4] stated that *Lacticaseibacillus casei* CC16 can colonize the intestines of common carp. Other papers have reported that *Pediococcus acidilactici* (Bactocell®, Lallemand Inc., Montreal, QC, Canada) [5], *Bacillus paralicheniformis* FA6 [6], *Lactiplantibacillus plantarum* G1 [7], *Lacticaseibacillus casei* ATCC 393 [8], *Latilactobacillus sakei* CLFP 202 [9], *Lactococcus lactis* CLFP 100 [9] and *Leuconostoc mesenteroides* CLFP 196 [9] can also colonize the GI tract of goldfish, grass carp, shabout fish and rainbow trout. However, some papers have shown that probiotic strains, including *Lactobacillus*, in the GI tract rapidly decreases following the withdrawal of supplementation [10–16], indicating their transient nature. Meanwhile, Ringø et al. [17] raised the following question: "Are probiotics permanently colonizing the GI tract?".

Colonization was defined by Conway and Cohen as the indefinite persistence of a particular bacterial population without the reintroduction of that bacterium [18]. Most bacterial cells are transiently present in the GI tract of aquatic animals, with the continuous intrusion of microbes from water and food [19]. Commercial feed or homemade feed are usually unsterile except for specific pathogen free (SPF) or gnotobiotic animals [20]. Considering the widespread existence of lactic acid bacteria (LAB) and *Bacillus*, it is rational to speculate on their existence in aquafeed. The transient microbes in the GI tract enter water with feces and can then be reintroduced to that same GI tract. However, in probiotic colonization-related studies, little attention has been paid to the influence of microbes originating from feed and water, resulting in a conclusion that ignores the prerequisite for colonization, i.e., that it occurs "without the reintroduction of that bacterium". In addition, the monitoring time for the persistence of probiotic microbes in the GI tract has often been insufficient, and there has been an absence of transit markers for evaluating the clearance time for transient microbes [21].

Colonization is a very important characteristic for screening additive strains and studying the mechanisms of probiotic action, but is associated with several significant challenges. First, the target bacteria being found in the water and diet can interfere with the colonization study. Second, lacking suitable methods for colonization study, some molecular methods such as 16S rRNA amplicon technology based on DNA samples cannot tell whether the bacteria are alive or dead. Third, once the probiotic supplementation has ceased, the proportion of the target strain may remain at a very low level [22], requiring a detection method with higher sensitivity for viable cells.

*L. casei* (Lc) is one of the species commonly used in aquaculture [4,17] and has shown some beneficial properties when applied to fish [23,24]. However, whether bacteria colonize the GI tract of fish has been unclear. To solve the issues above, the interfering microbes in feed and water were monitored and controlled, a transit marker was introduced, and an improved and highly sensitive selective culture method and high-throughput sequencing were both used to investigate whether *L. casei* can "truly" colonize the GI tract of crucian carp.

## 2. Materials and Methods

### 2.1. Bacteria Strains and Culture Condition

*Lacticaseibacillus casei* (Lc) ATCC 393 and *Geobacillus stearothermophilus* (Gs) ATCC 7953, were purchased from the China Center of Industrial Culture Collection and maintained with regular procedures.

Lc: The Lc strain was grown in MRS (De Man, Rogosa and Sharpe, Oxoid) broth at 37 °C overnight without agitation. The cells were harvested by centrifugation (5000× *g*, 5 min), resuspended in normal saline (0.85% (*w*/*v*) NaCl, pH 7.5) and adjusted to the necessary concentration.

Gs: The bacterial lawn grown on nutrient agar (NA, Aobox) supplemented with 18 μM/L $MnSO_4$ at 57 °C for 4 days was harvested and washed twice with normal saline and then resuspended in normal saline. After inactivation vegetative cells incubated in a water bath at 90 °C for 30 min, the Gs spore suspension was centrifuged, washed twice with normal saline again, and then adjusted to the necessary concentration.

### 2.2. Experiment Diet

For the sterilized diet (basal diet), five commercial aquafeeds were sterilized by $^{60}Co$ irradiation at 26.0 kGy, after which the efficacy of the sterilization was evaluated. The feed pellets with or without sterilization were homogenized and spread on nutrient agar and MRS agar with a pH of 5.4–5.5. The nutrient agar was incubated at 37 and 57 °C for 3 days to count the general heterotrophic bacteria and thermophiles, respectively. MRS agar was incubated at 37 °C for 3 days to count LAB. The colony number was counted to calculate the bacterial concentration in feed, and representative colonies with differing morphologies were selected for identification by 16S rRNA gene sequencing. Bacterial DNA was extracted using the TIANamp Bacteria DNA Kit (Tiangen) according to the manufacturer's protocol. The DNA samples were submitted to the Rui Biotech, Inc. (Beijing, China) for PCR amplification and sequencing. The 16S ribosomal RNA gene from each sample were amplified and sequenced using the bacterial universal primer 27F (5′-AGAGTTTGATCCTGGCTCAG-3′) and 1492R (5′-TACGGCTACCTTGTTACGA CTT-3′). Then the 16S sequences alignments were performed using BLAST based on 16S ribosomal RNA sequences database of NCBI. Sterilized diet No. 2 (Beijing Fangteqi Feed Co., Ltd., Beijing, China, Table A1) was used in the experiment.

For the Mixed diet, Lc and Gs suspension were prepared and sprayed on the sterile basal feed to achieve a final concentration of $1 \times 10^7$ cfu/g in Experiment 1. The final concentrations of Lc and Gs were $2 \times 10^9$ cfu/g and $1 \times 10^8$ cfu/g in Experiment 2, respectively. The experimental feed was air-dried in an oven for 10 min at 37 °C, and sealed and stored at 4 °C. The viable bacterial number in the feed was counted using the plate counting method at the beginning and end of the feeding experiments.

### 2.3. Experiment Design and Rearing Conditions

Two methods were used at two separate experimental phases. First, an improved and highly sensitive selective culture method (SCM) was established to compare the elimination kinetics between Lc and a transit marker (a Gs spore). Second, second-generation sequencing based on an 16S rRNA gene amplicon sequencing method (16S) was used to analyze the relative abundance of Lc and Gs. Meanwhile, the whole gastrointestines were sampled at the same time point, and their viable bacteria were monitored using the SCM. The flow chart of design of experiment see Figure A1.

Crucian carp (*Carassius auratus*) that weighed 20–40 g were obtained from the Beijing Longchi Aquaculture Farm. The fish were distributed into six separate glass aquariums (300 L) at a density of 24 fish per tank. Three glass aquariums were used for the treatment group (TG) and the others were used for the control group (CG). The study period was divided into three consecutive periods. First was the 21-day long, baseline period, during which the fish from both groups were fasted for 7 days and then acclimatized to the sterile pellet feed at 1.0–1.5% body weight once a day for 14 days. The last day of this period was defined as day −7. Next was the 7 day administration period (day −6 to day 0) and, finally, the post-administration period (19 days during Experiment 1 and 21 days during Experiment 2). During the administration period, the mixed diet was orally administered in both experiments for 7 days. The basal diet was used in all other periods, including the baseline period and the post-administration period. Meanwhile, the basal diet was used throughout the whole experiment in the control group. A total of nine fish with three in each tank were taken at days −7, 0, 7, 11, 15 and 19 during Experiment 1, whereas nine fish (six for the SCM and three for 16S) were collected at five time points during Experiment 2 (that is, days −7, 0, 7, 14 and 21).

During the baseline and post-administration periods, 100% of the water was renewed every 5 days in both experiments. Tap water was equilibrated to room temperature and aerated for 48 h before use. The physical parameters of the water were as follows: temperature 22–25 °C, pH 8.0–9.0, and dissolved oxygen > 6 mg/L.

*2.4. Monitoring Lc and Thermophiles in Water*

During the whole experimental period, 2 mL of water was sampled from the fish tanks every 3 days. A total of 1 mL water was spread on two MRS agar plates (pH 5.4–5.5, 500 μL on each plate), and the remaining 1 mL was spread on two nutrient agar plates (500 μL on each plate). MRS agar was incubated at 37 °C for 7 days. Nutrient agar was incubated at 57 °C for 2 days. The colonies were identified by 16S RNA gene sequencing.

*2.5. Experiment 1: The Improved, Highly Sensitive Selective Culture Method*

The pH of MRS medium was adjusted to 5.4–5.5 for the selective culture of Lc. The spore of Gs was used as the transit marker [21,25,26].

2.5.1. Gastrointestine Homogenate Preparations

The fish were sacrificed at the sampling point, and close to the entire GI tract, from the esophagus to the anus, was aseptically removed. Then, an ice-cold normal saline solution was added to make a 10% (*w*/*w*) homogenate using a glass homogenizer. Meanwhile, the effect of the 10% GI tract homogenate on Lc and Gs and their respective media were evaluated as described below.

The Lc suspension was inoculated into the 10% GI tract homogenate of the crucian carp and normal saline at 1% (*v*/*v*) to a final concentration of $5 \times 10^2$ cfu/mL. A 200 μL aliquot of homogenate containing Lc was spread on MRS agar with a pH of 5.4–5.5. A 200 μL aliquot of normal saline control containing Lc was spread on regular MRS agar. The plates were incubated at 37 °C for 7 days. Then, the colony number was counted to calculate the growth rate. The colony was identified at the species level by 16S RNA gene sequencing technology.

The Gs suspension was inoculated into the 10% GI tract homogenate of the crucian carp and the normal saline at 1% (*v*/*v*), achieving a final concentration of $1 \times 10^3$ cfu/mL. Aliquots (100 μL) of homogenate and normal saline containing Gs were spread on the nutrient agar. The plates were incubated at 57 °C for 2 days, and the colony number was then counted to calculate the growth rate. The colony was identified at the species level by 16S rRNA gene sequencing technology.

The growth rate was assessed by Equation(1).

$$\text{Growth rate} = (\text{the colony number of experiment group / the average colony number of control group}) \times 100\% \quad (1)$$

2.5.2. Dynamics of Lc and the Transit Marker in the Gastrointestinal Tract

The GI tract was removed at the appropriate sample time point, and homogenate was prepared as described above (2.5.1); half was spread on nutrient agar (100 μL per plate), and the other half was spread on MRS agar with a pH of 5.4–5.5 (200 μL per plate). The detection limit for Lc and Gs was 2 cfu/gastrointestine. In cases where no LAB grew on the MRS agar, an additional six fish were sacrificed, and all GI tract homogenates were spread on MRS with a pH of 5.4–5.5 to reach a detection limit of 1 cfu/gastrointestine.

Generally, 20–30 plates are required for a 5 mL GI tract homogenate to reach a detection limit of 2 cfu/gastrointestine. The colonies were identified by microscopic examination and/or 16S rRNA gene sequencing. The sum of the Lc colony number for each plate was the total viable bacteria in the GI tract when all the GI tract homogenate was spread on the plate.

*2.6. Experiment 2: 16S rRNA Gene Amplicon Sequencing Method (16S)*

Sample Collection, DNA Extraction and Bioinformatic Analysis.

Of the nine samples (three fish per replicate) that were randomly selected from each group at each sample time point, six fish were monitored using the SCM as described above (2.5.2) and the other three fish were used for second-generation sequencing. The gastrointestinal contents were removed under sterile conditions. Bacterial DNA was extracted using the E.Z.N.A

Mag-Bind Soil DNA Kit (Omega, Norcross, GA, USA). The DNA quality and concentrations were measured using a Qubit®3.0 spectrophotometer (Invitrogen, Waltham, MA, USA). The DNA samples were submitted to Sangon Biotech, Inc. (Shanghai, China) for PCR amplification and next-generation sequencing using an Illumina MiSeq platform. The primer sequences (341F (5′-CCTACACGACGCTCTTCCGATCTG(barcode) CCTACGGGNGGCWGCAG-3′) and 805R (5′-GACTGGAGTTCCTTGGCACCCGA GAATTCCAGACTACHVGGGTATCTAATCC-3′)), PCR cleanup, and sequencing were performed and a bioinformatic analysis was conducted as described in our previous study [27].

### 2.7. Data Analyses

Data were analyzed using *T*-test. A statistical analysis was performed using Microsoft Office Excel 2007(USA) with the level of significance set at $p < 0.05$.

## 3. Results

### 3.1. Effect of 100% Water Renewal on Interfering Bacteria

During the baseline period, no cultivable Lc or thermophiles were detected in the rearing water (<1 cfu/mL). During the administration period, 0–9 × $10^2$ cfu/mL of Lc and 0.1–8 × $10^3$ cfu/mL of Gs were detected in the rearing water. No Lc was detected following the cessation of bacterial supplementation and 100% water renewal up to the end of the experiments. Several Gs colonies were occasionally detected in the first week, whereas no Gs were detected after the second water renewal during the post-administration period.

### 3.2. Effect of Sterilizing the Feed with $^{60}$Co Irradiation

The bacterial content of the commercial aquafeed is shown in Table A2. There were general heterotrophic bacteria at $10^4$–$10^6$ cfu/g of the commercial diet, LAB at $10^2$–$10^4$ cfu/g and thermophiles at $10^2$–$10^4$ cfu/g. Using 16S rRNA gene sequencing identification, it was found that the general heterotrophic bacteria were mainly species of the genera *Bacillus* (including *Bacillus licheniformis* and *Bacillus subtilis*), and others include *Enterobacter, Parabacillus, Pantoea*, etc. The LAB were *Pediococcus, Enterococcus* and *Bacillus coagulans*. The thermophiles included mainly *Geobacillus, Parageobacillus*, and *Bacillus*. None of these bacteria were detected after $^{60}$Co irradiation sterilization.

Meanwhile, the concentration of Gs and Lc in the mixed diet did not attenuate at the end of either experiments (Table A3).

### 3.3. Selective Culture for LAB and Gs

The pH of MRS medium was adjusted to 5.4–5.5 for the selective culture of Lc. The MRS agar with a pH of 5.4–5.5 had high specificity for Lc growth, except for the occasional presence of some fungi and motile bacteria that failed to subculture in the rearing water and the gut at very low doses. There was no significant difference between the regular MRS and the 10% GI tract homogenate MRS (pH 5.4–5.5) (Figure 1). In other words, the improved MRS agar had a high specificity and sensitivity and was, thus, able to detect the LAB strains used in our study of the GI tract homogenate.

The growth rate of Gs at 57 °C was 83.78% ± 26.80% (Figure 2) when suspended in the 10% GI tract homogenate, which was slightly lower than that of the normal saline control. However, there were no significant differences between the two groups ($p > 0.05$).

### 3.4. The Concentration of Lc Changes in the GI Tract of Crucian Carp

The concentration of Lc and Gs in the GI tract decreased dramatically after the cessation of both bacteria supplements (Figure 3). In the first 3 days, the Lc concentration decreased from 2.6 × $10^5$ (5.43log) to 20.67 (1.32log) cfu/gastrointestine, and Lc could not be detected in the GI tracts of two out of nine fish. Seven days after the cessation of the mixed diet, Lc could not be detected in any of the sampled fish (< 2 cfu/gastrointestine), although Gs was remained detectable up to day 11 (7/9). As can be seen from Figure 3, Lc was eliminated from crucian carp gastrointestine faster than Gs.

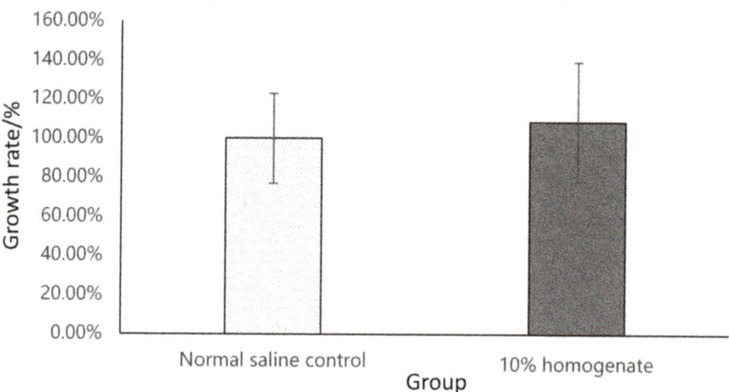


**Figure 1.** Comparison between the growth rate of *L. casei* in the normal saline control and 10% GI tract homogenate ($n = 9$) on the MRS plate.

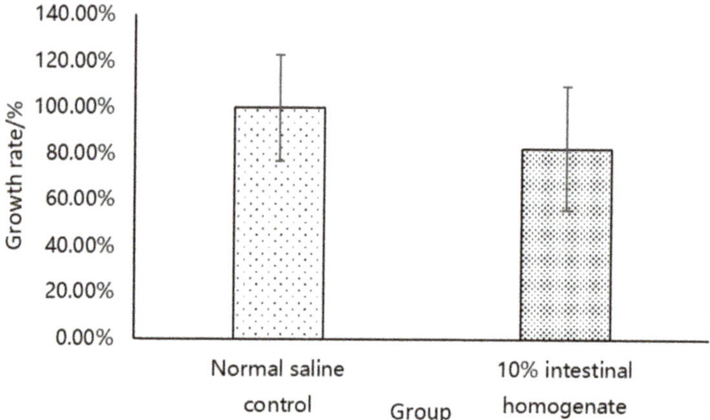


**Figure 2.** Comparison between the growth rate of Gs in the normal saline control and 10% GI tract homogenate ($n = 9$) on the NA plate.

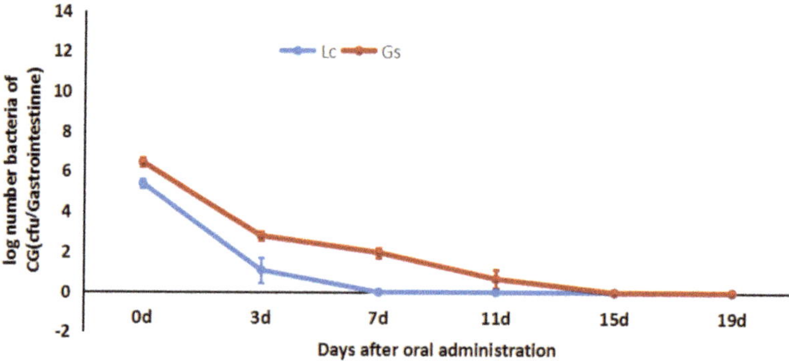


**Figure 3.** Kinetics of Lc and Gs elimination in the GI tract of crucian carp ($n = 9$).

### *3.5. Relative Abundance Changes of Lc and Gs in the Crucian Carp Gastrointestine*

Gastrointestinal content samples, collected at five time points during the three periods (from day −7 to day 21), were analyzed using a 16S RNA gene sequencing technique, and the results are shown in Figure 4. Lc was detected at very low abundance in the gastrointestine before the administration of the mixed diet (Day–7). It is not surprising that Lc became the major taxon in terms of abundance (36.75% ± 3.59%) after the administration of the mixed diet (day 0), whereas 7 days after the cessation of the mixed diet, the relative abundance of Lc decreased to 0.11% ± 0.03%. Fourteen days later, the relative abundance of Lc decreased to a very low level again, even lower than that of the control group (Figures 4 and 5).

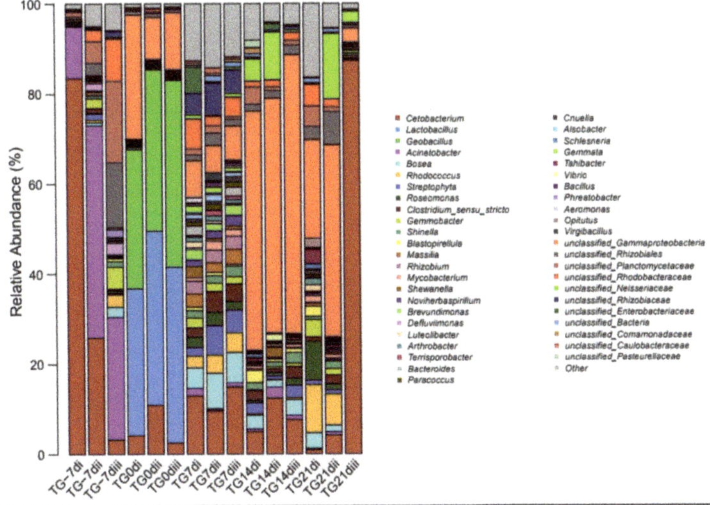


**Figure 4.** Bar plot illustrating the relative higher abundance bacterial genera for the individual fish. TG: treatment group: −7, 0, 7, 14 and 21 d represent the sample time points; i, ii, and iii represent individual triplicates within a group.

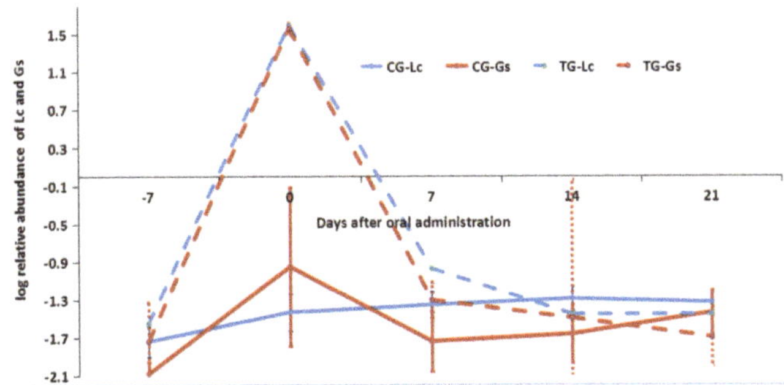


**Figure 5.** The changes in relative abundance of Lc and Gs in the CG and TG from day −7 to 21.

The relative abundance of Gs had the same trend as that of Lc (see Figures 4 and 5). At day 0, the relative abundance of Gs was 36.12% ± 5.31%, which was similar to that of Lc (Figure 5), but the number of viable Gs was eight times that of Lc (Figure 6). At day 7, although the relative abundance of Lc was 0.11% ± 0.03%, which was higher than other time points (except day 0), there was no viable Lc in the GI tract. We speculate that inactive Lc have reentered the GI tract because of the first incomplete replacement of the rearing water, and the same issue might also exist with the Gs. Viable Gs was detectable up to day 7, which is consistent with the results in Experiment 1. Regarding the control group, the relative Lc and Gs abundance remained at a very low level during the whole experiment, and no viable Lc and Gs were detected.

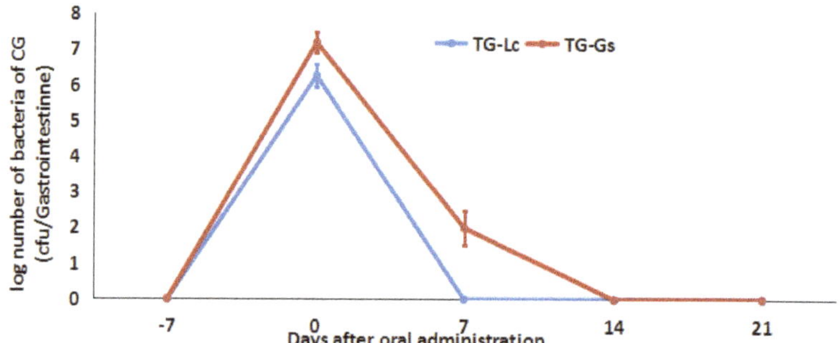


**Figure 6.** The changes in viable Lc and Gs bacteria in the TG from day −7 to 21.

## 4. Discussion

Here, an improved, highly sensitive selective culture method was used to monitor Lc in the GI tract of crucian carp whereby interference from nontarget bacteria was eliminated. Meanwhile, a transit marker was used to assess Lc colonization. In addition, a high-throughput sequencing technique was used to further understand changes in the relative abundance of Lc and Gs.

### 4.1. Elimination Interference Is Essential for Colonization

Compared with terrestrial animals and humans, the intestinal microbiota of fish is more easily affected by feed and rearing water [3,28], Moreover, it is inevitable that there will be *Lactobacillus* and *Bacillus* in fish diet. *Lactobacillus* and *Carnobacterium* could be detected in the gut of control groups in a probiotic feeding trial [13]. Merrifield et al. [29,30] also reported that *Enterococcus* and *Bacillus* could be detected in the gut of rainbow trout that were fed a diet without probiotic supplementation, and they considered that these bacteria may be indigenous species. In the five commercial feeds, we detected different species of LAB and *Bacillus* at different concentrations. One of the feeds contained *Pediococcus* at $1.4 \times 10^4$ cfu/g, and another feed contained *Bacillus* at over $10^6$ cfu/g (Table A2). Therefore, we proposed that sterile aquafeed should be used in GI microbe-related experiments. We therefore selected $^{60}$Co irradiation, which is a good sterilization method recommended for its wide use in SPF animal feed [20].

In the experiment, the target bacteria were more likely to reenter the gut via residual diet or feces. Merrifield et al. [29] found that $7.4 \times 10^3$ cfu/mL of *Bacillus* and $4.3 \times 10^3$ cfu/mL of *Enterococcus* were detected in the rearing water after feeding the diet supplemented with these bacteria, despite 15% water renewal per 72 h. Therefore, the authors suggested enhancing the water renewal rate to reduce background interference [29,30].

In rearing water with a pH of 8.0–9.0, the concentration of the Lc decreased dramatically from $1.0 \times 10^6$ cfu/mL at the beginning to <1 cfu/mL 7 days later (unpublished data). Considering their short life in water, 100% water renewal with an interval of 5 days is enough to control the amount of these Lc in the water. However, if a testing strain can endure the water environment (such as in the case of a Gs spore) or even proliferate, the persistence time would be overestimated, and the reintroduction of the testing strain would be obvious. Thus, a better method for controlling the testing strain in water is needed.

### 4.2. The Improved, Highly Sensitive Selective Culture Combined with a Transit Marker Is a Suitable Method for the Study of Colonization in Fish

Various methods have been developed to evaluate bacterial colonization in complex gut microbiota. Although tagging probiotic strains with fluorescence markers is an alternative, frequent plasmid loss during gut transition, low detection sensitivity and safety concerns hinder its further application. Species-specific PCR has also been developed to directly detect organisms in the extracted genome of fecal or GI tract samples. However, it cannot eliminate the baseline values of indigenous bacteria of the same species in their environments or diets [31]. At present, strain-specific PCR is used to detect and quantify strains; however, these strain-specific DNA fragments are based on a limited number of strains, making the strain-specificity robust only within a narrow confidence interval. These methods focus on humans and mice and are not suitable for colonization studies of aquatic animals such as fish. Although a selective medium method with colony identification is considered arduous and time-consuming, it is still a classic method in microbiology studies [32]. In particular, the method can tell whether the bacteria are alive or dead, whereas molecular methods cannot.

The MRS agar with a pH of 5.4–5.5 had high specificity and sensitivity for detecting acid-resistant bacterial species in the GI tract, such as the Lc strains used in our study. The weight of GI tract samples usually does not exceed 1 g after an appropriate starvation period when the bodyweight of the fish is less than 30 g. Then, a 10% homogenate of less than 10 mL can be entirely spread on agar on fewer than 50 plates at 200 μL/plate. The detection limit using this approach is 1 cfu/gastrointestine. Other culture-dependent methods have poor accuracy and a detection limit usually higher than 10 cfu/g [13,15], whereas our improved selective culture method is very suitable for fish colonization experiments.

Colonization was defined by Conway and Cohen as the indefinite persistence of a particular bacterial population without the reintroduction of that bacterium [18]. If a microbe can exit the GI tract in the extreme long term (such as its whole life) or extreme short term (such as a couple of days), then the conclusion of colonization is not easy to make. However, if a microbe merely exits the GI tract for "a period of time", how should we define the length of that time? Marteau and Vesa [21] indicated that using a transit marker is necessary when studying the colonization of potential probiotics, and the colonizer should persist for a longer period than the marker. A Gs spore is a good transit marker [21,25,26] for the following reasons: Firstly, its growing temperature ranges from 40 to 70 °C [33], so it usually cannot germinate, grow or reproduce in rearing water and fish gut. Secondly, the spores cannot be easily destroyed in the GI tract and feed preparation process. Thirdly, the spores can easily be counted based on high-temperature selective culture where other gastrointestinal bacteria usually cannot grow. Our study showed that the detection limit of Gs can reach 1 cfu/gastrointestine.

### 4.3. Monitored Relative Abundance Changes by High-Throughput Sequencing

With the second-generation sequencing technique for gut microbiome community analysis, we can identify bacterial components at the genus level. Some researchers employed 16S rRNA amplicon sequencing to study colonization [34,35]. Howitt compared traditional microbiological cultures and 16S polymerase chain reaction analyses for the identification of preoperative airway colonization in patients undergoing lung resection. The results showed that 16S PCR analyses identify colonizing bacteria in a similar proportion of preoperative BAL samples as traditional cultures [36]. An approach based on Illumina HiSeq 16S rRNA amplicon was used by Xia et al. [11], with results showing that *Lactococcus lactis* JCM5805 was below the detection level after the cessation of probiotics for 5 days, and they inferred that this strain could not colonize the gut; rather, the evaluation of colonization based on the 16S rRNA amplicon technology that they used is limited, for two reasons. First, the detection level of the method on a fish's gastrointestinal sample is unknown. Metagenomics is only able to distinguish bacteria with concentrations greater than $10^6$ bacteria per gram of feces [37]; thus, some low-abundance bacteria would be missed by metagenomic analysis. Second, the method is based on DNA samples and cannot determine the viability of bacteria, i.e., whether the bacteria are alive or dead, which could influence the interpretation of the results [2]. Of course, this method is feasible as an auxiliary means to understand changes in the abundance of the target bacteria.

*4.4. Lc ATCC 393 Cannot Colonize the Gastrointestinal Tract*

The persistence of probiotics in the gut is species-specific. In our previous study, even though an exogenous *Bacillus licheniformis* A1(Bli-A1) supplement was withdrawn, the concentration of Bli-A1 in the intestinal content was sustained at $3.3 \times 10^2$ cfu/g for at least 42 days with continuous sterile feed supplements [38]. In this study, when the detection limit was 1 cfu/gastrointestine, the elimination speed of Lc was even faster than that of the transit marker, indicating that Lc could not colonize in the gastrointestine of crucian carp. This is consistent with our previous studies of Lc on catfish [27]. We speculate that there are three reasons that Lc could not colonize in the gastrointestine of crucian carp. First, indigenous microbiomes drive colonization resistance to probiotics and/or additional bacteria [39]. Second, Gastrointestinal contents are not conducive to Lc reproduction. Third, Lc lacks the ability to adhere to the mucosa of the GI tract of crucian carp.

However, the supplement of Lc changed the gastrointestinal microbiota structure of crucian carp (Table S1), compared with day −7, the number of the high-abundant taxa (≥1%) increased from 9 (except other bacteria abundance) to 24 (except other bacteria abundance) on day 7, and recovered to the previous (day −7) microbiota structure until day 21.

## 5. Conclusions

The elimination speed of Lc was faster than the transit marker. Meanwhile, although Lc retained a low relative abundance from day 7 (0.11% ± 0.03%) to 21 in the crucian carp gastrointestine, they were not viable. The results indicate that the Lc ATCC 393 cannot colonize crucian carp. This study presents a method with a low detection limit for the colonization of LAB in fish and provides the idea of crucian carp to screen for beneficial probiotics.

**Supplementary Materials:** The following are available online at https://www.mdpi.com/article/10.3390/microorganisms9122547/s1. Table S1: The supplement of Lc changed the gastrointestinal microbiota structure of crucian carp.

**Author Contributions:** Conceptualization, H.Z., H.W. (Haibo Wang) and Y.M.; methodology, H.Z., H.W. (Haibo Wang), J.S., L.X. and H.W. (Hongwei Wang); software, H.Z., X.M. and H.W. (Hui Wang); validation, X.M. and Y.L.; formal analysis, H.W. (Hui Wang); investigation, H.Z., H.W. (Haibo Wang) and Y.M.; data curation, H.Z. and H.W. (Hongwei Wang); writing—original draft preparation, H.Z., H.W. (Haibo Wang) and J.S.; writing—review and editing, H.Z., H.W. (Haibo Wang), Y.M., X.M. and H.W. (Hui Wang); visualization, H.Z., H.W. (Haibo Wang) and Y.M.; project administration, H.Z., Y.L. and L.X.; funding acquisition, H.Z. and L.X. All authors have read and agreed to the published version of the manuscript.

**Funding:** This study was funded by the Central Public-Interest Scientific Institution Basal Research Fund, grant number CAFS (NO. 2019A006), grant number CAFS(NO. 2020TD11), National Key Research and Development Program of China (NO.2017YFC1600704).

**Institutional Review Board Statement:** This study was approved by the Animal Care and Use Committee of the Green Fish Drug Innovation Center at the Chinese Academy of Fishery Sciences (ACUCFD-2020-A06).

**Conflicts of Interest:** The authors declare no conflict of interest.

## Appendix A

**Table A1.** Proximate composition of No.2 diet used in the experiment.

| Proximate Composition | Proportion/% |
|---|---|
| Crude protein | not less than 33.0 |
| Crude lipid | not less than 5.0 |
| Crude fibre | not more than 8.0 |
| Crude ash | not more than 15.0 |
| Total phosphorus | not less than 1.1 |

## Appendix B

**Table A2.** Bacterial concentration of feed before $^{60}$Co irradiation. ($n$ = 3; cfu/g).

| Feed | General Heterotrophic Bacteria/Lg cell Concentration | Lactic Acid Bacteria/Lg cell Concentration | Thermophiles/Lg cell Concentration |
|---|---|---|---|
| No.1 | 4.84 ± 0.35 | 3.21 ± 0.57 | 4.00 ± 0.65 |
| No.2 | 5.50 ± 0.41 | 2.56 ± 0.21 | 4.22 ± 0.19 |
| No.3 | 4.29 ± 0.24 | 4.16 ± 0.15 | 3.80 ± 0.21 |
| No.4 | 6.48 ± 0.39 | 2.37 ± 0.10 | 4.37 ± 0.28 |
| No.5 | 4.70 ± 0.36 | 2.37 ± 0.10 | 2.94 ± 0.35 |

## Appendix C

**Table A3.** Bacterial concentration of feed at the beginning and end of the experiments (cfu/g).

| Feed | Lg cell Concentration | |
|---|---|---|
| | Beginning | End |
| *L.casei*/*G. stearothermophilus* (Experiment 1) | 7.0/6.9 | 6.8/6.8 |
| *L.casei*/*G. stearothermophilus* (Experiment 2) | 9.3/8.0 | 9.1/8.0 |

## Appendix D

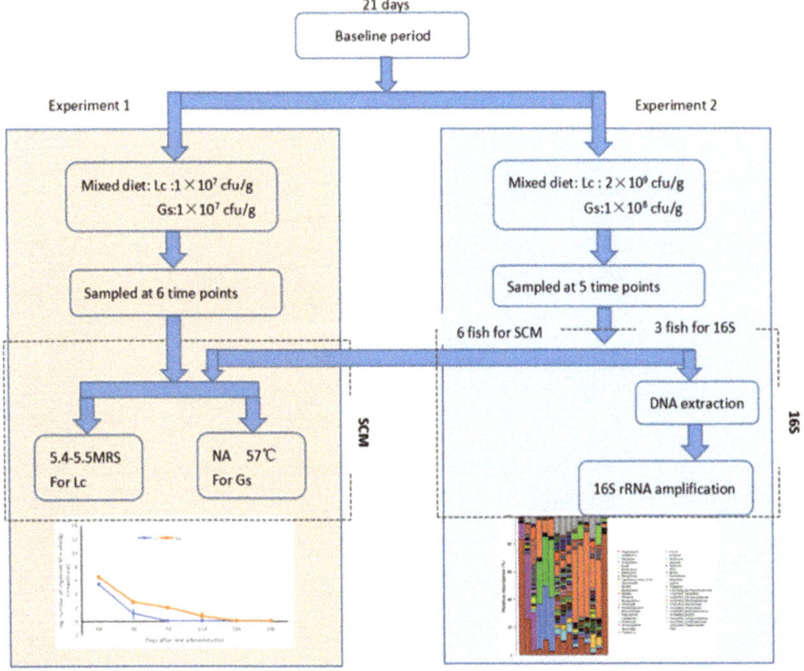


**Figure A1.** The flow chart of design of experiment.

## References

1. El-Saadony, M.T.; Alagawany, M.; Patra, A.K.; Kar, I.; Tiwari, R.; Dawood, M.A.O.; Dhama, K.; Abdel-Latif, H.M.R. The functionality of probiotics in aquaculture: An overview. *Fish Shellfish. Immun.* **2021**, *117*, 36–52. [CrossRef]
2. Vargas-Albores, F.; Martínez-Córdova, L.R.; Hernández-Mendoza, A.; Cicala, F.; Lago-Lestón, A.; Martínez-Porchas, M. Therapeutic modulation of fish gut microbiota, a feasible strategy for aquaculture? *Aquacultrue* **2021**, *544*, 737050. [CrossRef]
3. Li, X.; Ringø, E.; Hoseinifar, S.H.; Lauzon, H.L.; Birkbeck, H.; Yang, D. The adherence and colonization of microorganisms in fish gastrointestinal tract. *Rev. Aquacult.* **2019**, *11*, 603–618. [CrossRef]
4. Tian, J.; Kang, Y.; Chu, G.; Liu, H.; Kong, Y.; Zhao, L.; Kong, Y.; Shan, X.; Wang, G. Oral administration of lactobacillus casei expressing flagellin a protein confers effective protection against Aeromonas Veronii in common carp, Cyprinus Carpio. *Int. J. Mol. Sci.* **2020**, *21*, 33. [CrossRef] [PubMed]
5. Mehdinejad, N.; Imanpour, M.R.; Jafari, V. Combined or individual effects of dietary probiotic Pedicoccus acidilactici and nucleotide on growth performance, intestinal microbiota, hemato-biochemical parameters, and innate immune response in goldfish (*Carassius auratus*). *Probiot. Antimicrob. Protein* **2018**, *10*, 558–565. [CrossRef] [PubMed]
6. Zhao, D.; Wu, S.; Feng, W.; Jakovlić, I.; Tran, N.T.; Xiong, F. Adhesion and colonization properties of potentially probiotic Bacillus paralicheniformis strain FA6 isolated from grass carp intestine. *Fisheries Sci.* **2020**, *86*, 153–161. [CrossRef]
7. Mohammadian, T.; Alishahi, M.; Tabandeh, M.R.; Ghorbanpoor, M.; Gharibi, D. Changes in immunity, expression of some immune-related genes of shabot fish, Tor grypus, following experimental infection with Aeromonas hydrophila: Effects of autochthonous probiotics. *Probiot. Antimicrob. Protein* **2018**, *10*, 616–628. [CrossRef]
8. Zhao, L.L.; Liu, M.; Ge, J.W.; Qiao, X.Y.; Li, Y.J.; Liu, D.Q. Expression of infectious pancreatic necrosis virus (IPNV) VP2–VP3 fusion protein in Lactobacillus casei and immunogenicity in rainbow trouts. *Vaccine* **2012**, *30*, 1823–1829.
9. Balcázar, J.L.; Blas, I.D.B.; Ruiz-Zarzuela, I.; Vendrell, D.; Gironés, O.; Muzquiz, J.L. Enhancement of the immune response and protection induced by probiotic lactic acid bacteria against furunculosis in rainbow trout (*Oncorhynchus mykiss*). *FEMS Immunol. Med. Microbiol.* **2007**, *51*, 185–193. [CrossRef] [PubMed]
10. He, S.X.; Ran, C.; Qin, C.B.; Li, S.N.; Zhang, H.L.; Vos, W.M.D.; Ringø, E.; Zhou, Z.G. Anti-infective effect of adhesive probiotic lactobacillus in fish is correlated with their spatial distribution in the intestinal tissue. *Sci. Rep.* **2017**, *7*, 13195. [CrossRef]
11. Xia, Y.; Cao, J.M.; Wang, M.; Lu, M.X.; Chen, G.; Gao, F.Y.; Liu, Z.G.; Zhang, D.F.; Ke, X.L.; Yi, M.M. Effects of Lactococcus lactis subsp. lactis JCM5805 on colonization dynamics of gut microbiota and regulation of immunity in early ontogenetic stages of tilapia. *Fish Shellfish. Immun.* **2019**, *86*, 53–63. [CrossRef] [PubMed]
12. Huang, T.; Li, L.P.; Liu, Y.; Luo, Y.J.; Wang, R.; Tang, J.Y.; Chen, M. Spatiotemporal distribution of Streptococcus agalactiae attenuated vaccine strain YM001 in the intestinal tract of tilapia and its effect on mucosal associated immune cells. *Fish Shellfish. Immun.* **2019**, *87*, 714–720. [CrossRef]
13. Balcazar, J.L.; de Blas, I.; Ruiz-Zarzuela, I.; Vendrell, D.; Calvo, A.C.; Marquez, I.; Girones, O.; Muzquiz, J.L. Changes in intestinal microbiota and humoral immune response following probiotic administration in brown trout (*Salmo trutta*). *Br. J. Nutr.* **2007**, *97*, 522–527. [CrossRef] [PubMed]
14. Russo, P.; Iturria, I.; Mohedano, M.L.; Caggianiello, G.; Rainieri, S.; Fiocco, D.; Angel Pardo, M.; López, P.; Spano, G. Zebrafish gut colonization by mCherry-labelled lactic acid bacteria. *Appl. Microbiol. Biot.* **2015**, *99*, 3479–3490. [CrossRef]
15. Nikoskelainen, S.; Ouwehand, A.C.; Bylund, G.; Salminen, S.; Lilius, E. Immune enhancement in rainbow trout (Oncorhynchus mykiss) by potential probiotic bacteria (*Lactobacillus rhamnosus*). *Fish Shellfish. Immun.* **2003**, *15*, 443–452. [CrossRef]
16. Ringø, E.; Gatesoupe, F. Lactic acid bacteria in fish: A review. *Aquaculture* **1998**, *160*, 177–203. [CrossRef]
17. Ringø, E.; Van Doan, H.; Lee, S.H.; Soltani, M.; Hoseinifar, S.H.; Harikrishnan, R.; Song, S.K. Probiotics, lactic acid bacteria and bacilli: Interesting supplementation for aquaculture. *J. Appl. Microbiol.* **2020**, *129*, 116–136. [CrossRef] [PubMed]
18. Conway, T.; Cohen, P.S. Commensal and pathogenic Escherichia coli metabolism in the gut. *Microbiol. Spectr.* **2015**, *3*. [CrossRef] [PubMed]
19. Gatesoupe, F.J. The use of probiotics in aquaculture. *Aquaculture* **1999**, *180*, 147–165. [CrossRef]
20. Chen, Q.L.; Ha, Y.M.; Chen, Z.J. A study on radiation sterilization of SPF animal feed. *Radiat. Phys. Chem.* **2000**, *57*, 329–330. [CrossRef]
21. Marteau, P.; Vesa, T. Pharmacokinetics of probiotics and biotherapeutic agents in humans. *Biosci. Microflora* **1998**, *17*, 1–6. [CrossRef]
22. Banla, L.I.B.; Salzman, N.H.; Kristich, C.J. Colonization of the mammalian intestinal tract by enterococci. *Curr. Opin. Microbiol.* **2019**, *47*, 26–31. [CrossRef]
23. Safari, R.; Hoseinifar, S.H.; Nejadmoghadam, S.; Khalili, M. Apple cider vinegar boosted immunomodulatory and health promoting effects of Lactobacillus casei in common carp (*Cyprinus carpio*). *Fish Shellfish. Immunol.* **2017**, *67*, 441–448. [CrossRef] [PubMed]
24. Qin, C.B.; Xu, L.; Yang, Y.L.; He, S.X.; Dai, Y.Y.; Zhao, H.Y.; Zhou, Z.G. Comparison of fecundity and offspring immunity in zebrafish fed Lactobacillus rhamnosus CICC 6141 and Lactobacillus casei BL23. *Reproduction* **2013**, *147*, 53–64. [CrossRef]
25. Klijn, N.; Weerkamp, A.H.; de Vos, W.M. Genetic marking of Lactococcus lactis shows its survival in the human gastrointestinal tract. *Appl. Environ. Microbiol.* **1995**, *61*, 2771–2774. [CrossRef]
26. Vesa, T.; Pochart, P.; Marteau, P. Pharmacokinetics of Lactobacillus plantarum NCIMB 8826, Lactobacillus fermentum KLD, and Lactococcus lactis MG 1363 in the human gastrointestinal tract. *Aliment. Pharmacol. Ther.* **2000**, *14*, 823–828. [CrossRef]

27. Zhang, H.; Wang, H.; Hu, K.; Jiao, L.; Zhao, M.; Yang, X.; Xia, L. Effect of Dietary Supplementation of Lactobacillus Casei YYL3 and L. Plantarum YYL5 on Growth, Immune Response and Intestinal Microbiota in Channel Catfish. *Animals* **2019**, *9*, 1005. [CrossRef] [PubMed]
28. Ringø, E.; Zhou, Z.; Vecino, J.L.G.; Wadsworth, S.; Romero, J.; Krogdahl, Å.; Olsen, R.E.; Dimitroglou, A.; Foey, A.; Davies, S.; et al. Effect of dietary components on the gut microbiota of aquatic animals. A never-ending story? *Aquacult. Nutr.* **2016**, *22*, 219–282. [CrossRef]
29. Merrifield, D.L.; Dimitroglou, A.; Bradley, G.; Baker, R.T.M.; Davies, S.J. Probiotic applications for rainbow trout (*Oncorhynchus mykiss* Walbaum) I. Effects on growth performance, feed utilization, intestinal microbiota and related health criteria. *Aquacult. Nutr.* **2010**, *16*, 504–510. [CrossRef]
30. Merrifield, D.L.; Bradley, G.; Baker, R.T.M.; Davies, S.J. Probiotic applications for rainbow trout (*Oncorhynchus mykiss* Walbaum) II. Effects on growth performance, feed utilization, intestinal microbiota and related health criteria postantibiotic treatment. *Aquacult. Nutr.* **2010**, *16*, 496–503. [CrossRef]
31. Xiao, Y.; Zhao, J.X.; Zhang, H.; Zhai, Q.X.; Chen, W. Mining Lactobacillus and Bifidobacterium for organisms with long-term gut colonization potential. *Clin. Nutr.* **2020**, *39*, 1315–1323. [CrossRef]
32. Goossens, D.A.M.; Jonkers, D.M.A.E.; Russel, M.G.V.M.; Stobberingh, E.E.; Stockbrügger, R.W. The effect of a probiotic drink with Lactobacillus plantarum 299v on the bacterial composition in faeces and mucosal biopsies of rectum and ascending colon. *Aliment. Pharmacol. Ther.* **2006**, *23*, 255–263. [CrossRef] [PubMed]
33. De Vos, P.; Garrity, G.M.; Jones, D.; Krieg, N.R.; Ludwig, W.; Rainey, F.A.; Schleifer, K.-H.; Whitman, W.B. *Bergey's Manual of Systematic Bacteriology*; Springer: Berlin/Heidelberg, Germany, 2009; Volume 3.
34. Zhang, X.; Wu, J.; Zhou, C.; Tan, Z.; Jiao, J. Spatial and temporal organization of jejunal microbiota in goats during animal development process. *J. Appl. Microbiol.* **2021**, *131*, 68–79. [CrossRef]
35. Crobach, M.J.T.; Ducarmon, Q.R.; Terveer, E.M.; Harmanus, C.; Sanders, I.M.J.G.; Verduin, K.M.; Kuijper, E.J.; Zwittink, R.D. The bacterial gut microbiota of adult patients infected, colonized or noncolonized by clostridioides difficile. *Microorganisms* **2020**, *8*, 677. [CrossRef] [PubMed]
36. Howitt, S.H.; Blackshaw, D.; Fontaine, E.; Hassan, I.; Malagon, I. Comparison of traditional microbiological culture and 16S polymerase chain reaction analyses for identification of preoperative airway colonization for patients undergoing lung resection. *J. Crit. Care* **2018**, *46*, 84–87. [CrossRef] [PubMed]
37. Amrane, S.; Lagier, J. Metagenomic and clinical microbiology. *Hum. Microbiome J.* **2018**, *9*, 1–6. [CrossRef]
38. Zhang, H.Y.; Wang, H.B.; Zhao, M.J.; Jiao, L.T.; Hu, K.; Yang, X.L.; Xia, L. Colonization of Bacillus licheniformis A1 in intestine of Ictalurus punctatus. *J. Fish. Sci. China* **2019**, *26*, 1136–1143.
39. Zmora, N.; Zilberman-Schapira, G.; Suez, J.; Mor, U.; Dori-Bachash, M.; Bashiardes, S.; Kotler, E.; Zur, M.; Regev-Lehavi, D.; Ben-Zeev Brik, R.; et al. Personalized gut mucosal colonization resistance to empiric probiotics is associated with unique host and microbiome features. *Cell* **2018**, *174*, 1388–1405. [CrossRef] [PubMed]

*Article*

# Probiotics Improve Eating Disorders in Mandarin Fish (*Siniperca chuatsi*) Induced by a Pellet Feed Diet via Stimulating Immunity and Regulating Gut Microbiota

**Xiaoli Chen [1,2], Huadong Yi [1,2], Shuang Liu [1,2], Yong Zhang [1], Yuqin Su [1,2], Xuange Liu [1,2], Sheng Bi [1,2], Han Lai [1,2], Zeyu Zeng [1,2] and Guifeng Li [1,2,*]**

[1] State Key Laboratory of Biocontrol, Guangdong Provincial Key Laboratory for Aquatic Economic Animals, School of Life Sciences, Sun Yat-Sen University, Guangzhou 510275, China; chenxli27@mail2.sysu.edu.cn (X.C.); yihd3@mail2.sysu.edu.cn (H.Y.); liush276@mail2.sysu.edu.cn (S.L.); lsszy@mail.sysu.edu.cn (Y.Z.); suyq9@mail2.sysu.edu.cn (Y.S.); liuxg27@mail2.sysu.edu.cn (X.L.); bish@mail2.sysu.edu.cn (S.B.); laih5@mail2.sysu.edu.cn (H.L.); zengzy5@mail3.sysu.edu.cn (Z.Z.)

[2] Guangdong Provincial Engineering Technology Research Center for Healthy Breeding of Important Economic Fish, Guangzhou 510006, China

* Correspondence: liguif@mail.sysu.edu.cn

**Citation:** Chen, X.; Yi, H.; Liu, S.; Zhang, Y.; Su, Y.; Liu, X.; Bi, S.; Lai, H.; Zeng, Z.; Li, G. Probiotics Improve Eating Disorders in Mandarin Fish (*Siniperca chuatsi*) Induced by a Pellet Feed Diet via Stimulating Immunity and Regulating Gut Microbiota. *Microorganisms* **2021**, *9*, 1288. https://doi.org/10.3390/microorganisms9061288

Academic Editor: Konstantinos Ar. Kormas

Received: 24 May 2021
Accepted: 9 June 2021
Published: 12 June 2021

**Publisher's Note:** MDPI stays neutral with regard to jurisdictional claims in published maps and institutional affiliations.



**Abstract:** Eating disorders are directly or indirectly influenced by gut microbiota and innate immunity. Probiotics have been shown to regulate gut microbiota and stimulate immunity in a variety of species. In this study, three kinds of probiotics, namely, *Lactobacillus plantarum*, *Lactobacillus rhamnosus* and *Clostridium butyricum*, were selected for the experiment. The results showed that the addition of three probiotics at a concentration of $10^8$ colony forming unit/mL to the culture water significantly increased the ratio of the pellet feed recipients and survival rate of mandarin fish (*Siniperca chuatsi*) under pellet-feed feeding. In addition, the three kinds of probiotics reversed the decrease in serum lysozyme and immunoglobulin M content, the decrease in the activity of antioxidant enzymes glutathione and catalase and the decrease in the expression of the appetite-stimulating regulator agouti gene-related protein of mandarin fish caused by pellet-feed feeding. In terms of intestinal health, the three probiotics reduced the abundance of pathogenic bacteria *Aeromonas* in the gut microbiota and increased the height of intestinal villi and the thickness of foregut basement membrane of mandarin fish under pellet-feed feeding. In general, the addition of the three probiotics can significantly improve eating disorders of mandarin fish caused by pellet feeding.

**Keywords:** eating disorders; feeding behavior; gut microbiota; *Siniperca chuatsi*; innate immunity; appetite; *Lactobacillus plantarum*; *Lactobacillus rhamnosus*; *Clostridium butyricum*

## 1. Introduction

Eating disorders mainly refer to a group of syndromes characterized by abnormal feeding behaviors, accompanied by significant weight changes or physiological dysfunctions [1–3]. The main clinical types include anorexia nervosa, bulimia nervosa, binge eating disorder and avoidance/restrictive food intake disorder. Moreover, eating disorders occur throughout the age groups and have an essential impact on physical and mental health [4]. They increase the likelihood of anxiety, obesity, suicidal intentions, depression, drug abuse and health problems [5]. Eating disorders are associated with the establishment of food preferences and aversions and are influenced by the sensorial characteristics of food [6]. A better understanding of food preferences and aversions can improve the prevention and treatment of eating disorders [7].

Food preference is an innate behavioral trait which is affected by both genes and the environment [8,9]. The hypothalamus contains orexigenic neurons that express neuropeptide Y (NPY) and agouti-related peptide (AgRP), which participate in food intake control and are regulated by the peripheral hormone leptin and ghrelin [10,11]. NPY is a peptide composed of 36 amino acids. As an appetite-stimulating factor, it plays a crucial role in regulating energy homeostasis and food intake [12]. AgRP increases food intake by antagonizing the effect of the anorexigenic POMC product, α-melanocyte stimulating hormone (α-MSH) [13,14]. There seems to be a species-specific variability in the functions of leptin and ghrelin with regards to the regulation of feeding and metabolism in fish [15]. Ghrelin acts as an appetite stimulant in a variety of fish species, but there is also conflicting evidence, such as in *Salmoniformes* [16].

As an essential modulator of host physiology and behavior, intestinal bacteria have been shown to influence feeding behavior and food choice [17–25]. Gut microbiota can influence host eating behavior by directly affecting nutrient sensing, appetite and satiety-regulating systems through the production of neuroactive substances and short-chain fatty acids or indirectly manipulating intestinal barrier function, interacting with bile acid metabolism, modulating the immune system and influencing host antigen production [26]. Gut microbiota play a vital role in regulating host eating disorders' behavioral comorbidities, such as obesity, anorexia nervosa and severe acute malnutrition. A growing body of evidence links the gut microbiota with nutrition, immune, anti-oxidative stress and appetite. Influencing one of these factors will most likely lead to changes in the others, thereby making the gut microbiota easily accessible and manipulable for targeting host food preferences [26].

Administration of probiotics is an effective strategy to maintain the balance of the gut microbiota [27]. Probiotics are defined as microbial cells or compounds that have a beneficial effect on the health of the host. In aquaculture, probiotics can prevent the spread of diseases, increase food conversion efficiency and stimulate growth by improving the composition of the gastrointestinal microbiota, strengthening the immune system and increasing the resistance to farmed stressors [28–30]. In addition, probiotics have become an alternative to antibiotics and other drug treatments in the aquaculture industry and are considered a new tool for disease control [28,31,32]. Microorganisms commonly used as probiotics in aquaculture include bacteria, yeast and algae [33].

Among several probiotic bacterial species, numerous reports have been published on the beneficial role of *Lactobacillus plantarum*, *Lactobacillus rhamnosus* and *Clostridium butyricum* as probiotics in aquaculture [34–40]. *L. plantarum* is a rod-shaped, gram-positive, non-spore-forming facultative anaerobic bacteria that belong to the *Lactobacillaceae* family. It has been reported to reduce the adhesion and growth of harmful bacteria via producing antimicrobial compounds [41–43], improve the growth and feed efficiency of carp (*Catla catla*) [44,45], grouper (*Epinephelus coioides*) [46], tilapia (*Oreochromis niloticus*) [47], shrimp (*Penaeus indicus*) [48] and pacific white shrimp (*Litopenaeus vannamei*) [49] and enhance the immunity and survival rate of pacific white shrimp (*Litopenaeus vannamei*) [50,51] and tilapia [52]. Previous studies have shown that *L. rhamnose* can affect the appetite and energy metabolism of the host by regulating the expression of γ-aminobutyric acid and its receptors in the central nervous system [53–57]. *C. butyricum* is a spore-forming bacterium belonging to Gram-positive anaerobe that can produce butyric acid and exists in the intestine of healthy animals and human [58–60]. Compared with other probiotics, *C. butyricum* has a more vital tolerance ability to higher temperature environments, lower pH, bile salt and several antibiotics. Therefore, *C. butyricum* has always been regarded as a good and safe food additive [58]. *C. butyricum* has a positive effect on immune function and is connected with increased population of *Bifidobacterium* and *Lactobacillus* and decreased concentration of pathogenic bacteria in the intestinal tract of humans, mice, piglets and broiler chickens [61,62]. *C. butyricum* can inhibit intestinal inflammation and regulate gut microbiota through the immune pathway [63–65].

Mandarin fish (*Siniperca chuatsi*) is a precious freshwater farmed fish with unique live bait feeding habits, and it does not easily accept dead bait or pellet feed [10,66]. The preference for a live bait diet increases the cost of mandarin fish farming and the risk of infectious diseases, limiting the development of mandarin fish farming. For this problem, previous studies mainly focused on optimizing the domestication process and breeding conditions (such as temperature), strengthening the training of learning and memory and using attractants, which promoted the development of pellet feed for mandarin fish [67–70]. However, there are still problems such as high mortality and slow growth of mandarin fish fed with pellet feed. Recently, relationships among gut microbiota, host immunity and feeding preference behavior have attracted research attention [71]. Probiotics intervention is an effective way to regulate the gut microbiota [27]. In this study, three probiotics that have been shown to be safe for aquatic animals, *L. plantarum*, *L. rhamnosus* and *C. butyricum*, were selected to investigate whether probiotics can improve the eating disorders of mandarin fish caused by pellet feed diet by modulating the gut microbiota, immune parameters, appetite and intestinal morphology, which may contribute to the theoretical foundation of probiotics intervention in the treatment of dietary disorders.

## 2. Materials and Methods

### 2.1. Bacteria Strains

The three probiotic strains, *L. plantarum* (ATCC 8014), *L. rhamnosus* (ATCC 7469) and *C. butyricum* (ATCC 19398), were purchased from Guangdong Microbial Culture Collection Center (GDMCC). The bacteria were cultured as described previously [36,72,73]. Briefly, the two activated bacterial suspensions of *L. plantarum* and *L. rhamnosus* were separately incubated into MRS liquid broth (Merck, Darmstadt, Germany). The activated bacterial suspension of *C. butyricum* was incubated into the reinforced clostridial medium (RCM) and then placed in an anaerobic workstation at 37 °C for 12 h. The bacterial titers were measured by making tenfold dilution series in triplicate on agar plates. Optical densities (OD) were measured using a spectrophotometer (Spectroscan UV 2600, Thermo Scientific, Waltham, MA, USA) at 600 nm. The strains were harvested via centrifugation at 4000× *g* for 10 min, washed twice with normal saline (0.9% NaCl) and resuspended at $2 \times 10^{10}$ colony forming unit (CFU)/mL in sterile normal saline. Culture bacterial cells were afterward kept at 4 °C until usage.

### 2.2. Animal Treatments

All experimental procedures were approved by the Institutional Animal Care and Use Committee of Sun Yat-sen University and performed according to the guidelines for experimental animals established by this committee. One thousand and five hundred healthy mandarin fish were obtained from a fish farm in Foshan, Guangdong, China. All experimental fish were acclimatized for two weeks in 3200 L rectangular aquaria to laboratory conditions before pellet-feed feeding.

After the adaptive feeding, a total of 1350 healthy mandarin fish weighing 2.5 ± 0.1 g (mean ± standard error of mean (SEM)) were randomly allocated into one of five groups (270 fish per group): live bait fish feeding group (LBFD), pellet-feed feeding group with probiotics free (PFD), pellet-feed feeding group with *L. plantarum* plus (PFDLP), pellet-feed feeding group with *L. rhamnosus* plus (PFDLR) and pellet-feed feeding group with *C. butyricum* plus (PFDCB). Each group of experimental fish was randomly assigned to three 800 L replicated water tanks (90 fish per tank). Mandarin fish in the PFDLP, PFDLR and PFDCB groups were treated with *L. plantarum*, *L. plantarum* and *C. butyricum* at a final concentration of $10^8$ CFU/mL for one week, while the remaining two groups, LBFD and PFD, were not treated. In this time, all fish received a live bait fish diet twice a day (at 06:00 and 18.00 h) at 5% of initial body weight. Mud carp (*Cirrhinus molitorella*) was used as the live bait fish in this study.

During the period of pellet-feed feeding, the PFD, PFDCB, PFDLR and PFDLP groups of experimental fish were overfed from dead fish (1 week) to commercial feed (4 weeks) following the domestication process established by Liang et al. [67], while the LBFD group of experimental fish maintained a live bait diet. Each group of experimental fish was fed twice a day (at 06:00 and 18.00 h) at 5% of initial body weight to approximate satiation. The main nutritional composition of the commercial feed purchased from Foshan Nanhai Jieda Feed Co., LTD. (Lishui, China), is 48% crude protein, 5% crude fat, 3% crude fiber,

19% crude ash, 10% water, 4% calcium, 2% total phosphorus, 3%NaCl and 2.7% lysine. The soft pellet feed with a diameter of 50 mm was made with a feed machine and stored at −20 °C until use. Part of the water tank was replaced daily to remove waste and feces. When partially replacing the aquaculture water, an appropriate amount of *L. plantarum*, *L. plantarum* and *C. butyricum* was added to the PFDLP, PFDLR and PFDCB groups to maintain the concentration at $1 \times 10^8$ CFU/mL. The water quality of each tank was kept within the best physical parameter range, temperature (24.13 ± 0.52 °C), pH (7.41 ± 0.15), ammonia-nitrogen (0.27 ± 0.05 mg/L) and dissolved oxygen (7.52 ± 0.15 mg/L), during the experiment.

### 2.3. Proportion of Pellet Feed Recipients and Survival Analysis

The number of pellet feed recipients in groups PFD, PFDLP, PFDLR and PFDCB were counted on days 7, 14 and 28 after pellet-feed feeding, and the proportion of pellet feed recipients (POPFR) was calculated according to the following formula: POPFR (%) = [Number of pellet feed recipients/Number of initial mandarin fish] × 100. During the feeding trial, the number of deaths in each group was recorded every day, and Kaplan Meyer's (KM) survival analysis was used to evaluate the survival differences between groups.

### 2.4. Sample Collection

On days 7, 14 and 28 of pellet-feed feeding, twelve mandarin fish were randomly collected from each tank and then anesthetized with tricaine methanesulfonate (MS-222) for subsequent sampling. Blood samples collected from the tail vascular vein of each fish were placed in centrifuge tubes and centrifuged at 4 °C and 4000 rpm for 15 min to separate the serum. The separated serum was stored at −80 °C for further determination of immune parameters. Brain and gut samples were collected and placed in RNA Later® (Qiagen, Hilden, Germany) at 4 °C overnight and then stored at −80 °C for gene expression analysis. A separate liver, intestine and gills were homogenized with cold phosphate buffer saline (PH 7.5). The homogenate was then centrifuged at 4 °C and 8000 rpm for 10 min, and the supernatant was taken and stored at −20 °C for analysis of antioxidants and oxidative stress parameters. Intestinal samples containing the inclusion were collected and placed in sterile Eppendorf tubes, immediately frozen in liquid nitrogen, and then stored at −80 °C for microbiome analysis. Intestinal tissue was collected and fixed in Bouin's solution for 24 h before histological analysis was performed.

### 2.5. Serum Parameter Analysis

#### 2.5.1. Serum Lysozyme Content

According to the instruction manual, lysozyme content in serum was strictly analyzed (Nanjing Jiancheng Bioengineering Institute, Nanjing, China).

#### 2.5.2. Measurement of IgM and CRP

Reagent kits for immunoglobulin M (IgM) and C-reactive protein (CRP) were obtained from Shanghai Enzyme-linked Biotechnology Co., Ltd., Shanghai, China. Each parameter was strictly analyzed in accordance using a double-antibody sandwich ELISA with the manufacturer's instructions.

### 2.6. Antioxidant and Oxidative Stress Parameters

The superoxide dismutase (SOD) activity, CAT activity, glutathione (GSH) content and malondialdehyde (MDA) content were determined according to the instructions provided in the commercial kits (Nanjing Jiancheng Bioengineering Institute, Nanjing, China). SOD, GSH, CAT and MDA measurements were based on the WST-1 method [74], xanthine oxidase method [75], ammonium molybdate colorimetric method [76] and thiobarbituric acid method [77], respectively.

### 2.7. Gene Expression Analysis

#### 2.7.1. Extraction of total RNA and Reverse Transcription

According to the manufacturer's instructions, total RNAs were extracted from each tissue sample (50–100 mg) using RNAiso Plus reagent (Takara, Shiga, Japan). RNA concentrations and purity were determined using a Nanodrop 2000 c spectrophotometer (Thermo Fisher, Waltham, MA, USA). RNA was used as a templet for cDNA synthesis using PrimeScript$^{TM}$ reverse transcription (RT) reagent kit (TaKaRa, Shiga, Japan) following the manufacturer's guidelines and stored at −80 °C until analysis.

#### 2.7.2. Real-Time Quantitative PCR (RT-qPCR)

Total RNA was isolated from different tissues by using RNAiso Plus reagent (Takara, Shiga, Japan) according to the manufacturer's instructions. First-strand complementary DNAs (cDNAs) were synthesized using PrimeScript$^{TM}$ RT reagent kit (Takara, Shiga, Japan) following the manufacturer's guidelines. The expression levels of *ghrelin, leptin, npy, agrp* and *β-actin* were detected using the corresponding forward and reverse primers, which were designed using Primer Express software (Applied Biosystems, Waltham, MA, USA) (Table 1). *β-actin* served as a housekeeping gene in order to normalize the expression levels. Quantitative PCR (qPCR) was performed on a total reaction volume of 10 µL, containing 0.2 µM primers, 1µL of cDNA, 5 µL of 2 × SYBR premix ExTaq$^{TM}$ (Takara, Shiga, Japan) and 3.6 µL of ultrapure water using the following setting: 40 cycles of amplification (5 s at 95 °C, 40 s at 60 °C and 1 s at 70 °C). All RT-qPCR reactions were performed in triplicate on a LightCycler 480 instrument (Roche Diagnostics, Rotkreuz, Switzerland). Data were analyzed using the $2^{-\Delta\Delta Ct}$ method [78].

**Table 1.** Sequences of primer pairs used for real-time quantitative PCR in this study.

| Gene | Primer Name | Primer Sequence (5′-3′) | Annealing Temp (°C) |
|---|---|---|---|
| *ghrelin* | Scghrelin-F | GCTTTCTCAGCCCTTCAC | 60 |
| | Scghrelin-R | GGTTGTCCTCAGTGGGTTG | |
| *leptin* | scleptinB-F | CGAGAGTCACCTTTACCTG | 58 |
| | scleptinB-R | GTGCAAATAAGCCTCTAAGTG | |
| *npy* | scNPY-F | GCAAATCTCCCTCTGACAATC | 60 |
| | scNPY-R | GGTTTCACCGGGTATCCTT | |
| *agrp* | scAgRP-F | GAGCCAAGCGAAGACCAGA | 58 |
| | scAgRP-R | GCAGCACGGCAAATGAGAG | |
| *β-actin* | β-actin-F | CCCTCTGAACCCCAAAGCCA | 59 |
| | β-actin-R | CAGCCTGGATGGCAACGTACA | |

### 2.8. Gut Microbiota Analysis

Total bacterial DNA of the intestine samples with retained contents was extracted using an E.Z.N.A. ®Stool DNA Kit (Omega, Norcross, GA, USA). After measurement of the concentration and quality of the extracted DNA using a Nanodrop 2000c spectrophotometer (Thermo Fisher, Waltham, MA, USA), the V4-V5 region of the bacterial 16S DNA gene was amplified via the PCR method using the primers of 515F (5′-GTGCCAGCMGCCGCGGTAA-3′) and 806R (5′-CCGTCAATTCCTTTG AGTTT-3′). The high throughput sequencing for the qualified amplicon was performed on the Illumina NovaSeq6000 platform at Novogene Biotech Co., Ltd. (Beijing, China). Paired-end reads were assigned to samples based on a unique barcode and truncated by cutting off the barcode and primer sequence. The raw tags were then produced via FLASH (V1.2.7) [79]. Sequences were analyzed with the UCHIME algorithm [80] and QIIME [81]. The effective tags were filtered and clustered into operational taxonomic units (OTUs) under a 97% nucleotide similarity level. The taxonomic annotation of OTUs was performed using Uparse software [82]. The alpha diversity, including the observed species, Chao 1, abundance-based coverage estimator (ACE), Simpson, Shannon and PD whole tree, was calculated using QIIME (Version 1.9.1) to analyze the abundance and diversity. A Venn diagram was constructed to describe the core components of the genera. Beta diversity was evaluated using principal coordinates analysis (PCoA). Linear discriminant analysis effect size (LEfSe) was used to identify significant differences in the relative abundance of bacterial taxa [83]. Predicted functional pathways were annotated using the Kyoto encyclopedia of genes and genomes (KEGG) at level 1. Tax4Fun was used to predict the functional profile of the intestinal microbiota [84]. All figures were drawn using R software (Version 2.15.3).

### 2.9. Intestinal Histological Assessment

The foregut, midgut and hindgut tissues were fixed in Bouin's solution for 24 h and then dehydrated, embedded in paraffin and sectioned into 4-μm transverse cuts following the axis of the gut lumen. Hematoxylin and eosin (H.E.) were applied for the staining, and histological examination of the samples was carried out using an optic microscope (Nikon, Tokyo, Japan) with a digital camera (Nikon, Tokyo, Japan). The intestinal villi height and basement membrane thickness of each segment was measured with Image-Pro software.

### 2.10. Statistical Analysis

All the experimental data were tested for normality and homogeneity of variances using the Shapiro-Wilk's test and Levene's test, respectively, and presented as the mean ± SEM. Significant differences were determined using the one-way analysis of variance (ANOVA) test, followed by Fisher's least significant difference post hoc test and Duncan's multiple range tests, after confirming data normality and homogeneity of variances. Statistical analysis was performed using SPSS software 19.0 (SPSS Inc., New York, NY, USA) and the Windows-based Graph pad prism statistical software (San Diego, CA, USA). A $p$ value less than 0.05 was accepted as statistically significant.

## 3. Results

### 3.1. Proportion of Pellet Feed Recipients

The POPFR of mandarin fish in different feeding groups (PFD, PFDCB, PFDLR and PFDLP) was tested on the 7th, 14th and 28th day of feeding. As shown in Figure 1, on the 28th day of feeding, the POPFR of mandarin fish in the PFDLP, PFDLR and PFDCB groups was higher than that in the PFD group, and the PFDLP and PFDCB groups reached a significant level of difference ($p < 0.05$). The highest POPFR of mandarin fish was recorded in PFDLP (81%) compared to PFD (68%) on the 28th day of feeding.

### 3.2. Survival Analysis

Mandarin fish fed with pellet feed without probiotics supplemented had a lower survival rate than those fed with live bait at the end of the experiment (Figure 2). Application of *L. plantarum*, *L. rhamnosus* and *C. butyricum* significantly reduced the decrease of the survival rate of mandarin fish caused by the pellet feed diet at the end of the experiment (Figure 2).

### 3.3. Serum Parameter Analysis

#### 3.3.1. Serum Lysozyme Content

Mandarin fish in the PFD group had lower serum lysozyme content than that in the LBFD group at days 7, 14 and 28 of feeding (Figure 3). The effects of *L. plantarum*, *L. rhamnosus* and *C. butyricum* on serum lysozyme content are shown in Figure 3. Application of *L. plantarum*, *L. rhamnosus* and *C. butyricum* reduced the decrease of the serum lysozyme content of mandarin fish caused by the pellet feed diet (Figure 3). Compared with the PFD group, the content of serum lysozyme increased significantly on the 7th and 28th day in the PFDLP group and on the 14th day in the PFDLR group ($p < 0.05$) (Figure 3). The highest serum lysozyme content of mandarin fish was noticed in PFDLP after being fed for 28 days (Figure 3).

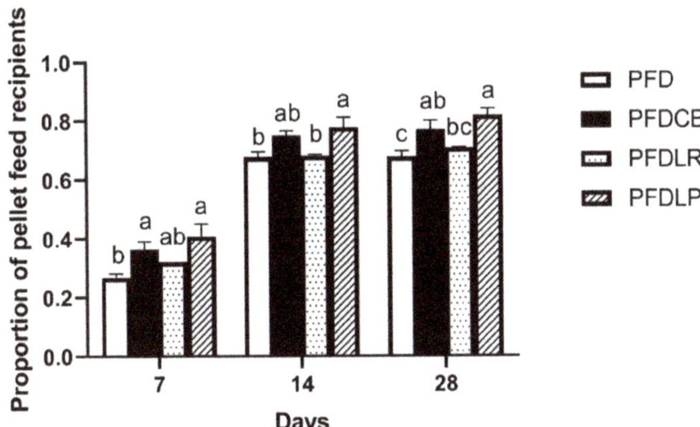


**Figure 1.** POPFR of mandarin fish in different feeding groups (PFD, PFDCB, PFDLR and PFDLP) at days 7, 14 and 28 of feeding. Data are presented as mean ± SEM ($n$ = 3). Abbreviations: PFD, pellet-feed feeding group with probiotics free; PFDCB, pellet-feed feeding group with *C. butyricum* plus; PFDLR, pellet-feed feeding group with *L. rhamnosus* plus; PFDLP, pellet-feed feeding group with *L. plantarum* plus. A value followed by a lowercase superscript (a–c) differs significantly from all other values not followed by the same lowercase superscript at the same time point based on ANOVA followed by the post hoc test ($p < 0.05$).

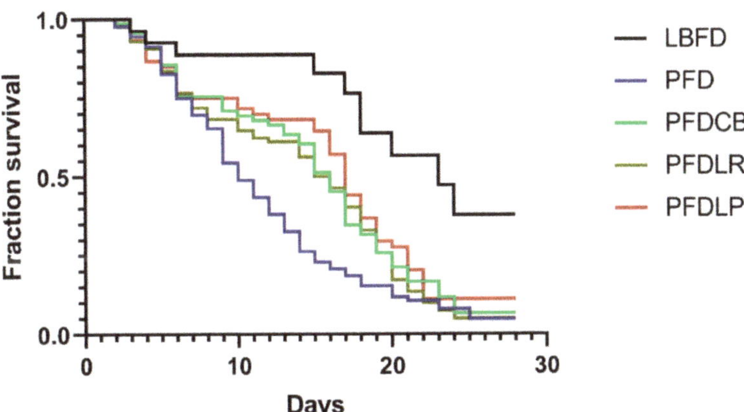


**Figure 2.** Kaplan Meyer's (KM) survival analysis of mandarin fish in different feeding groups (LBFD, PFD, PFDCB, PFDLR and PFDLP) during 28 days of feeding. Abbreviations: LBFD, live bait fish feeding group; PFD, pellet-feed feeding group with probiotics free; PFDCB, pellet-feed feeding group with *C. butyricum* plus; PFDLR, pellet-feed feeding group with *L. rhamnosus* plus; PFDLP, pellet-feed feeding group with *L. plantarum* plus.

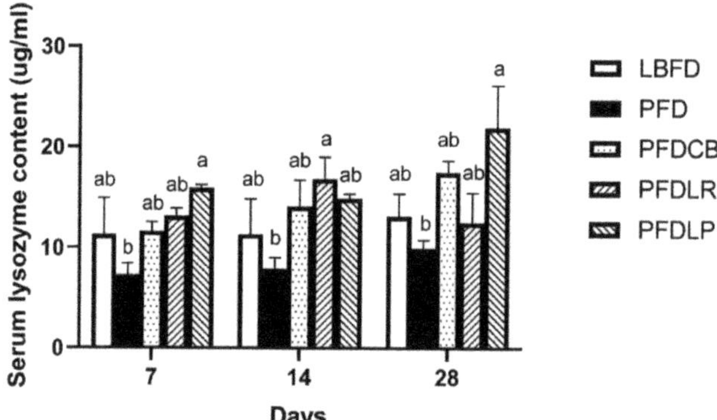


**Figure 3.** Serum lysozyme content of mandarin fish in different feeding groups (LBFD, PFD, PFDCB, PFDLR and PFDLP) at day 7, 14 and 28 of feeding. Data are presented as mean ± SEM ($n$ = 9). Abbreviations: LBFD, live bait fish feeding group; PFD, pellet-feed feeding group with probiotics free; PFDCB, pellet-feed feeding group with *C. butyricum* plus; PFDLR, pellet-feed feeding group with *L. rhamnosus* plus; PFDLP, pellet-feed feeding group with *L. plantarum* plus. A value followed by a lowercase superscript (a–b) differs significantly from all other values not followed by the same lowercase superscript at the same time point based on ANOVA followed by the post hoc test ($p < 0.05$).

3.3.2. Measurement of IgM and CRP

The effects of *L. plantarum*, *L. rhamnosus* and *C. butyricum* on serum IgM and CRP content are shown in Figure 4. Although the application of *L. plantarum*, *L. rhamnosus* and *C. butyricum* reduced the decrease of the serum IgM level of mandarin fish at days 14 and 28 of feeding (Figure 4A), serum CRP content was not significantly affected by pellet feed and probiotics application (Figure 4B). Compared with the PFD group, the serum IgM content of the PFDLP group supplemented with *L. plantarum* was significantly increased at days 14 and 28 of feeding ($p < 0.05$) (Figure 4A).

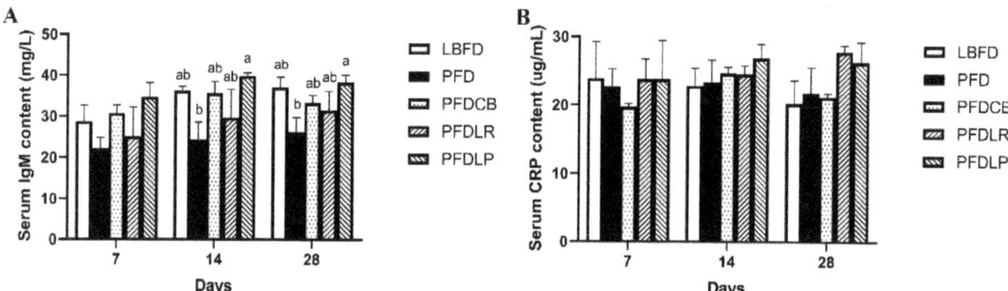


**Figure 4.** Serum content of IgM (**A**) and CRP (**B**) of mandarin fish in different feeding groups (LBFD, PFD, PFDCB, PFDLR and PFDLP) at day 7, 14 and 28 of feeding. Data are presented as mean ± SEM ($n$ = 9). Abbreviations: IgM, immunoglobulin M; CRP, C-reactive protein; LBFD, live bait fish feeding group; PFD, pellet-feed feeding group with probiotics free; PFDCB, pellet-feed feeding group with *C. butyricum* plus; PFDLR, pellet-feed feeding group with *L. rhamnosus* plus; PFDLP, pellet-feed feeding group with *L. plantarum* plus. A value followed by a lowercase superscript (a–b) differs significantly from all other values not followed by the same lowercase superscript at the same time point based on ANOVA followed by the post hoc test ($p < 0.05$).

*3.4. Antioxidant and Oxidative Stress Parameters*

GSH content and CAT activity in liver, gut and gill of mandarin fish in the PFD group were decreased compared with that in the LBFD group (Figure 5B,C), while the MDA level in liver, gut and gill was increased (Figure 5D). Compared with the LBFD group, the decrease of GSH content in gut, the decrease of CAT activity in gill and the increase of MDA content in gill in the PFD group reached significant difference levels ($p < 0.05$) (Figure 5B–D). The application of *L. plantarum*, *L. rhamnosus* and *C. butyricum* motivated an elevation of GSH content and CAT activity (Figure 5B,C) and a reduced MDA content in the liver, gut and gill of mandarin fish in the PFDLP, PFDLR and PFDCB groups when compared to the PFD group (Figure 5D). The content of GSH in liver and gill of mandarin fish in the PFDLP group treated with *L. plantarum* was significantly higher than that in the PFD group ($p < 0.05$) (Figure 5B). Compared with the mandarin fish in the PFD group, application of *L. plantarum*, *L. rhamnosus* and *C. butyricum* significantly increased CAT activity in liver and MDA content in gill ($p < 0.05$) (Figure 5C,D).

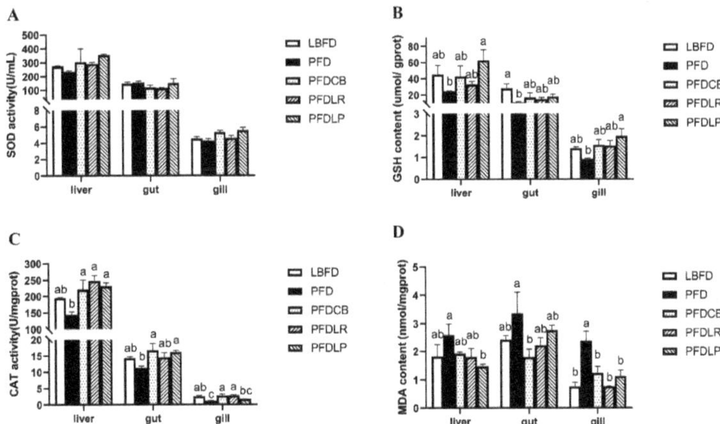


**Figure 5.** Activities of SOD (**A**), GSH (**B**) and CAT (**C**) and content of MDA (**D**) in the gut, liver and gills of mandarin fish in different feeding groups (LBFD, PFD, PFDCB, PFDLR and PFDLP) after being fed for 28 days. Data are presented as mean ± SEM ($n = 9$). Abbreviations: SOD, superoxide dismutase; GSH, glutathione; CAT, Catalase; MDA, malondialdehyde; LBFD, live bait fish feeding group; PFD, pellet-feed feeding group with probiotics free; PFDCB, pellet-feed feeding group with *C. butyricum* plus; PFDLR, pellet-feed feeding group with *L. rhamnosus* plus; PFDLP, pellet-feed feeding group with *L. plantarum* plus. A value followed by a lowercase superscript (a–c) differs significantly from all other values not followed by the same lowercase superscript at the same time point based on ANOVA followed by the post hoc test ($p < 0.05$).

### *3.5. Expression of Appetite-Related Genes*

For appetite control genes expression in the brain and gut of mandarin fish in different feeding groups (LBFD, PFD, PFDCB, PFDLR and PFDLP) after being fed for 28 days, we found a significantly increased mRNA level of *leptin* in the gut of mandarin fish and a significantly decreased mRNA level of *npy* and *agrp* in the brain of mandarin fish in the PFD group compared to the LBFD group ($p < 0.05$) (Figure 6A,B). After applying *L. plantarum*, *L. rhamnosus* and *C. butyricum*, the *leptin* expression levels in the mandarin fish gut were significantly down-regulated in the PFDLP and PFDCB groups compared with that in the PFD group ($p < 0.05$) (Figure 6A,B). The *agrp* expression levels in the mandarin fish brain were significantly up-regulated in the PFDLP, PFDLR and PFDCB groups compared with that in the PFD group ($p < 0.05$) (Figure 6A,B).

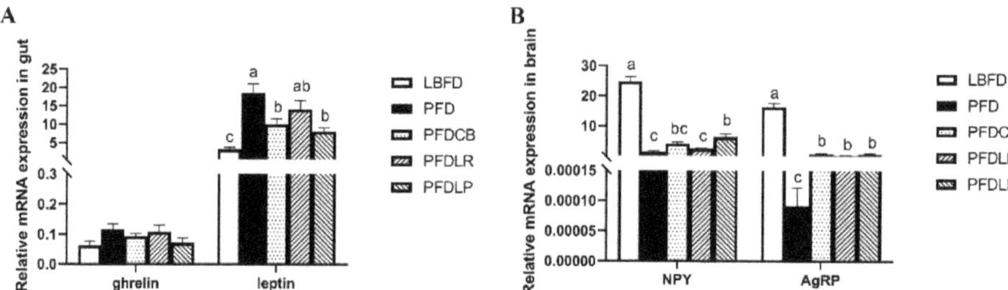


**Figure 6.** Relative mRNA expressions of appetite control genes in the gut (**A**) and brain (**B**) of mandarin fish in different feeding groups (LBFD, PFD, PFDCB, PFDLR and PFDLP) after being fed for 28 days. Data are presented as mean ± SEM ($n$ = 9). Abbreviations: NPY, *nerve peptide y*; AgRP, *agouti gene-related protein*; LBFD, live bait fish feeding group; PFD, pellet-feed feeding group with probiotics free; PFDCB, pellet-feed feeding group with *C. butyricum* plus; PFDLR, pellet-feed feeding group with *L. rhamnosus* plus; PFDLP, pellet-feed feeding group with *L. plantarum* plus. Data are presented as mean ± SEM ($n$ = 9). A value followed by a lowercase superscript (a–c) differs significantly from all other values not followed by the same lowercase superscript at the same time point based on ANOVA followed by the post hoc test ($p < 0.05$).

## *3.6. Gut Microbiota Analysis*

### 3.6.1. Richness and Diversity

The alpha diversity index, including observed species, Shannon, Simpson, Chao 1, ACE and PD whole tree, was calculated to assess the diversity and richness of intestinal microbiota of mandarin fish in different groups. No significant difference was observed in the Shannon and Simpson indices between groups ($p < 0.05$) (Table 2). The observed species, Chao1, ACE and PD whole tree indices of the PFDLP group were higher than that of other groups, and there was significant difference compared with the LBFD and PFDCB groups ($p < 0.05$) (Table 2). A Venn diagram was constructed to identify the core and different OTUs existing in mandarin fish under different feeding strategies. In this regard, 168 OTUs were shared among all mandarin fish gut samples. In contrast, 430 OTUs, 534 OTUs, 661 OTUs, 269 OTUs and 176 OTUs were unique to LBFD, PFD, PFDLP, PFDLR and PFDCB groups, respectively (Figure 7). Simultaneously, the intestinal microbiota community structure was further investigated using PCoA based on the binary jaccard distance (Figure 8). PCoA analysis showed 16.8% and 12.16% explained variance of principal component analysis PCoA1 and PCoA2, respectively. PCoA cluster analysis indicated that three clusters were formed and separated between the bait fish diet group (LBFD), pellet feed group (PFD) and probiotic-treated pellet feed group (PFDLP, PFDLR and PFDCB) after being fed for 28 days (Figure 8). This suggested that different feeding strategies of mandarin fish led to different intestinal community structures (Figure 8).

**Table 2.** Richness and diversity indices of mandarin fish intestinal microbial populations in different feeding groups (LBFD, PFD, PFDCB, PFDLR and PFDLP) after being fed for 28 days.

| Index | LBFD | PFD | PFDCB | PFDLR | PFDLP |
|---|---|---|---|---|---|
| Observed species | 418.33 ± 107.84 $^{c}$ | 721.00 ± 66.55 $^{a,b}$ | 435.75 ± 18.91 $^{c}$ | 493.75 ± 85.17 $^{b,c}$ | 777.25 ± 35.03 $^{a}$ |
| Shannon | 2.61 ± 125.16 $^{a}$ | 2.85 ± 39.55 $^{a}$ | 1.88 ± 22.37 $^{a}$ | 1.62 ± 126.45 $^{a}$ | 2.85 ± 12.40 $^{a}$ |
| Simpson | 0.71 ± 151.35 $^{a}$ | 0.59 ± 47.19 $^{a}$ | 0.49 ± 16.06 $^{a}$ | 0.41 ± 131.67 $^{a}$ | 0.55 ± 29.14 $^{a}$ |
| Chao 1 | 579.12 ± 0.51 $^{b,c}$ | 859.22 ± 0.10 $^{a,b}$ | 556.92 ± 0.34 $^{c}$ | 692.79 ± 0.35 $^{a,b,c}$ | 918.94 ± 0.33 $^{a}$ |
| ACE | 612.28 ± 0.07 $^{b}$ | 902.63 ± 0.07 $^{a,b}$ | 588.27 ± 0.12 $^{b}$ | 722.63 ± 0.12 $^{a,b}$ | 993.75 ± 0.05 $^{a}$ |
| PD whole tree | 68.41 ± 15.70 $^{b}$ | 85.20 ± 5.46 $^{b}$ | 80.21 ± 16.99 $^{b}$ | 169.72 ± 44.11 $^{a,b}$ | 199.00 ± 36.99 $^{a}$ |

ACE: abundance-based coverage estimator; LBFD, live bait fish feeding group; PFD, pellet-feed feeding group with probiotics free; PFDCB, pellet-feed feeding group with *C. butyricum* plus; PFDLR, pellet-feed feeding group with *L. rhamnosus* plus; PFDLP, pellet-feed feeding group with *L. plantarum* plus. The numbers represent the mean ± SEM ($n$ = 3). A value followed by a lowercase superscript (a–c) differs significantly from all other values not followed by the same lowercase superscript at the same time point based on ANOVA followed by the post hoc test ($p < 0.05$).

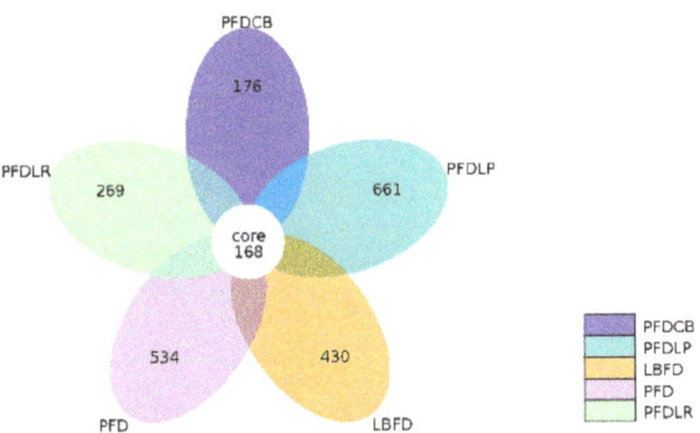


**Figure 7.** Venn diagram analysis depicting the numbers of shared and unique OTUs of mandarin fish intestinal microbial populations in different feeding groups (LBFD, PFD, PFDCB, PFDLR and PFDLP) after being fed for 28 days. Abbreviations: LBFD, live bait fish feeding group; PFD, pellet-feed feeding group with probiotics free; PFDCB, pellet-feed feeding group with *C. butyricum* plus; PFDLR, pellet-feed feeding group with *L. rhamnosus* plus; PFDLP, pellet-feed feeding group with *L. plantarum* plus.

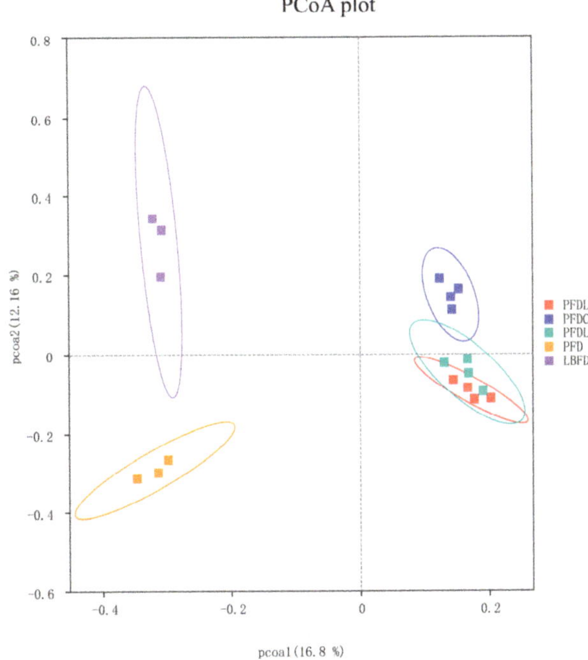


**Figure 8.** PCoA based on the binary jaccard distance of the intestinal bacterial communities of mandarin fish in different feeding groups (LBFD, PFD, PFDCB, PFDLR and PFDLP) after being fed for 28 days. Abbreviations: LBFD, live bait fish feeding group; PFD, pellet-feed feeding group with probiotics free; PFDCB, pellet-feed feeding group with *C. butyricum* plus; PFDLR, pellet-feed feeding group with *L. rhamnosus* plus; PFDLP, pellet-feed feeding group with *L. plantarum* plus.

### 3.6.2. Community Composition and Biomarker Analysis

The gut microbiota of mandarin fish in different feeding groups (LBFD, PFD, PFDCB, PFDLR and PFDLP) showed their unique microbial population structure. At the phylum and genus level, the top 10 abundant microbiota composition in the intestine of mandarin fish in different feeding groups (LBFD, PFD, PFDCB, PFDLR and PFDLP) after being fed for 28 days is represented in Figure 9. The gut microbiota of mandarin fish in the LBFD group was dominated by *Fusobacteriota* and *Proteobacteria* at the phylum level, and *Proteobacteria* was the dominant phylum in the gut microbiota of the PFD, PFDLP, PFDLR and PFDCB groups (Figure 9A). The abundance of *Aeromonas* in the PFDLP, PFDLR and PFDCB groups was significantly lower than in the PFD group (Figure 9B). LEfSe analysis revealed 19, 24, 17, 7 and 1 biomarkers with significantly higher relative abundance in the LBFD, PFD, PFDLP, PFDLR and PFDCB groups, respectively (Figure 10A). *Aeromonas* was a biomarker for PFD compared with other groups (Figure 10B).

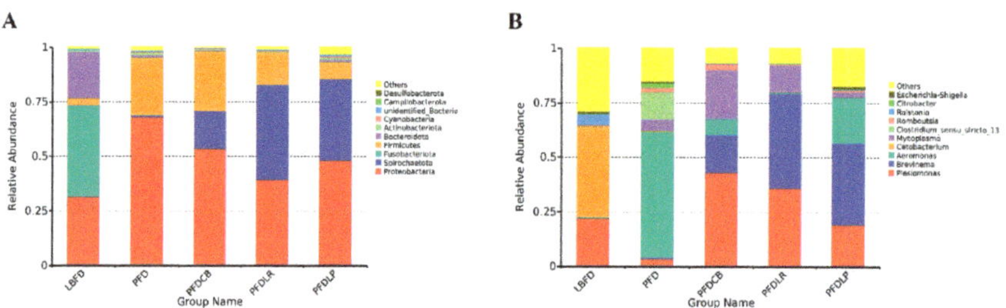


**Figure 9.** The abundance of composition at phylum (**A**) and genus (**B**) level in mandarin fish intestinal microbial populations in different feeding groups (LBFD, PFD, PFDCB, PFDLR and PFDLP) after being fed for 28 days. Abbreviations: LBFD, live bait fish feeding group; PFD, pellet-feed feeding group with probiotics free; PFDCB, pellet-feed feeding group with *C. butyricum* plus; PFDLR, pellet-feed feeding group with *L. rhamnosus* plus; PFDLP, pellet-feed feeding group with *L. plantarum* plus.

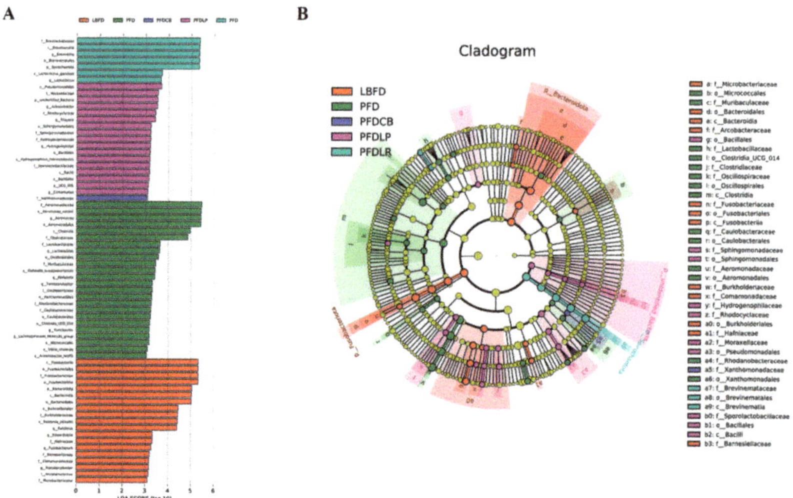


**Figure 10.** Intergroup variation in the relative abundance of the intestinal microbial communities. (**A**) Cladogram of LEfSe. (**B**) Bacterial taxa differentially displayed in the mandarin fish intestinal microbial populations in different feeding groups (LBFD, PFD, PFDCB, PFDLR and PFDLP) after being fed for 28 days were identified via LEfSe using an LDA score threshold of >3. Abbreviations: LBFD, live bait fish feeding group; PFD, pellet-feed feeding group with probiotics free; PFDCB, pellet-feed feeding group with *C. butyricum* plus; PFDLR, pellet-feed feeding group with *L. rhamnosus* plus; PFDLP, pellet-feed feeding group with *L. plantarum* plus.

### 3.6.3. Functional Prediction

Functional prediction on the KEGG database was annotated based on 16S sequencing data. As shown in Figure 11, the abundance of functional categories based on KEGG (level 1) between different feeding groups (LBFD, PFD, PFDCB, PFDLR and PFDLP) after being fed for 28 days were analyzed. The abundance of human pathogens pneumonia and human pathogens nosocomial significantly increased in the PFD group compared with other groups ($p < 0.05$) (Figure 11).

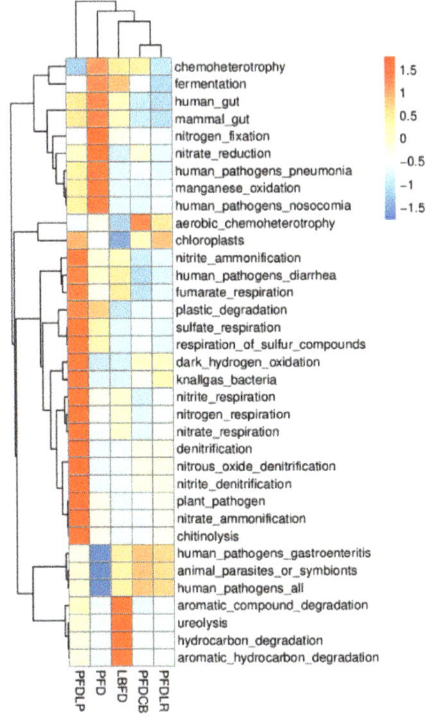


**Figure 11.** Heatmap showing the relative abundances of KEGG ortholog groups of mandarin fish intestinal microbial populations in different feeding groups (LBFD, PFD, PFDCB, PFDLR and PFDLP) after being fed for 28 days. The heatmap was made based on Tax4Fun functional annotations, and the color intensity indicates the abundance information. Abbreviations: LBFD, live bait fish feeding group; PFD, pellet-feed feeding group with probiotics free; PFDCB, pellet-feed feeding group with *C. butyricum* plus; PFDLR, pellet-feed feeding group with *L. rhamnosus* plus; PFDLP, pellet-feed feeding group with *L. plantarum* plus.

### 3.7. *Intestinal Histological Assessment*

Histological changes of the intestinal tract were observed in different feeding groups (LBFD, PFD, PFDCB, PFDLR and PFDLP) after being fed for 28 days (Figure 12). By comparing LBFD, PFD, PFDCB, PFDLR and PFDLP, the result showed that *C. butyricum*, *L. rhamnosus* and *L. plantarum* could significantly increase the villi height of the foregut, midgut and hindgut of mandarin fish fed with pellet feed ($p < 0.05$; Figure 13A) and significantly reverse the decrease in the thickness of foregut basement membrane caused by pellet-feed feeding ($p < 0.05$; Figure 13B).

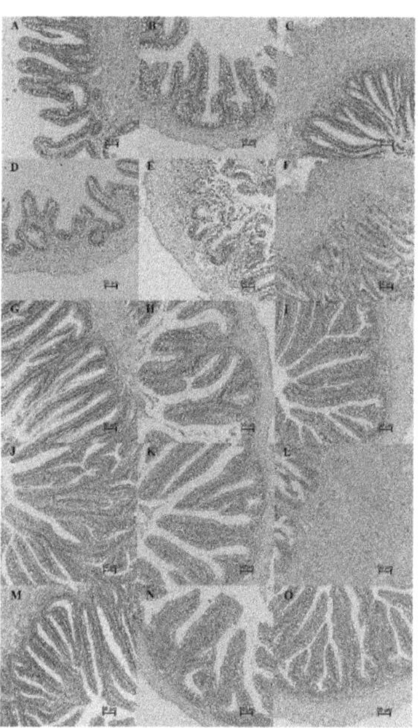

**Figure 12.** Photomicrographs showing histological sections of the intestinal tract of mandarin fish in different feeding groups (LBFD, PFD, PFDCB, PFDLR and PFDLP) after being fed for 28 days. (H.E. staining; scale bar: 50 μm; magnification ×200). (**A–C**) Foregut, midgut and hindgut of mandarin fish in LBFD group. (**D–F**) Foregut, midgut and hindgut of mandarin fish in PFD group. (**G–I**) Foregut, midgut and hindgut of mandarin fish in PFDCB group. (**J–L**) Foregut, midgut and hindgut of mandarin fish in PFDLR group. (**M–O**) Foregut, midgut and hindgut of mandarin fish in PFDLP group. Abbreviations: H.E., hematoxylin and eosin staining; LBFD, live bait fish feeding group; PFD, pellet-feed feeding group with probiotics free; PFDCB, pellet-feed feeding group with *C. butyricum* plus; PFDLR, pellet-feed feeding group with *L. rhamnosus* plus; PFDLP, pellet-feed feeding group with *L. plantarum* plus.

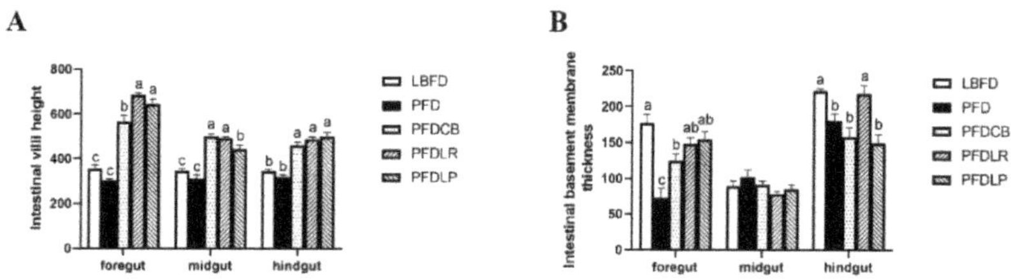


**Figure 13.** Intestinal villi height (**A**) and basement membrane thickness (**B**) of mandarin fish in different feeding groups (LBFD, PFD, PFDCB, PFDLR and PFDLP) after being fed for 28 days. Data are presented as mean ± SEM ($n$ = 3). Abbreviations: LBFD, live bait fish feeding group; PFD, pellet-feed feeding group with probiotics free; PFDCB, pellet-feed feeding group with *C. butyricum* plus; PFDLR, pellet-feed feeding group with *L. rhamnosus* plus; PFDLP, pellet-feed feeding group with *L. plantarum* plus. A value followed by a lowercase superscript (a–c) differs significantly from all other values not followed by the same lowercase superscript at the same time point based on ANOVA followed by the post hoc test ($p < 0.05$).

## 4. Discussion

Mandarin fish have a food preference for live bait and show certain eating disorders with dead bait fish or pellet feed. The increased mortality rate of mandarin fish under pellet feeding conditions seriously affects its economic benefits [10,66]. The eating disorder is characterized by abnormal feeding behaviors associated with the establishment of food preference [1–3,6]. Gut microbiota can regulate host food preferences through interactions with nutritional, immune, antioxidant stress and appetite levels [26,85]. Previous studies have shown that probiotics can influence the feeding behavior of the host by regulating the microbiota [26,27,86]. However, few studies have been done on the regulatory effect of probiotics on eating disorders, especially on the pellet feed intake of mandarin fish [87]. Therefore, the present study was conducted to assess the effects of *L. plantarum*, *L. rhamnosus* and *C. butyricum* on the POPFR, survival, appetite, gut microbiota, innate immunity, antioxidant capacity and intestinal histology in mandarin fish and to explore the role of probiotics in regulating feeding behavior in vivo.

Acceptance of pelleted feed and survival rate are direct indicators of the improvement of eating disorders during the feeding process of mandarin fish with pellet feed. In this study, we observed that supplementation with either of the three probiotics effectively increased the POPFR in mandarin fish compared to those fed the same diet but without probiotics supplementation. Moreover, pellet feed diet can lead to the reduction of survival rate of mandarin fish, which is consistent with the previous report that the dietary conversion of *Sparus aurata* larvae and *Solea senegalensis* larvae from live bait to alginate microdiets resulted in a significant decrease in survival rate, which may be related to the changes of physiological stress and nutritional status of the larvae fish [88–90]. Results showed that *L. plantarum*, *L. rhamnosus* and *C. butyricum* can significantly reverse the increase in mortality of mandarin fish caused by feeding pellets at the end of the 28-day experiment. This finding is consistent with a previous study in which the administration of *L. plantarum* to the rainbow trout at a dose of $10^6$ CFU/g for 36 consecutive days significantly improved the survival rate of rainbow trout when attacked by *Lactococcus garvieae* [91]. Similarly, Hooshyar reported that *L. rhamnosus* ATCC 7469 significantly increased the survival rate of rainbow trout (*Oncorhynchus mykiss*) when attacked by *Yersinia ruckeri* [36]. Duan reported that supplementation of *C. butyricum* ($1 \times 10^9$ CFU/g) for 56 days improved the survival of black tiger shrimp (*Penaeus monodon*) after exposure to nitrite stress for 24 and 48 h [92]. Proper nutrition can affect intestinal health through several pathways, including intestinal morphology, microbial diversity, intestinal barriers and oxidative status [93]. The improvement of survival of cultured animals after applying *L. rhamnosus*, *L. plantarum* and *C. butyricum* may result from their positive regulation of nutritional status, intestinal morphology, gut microbiota, oxidative status and immune system [38,94,95]. Therefore, the administration of probiotics may be a potential method to improve the eating disorders of mandarin fish caused by pellet feed and increase the POPFR of mandarin fish without side effects because probiotics such as *L. plantarum*, *L. rhamnosus* and *C. butyricum* are generally regarded as safe for aquatic animals.

Appetite is one reason influencing the eating preference of mandarin fish [96]. Feeding behavior is ultimately regulated by central feeding centers of the brain, which receive and process information from endocrine signals from both the brain and periphery. These signals, such as hormones that inhibit (e.g., leptin) or increase (e.g., Agrp) ingestion, provide information about nutritional status and ingestion [97–99]. Npy is considered the most potent orexigenic molecule in fish, mediated by gut microbiota changes [100,101]. Agrp is one of the most potent appetite stimulants within the hypothalamus and mediates the peripheral body weight regulators such as ghrelin and leptin [100,102]. In the present study, we observed that *L. rhamnosus*, *L. plantarum* and *C. butyricum* could reverse the decrease of *agrp* expression in the brain tissue of mandarin fish caused by pellet-feed feeding. At the peripheral level, ghrelin is a potent appetite stimulant and is highly expressed in the fish gut [103,104]. Furthermore, the gastrointestinal hormone ghrelin is a vital molecule that regulates intestinal motility and secretion [105,106]. Leptin plays an anorexic role by down-regulating orexigenic signals such as Npy [107]. This study showed that the treatment of *L. rhamnosus, L. plantarum* and *C. butyricum* can reverse the high expression of the peripheral hormone leptin in the intestinal tissue of mandarin fish caused by feed-pellet feeding. These results agree with the previous findings on the regulation of appetite of *L. rhamnosus* on larval Nile tilapia [38]. All this indicates that probiotics treatment can promote the appetite of pellet feeding mandarin fish through reducing the expression of the peripheral appetite-suppressing hormone leptin and increasing the expression of the central appetite-promoting factor Npy/Agrp. Previous studies have shown that the gut microbiota can affect host appetite and eating behavior by directly affecting nutrient sensing and the satiety regulation system [26]. In this study, the appetite-promoting effect of *L. rhamnosus, L. plantarum* and *C. butyricum* may be mediated by their regulation on the gut microbiota of mandarin fish.

The immune system can influence eating behavior through interactions with gut bacteria and appetite [108,109]. As lower vertebrates, fish mainly rely on the innate immune system to resist pathogens [110]. Lysozyme is responsible for bacterial lysis and activation of phagocytes and complement systems [111]. IgM mainly exists in the serum, which is the most essential component of teleost humoral immunity, and it can recognize, bind and precipitate antigens and activate the complement system [112]. To assess if *L. rhamnosus, L. plantarum, C. butyricum* affects the immune system of feed-fed mandarin fish, we measured the levels of lysozyme and IgM in the serum. We found that at the end of 28 days of cultivation, the three probiotics can increase the reduction of mandarin fish serum lysozyme and IgM content caused by pellet feed domestication, and *L. plantarum* is the most significant. All this is similar to the finding in a previous publication suggesting that the feed supplement of *L. plantarum* CCFM8661 restored the decrease in serum lysozyme of Nile tilapia caused by waterborne Pb exposure [113]. In Wang's study, administration of *C. butyricum* significantly increased the serum IgM levels in piglets on day 28 [114]. Liao et al. have confirmed that a diet supplemented with *C. butyricum* increased the IgM concentration compared with that of chicks in the control group at 21 and 42 days old [115]. This study proved that the addition of *L. plantarum, L. rhamnosus* and *C. butyricum* reversed the decrease in serum lysozyme and IgM content caused by pellet-feed feeding, which may further ameliorate eating disorders by regulating the appetite and gut microbiota of mandarin fish.

Anti-oxidative enzymes are the major components of anti-oxidative defense systems in living organisms [116]. The host gut microbiota directly or indirectly influences the central nervous system by affecting local OS levels and the permeability of the gut and then influences the behavioral characteristics of the host. SOD, CAT and GSH are considered the three main antioxidant enzymes in the primary antioxidant defense system, eliminating ROS in the body during oxidative damage [117]. MDA is an essential product of membrane lipid peroxidation and a well-known aging indicator reflecting the degree of oxidative stress in cells [118]. In this study, compared with the LBFD group, the decrease of GSH content in gut, the decrease of CAT activity in gill and the increase of MDA content in gill in the PFD group reached significant difference levels ($p < 0.05$), indicating that the pellet diet induced oxidative stress in mandarin fish, which is in accord with the results found in *Solea senegalensis* larvae and hybrid mandarin fish [88,119]. The low nutritional status and stress caused by the pellet diet may decrease antioxidant capacity in mandarin fish [88,120–124]. Furthermore, compared with the mandarin fish in the PFD group, application of *L. plantarum, L. rhamnosus* and *C. butyricum* significantly increased CAT activity in liver and MDA content in gill ($p < 0.05$). Increased CAT activity and GSH content accompanied by decreased MDA levels was observed after the application of three probiotics compared with the PFD group, which indicates that *L. rhamnosus, L. plantarum* and *C. butyricum* could enhance the antioxidant capacity of the host, which is consistent with the findings in rainbow trout, the black tiger shrimp (*Penaeus monodon*) and Nile tilapia [36,92,113,125]. The three kinds of probiotics showed an excellent free radical scavenging ability in the oxidative damage of the liver, intestine, and gill tissues, which may be attributed to its ability in gut microbiota and immune system regulation.

It has been reported that gut microbiota plays a causal role in regulating the feeding behavior of the host and can directly or indirectly affect the appetite and food intake of the host [26,126]. The composition of the intestinal microbiome is influenced by both host genotype and environment. Previous studies have shown that the gut microbiota of aquatic species is influenced by several abiotic factors [127,128]. Diet is considered one way to change the gut microbiota and the exogenous factors affecting the gut microbiota [129–138]. In this study, compared with the live bait fish diet, the pellet feed diet changed the intestinal colony structure of mandarin fish, which may be mainly caused by changes in the dietary structure and also be affected by environmental stress (including dietary stress) [139]. Disturbance of gut microbiota balance could lead to the establishment of harmful bacteria, causing disease problems [140–143]. In addition to diet, probiotic treatments can also affect the gut microbiome [144,145]. Probiotics play an essential role in the welfare of the host by maintaining a healthier balance of intestinal microbiota, which provides a defensive barrier against colonization of harmful bacteria and stimulates the immune system [146–148]. In this study, the addition of three probiotics significantly reduced the increased abundance of pathogenic bacteria *Aeromonas* caused by pellet-feed feeding, which may be achieved through the direct competition of probiotics on the abundance of pathogenic bacteria and indirect regulation of host immunity. According to reports, lactic acid bacteria inhibit the growth of harmful bacteria by producing antimicrobial compounds and competing for nutrients and attachment sites [41,149]. The present result agrees with earlier findings where a similar decrease in pathogenic bacteria (*Aeromonas* sp. and *Pseudomonas* sp.) was reported in giant freshwater prawn (*Macrobrachium rosenbergii*) feeding with a diet supplemented with *L. plantarum* [150]. This result is also consistent with the early discovery which reported that *L. rhamnosus* micro-granules administered for 30 days to tilapia larvae could significantly reduce the proportion of potentially pathogenic bacteria [38]. In addition, *C. butyricum* treatment reversed the increased abundance of intestinal pathogens in mice induced by severe acute pancreatitis and intra-abdominal hypertension [73]. All this indicates that these three probiotics can inhibit the abundance of harmful intestinal bacteria *Aeromonas* in the in vivo model resulting from direct competition between probiotics and pathogenic bacteria and host immunity regulation.

The intestine is the leading site of nutrient absorption, and the health of villi is a crucial factor influencing nutrient absorption. Consistent with the description of Wu et al. on the histological and histochemical characterization of mandarin fish tissues and organs, in our study, mandarin fish fed with live bait showed a conventional histological pattern of intestinal tissue [151]. In contrast, histological changes were detected in mandarin fish fed with pellet feed. Compared to mandarin fish fed on the live feed, the thickness of the foregut basement membrane in pellet feed-fed mandarin fish was significantly reduced, with similar results in other fish [88,152,153]. In addition, our results indicated that dietary supplement of *L. plantarum*, *L. rhamnosus* and *C. butyricum* enhanced the intestinal health development in mandarin fish by increasing the height of intestinal villi and the thickness of foregut basement membrane. Similarly, *L. plantarum* favorably recovered the cyclophosphamide-induced abnormal intestinal morphology in mice by improving the villus height [154]. Pangasius catfish (*Pangasius bocourti*) fed a diet supplemented with *L. plantarum* for 90 days exhibited a greater villus height in all intestines, with significant differences in the proximal intestine [155]. Wang et al. reported that *C. butyricum* increased the jejunal villus length and jejunal villus height to crypt depth ratio, while they decreased the jejunal crypt depth compared with those of the control and protected the intestinal villi morphology in a piglet model [114]. According to Sewaka et al., *L. rhamnosus* increased the villous height in the proximal, middle and distal parts of the intestine of juvenile red tilapia (*Oreochromis* spp.) [37]. Moreover, Casas et al. reported that the intestinal villus height of weanling pigs tended to increase as the dose of *C. butyricum* increased in the diet [94]. Our findings indicate that the application of probiotics could effectively promote the intestinal health of mandarin fish fed with pellet feed, which may benefit from repair of the intestinal microbial barrier. At the same time, the promoting effect of probiotics on intestinal health may be one of the reasons for the improvement of survival rate of mandarin fish fed with pellet feed.

## 5. Conclusions

In summary, the present results confirmed that the application of *L. plantarum*, *L. rhamnosus* and *C. butyricum* could significantly improve the eating disorders of mandarin fish caused by pellet-feed feeding, which expressed as significantly increased POPFR and survival rate. All of these may be related to the ability of probiotics to regulate gut microbiota, activate immunity, boost appetite, improve antioxidant capacity and protect intestinal tissues. This study explores the problem of eating disorders in non-mammals and tried to solve the eating disorders caused by pellet-feed feeding of mandarin fish by regulating gut microbiota using probiotics. In this study, the influence of probiotics intervention on eating disorders and its mechanism were studied using mandarin fish fed with pellet feed as a model. Due to the complex interactions between the gut microbiota, immune system, appetite and oxidative stress, the causal relationship between them needs to be further investigated. The conversion of pellet feed for mandarin fish has always been considered a global problem, and this study provides a new train of thought. More solutions, such as the application of other probiotics, prebiotics or immunostimulants, are worth investigating.

**Author Contributions:** Conceptualization, X.C. and G.L.; methodology, X.C.; software, X.C.; validation, X.C., H.Y. and S.L.; formal analysis, X.C. and Y.Z.; investigation, X.C. and Y.S.; resources, X.C. and X.L.; data curation, X.C. and S.B.; writing—original draft preparation, X.C.; writing—review and editing, X.C. and H.L.; visualization, X.C. and Z.Z.; supervision, X.C.; project administration, X.C.; funding acquisition, G.L. All authors have read and agreed to the published version of the manuscript.

**Funding:** This research was funded by the Guangdong Province Key Field R&D Program Project, grant number 20200202; Guangdong Basic and Applied Basic Research Foundation, grant number 2019B1515120072; the State Guides the Local Science and Technology Development Special Funding Project in 2021 (Construction of Germ plasm Resource Bank of *Micropterus salmoides* and *Siniperca chuatsi* in Guangdong Province); the Research in Spreading and Breeding Feedstuff of a New Breed of *Siniperca chuatsi* × *Siniperca scherzeri*, grant number Yuenong 2019A2.

**Institutional Review Board Statement:** The study was conducted according to the guidelines of the Declaration of Helsinki, and approved by the Institutional Animal Care and Use Committee of Sun Yat-sen University (SYSU-IACUC-2020-B0423).

**Informed Consent Statement:** Not applicable.

**Data Availability Statement:** The data presented in this study are available on request from the corresponding author.

**Acknowledgments:** The authors gratefully acknowledge the aquaculture farm (Foshan, Guangdong) for providing fry and breeding help. The author would like to thank the laboratory instrument management teacher for help in using the instrument.

**Conflicts of Interest:** The authors declare no conflict of interest.

## References

1. Cooper, Z.; Grave, R.D. Eating Disorders: Transdiagnostic Theory and Treatment. In *The Science of Cognitive Behavioral Therapy*; Academic Press: Cambridge, MA, USA, 2017; pp. 337–357.
2. Smink, F.R.E.; Van Hoeken, D.; Hoek, H.W. Epidemiology of Eating Disorders: Incidence, Prevalence and Mortality Rates. *Curr. Psychiatry Rep.* **2012**, *14*, 406–414. [CrossRef]
3. Hoek, H.W.; Van Hoeken, D. Review of the prevalence and incidence of eating disorders. *Int. J. Eat. Disord.* **2003**, *34*, 383–396. [CrossRef] [PubMed]
4. Hay, P. Current approach to eating disorders: A clinical update. *Intern. Med. J.* **2020**, *50*, 24–29. [CrossRef]
5. Johnson, J.G.; Cohen, P.; Kasen, S.; Brook, J.S. Eating Disorders during Adolescence and the Risk for Physical and Mental Disorders During Early Adulthood. *Arch. Gen. Psychiatry* **2002**, *59*, 545–552. [CrossRef] [PubMed]

6. Clouard, C.; Meunier-Salaün, M.; Val-Laillet, D. Food preferences and aversions in human health and nutrition: How can pigs help the biomedical research? *Animal* **2012**, *6*, 118–136. [CrossRef] [PubMed]
7. Yanovski, S. Sugar and fat: Cravings and aversions. *J. Nutr.* **2003**, *133*, 835S–837S. [CrossRef]
8. He, S.; You, J.-J.; Liang, X.-F.; Zhang, Z.-L.; Zhang, Y.-P. Transcriptome sequencing and metabolome analysis of food habits domestication from live prey fish to artificial diets in mandarin fish (*Siniperca chuatsi*). *BMC Genom.* **2021**, *22*, 1–12. [CrossRef]
9. Ventura, A.K.; Worobey, J. Early Influences on the Development of Food Preferences. *Curr. Biol.* **2013**, *23*, R401–R408. [CrossRef]
10. Chen, K.; Zhang, Z.; Li, J.; Xie, S.; Shi, L.-J.; He, Y.-H.; Liang, X.-F.; Zhu, Q.-S.; He, S. Different regulation of branched-chain amino acid on food intake by TOR signaling in Chinese perch (*Siniperca chuatsi*). *Aquaculture* **2021**, *530*, 735792. [CrossRef]
11. Cota, D.; Yoon, A.; Peng, G.; Brandenburg, Y.; Zollo, O.; Xu, W.; Rego, E.; Ruggero, D. Hypothalamic mTOR Signaling Regulates Food Intake. *Science* **2006**, *312*, 927–930. [CrossRef]
12. Valassi, E.; Scacchi, M.; Cavagnini, F. Neuroendocrine control of food intake. *Nutr. Metab. Cardiovasc. Dis.* **2008**, *18*, 158–168. [CrossRef]
13. Sohn, J.-W. Network of hypothalamic neurons that control appetite. *BMB Rep.* **2015**, *48*, 229–233. [CrossRef]
14. Volkoff, H. The Neuroendocrine Regulation of Food Intake in Fish: A Review of Current Knowledge. *Front. Neurosci.* **2016**, *10*, 540. [CrossRef]
15. Gorissen, M.; Flik, G. Leptin in teleostean fish, towards the origins of leptin physiology. *J. Chem. Neuroanat.* **2014**, *61-62*, 200–206. [CrossRef]
16. Jönsson, E. The role of ghrelin in energy balance regulation in fish. *Gen. Comp. Endocrinol.* **2013**, *187*, 79–85. [CrossRef] [PubMed]
17. Clemente, J.C.; Ursell, L.K.; Parfrey, L.W.; Knight, R. The Impact of the Gut Microbiota on Human Health: An Integrative View. *Cell* **2012**, *148*, 1258–1270. [CrossRef]
18. Subramanian, S.; Blanton, L.V.; Frese, S.A.; Charbonneau, M.; Mills, D.A.; Gordon, J.I. Cultivating Healthy Growth and Nutrition through the Gut Microbiota. *Cell* **2015**, *161*, 36–48. [CrossRef]
19. Vuong, H.E.; Yano, J.M.; Fung, T.C.; Hsiao, E.Y. The Microbiome and Host Behavior. *Annu. Rev. Neurosci.* **2017**, *40*, 21–49. [CrossRef] [PubMed]
20. Cryan, J.F.; O'Riordan, K.J.; Cowan, C.S.M.; Sandhu, K.V.; Bastiaanssen, T.F.S.; Boehme, M.; Codagnone, M.G.; Cussotto, S.; Fulling, C.; Golubeva, A.V.; et al. The Microbiota-Gut-Brain Axis. *Physiol. Rev.* **2019**, *99*, 1877–2013. [CrossRef] [PubMed]
21. Martino, M.E.; Ma, D.; Leulier, F. Microbial influence on Drosophila biology. *Curr. Opin. Microbiol.* **2017**, *38*, 165–170. [CrossRef] [PubMed]
22. Ezra-Nevo, G.; Henriques, S.F.; Ribeiro, C. The diet-microbiome tango: How nutrients lead the gut brain axis. *Curr. Opin. Neurobiol.* **2020**, *62*, 122–132. [CrossRef] [PubMed]
23. Vijay-Kumar, M.; Aitken, J.D.; Carvalho, F.A.; Cullender, T.C.; Mwangi, S.; Srinivasan, S.; Sitaraman, S.V.; Knight, R.; Ley, R.E.; Gewirtz, A.T. Metabolic Syndrome and Altered Gut Microbiota in Mice Lacking Toll-Like Receptor 5. *Science* **2010**, *328*, 228–231. [CrossRef]
24. Breton, J.; Tennoune, N.; Lucas, N.; Francois, M.; Legrand, R.; Jacquemot, J.; Goichon, A.; Guérin, C.; Peltier, J.; Pestel-Caron, M.; et al. Gut Commensal E. coli Proteins Activate Host Satiety Pathways following Nutrient-Induced Bacterial Growth. *Cell Metab.* **2016**, *23*, 324–334. [CrossRef] [PubMed]
25. Leitão-Gonçalves, R.; Santos, Z.; Francisco, A.P.; Fioreze, G.T.; Anjos, M.; Baltazar, C.; Elias, A.P.; Itskov, P.; Piper, M.; Ribeiro, C. Commensal bacteria and essential amino acids control food choice behavior and reproduction. *PLoS Biol.* **2017**, *15*, e2000862. [CrossRef]
26. Van De Wouw, M.; Schellekens, H.; Dinan, T.G.; Cryan, J.F. Microbiota-Gut-Brain Axis: Modulator of Host Metabolism and Appetite. *J. Nutr.* **2017**, *147*, 727–745. [CrossRef]
27. De Almada, C.N.; Almada, C.N.; Martinez, R.C.; Sant'Ana, A.S. Paraprobiotics: Evidences on their ability to modify biological responses, inactivation methods and perspectives on their application in foods. *Trends Food Sci. Technol.* **2016**, *58*, 96–114. [CrossRef]
28. Echeng, G.; Ehao, H.; Exie, S.; Ewang, X.; Edai, M.; Ehuang, L.; Eyuan, Z.-H. Antibiotic alternatives: The substitution of antibiotics in animal husbandry? *Front. Microbiol.* **2014**, *5*, 217. [CrossRef]
29. Watts, M.; Munday, B.L.; Burke, C.M. Immune responses of teleost fish. *Aust. Veter J.* **2001**, *79*, 570–574. [CrossRef]
30. Gatesoupe, F. The use of probiotics in aquaculture. *Aquaculture* **1999**, *180*, 147–165. [CrossRef]
31. Chauhan, A.; Singh, R. Probiotics in aquaculture: A promising emerging alternative approach. *Symbiosis* **2019**, *77*, 99–113. [CrossRef]
32. Ringø, E. Probiotics in shellfish aquaculture. *Aquac. Fish.* **2020**, *5*, 1–27. [CrossRef]
33. Balcázar, J.L.; de Blas, I.; Ruiz-Zarzuela, I.; Cunningham, D.; Vendrell, D.; Múzquiz, J.L. The role of probiotics in aquaculture. *Veter Microbiol.* **2006**, *114*, 173–186. [CrossRef]
34. Wang, L.; Jian, Y.; Liu, Q.; Li, F.; Chang, L. Electromagnetohydrodynamic flow and heat transfer of third grade fluids between two micro-parallel plates. *Colloids Surf. A Physicochem. Eng. Asp.* **2016**, *494*, 87–94. [CrossRef]
35. Van Nguyen, N.; Onoda, S.; Van Khanh, T.; Hai, P.D.; Trung, N.T.; Hoang, L.; Koshio, S. Evaluation of dietary Heat-killed Lactobacillus plantarum strain L-137 supplementation on growth performance, immunity and stress resistance of Nile tilapia (*Oreochromis niloticus*). *Aquaculture* **2019**, *498*, 371–379. [CrossRef]

36. Hooshyar, Y.; Kenari, A.A.; Paknejad, H.; Gandomi, H. Effects of Lactobacillus rhamnosus ATCC 7469 on Different Parameters Related to Health Status of Rainbow Trout (*Oncorhynchus mykiss*) and the Protection Against Yersinia ruckeri. *Probiotics Antimicrob. Proteins* **2020**, *12*, 1370–1384. [CrossRef]
37. Sewaka, M.; Trullas, C.; Chotiko, A.; Rodkhum, C.; Chansue, N.; Boonanuntanasarn, S.; Pirarat, N. Efficacy of synbiotic Jerusalem artichoke and Lactobacillus rhamnosus GG-supplemented diets on growth performance, serum biochemical parameters, intestinal morphology, immune parameters and protection against Aeromonas veronii in juvenile red tilapia (*Oreochromis* spp.). *Fish Shellfish Immunol.* **2019**, *86*, 260–268. [CrossRef] [PubMed]
38. Giorgia, G.; Elia, C.; Andrea, P.; Cinzia, C.; Stefania, S.; Ana, R.; Daniel, M.L.; Ike, O.; Oliana, C. Effects of Lactogen 13, a New Probiotic Preparation, on Gut Microbiota and Endocrine Signals Controlling Growth and Appetite of Oreochromis niloticus Juveniles. *Microb. Ecol.* **2018**, *76*, 1063–1074. [CrossRef] [PubMed]
39. Poolsawat, L.; Li, X.; He, M.; Ji, D.; Leng, X.J.A.N. Clostridium butyricum as probiotic for promoting growth performance, feed utilization, gut health and microbiota community of tilapia (*Oreochromis niloticus* × *O. aureus*). *Aquac. Nutr.* **2020**, *26*, 657–670. [CrossRef]
40. Zhang, M.; Dong, B.; Lai, X.; Chen, Z.; Hou, L.; Shu, R.; Huang, Y.; Shu, H. Effects of Clostridium butyricum on growth, digestive enzyme activity, antioxidant capacity and gut microbiota in farmed tilapia (*Oreochromis niloticus*). *Aquac. Res.* **2021**, *52*, 1573–1584. [CrossRef]
41. Balcázar, J.L.; Vendrell, D.; de Blas, I.; Ruiz-Zarzuela, I.; Muzquiz, J.L.; Girones, O. Characterization of probiotic properties of lactic acid bacteria isolated from intestinal microbiota of fish. *Aquaculture* **2008**, *278*, 188–191. [CrossRef]
42. Henriques, S.F.; Dhakan, D.B.; Serra, L.; Francisco, A.P.; Carvalho-Santos, Z.; Baltazar, C.; Elias, A.P.; Anjos, M.; Zhang, T.; Maddocks, O.D.K.; et al. Metabolic cross-feeding in imbalanced diets allows gut microbes to improve reproduction and alter host behaviour. *Nat. Commun.* **2020**, *11*, 4236. [CrossRef] [PubMed]
43. Lash, B.W.; Mysliwiec, T.H.; Gourama, H. Detection and partial characterization of a broad-range bacteriocin produced by Lactobacillus plantarum (ATCC 8014). *Food Microbiol.* **2005**, *22*, 199–204. [CrossRef]
44. Giri, S.S.; Sukumaran, V.; Oviya, M. Potential probiotic Lactobacillus plantarum VSG3 improves the growth, immunity, and disease resistance of tropical freshwater fish, Labeo rohita. *Fish Shellfish. Immunol.* **2013**, *34*, 660–666. [CrossRef] [PubMed]
45. Parthasarathy, R.; Ravi, D. Probiotic bacteria as growth promoter and biocontrol agent against *Aeromonas hydrophila* in *Catla catla* (Hamilton, 1822). *Indian J. Fish.* **2011**, *58*, 87–93.
46. Son, V.M.; Chang, C.-C.; Wu, M.-C.; Guu, Y.-K.; Chiu, C.-H.; Cheng, W. Dietary administration of the probiotic, Lactobacillus plantarum, enhanced the growth, innate immune responses, and disease resistance of the grouper Epinephelus coioides. *Fish Shellfish Immunol.* **2009**, *26*, 691–698. [CrossRef]
47. Jatobá, A.; Vieira, F.D.N.; Buglione-Neto, C.C.; Mouriño', J.L.P.; Silva, B.C.; Seiftter, W.Q.; Andreatta, E.R. Diet supplemented with probiotic for Nile tilapia in polyculture system with marine shrimp. *Fish Physiol. Biochem.* **2011**, *37*, 725–732. [CrossRef]
48. Uma, A.; Abraham, T.J.; Sundararaj, V. Effect of a probiotic bacterium, Lactobacillus plantarum on disease resistance of Penaeus indicus larvae. *Indian J. Fish.* **1999**, *46*, 367–373.
49. Kongnum, K.; Hongpattarakere, T. Effect of Lactobacillus plantarum isolated from digestive tract of wild shrimp on growth and survival of white shrimp (*Litopenaeus vannamei*) challenged with Vibrio harveyi. *Fish Shellfish Immunol.* **2012**, *32*, 170–177. [CrossRef]
50. Chiu, C.-H.; Guu, Y.-K.; Liu, C.-H.; Pan, T.-M.; Cheng, W. Immune responses and gene expression in white shrimp, Litopenaeus vannamei, induced by Lactobacillus plantarum. *Fish Shellfish Immunol.* **2007**, *23*, 364–377. [CrossRef]
51. Vieira, F.N.; Buglione, C.C.; Mouriño, J.P.L.; Jatobá, A.; Martins, M.L.; Schleder, D.D.; Andreatta, E.R.; Barraco, M.A.; Vinatea, L.A. Effect of probiotic supplemented diet on marine shrimp survival after challenge with Vibrio harveyi. *Arquivo Brasileiro de Medicina Veterinária e Zootecnia* **2010**, *62*, 631–638. [CrossRef]
52. Iman, M.K.A.; Wafaa, T.A.; Elham, S.A.; Mohammad, M.N.A.; El-Shafei, K.; Osama, M.S.; Gamal, A.I.; Zeinab, I.S.; El-Sayed, H.S. Evaluation of Lactobacillus plantarum as a probiotic in aquaculture: Emphasis on growth performance and innate immunity. *J. Appl. Sci. Res.* **2013**, *9*, 572–582.
53. Bravo, J.A.; Forsythe, P.; Chew, M.V.; Escaravage, E.; Savignac, H.M.; Dinan, T.G.; Bienenstock, J.; Cryan, J.F. Ingestion of Lactobacillus strain regulates emotional behavior and central GABA receptor expression in a mouse via the vagus nerve. *Proc. Natl. Acad. Sci. USA* **2011**, *108*, 16050–16055. [CrossRef]
54. Liang, S.; Wang, T.; Hu, X.; Luo, J.; Li, W.; Wu, X.; Duan, Y.; Jin, F. Administration of Lactobacillus helveticus NS8 improves behavioral, cognitive, and biochemical aberrations caused by chronic restraint stress. *Neuroscience* **2015**, *310*, 561–577. [CrossRef]
55. Janik, R.; Thomason, L.A.; Stanisz, A.M.; Forsythe, P.; Bienenstock, J.; Stanisz, G.J. Magnetic resonance spectroscopy reveals oral Lactobacillus promotion of increases in brain GABA, N-acetyl aspartate and glutamate. *NeuroImage* **2016**, *125*, 988–995. [CrossRef] [PubMed]
56. Lyte, M. Probiotics function mechanistically as delivery vehicles for neuroactive compounds: Microbial endocrinology in the design and use of probiotics. *BioEssays* **2011**, *33*, 574–581. [CrossRef] [PubMed]
57. Lyte, M. Microbial endocrinology. *Gut Microbes* **2014**, *5*, 381–389. [CrossRef] [PubMed]
58. Kong, Q.; He, G.-Q.; Jia, J.-L.; Zhu, Q.-L.; Ruan, H. Oral Administration of Clostridium butyricum for Modulating Gastrointestinal Microflora in Mice. *Curr. Microbiol.* **2010**, *62*, 512–517. [CrossRef]

59. Yang, C.; Cao, G.; Xiao, Y.; Chen, A.; Liu, T.; Zhou, L.; Zhang, L.; Ferket, P. Effects of Clostridium butyricum on Growth Performance, Nitrogen Metabolism, Intestinal Morphology and Cecal Microflora in Broiler Chickens. *J. Anim. Veter Adv.* **2012**, *11*, 2665–2671. [CrossRef]
60. Juan, Z.; Chun, W.; Zhao-Ling, S.; Ming-Hua, Z.; Hai-Xia, W.; Meng-Yun, L.; Jian-Qiong, H.; Yue-Jie, Z.; Xin, S. Oral administration of Clostridium butyricum CGMCC0313-1 reduces ovalbumin-induced allergic airway inflammation in mice. *Respirology* **2017**, *22*, 898–904. [CrossRef]
61. Yang, C.M.; Cao, G.T.; Ferket, P.; Liu, T.T.; Zhou, L.; Zhang, L.; Xiao, Y.P.; Chen, A.G. Effects of probiotic, Clostridium butyricum, on growth performance, immune function, and cecal microflora in broiler chickens. *Poult. Sci.* **2012**, *91*, 2121–2129. [CrossRef]
62. Chen, L.; Li, S.; Zheng, J.; Li, W.; Jiang, X.; Zhao, X.; Wu, D. Effects of dietary Clostridium butyricum supplementation on growth performance,intestinal development, and immune response of weaned piglets challenged with lipopolysaccharide. *J. Anim. Sci. Biotechnol.* **2018**, *9*, 1–14. [CrossRef]
63. Sisson, G.; Ayis, S.; Sherwood, R.A.; Bjarnason, I. Randomised clinical trial: A liquid multi-strain probiotic vs. placebo in the irritable bowel syndrome—A 12 week double-blind study. *Aliment. Pharmacol. Ther.* **2014**, *40*, 51–62. [CrossRef] [PubMed]
64. Sun, Y.-Y.; Li, M.; Li, Y.-Y.; Li, L.-X.; Zhai, W.-Z.; Wang, P.; Yang, X.-X.; Gu, X.; Song, L.-J.; Li, Z.; et al. The effect of Clostridium butyricum on symptoms and fecal microbiota in diarrhea-dominant irritable bowel syndrome: A randomized, double-blind, placebo-controlled trial. *Sci. Rep.* **2018**, *8*, 2964. [CrossRef] [PubMed]
65. Hudson, L.E.; Anderson, S.E.; Corbett, A.H.; Lamb, T.J. Gleaning Insights from Fecal Microbiota Transplantation and Probiotic Studies for the Rational Design of Combination Microbial Therapies. *Clin. Microbiol. Rev.* **2017**, *30*, 191–231. [CrossRef] [PubMed]
66. Lv, L.; Liang, X.F.; Huang, K.; He, S. Effect of agmatine on food intake in mandarin fish (*Siniperca chuatsi*). *Fish Physiol. Biochem.* **2019**, *45*, 1709–1716. [CrossRef]
67. Liang, X.F.; Oku, H.; Ogata, H.Y.; Liu, J.; He, X. Weaning Chinese perch *Siniperca chuatsi* (Basilewsky) onto artificial diets based upon its specific sensory modality in feeding. *Aquac. Res.* **2001**, *32*, 76–82. [CrossRef]
68. Zhang, X.; Xiang, J.; Yin, M. The influences of temperature on the feeding,growth and development oflarval siniperca chuatsi. *J. Fish. China* **1999**, *1*, 91–94.
69. Dou, Y.; He, S.; Liang, X.-F.; Cai, W.; Wang, J.; Shi, L.; Li, J. Memory Function in Feeding Habit Transformation of Mandarin Fish (*Siniperca chuatsi*). *Int. J. Mol. Sci.* **2018**, *19*, 1254. [CrossRef] [PubMed]
70. Li, L.; Fang, J.; Liang, X.; Alam, M.S.; Liu, L.; Yuan, X. Effect of feeding stimulants on growth performance, feed intake and appetite regulation of mandarin fish,Siniperca chuatsi. *Aquac. Res.* **2019**, *50*, 3684–3691. [CrossRef]
71. Cao, G.; Tao, F.; Hu, Y.; Li, Z.; Zhang, Y.; Deng, B.; Zhan, X. Positive effects of a Clostridium butyricum-based compound probiotic on growth performance, immune responses, intestinal morphology, hypothalamic neurotransmitters, and colonic microbiota in weaned piglets. *Food Funct.* **2019**, *10*, 2926–2934. [CrossRef] [PubMed]
72. Zheng, X.; Duan, Y.; Dong, H.; Zhang, J. Effects of Dietary Lactobacillus plantarum on Growth Performance, Digestive Enzymes and Gut Morphology of Litopenaeus vannamei. *Probiotics Antimicrob. Proteins* **2018**, *10*, 504–510. [CrossRef] [PubMed]
73. Zhao, H.-B.; Jia, L.; Yan, Q.-Q.; Deng, Q.; Wei, B. Effect of Clostridium butyricum and Butyrate on Intestinal Barrier Functions: Study of a Rat Model of Severe Acute Pancreatitis With Intra-Abdominal Hypertension. *Front. Physiol.* **2020**, *11*, 561061. [CrossRef]
74. Ukeda, H. Novel assay methods of enzyme superoxide dismutase (SOD). *Nippon Nogeikagaku Kaishi J. Japan Soc. Biosci. Biotechnol. Agrochem.* **2001**, *75*, 562–565.
75. Buege, J.A.; Aust, S.D. Microsomal lipid peroxidation. *Methods Enzymol.* **1978**, *52*, 302–310. [CrossRef] [PubMed]
76. Nebot, C.; Moutet, M.; Huet, P.; Xu, J.; Yadan, J.; Chaudière, J. Spectrophotometric Assay of Superoxide Dismutase Activity Based on the Activated Autoxidation of a Tetracyclic Catechol. *Anal. Biochem.* **1993**, *214*, 442–451. [CrossRef]
77. Reiners, J.J.; Thai, G.; Rupp, T.; Cantu, A.R. Assessment of the antioxidant/prooxidant status of murine skin following topical treatment with 12- O -tetradecanoylphorbol-13-acetate and throughout the ontogeny of skin cancer. Part I: Quantitation of superoxide dismutase, catalase, glutathione peroxidase and xanthine oxidase. *Carcinogenesis* **1991**, *12*, 2337–2343. [CrossRef] [PubMed]
78. Schmittgen, T.D.; Livak, K.J. Analyzing real-time PCR data by the comparative $C_T$ method. *Nat. Protoc.* **2008**, *3*, 1101–1108. [CrossRef]
79. Magoč, T.; Salzberg, S.L. FLASH: Fast Length Adjustment of Short Reads to Improve Genome Assemblies. *Bioinformatics* **2011**, *27*, 2957–2963. [CrossRef] [PubMed]
80. Edgar, R.C.; Haas, B.J.; Bioinformatics, C.J. UCHIME improves sensitivity and speed of chimera detection. *Bioinformatics* **2011**, *27*, 2194–2200. [CrossRef] [PubMed]
81. Caporaso, J.G.; Kuczynski, J.; Stombaugh, J.; Bittinger, K.; Bushman, F.D.; Costello, E.K.; Fierer, N.; Peña, A.G.; Goodrich, J.K.; Gordon, J.I.; et al. QIIME Allows Analysis of High-Throughput Community Sequencing data. *Nat. Methods* **2010**, *7*, 335–336. [CrossRef]
82. Edgar, R.C. UPARSE: Highly accurate OTU sequences from microbial amplicon reads. *Nat. Methods* **2013**, *10*, 996–998. [CrossRef]
83. Segata, N.; Izard, J.; Waldron, L.; Gevers, D.; Miropolsky, L.; Garrett, W.S.; Huttenhower, C. Metagenomic biomarker discovery and explanation. *Genome Biol.* **2011**, *12*, R60. [CrossRef] [PubMed]
84. Aßhauer, K.P.; Wemheuer, B.; Daniel, R.; Meinicke, P. Tax4Fun: Predicting functional profiles from metagenomic 16S rRNA data: Fig. 1. *Bioinformatics* **2015**, *31*, 2882–2884. [CrossRef]

85. Seitz, J.; Trinh, S.; Herpertz-Dahlmann, B. The Microbiome and Eating Disorders. *Psychiatr. Clin. N. Am.* **2019**, *42*, 93–103. [CrossRef] [PubMed]
86. Dinan, T.G.; Cryan, J.F. Melancholic microbes: A link between gut microbiota and depression? *Neurogastroenterol. Motil.* **2013**, *25*, 713–719. [CrossRef]
87. Chen, X.; Yi, H.; Liu, S.; Zhang, Y.; Su, Y.; Liu, X.; Bi, S.; Lai, H.; Zeng, Z.; Li, G. Promotion of pellet-feed feeding in mandarin fish (Siniperca chuatsi) by Bdellovibrio bacteriovorus is influenced by immune and intestinal flora. *Aquaculture* **2021**, *542*, 736864. [CrossRef]
88. Fernandez-Diaz, C.; Kopecka, J.; Cañavate, J.P.; Sarasquete, C.; Sole, M. Variations on development and stress defences in Solea senegalensis larvae fed on live and microencapsulated diets. *Aquaculture* **2006**, *251*, 573–584. [CrossRef]
89. Dawood, M.A. Nutritional immunity of fish intestines: Important insights for sustainable aquaculture. *Rev. Aquac.* **2021**, *13*, 642–663. [CrossRef]
90. Yúfera, M.; Fernandez-Diaz, C.; Pascual, E. Food microparticles for larval fish prepared by internal gelation. *Aquaculture* **2005**, *248*, 253–262. [CrossRef]
91. Pérez-Sánchez, T.; Balcazar, J.L.; Merrifield, D.; Carnevali, O.; Gioacchini, G.; de Blas, I.; Ruiz-Zarzuela, I. Expression of immune-related genes in rainbow trout (*Oncorhynchus mykiss*) induced by probiotic bacteria during Lactococcus garvieae infection. *Fish Shellfish Immunol.* **2011**, *31*, 196–201. [CrossRef]
92. Duan, Y.; Zhang, J.; Huang, J.; Jiang, S. Effects of Dietary Clostridium butyricum on the Growth, Digestive Enzyme Activity, Antioxidant Capacity, and Resistance to Nitrite Stress of Penaeus monodon. *Probiotics Antimicrob. Proteins* **2018**, *11*, 938–945. [CrossRef] [PubMed]
93. Oliva-Teles, A. Nutrition and health of aquaculture fish. *J. Fish Dis.* **2012**, *35*, 83–108. [CrossRef] [PubMed]
94. A Casas, G.; Blavi, L.; Cross, T.-W.L.; Lee, A.H.; Swanson, K.; Stein, H.H. Inclusion of the direct-fed microbial Clostridium butyricum in diets for weanling pigs increases growth performance and tends to increase villus height and crypt depth, but does not change intestinal microbial abundance. *J. Anim. Sci.* **2020**, *98*. [CrossRef]
95. Zhang, F.; Li, Y.; Wang, X.; Wang, S.; Bi, D. The Impact ofLactobacillus plantarumon the Gut Microbiota of Mice with DSS-Induced Colitis. *BioMed Res. Int.* **2019**, *2019*, 1–10. [CrossRef]
96. He, S.; Liang, X.F.; Sun, J.; Li, L.; Yu, Y.; Huang, W.; Qu, C.M.; Cao, L.; Bai, X.L.; Tao, Y. Insights into food preference in hybrid F1 of Siniperca chuatsi (♀)× Siniperca scherzeri (♂) mandarin fish through transcriptome analysis. *BMC Genom.* **2013**, *14*, 1–11. [CrossRef] [PubMed]
97. Biochemistry, H.V.J.C.; Molecular, P.P.A.; Physiology, I. The role of neuropeptide Y, orexins, cocaine and amphetamine-related transcript, cholecystokinin, amylin and leptin in the regulation of feeding in fish. *Biochem. Physiol. Part A Mol. Integr. Physiol.* **2006**, *144*, 325–331.
98. Rui, L. Brain regulation of energy balance and body weight. *Rev. Endocr. Metab. Disord.* **2013**, *14*, 387–407. [CrossRef]
99. Crespo, C.S.; Cachero, A.P.; Jimã Nez, L.P.; Barrios, V.; Ferreiro, E.A. Peptides and Food Intake. *Front. Endocrinol.* **2014**, *5*, 58. [CrossRef]
100. Volkoff, H.; Canosa, L.; Unniappan, S.; Cerdá-Reverter, J.; Bernier, N.; Kelly, S.; Peter, R. Neuropeptides and the control of food intake in fish. *Gen. Comp. Endocrinol.* **2005**, *142*, 3–19. [CrossRef]
101. Holzer, P.; Reichmann, F.; Farzi, A. Neuropeptide Y, peptide YY and pancreatic polypeptide in the gut–brain axis. *Neuropeptides* **2012**, *46*, 261–274. [CrossRef]
102. Cerdá-Reverter, J.M.; Agulleiro, M.J.; Guillot, R.; Sánchez, E.; Ceinos, R.M.; Rotllant, J. Fish melanocortin system. *Eur. J. Pharmacol.* **2011**, *660*, 53–60. [CrossRef]
103. Riley, L.G.; Fox, B.K.; Kaiya, H.; Hirano, T.; Grau, E.G. Long-term treatment of ghrelin stimulates feeding, fat deposition, and alters the GH/IGF-I axis in the tilapia, Oreochromis mossambicus. *Gen. Comp. Endocrinol.* **2005**, *142*, 234–240. [CrossRef] [PubMed]
104. Kaiya, H.; Miyazato, M.; Kangawa, K.; Peter, R.E.; Unniappan, S. Ghrelin: A multifunctional hormone in non-mammalian vertebrates. *Comp. Biochem. Physiol. Part A Mol. Integr. Physiol.* **2008**, *149*, 109–128. [CrossRef]
105. Tan, Y.-Y.; Ji, Z.-L.; Zhao, G.; Jiang, J.-R.; Wang, N.; Wang, J.-M. Decreased SCF/c-kit signaling pathway contributes to loss of interstitial cells of Cajal in gallstone disease. *Int. J. Clin. Exp. Med.* **2014**, *7*, 4099–4106. [PubMed]
106. ThanThan, S.; Mekaru, C.; Seki, N.; Hidaka, K.; Ueno, A.; ThidarMyint, H.; Kuwayama, H. Endogenous ghrelin released in response to endothelin stimulates growth hormone secretion in cattle. *Domest. Anim. Endocrinol.* **2010**, *38*, 1–12. [CrossRef] [PubMed]
107. Won, E.T.; Douros, J.D.; Hurt, D.A.; Borski, R.J. Leptin stimulates hepatic growth hormone receptor and insulin-like growth factor gene expression in a teleost fish, the hybrid striped bass. *Gen. Comp. Endocrinol.* **2016**, *229*, 84–91. [CrossRef]
108. Browning, K.N.; Verheijden, S.; Boeckxstaens, G.E. The Vagus Nerve in Appetite Regulation, Mood, and Intestinal Inflammation. *Gastroenterology* **2017**, *152*, 730–744. [CrossRef]
109. Caricilli, A. Intestinal barrier: A gentlemen's agreement between microbiota and immunity. *World J. Gastrointest. Pathophysiol.* **2014**, *5*, 18. [CrossRef]
110. Uribe, C.; Folch, H.; Enriquez, R.; Moran, G. Innate and adaptive immunity in teleost fish: A review. *Veterinární Medicína* **2011**, *56*, 486–503. [CrossRef]

111. Singh, P.B.; Singh, V. Cypermethrin induced histological changes in gonadotrophic cells, liver, gonads, plasma levels of estradiol-17β and 11-ketotestosterone, and sperm motility in *Heteropneustes fossilis* (Bloch). *Chemosphere* **2008**, *72*, 422–431. [CrossRef] [PubMed]
112. Dominguez, M.; Takemura, A.; Tsuchiya, M.; Nakamura, S. Impact of different environmental factors on the circulating immunoglobulin levels in the Nile tilapia, Oreochromis niloticus. *Aquaculture* **2004**, *241*, 491–500. [CrossRef]
113. Zhai, Q.; Wang, H.; Tian, F.; Zhao, J.; Zhang, H.; Chen, W. Dietary Lactobacillus plantarum supplementation decreases tissue lead accumulation and alleviates lead toxicity in Nile tilapia (*Oreochromis niloticus*). *Aquac. Res.* **2017**, *48*, 5094–5103. [CrossRef]
114. Wang, K.; Cao, G.; Zhang, H.; Li, Q.; Yang, C. Effects of Clostridium butyricum and Enterococcus faecalis on growth performance, immune function, intestinal morphology, volatile fatty acids, and intestinal flora in a piglet model. *Food Funct.* **2019**, *10*, 7844–7854. [CrossRef] [PubMed]
115. Liao, X.D.; Ma, G.; Cai, J.; Fu, Y.; Yan, X.Y.; Wei, X.B.; Zhang, R.J. Effects ofClostridium butyricum on growth performance, antioxidation, and immune function of broilers. *Poult. Sci.* **2015**, *94*, 662–667. [CrossRef]
116. Jia, R.; Gu, Z.; He, Q.; Du, J.; Cao, L.; Jeney, G.; Xu, P.; Yin, G. Anti-oxidative, anti-inflammatory and hepatoprotective effects of Radix Bupleuri extract against oxidative damage in tilapia (*Oreochromis niloticus*) via Nrf2 and TLRs signaling pathway. *Fish Shellfish Immunol.* **2019**, *93*, 395–405. [CrossRef]
117. Dong, Y.; Tu, J.; Zhou, Y.; Zhou, X.H.; Xu, B.; Zhu, S.Y. Silybum marianum oil attenuates oxidative stress and ameliorates mitochondrial dysfunction in mice treated with D-galactose. *Pharmacogn. Mag.* **2014**, *10*, S92–S99. [CrossRef] [PubMed]
118. Chen, P.; Chen, F.; Zhou, B. Antioxidative, anti-inflammatory and anti-apoptotic effects of ellagic acid in liver and brain of rats treated by D-galactose. *Sci. Rep.* **2018**, *8*, 1–10. [CrossRef] [PubMed]
119. Li, Y.; Li, J.; Lu, J.; Li, Z.; Shi, S.; Liu, Z. Effects of live and artificial feeds on the growth, digestion, immunity and intestinal microflora of mandarin fish hybrid (*Siniperca chuatsi* ♀× *Siniperca scherzeri* ♂). *Aquac. Res.* **2017**, *48*, 4479–4485. [CrossRef]
120. Peters, L.D.; Livingstone, D. Antioxidant enzyme activities in embryologic and early larval stages of turbot. *J. Fish Biol.* **1996**, *49*, 986–997. [CrossRef]
121. Mourente, G.; Tocher, D.; Diaz, E.; Grau, A.; Pastor, E. Relationships between antioxidants, antioxidant enzyme activities and lipid peroxidation products during early development in Dentex dentex eggs and larvae. *Aquaculture* **1999**, *179*, 309–324. [CrossRef]
122. Dandapat, J.; Chainy, G.B.; Rao, K.J. Lipid peroxidation and antioxidant defence status during larval development and metamorphosis of giant prawn, Macrobrachium rosenbergii. *Comp. Biochem. Physiol. Part C Toxicol. Pharmacol.* **2003**, *135*, 221–233. [CrossRef]
123. Rueda-Jasso, R.; Conceição, L.; Dias, J.; De Coen, W.; Gomes, E.; Rees, J.; Soares, F.; Dinis, M.T.; Sorgeloos, P. Effect of dietary non-protein energy levels on condition and oxidative status of Senegalese sole (*Solea senegalensis*) juveniles. *Aquaculture* **2004**, *231*, 417–433. [CrossRef]
124. Solé, M.; Potrykus, J.; Fernandez-Diaz, C.; Blasco, J. Variations on stress defences and metallothionein levels in the Senegal sole, Solea senegalensis, during early larval stages. *Fish Physiol. Biochem.* **2004**, *30*, 57–66. [CrossRef]
125. Liu, J.; Fu, Y.; Zhang, H.; Wang, J.; Zhu, J.; Wang, Y.; Guo, Y.; Wang, G.; Xu, T.; Chu, M.; et al. The hepatoprotective effect of the probiotic Clostridium butyricum against carbon tetrachloride-induced acute liver damage in mice. *Food Funct.* **2017**, *8*, 4042–4052. [CrossRef] [PubMed]
126. Alcock, J.; Maley, C.C.; Aktipis, C.A. Is eating behavior manipulated by the gastrointestinal microbiota? Evolutionary pressures and potential mechanisms. *BioEssays* **2014**, *36*, 940–949. [CrossRef] [PubMed]
127. Soriano, E.L.; Ramírez, D.T.; Araujo, D.R.; Gómez-Gil, B.; Castro, L.I.; Sánchez, C.G. Effect of temperature and dietary lipid proportion on gut microbiota in yellowtail kingfish Seriola lalandi juveniles. *Aquaculture* **2018**, *497*, 269–277. [CrossRef]
128. Dehler, C.E.; Secombes, C.J.; Martin , S.A.M. Environmental and physiological factors shape the gut microbiota of Atlantic salmon parr (*Salmo salar* L.). *Aquaculture* **2017**, *467*, 149–157. [CrossRef] [PubMed]
129. Benson, A.K.; Kelly, S.A.; Legge, R.; Ma, F.; Low, S.J.; Kim, J.; Zhang, M.; Oh, P.L.; Nehrenberg, D.; Hua, K.; et al. Individuality in gut microbiota composition is a complex polygenic trait shaped by multiple environmental and host genetic factors. *Proc. Natl. Acad. Sci. USA* **2010**, *107*, 18933–18938. [CrossRef] [PubMed]
130. Kovacs, A.; Ben-Jacob, N.; Tayem, H.; Halperin, E.; Iraqi, F.A.; Gophna, U. Genotype Is a Stronger Determinant than Sex of the Mouse Gut Microbiota. *Microb. Ecol.* **2010**, *61*, 423–428. [CrossRef]
131. Rawls, J.; Mahowald, M.A.; Ley, R.E.; Gordon, J.I. Reciprocal Gut Microbiota Transplants from Zebrafish and Mice to Germ-free Recipients Reveal Host Habitat Selection. *Cell* **2006**, *127*, 423–433. [CrossRef]
132. Arkoosh, M.R.; Clemons, E.; Kagley, A.N.; Stafford, C.; Glass, A.C.; Jacobson, K.; Reno, P.; Myers, M.S.; Casillas, E.; Loge, F.; et al. Survey of Pathogens in Juvenile SalmonOncorhynchusSpp. Migrating through Pacific Northwest Estuaries. *J. Aquat. Anim. Heal.* **2004**, *16*, 186–196. [CrossRef]
133. De Filippo, C.; Cavalieri, D.; Di Paola, M.; Ramazzotti, M.; Poullet, J.B.; Massart, S.; Collini, S.; Pieraccini, G.; Lionetti, P. Impact of diet in shaping gut microbiota revealed by a comparative study in children from Europe and rural Africa. *Proc. Natl. Acad. Sci. USA* **2010**, *107*, 14691–14696. [CrossRef] [PubMed]
134. A Koeth, R.; Wang, Z.; Levison, B.S.; A Buffa, J.; Org, E.; Sheehy, B.T.; Britt, E.B.; Fu, X.; Wu, Y.; Li, L.; et al. Intestinal microbiota metabolism of l-carnitine, a nutrient in red meat, promotes atherosclerosis. *Nat. Med.* **2013**, *19*, 576–585. [CrossRef] [PubMed]

135. Muegge, B.D.; Kuczynski, J.; Knights, D.; Clemente, J.C.; González, A.; Fontana, L.; Henrissat, B.; Knight, R.; Gordon, J.I. Diet Drives Convergence in Gut Microbiome Functions Across Mammalian Phylogeny and Within Humans. *Science* **2011**, *332*, 970–974. [CrossRef] [PubMed]
136. Bolnick, D.I.; Snowberg, L.K.; Hirsch, P.E.; Lauber, C.L.; Org, E.; Parks, B.; Lusis, A.J.; Knight, R.; Caporaso, J.G.; Svanbäck, R. Individual diet has sex-dependent effects on vertebrate gut microbiota. *Nat. Commun.* **2014**, *5*, 4500. [CrossRef]
137. David, L.A.; Maurice, C.F.; Carmody, R.N.; Gootenberg, D.; Button, J.E.; Wolfe, B.E.; Ling, A.V.; Devlin, A.S.; Varma, Y.; Fischbach, M.A.; et al. Diet rapidly and reproducibly alters the human gut microbiome. *Nat. Cell Biol.* **2014**, *505*, 559–563. [CrossRef]
138. Turnbaugh, P.J.; Hamady, M.; Yatsunenko, T.; Cantarel, B.L.; Duncan, A.; Ley, R.E.; Sogin, M.L.; Jones, W.J.; Roe, B.A.; Affourtit, J.P.; et al. A core gut microbiome in obese and lean twins. *Nat. Cell Biol.* **2008**, *457*, 480–484. [CrossRef]
139. Burns, A.R.; Stephens, W.Z.; Stagaman, K.; Wong, S.; Rawls, J.; Guillemin, K.; Bohannan, B.J. Contribution of neutral processes to the assembly of gut microbial communities in the zebrafish over host development. *ISME J.* **2016**, *10*, 655–664. [CrossRef]
140. Nayak, S.K. Role of gastrointestinal microbiota in fish. *Aquac. Res.* **2010**, *41*, 1553–1573. [CrossRef]
141. Sekirov, I.; Finlay, B.B. The role of the intestinal microbiota in enteric infection. *J. Physiol.* **2009**, *587*, 4159–4167. [CrossRef]
142. Olsen, R.E.; Sundell, K.; Mayhew, T.M.; Myklebust, R.; Ringø, E. Acute stress alters intestinal function of rainbow trout, *Oncorhynchus mykiss* (Walbaum). *Aquaculture* **2005**, *250*, 480–495. [CrossRef]
143. Gareau, M.G.; Wine, E.; Sherman, P.M. Early Life Stress Induces Both Acute and Chronic Colonic Barrier Dysfunction: Figure. *NeoReviews* **2009**, *10*, e191–e197. [CrossRef]
144. Stecher, B.; Robbiani, R.; Walker, A.W.; Westendorf, A.M.; Barthel, M.; Kremer, M.; Chaffron, S.; MacPherson, A.J.; Buer, J.; Parkhill, J.; et al. Salmonella enterica Serovar Typhimurium Exploits Inflammation to Compete with the Intestinal Microbiota. *PLoS Biol.* **2007**, *5*, e244–e2189. [CrossRef] [PubMed]
145. Dethlefsen, L.; Huse, S.; Sogin, M.L.; Relman, D.A. The Pervasive Effects of an Antibiotic on the Human Gut Microbiota, as Revealed by Deep 16S rRNA Sequencing. *PLoS Biol.* **2008**, *6*, e280. [CrossRef] [PubMed]
146. Cain, K.; Swan, C. Barrier function and immunology. In *The Multifunctional Gut of Fish*; Fish Physiology; Grosell, M., Farrell, A.P., Brauner, C.J., Eds.; Academic Press: Cambridge, MA, USA, 2010; Volume 30, pp. 111–134.
147. Gaggìa, F.; Mattarelli, P.; Biavati, B. Probiotics and prebiotics in animal feeding for safe food production. *Int. J. Food Microbiol.* **2010**, *141*, S15–S28. [CrossRef] [PubMed]
148. Gómez, G.D.; Balcázar, J.L. A review on the interactions between gut microbiota and innate immunity of fish: Table 1. *FEMS Immunol. Med. Microbiol.* **2008**, *52*, 145–154. [CrossRef]
149. Nikoskelainen, S.; Salminen, S.; Bylund, G.; Ouwehand, A.C. Characterization of the Properties of Human- and Dairy-Derived Probiotics for Prevention of Infectious Diseases in Fish. *Appl. Environ. Microbiol.* **2001**, *67*, 2430–2435. [CrossRef]
150. Dash, G.; Raman, R.P.; Prasad, K.P.; Makesh, M.; Pradeep, M.; Sen, S. Evaluation of Lactobacillus plantarum as feed supplement on host associated microflora, growth, feed efficiency, carcass biochemical composition and immune response of giant freshwater prawn, *Macrobrachium rosenbergii* (de Man, 1879). *Aquaculture* **2014**, *432*, 225–236. [CrossRef]
151. Xue-Feng, W.U.; Zhao, J.L.; Qian, Y.Z.; Chao, W. Histological Study of the Digestive System Organogenesis for the Mandarin Fish, Siniperca chuatsi. *Zool. Res.* **2007**, *28*, 511–518.
152. Yúfera, M.; Pascual, E.; Fernandez-Diaz, C. A highly efficient microencapsulated food for rearing early larvae of marine fish. *Aquaculture* **1999**, *177*, 249–256. [CrossRef]
153. Olsen, A.I.; Attramadal, Y.; Reitan, K.I.; Olsen, Y. Food selection and digestion characteristics of Atlantic halibut (Hippoglossus hippoglossus) larvae fed cultivated prey organisms. *Aquaculture* **2000**, *181*, 293–310. [CrossRef]
154. Meng, Y.; Wang, J.; Wang, Z.; Zhang, G.; Liu, L.; Huo, G.; Li, C. Lactobacillus plantarum KLDS1.0318 Ameliorates Impaired Intestinal Immunity and Metabolic Disorders in Cyclophosphamide-Treated Mice. *Front. Microbiol.* **2019**, *10*, 731. [CrossRef] [PubMed]
155. Doan, H.V.; Doolgindachbaporn, S.; Suksri, A.J.A.N. Effect of Lactobacillus plantarum and Jerusalem artichoke (*Helianthus tuberosus*) on growth performance, immunity and disease resistance of Pangasius catfish (*Pangasius bocourti*, Sauvage 1880). *Aquac. Nutr.* **2016**, *22*, 444–456. [CrossRef]

 *microorganisms* 

*Article*

# Evaluating Methods of Preserving Aquatic Invertebrates for Microbiome Analysis

**Stephanie N. Vaughn and Colin R. Jackson ***

Department of Biology, University of Mississippi, University, MS 38677, USA; snvaughn@go.olemiss.edu
* Correspondence: cjackson@olemiss.edu

**Abstract:** Research on the microbiomes of animals has increased substantially within the past decades. More recently, microbial analyses of aquatic invertebrates have become of increased interest. The storage method used while collecting aquatic invertebrates has not been standardized throughout the scientific community, and the effects of common storage methods on the microbial composition of the organism is unknown. Using crayfish and dragonfly nymphs collected from a natural pond and crayfish maintained in an aquarium, the effects of two common storage methods, preserving in 95% ethanol and freezing at −20 °C, on the invertebrate bacterial microbiome was evaluated. We found that the bacterial community was conserved for two sample types (gut and exoskeleton) of field-collected crayfish stored either in ethanol or frozen, as was the gut microbiome of aquarium crayfish. However, there were significant differences between the bacterial communities found on the exoskeleton of aquarium crayfish stored in ethanol compared to those that were frozen. Dragonfly nymphs showed significant differences in gut microbial composition between species, but the microbiome was conserved between storage methods. These results demonstrate that preserving field-collected specimens of aquatic invertebrates in 95% ethanol is likely to be a simple and effective sample preservation method for subsequent gut microbiome analysis but is less reliable for the external microbiome.

**Keywords:** invertebrate; microbiome; sample preservation; crayfish; dragonfly

**Citation:** Vaughn, S.N.; Jackson, C.R. Evaluating Methods of Preserving Aquatic Invertebrates for Microbiome Analysis. *Microorganisms* **2022**, *10*, 811. https://doi.org/10.3390/microorganisms10040811

Academic Editors: Konstantinos Ar. Kormas and James S. Maki

Received: 23 January 2022
Accepted: 12 April 2022
Published: 13 April 2022



## 1. Introduction

Over the past twenty years, the human microbiome has been at the forefront of health-related research [1]. This has largely been because of an increase in technologies allowing for next generation 16S rRNA gene sequencing, and various human diseases are now known to be the result of gut dysbiosis [2–4]. Advances in more efficient DNA sequencing methods, such as next generation sequencing and Illumina technology, have enabled scientists to pursue microbiome research beyond that of humans [2,5–7]. Substantial interest over the past decade has focused on host-related microbiomes of other animals. However, throughout this increase in animal microbiome studies ranging from humans to invertebrates, there have been inconsistencies between findings, partly because of differences in sample storage methods [4,5,8–12]. This is especially pronounced in aquatic invertebrates where field collection of samples often requires immediate storage, yet the most suitable method for conserving the microbiome of samples has not been defined and has seldom been investigated [13–15].

Preserving aquatic invertebrate samples is crucial to accurately analyzing the bacterial community associated with the specimens of interest. Many studies have incorporated some method of sample preservation prior to later analyses, yet the impact of storage methods on the microbial community of these samples is poorly understood [8,13,14,16,17]. This is important for samples collected in the field, where there may be substantial travel time between the sampling site and laboratory. Common forms of sample preservation for field-collected invertebrates include flash freezing with liquid nitrogen, freezing, and storage in ethanol or RNAlater [13,14,16]. When the goal of the storage is to preserve the invertebrate specimen itself, these methods may be sufficient, but when wanting to conserve the bacterial community associated with these specimens, the impact of these methods are not well understood. While a storage method must preserve the microbiome of a particular sample, there are also logistical considerations, especially in the context of fieldwork, and approaches differ in their availability, ease of use in the field, and cost.

The aim of this study was to determine the effects that two common storage methods, preserving samples in ethanol or freezing, have on the microbiome aquatic invertebrates. Aquatic invertebrates are of increasing interest for microbiome studies because of their significant roles in marine and freshwater ecosystems [18,19]. Ecosystem services provided by aquatic invertebrates include bioturbation, filter feeding, nutrient and chemical retention, and food web interactions [18,20]. Here, we determine how storing samples in 95% ethanol and freezing at −20 °C affected the bacterial composition of gut and exoskeleton samples from one species of crayfish (*Procambarus vioscai paynei*) and three species of dragonfly nymphs (*Libellula luctuosa*, *Pachydiplax longipennis*, and *Erythemis simplicicollis*) collected from a natural pond, as well as a second species of crayfish (*Faxonius virilis*) maintained in an aquarium to help standardize their microbiome prior to collection. Partial 16S rRNA gene sequences obtained from high throughput sequencing were classified into amplicon sequence variants (ASVs) to assess bacterial microbiome composition and alpha diversity of each specimen, and beta diversity between specimens. We show that preservation in 95% ethanol, as is commonly used to preserve invertebrate specimens for other purposes, is a valid method for the preservation of gut microbiomes of aquatic invertebrates, and potentially suitable for the preservation of the external, exoskeleton microbiome.

## 2. Materials and Methods

### 2.1. Specimen Collection and Processing

Multiple experiments were conducted to assess the effects of storage method on the microbiomes of aquatic invertebrates. The first experiment used field-collected aquatic invertebrates: ten crayfish (*Procambarus vioscai paynei*) and 18 dragonfly nymphs (six each of *Libellula luctuosa*, *Pachydiplax longipennis*, *Erythemis simplicicollis*). All organisms were collected on 5/5/2021 from ponds at the University of Mississippi Field Station (UMFS; Lafayette County, MS, USA). Numbers of each species of invertebrate were determined from what was caught. Immediately after collection, specimens were placed into buckets of pond water and transported (1 h) to the laboratory at University of Mississippi main campus. At the laboratory, five crayfish were placed, individually, into 95% ethanol while five were sealed, individually, in sterile bags and frozen in a −20 °C freezer. Similarly, three dragonfly nymphs of each species were placed, individually, in 95% ethanol and three were placed in sterile bags in a −20 °C freezer. Specimens were preserved for almost three months (83 days) before being sampled for microbiome composition.

In a second experiment, a group of commercially acquired crayfish (*Faxonius virilis*) were housed in a 30-gallon aquarium in the laboratory for 24 days in an attempt to reduce individual to individual variation in their microbiome. Aquarium crayfish were fed a standardized diet of commercial food pellets (Hikari Crab Cuisine, Kyorin Co., Ltd., Himeji City, Japan), and Pro PlecoWafers, Tetra, Melle, Germany). After 24 d, 14 visibly healthy crayfish were removed and seven were placed, individually, into 95% ethanol while the other seven were sealed in sterile bags and frozen at −20 °C, as per the field-collected invertebrates. Specimens from the aquarium experiment were preserved for two months (60 days) before being dissected.

For all crayfish, exoskeleton samples were collected by gently rinsing each crayfish quickly in sterile water to remove non-attached microorganisms. This rinsing also served to partly thaw frozen specimens and removed residual ethanol from ethanol-preserved samples. Samples were then scrubbed gently three times for 30 s each using a sterile toothbrush. Material that was scrubbed off was placed into the initial buffer solution (CD1) from a PowerSoil Pro kit (Qiagen, Germantown, MD, USA). Following exoskeleton scraping, crayfish were dissected by making an incision on the dorsal side of the telson and up the abdomen and the gut extracted. The extracted gut samples were placed directly into bead beating tubes containing buffer solution (CD1) from the PowerSoil Pro kit. Dragonfly nymphs were too small to assess for exoskeleton microbiome composition so only the gut microbiome was examined. The guts of dragonfly nymphs were obtained by cutting through the dorsal portion of the abdominal segments and placing the gut into bead beating tubes containing buffer solution (CD1) from PowerSoil Pro kit.

### 2.2. DNA Extraction, Amplification, and Sequencing

DNA was extracted from all sample types using the PowerSoil Pro kit and following manufacturer's instructions. A 250 bp portion of the V4 region of the bacterial 16S rRNA gene in each sample was sequenced using a dual-index 8-nucleotide barcoding approach [21]. This approach uses a single round of PCR, reducing the risk of amplification artifacts. Following amplification, the presence of amplicons was verified using agarose gels, amplification products standardized using SequalPrep plates (Life Technologies, Grand Island, NY, USA), and barcoded products pooled prior to sequencing. The assembled library was spiked with 20% PhiX [22,23] and sequenced on an Illumina MiSeq at the University of Mississippi Medical Center (UMMC) Molecular and Genomics Core Facility.

Raw sequence files (fastq) were processed using the standard 16S rRNA pipeline of the DADA2 package version 1.12.1 [24] within R version 1.3.1073 [25]. At least 80% of sequences from each sample were retained following quality trimming: truncLen = c(240,160), maxN = 0, maxEE = c(2,2), truncQ = 2. Quality profile plots were inspected to ensure proper quality of trimmed reads. During merging of reads, sequences were trimmed further to account for any overhang (trimOverhang = TRUE) and sequences shorter than 250 base pairs (bp) and longer than 256 bp were trimmed. Chimeras were removed using the "consensus" method. Sequences were classified against the RDP v.18 database [26]. Final amplicon sequence variant (ASV) data was transformed into relative abundance (% sequence reads) of microbial taxa for further compositional analysis.

### 2.3. Statistical Analyses

Alpha diversity was assessed using the Inverse Simpson's Index to measure overall bacterial species diversity and Observed Species Richness (richness based on repeated subsampling of the rarefied number of sequences) to determine richness of ASVs. Two-way analysis of variance (ANOVA) tests were performed on samples to determine differences in mean diversity and richness between storage method (frozen or ethanol) and sample type (gut or exoskeleton) for crayfish, or storage method and species for dragonfly nymphs. One-way ANOVAs were performed to further asses the differences in evenness and richness estimates based on crayfish separated by their storage method and corresponding sample type (gut, exoskeleton). Effect sizes were calculated using the pwr package of R to assess statistical importance of ANOVA results. No *a priori* hypothesis were stated, therefore, TukeyHSD post hoc tests were performed to further assess the differences among group means of significant variables. Multivariate analysis of variance (MANOVA) tests were used to assess if bacterial phyla differed between storage method for each invertebrate/experiment (aquarium crayfish, pond crayfish, and dragonfly nymphs). Bray–Curtis dissimilarity matrices compared structural differences of bacterial communities by storage method, and sample type for crayfish, or species for dragonflies. Permutational multivariate analysis (PERMANOVA) tests using Bray–Curtis distance matrices were performed to determine whether storage method, sample type, and/or species significantly affected the composition of the microbiome. Non-metric multidimensional scaling (NMDS) ordinations were created using the metaMDS function in the Vegan package [27] of R to visualize these differences. The most frequent ASVs in ethanol-preserved and frozen gut and exoskeletons of crayfish samples was determined using the "microbiome" package version 1.12.0 [28] in R where "core" AVSs were specified as those most commonly found in samples of each category.

## 3. Results

### 3.1. Sequence Counts

Initial DADA2 analysis yielded 3810 ASVs from a total of 815,362 16S rRNA sequence reads of the V4 region. Following trimming, merging, chimera removal, and classification against RDP (version 18), 3693 ASVs from 671,032 sequences were retained for the full dataset. Independent *t*-tests were run to determine any potential effect that storage method may have on the amount of sequence reads retained per sample. Aquarium crayfish showed a significantly higher number of sequence reads for gut samples from ethanol-preserved crayfish (15,057 $\pm$ 7782 sequences) compared to those from frozen crayfish (7227 $\pm$ 3735; $p < 0.01$, $t(13) = -2.374$). Exoskeleton samples from frozen field-collected crayfish showed a significantly higher number of reads compared to exoskeleton samples from ethanol-preserved field-collected crayfish (20,889 $\pm$ 2678 and 7403 $\pm$ 2861, respectively; $p < 0.001$, $t(7) = 11.41$). Rarefaction parameters were set to retain samples containing more than 2000 sequences for crayfish, which subsequently removed four samples: one frozen aquarium crayfish gut sample, two field-collected ethanol-preserved crayfish gut samples, and one field-collected ethanol-preserved crayfish exoskeleton samples. Dragonfly nymphs showed lower overall numbers of sequence reads retained compared to that of crayfish. Thus, rarefaction parameters for dragonfly nymph samples were set to 1000 sequences which subsequently removed six dragonfly nymphs. Dragonfly nymph samples showed no significant difference between the number of sequence reads retained in ethanol-preserved compared to that of frozen samples.

### 3.2. Differences in the Crayfish Microbiome between Sample Types and Preservation Method

There were significant differences in overall microbiome composition between gut and exoskeleton samples for both aquarium (*F. virilis*) and field-collected crayfish (*P. vioscai paynei*; Adonis PERMANOVA analyses based on Bray–Curtis distances showed $p < 0.001$, $F = 11.554$ and $p < 0.021$, $F = 3.348$, respectively). The gut microbiome of aquarium crayfish showed no significant difference in overall bacterial composition based on storage method (ethanol or frozen; Figure 1A); however, there was a significant difference in overall bacterial community composition between the ethanol-preserved and frozen exoskeleton samples of aquarium crayfish ($p < 0.01$, $F = 4.837$; Figure 1B). Neither gut nor exoskeleton microbiomes of field-collected crayfish differed in terms of overall bacterial composition when comparing storage method (Figure 1C,D).

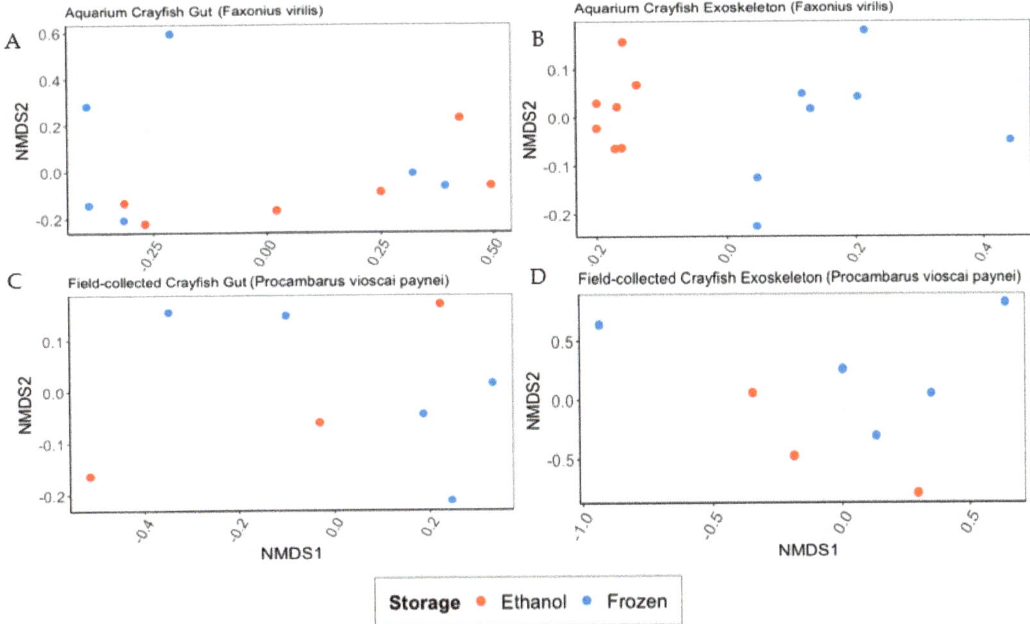


**Figure 1.** NMDS ordinations based on Bray–Curtis dissimilarity scores for bacterial communities of aquarium (*Faxonius virilis*; (**A**,**B**)) and field-collected (*Procambarus vioscai paynei*; (**C**,**D**)) crayfish based on sample preservation method (95% ethanol or frozen at −20 °C) and separated by sample type (gut, (**A**,**C**), or exoskeleton, (**B**,**D**)). Sample preservation method within each plot is represented by color. Gut and exoskeleton communities were significantly different for both aquarium crayfish ($p < 0.001$, F = 11.554) and field-collected crayfish ($p < 0.05$, F = 3.3). Sample preservation method only produced a significant difference in the bacterial community for exoskeleton samples from aquarium crayfish ($p < 0.01$, F = 4.837; (**B**)).

There was a significant difference in the Inverse Simpson's Index and Observed Species Richness based on microbiome location for the aquarium-maintained *F. virilis*, with the exoskeleton microbiome being richer ($p < 0.001$, F = 56.312) and more diverse ($p < 0.01$, F = 13.522) than the gut microbiome. This was particularly pronounced for Species Observed, where exoskeleton samples predicted approximately 400 observed bacterial species compared to 150–300 in the gut community (Figure 2A). The Inverse Simpson's Index was significantly higher in exoskeleton microbiomes of ethanol-preserved of *F. virilis* compared to those from frozen crayfish ($p < 0.01$, F = 11.537; Figure 2B), although storage method did not affect the species diversity of gut microbiomes for these samples (Figure 2B). Field-collected *P. vioscai paynei* showed significant differences in Observed Species Richness and the Inverse Simpson's Index between gut and exoskeleton samples, with gut microbiomes being higher for both indices ($p < 0.01$, F = 15.87 and $p < 0.05$, F = 8.246, respectively). Neither gut nor exoskeleton samples of field-collected crayfish showed significant differences in diversity indices based on sample storage method (Figure 2C,D). Cohen's effect size was medium to large (0.33–0.91) for all comparisons between frozen and ethanol-preserved samples, with the exception of aquarium-maintained *F. virilis* (0.06).

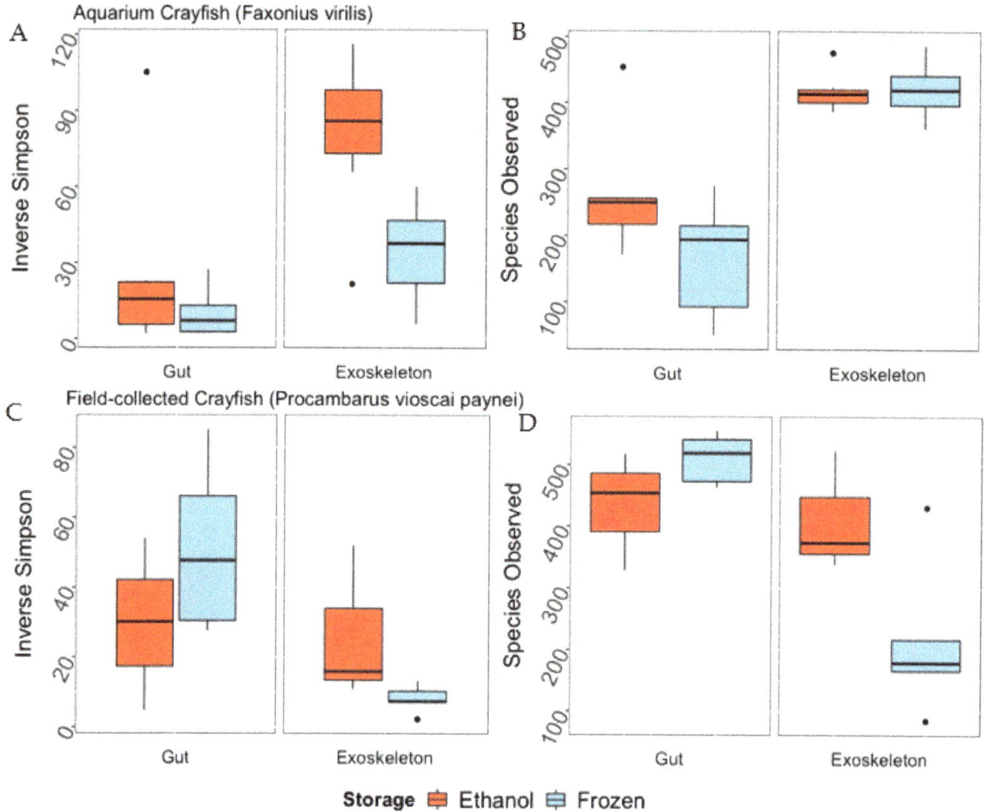


**Figure 2.** Alpha diversity metrics (Inverse Simpson's Index, (**A**,**C**); Observed Species Richness, (**B**,**D**)) derived from gut or exoskeleton bacterial communities of aquarium-maintained (*Faxonius virilis*; (**A**,**B**)) and field-collected (*Procambarus vioscai paynei*; (**C**,**D**)) crayfish collected and stored under different conditions. Samples are separated into their corresponding storage method (95% ethanol, frozen). Boxes show the interquartile range/distribution of values measured in each metric with the black solid line representing the median value from sample type. Vertical lines represent the highest and lowest values associated with each sample type. Dots represent outliers from each group. Observed Species Richness was significantly different between exoskeleton and gut samples for aquarium and field-collected crayfish ($p < 0.001$, F = 56.312 and $p < 0.01$, F = 15.874, respectively), as was the Inverse Simpson's Index ($p < 0.01$, F = 13.522 for aquarium and $p < 0.05$, F = 8.246 for field-collected). Sample preservation method was only significant for the Inverse Simpson's Index of exoskeleton samples from aquarium crayfish ($p < 0.01$, F = 11.537; (**B**)).

There were significant differences in the major bacterial phyla found in gut and exoskeleton samples of aquarium *F. virilis* crayfish (MANOVA; $p < 0.01$, F = 11.554; Figure 3). Based on the proportions of 16S rRNA gene sequences, major bacterial phyla (or subphyla of Proteobacteria) found in the guts of *F. virilis* were the Firmicutes (35.6% of sequences), Bacteroidetes (12.0%), Actinobacteria (10.3%), Gammaproteobacteria (9.50%), Alphaproteobacteria (9.10%), Betaproteobacteria (8.58%), and Planctomycetes (4.35%). Major bacterial phyla/subphyla in exoskeleton samples of aquarium-maintained *F. virilis* were the Bacteroidetes (20.3%), Betaproteobacteria (16.4%), Actinobacteria (15.0%), Alphaproteobacteria (13.9%), Planctomycetes (9.01%), Verrucomicrobia (3.51%), and Deltaproteobacteria (3.19%). Bacterial phyla that differed significantly in their representation between gut and exoskeleton samples were the Firmicutes (MANOVA; $p < 0.001$, F = 19.154) which were proportionally more abundant in gut samples (35.0% more) and Alphaproteobacteria ($p < 0.05$, F = 7.168) which were proportionally more abundant in exoskeleton samples (4.8% more). While there was some variability in the proportions of major bacterial phyla in the gut microbiomes of *F. virilis* between ethanol-preserved and frozen samples, none of this variability was significant (MANOVA; $p > 0.05$). The exoskeleton microbiomes of aquarium crayfish did show differences in the composition of major bacterial phyla based on sample storage method, with the percentage representation of Betaproteobacteria (MANOVA; $p < 0.001$, F = 2.812), and Bacteroidetes ($p < 0.001$, F = 26.264), being significantly higher in ethanol-preserved samples (+8.19% and +10.7%, respectively) and the percentage of Actinobacteria being +16.2% higher in frozen samples ($p < 0.01$, F = 11.522).

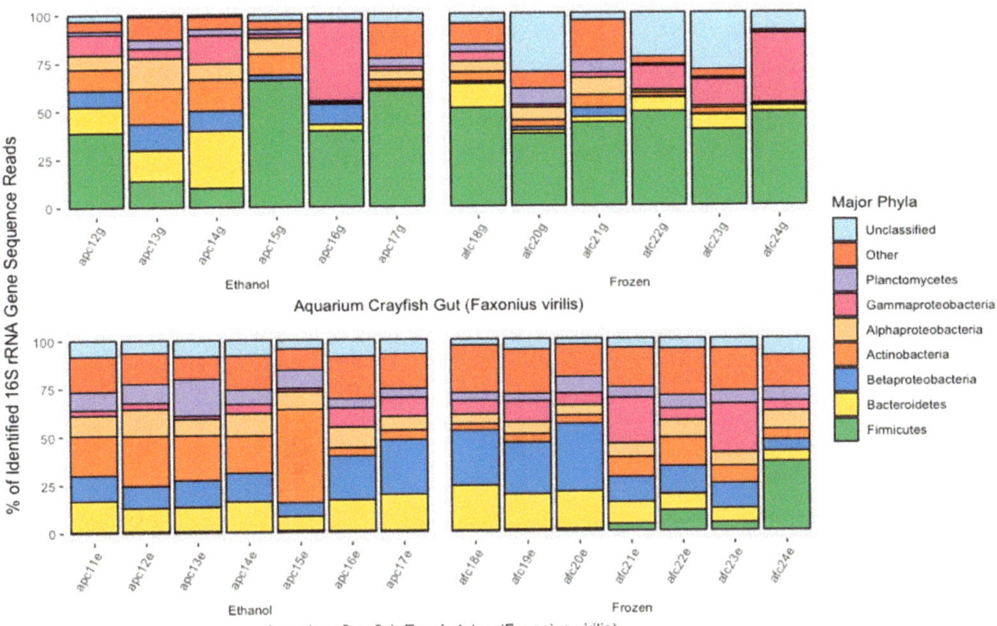


**Figure 3.** Major bacterial phyla found in the gut and exoskeleton microbiomes of aquarium-maintained crayfish (*Faxonius virilis*) as determined from percent of 16S rRNA gene sequence reads. Each bar represents one individual and are separated by sample type (gut or exoskeleton) and sample storage method (in 95% ethanol or frozen at −20 °C). Sample names are located on the *x*-axis and correspond to the location (i.e., a = aquarium), storage method (i.e., p = ethanol-preserved, f = frozen), the number order of crayfish collection, storage, and subsequent dissection, and the sample type being analyzed (i.e., g = gut, e = exoskeleton).

As with aquarium crayfish, the major bacterial phyla/subphyla in the microbiomes of field-collected *P. vioscai paynei* crayfish were significantly different between exoskeleton and gut samples (MANOVA; $p < 0.05$, F = 3.48; Figure 4). The gut microbiome (Figure 4) was primarily composed of Firmicutes (49.4% of sequences), Cyanobacteria (6.12%), Alphaproteobacteria (5.72%), Planctomycetes (5.47%), Bacteroidetes (4.87%%), and Actinobacteria (4.45%). The major bacterial phyla making up the exoskeleton microbiome were Betaproteobacteria (22.8%), Bacteroidetes (15.5%), Verrucomicrobia (12.5%), Gammaproteobacteria (12.4%), Alphaproteobacteria (7.14%), Actinobacteria (6.98%), and Planctomycetes (5.61%). Gut and exoskeleton samples from field-collected crayfish differed in their percentage representation of Actinobacteria (MANOVA; $p < 0.05$, F = 5.135, +2.53% in exoskeleton) and Verrucomicrobia ($p < 0.01$, F = 11.280, +11.53% in exoskeleton). Storage method had no significant effect on proportions of any of the major bacterial phyla/subphlya in the gut or exoskeleton microbiome for field-collected crayfish.

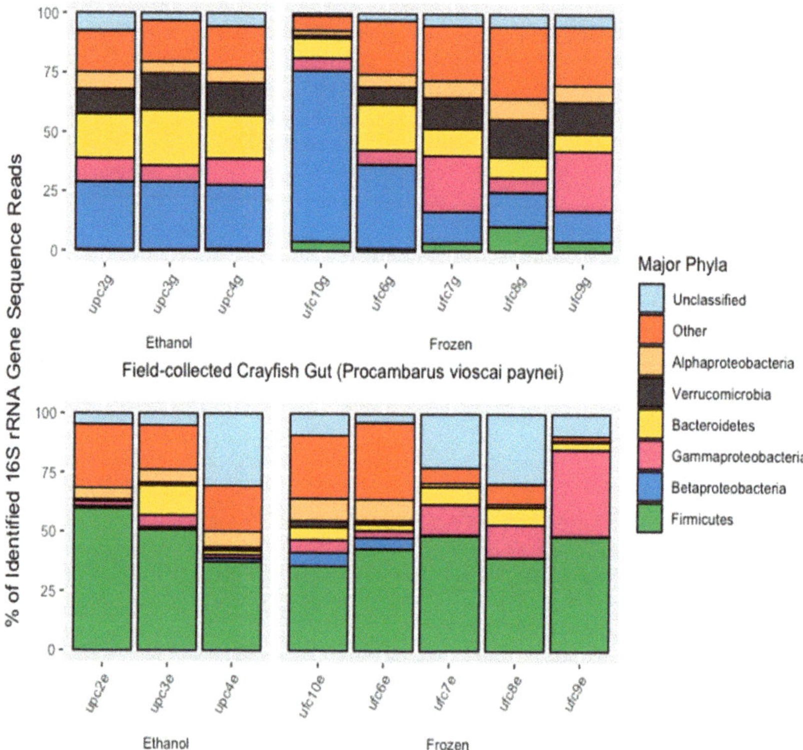


**Figure 4.** Major bacterial phyla found in the gut and exoskeleton microbiomes of field-collected crayfish (*Procambarus vioscai paynei*) as determined from percent of 16S rRNA gene sequence reads. Each bar represents one individual and are separated by sample type (gut or exoskeleton) and sample storage method (in 95% ethanol or frozen at −20 °C). Sample names are located on the *x*-axis and correspond to the location (i.e., a = aquarium), storage method (i.e., *p* = ethanol-preserved, f = frozen), the number order of crayfish collection, storage, and subsequent dissection, and the sample type being analyzed (i.e., g = gut, e = exoskeleton).

### *3.3. Dominant ASVs by Sample Type and Preservation Method*

The most frequently observed ASVs from aquarium and field-collected crayfish of each sample type preserved in ethanol of frozen were determined and classified by their finest identified taxonomic level. For gut microbiome samples from aquarium crayfish (*F. virilis*), four of the six most abundant ASVs were the same regardless of the method of sample preservation (Table 1). Those that were not specifically identified as the same ASV all classified within the Proteobacteria phylum (ASV34, ASV69, ASV25, and ASV27). ASV1, ASV4, and ASV9 were the three most abundant ASVs within both frozen and ethanol-preserved gut samples; however, the most abundant in these samples, ASV1, could not be identified further than the phylum level (Firmicutes). Consistency in dominant ASVs between sample storage procedures was much less for the exoskeleton samples from aquarium crayfish, with only one of the six most frequent ASVs being in the core microbiome of both ethanol-preserved and frozen samples (ASV9, identified as a member of *Mycobacterium*).

**Table 1.** The core microbiome (most frequently identified ASVs from each sample) of aquarium crayfish (*Faxonius virilis*) gut and exoskeleton samples, separated into those preserved in 95% ethanol or frozen at −20 °C.

| Aquarium Crayfish | ASV | Identification | Frequency [a] | Relative Abundance [b] | CI (+/−) |
|---|---|---|---|---|---|
| Gut Ethanol | ASV 1 | Firmicutes (Firmicutes) | 6/6 | 24.2% | 8.39% |
| | ASV 4 | *Flavobacterium* (Bacteroidetes) | 6/6 | 7.10% | 2.55% |
| | ASV 9 | *Mycobacterium* (Actinobacteria) | 6/6 | 3.67% | 1.22% |
| | ASV 34 | *Gemmobacter* (Alpharoteobacteria) | 6/6 | 2.32% | 0.81% |
| | ASV 33 | *Mycobacterium* (Actinobacteria) | 6/6 | 1.77% | 0.60% |
| | ASV 69 | *Dechloromonas* (Betaproteobacteria) | 5/6 | 1.28% | 0.44% |
| Gut Frozen | ASV 1 | Firmicutes (Firmicutes) | 6/6 | 21.8% | 5.01% |
| | ASV 4 | *Flavobacterium* (Bacteroidetes) | 6/6 | 5.58% | 1.07% |
| | ASV 9 | *Mycobacterium* (Actinobacteria) | 6/6 | 5.23% | 0.63% |
| | ASV 27 | *Hydromonas* (Betaproteobacteria) | 6/6 | 3.93% | 1.48% |
| | ASV 25 | *Citrobacter* (Gammaproteobacteria) | 5/6 | 2.10% | 0.12% |
| | ASV 33 | *Mycobacterium* (Actinobacteria) | 6/6 | 1.82% | 0.25% |
| Exoskeleton Ethanol | ASV 3 | Kineosporiaceae (Actinobacteria) | 7/7 | 15.5% | 1.93% |
| | ASV 31 | Bacteroidetes (Bacteroidetes) | 7/7 | 2.14% | 0.18% |
| | ASV 19 | Phycisphaeraceae (Planctomycetes) | 6/7 | 1.98% | 0.60% |
| | ASV 21 | Pirellulaceae (Planctomycetes) | 7/7 | 1.84% | 0.13% |
| | ASV 28 | *Fimbriiglobus* (Planctomycetes) | 6/7 | 1.26% | 0.07% |
| | ASV 9 | *Mycobacterium* (Actinobacteria) | 7/7 | 1.11% | 0.10% |
| Exoskeleton Frozen | ASV 1 | Firmicutes (Firmicutes) | 7/7 | 19.1% | 4.04% |
| | ASV 4 | *Flavobacterium* (Bacteroidetes) | 7/7 | 4.81% | 0.89% |
| | ASV 9 | *Mycobacterium* (Actinobacteria) | 7/7 | 4.56% | 0.55% |
| | ASV 27 | *Hydromonas* (Betaproteobacteria) | 5/7 | 3.36% | 1.18% |
| | ASV 25 | *Citrobacter* (Gammaproteobacteria) | 6/7 | 1.81% | 0.24% |
| | ASV 33 | *Mycobacterium* (Actinobacteria) | 7/7 | 1.56% | 0.22% |

[a] Frequency was determined from the number of individuals found with that ASV. [b] Relative abundance was determined from the total number of each ASV identified within each storage group (i.e., ethanol and frozen).

The most frequently detected ASVs in the gut microbiome of field-collected *P. vioscai paynei* were generally the same regardless of sample storage method, with five of the six most common ASVs being found in both ethanol-preserved and frozen gut samples (Table 2). Sample storage method had a greater impact on the exoskeleton microbiome of field-collected crayfish, with only two of six common ASVs (ASV12 identified as *Sphaerotilus*, and ASV 16 identified as *Verrucomicrobium*) being the same for ethanol-preserved and frozen samples (Table 2).

**Table 2.** The core microbiome (most frequently identified ASVs from each sample) of field-collected crayfish (*Procambarus vioscai paynei*) gut and exoskeleton samples, separated into those preserved in ethanol or frozen at −20 °C.

| Field-Collected Crayfish | ASV | Identification | Frequency [a] | Relative Abundance [b] | CI (+/−) |
|---|---|---|---|---|---|
| Gut Ethanol | ASV 7 | *Catenococcus* (Gammaproteobacteria) | 3/3 | 16.6% | 2.96% |
| | ASV 1 | *Rhodobacter* (Firmicutes) | 3/3 | 12.7% | 3.22% |
| | ASV 15 | Bacilli (Firmicutes) | 3/3 | 10.3% | 1.86% |
| | ASV 11 | *Clostridium_XIVb* (Firmicutes) | 3/3 | 7.72% | 1.81% |
| | ASV 22 | Firmicutes (Fimicutes) | 3/3 | 5.73% | 1.65% |
| | ASV 32 | *Dysgonomonas* (Bacteroidetes) | 3/3 | 4.73% | 0.75% |
| Gut Frozen | ASV 17 | Firmicutes (Firmicutes) | 5/5 | 14.5% | 3.51% |
| | ASV 1 | Firmicutes (Firmicutes) | 5/5 | 12.5% | 1.53% |
| | ASV 22 | Firmicutes (Firmicutes) | 3/5 | 4.77% | 1.03% |
| | ASV 11 | *Clostridium_XIVb* (Firmicutes) | 5/5 | 4.53% | 0.86% |
| | ASV 32 | *Dysgonomonas* (Bacteroidetes) | 3/5 | 3.41% | 1.01% |
| | ASV 15 | Bacilli (Firmicutes) | 3/5 | 2.09% | 0.39% |
| Exoskeleton Ethanol | ASV 83 | Methylococcaceae (Gammaproteobacteria) | 3/3 | 3.78% | 1.44% |
| | ASV 16 | *Verrucomicrobium* (Verrucomicrobia) | 3/3 | 3.37% | 1.24% |
| | ASV 68 | Kineosporiaceae (Actinobacteria) | 3/3 | 3.32% | 0.83% |
| | ASV 115 | Verrucomicrobiaceae (Verrucomicrobia) | 3/3 | 2.73% | 0.54% |
| | ASV 171 | Verrucomicrobia (Verrucomicrobia) | 3/3 | 1.62% | 0.13% |
| | ASV 193 | Micrococcales (Actinobacteria) | 3/3 | 1.05% | 0.08% |
| Exoskeleton Frozen | ASV 8 | Comamonadaceae (Proteobacteria) | 5/5 | 9.49% | 1.06% |
| | ASV 16 | *Verrucomicrobium* (Verrucomicrobia) | 5/5 | 4.87% | 0.76% |
| | ASV 24 | *Methylobacter* (Gammaproteobacteria) | 5/5 | 3.84% | 0.35% |
| | ASV 12 | *Sphaerotilus* (Betaproteobacteria) | 5/5 | 3.73% | 0.20% |
| | ASV 29 | Comamonadaceae (Betaproteobacteria) | 5/5 | 2.80% | 0.57% |
| | ASV 35 | *Aquabacterium* (Betaproteobacteria) | 5/5 | 2.73% | 0.19% |

[a] Frequency was determined from the number of individuals found with that ASV. [b] Relative abundance was determined from the total number of each ASV identified within each storage group (i.e., ethanol and frozen).

### 3.4. Patterns in the Dragonfly Nymph Microbiome by Species and Preservation Method

Gut microbiomes of the three species of dragonfly nymphs (*E. simplicicollis*, *L. luctuosa*, *P. longipennis*) were significantly different from each other based on species (Adonis PERMANOVA analyses based on Bray–Curtis distances; $p < 0.05$, F = 1.844; Figure 5A). There was, however, no difference in overall microbiome composition based on sample preservation method (Figure 5A). Similarly, there were no significant differences in the alpha diversity indices (Inverse Simpson's Index, Observed Species Richness) of dragonfly gut microbiomes based on sample preservation method or, for that matter, by host species (Figure 5B). Dominant bacterial phyla (subphyla for Proteobacteria) in the 16S rRNA gene sequence dataset recovered from dragonfly nymphs were the Betaproteobacteria (32.7% of recovered sequences), Gammaproteobacteria (16.6%), Firmicutes (9.61%), Alphaproteobacteria (8.90%), Bacteroidetes (6.18%), and Planctomycetes (4.35%) (Figure 5C). The only phyla that showed a significant difference in relative abundance based on sample storage method, were the Bacteroidetes (MANOVA; $p < 0.05$, F = 7.242), which were found at a higher proportion in the frozen *P. longipennis* samples (22.6% more abundant) compared to ethanol-preserved samples of the same species.

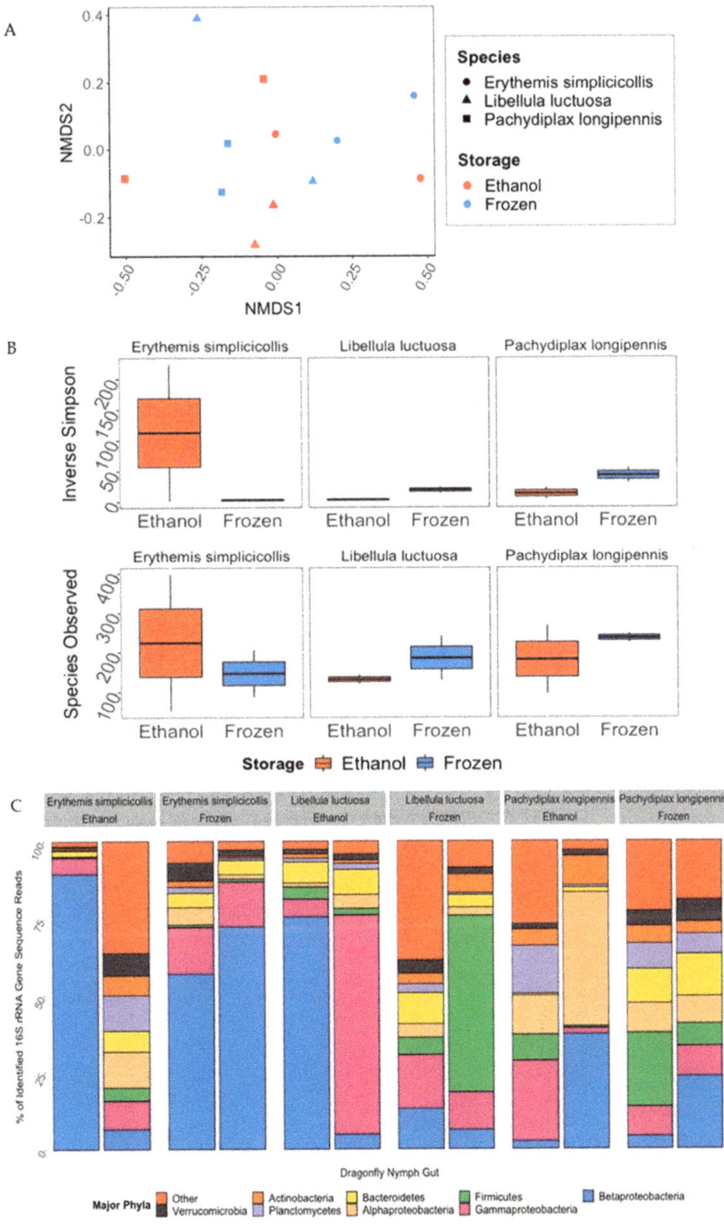


**Figure 5.** Diversity patterns in gut microbiome of three species of dragonfly nymphs (*E. simplicicollis, L. luctuosa,* and *P. longipennis*) that were preserved in 95% ethanol or frozen at −20 °C. (**A**) NMDS ordination based on Bray–Curtis dissimilarity scores (**B**) Alpha diversity plots of Inverse Simpson's Index and Observed Species Richness separated by host species and preservation method. There were no significant differences in diversity indices between preservation methods for any species. (**C**) Major bacterial phyla found in the gut of dragonfly nymphs as determined from percent of 16S rRNA gene sequence reads. Each bar represents one individual and are separated by storage method (gut or exoskeleton) and nymph species.

## 4. Discussion

While the number of studies analyzing the host-associated bacterial communities of aquatic invertebrates is increasing, there are few studies analyzing the effects that preservation has on stored specimen's microbiome. Of the few studies previously analyzing the effects that preservation has on any microbiome sample [8,16,17,29], they have primarily focused on preserving fecal specimens of vertebrates rather than preserving the entire host as we did for the aquatic invertebrates sampled in this study. Furthermore, the results of the previous studies were inconclusive as to which storage method would be ideal for microbiome preservation of their samples, leaving the decision to the investigator. However, given that ethanol is one of the most commonly used preservation methods for storing aquatic invertebrates [30–33], it is critical to understand the effects ethanol has on the bacterial community of host species before choosing and standardizing field-preservation methods or analyzing invertebrates stored for the long-term in collections.

Consistent in all analyses were the differences between the gut and exoskeleton microbiomes of both crayfish species and the differences between species for dragonfly nymphs. In the current study, these differences were apparent regardless of sample storage method (freezing, preservation in 95% ethanol) suggesting that broad ecological patterns are likely to be detected regardless of how samples are preserved. The bacterial communities associated with aquatic macroinvertebrates (e.g., crayfish) has often been found to differ based on the locality of the sample [34–37]. Skelton et al. [36] characterized the carapace and gill microbiomes of the crayfish species, *Cambarus sciotensis*, the first characterization of any crayfish microbiome to their knowledge. They found that the bacterial community of the exoskeleton was largely influenced by the water column that crayfish were collected from [36]. That study, along with more recent studies [34,35,37], and the results of the current study show the differences in bacterial diversity and major bacterial taxa between different parts of the crayfish body, and suggest that each area may have its own functional role for the well-being of the host.

When investigating multiple insect species (*Pieris rapae* (Lepidoptera), *Arphia conspersa* (Orthoptera), *Epilachna varivestis* (Coleoptera), *Apis mellifera* (Hymenoptera)) preserved by various methods, Hammer et al. [15] found similar results to our study, in that they were able to distinguish the microbiomes between different species, regardless of storage method [13]. However, they declared that no single storage method had a significantly greater preservation effect on the bacterial community of the insects than any other and suggest that storage method be determined by the investigator based on cost and efficiency (i.e., travel time from field to laboratory). Along with our findings that 95% ethanol was a suitable sample preservation method for microbiome analyses of crayfish and other aquatic invertebrates, this suggests the potential that samples that have been stored long-term in ethanol, as is common in collections, could be characterized to assess their microbiomes. That said, assessing the effects of longer-term storage in ethanol should be a priority, although such studies would, by nature, take a much longer period of time.

The most dominant taxa in the gut microbiome of aquarium crayfish, both ethanol-preserved and frozen, were Firmicutes, consistent with previous studies analyzing gut bacteria of crayfish [34,35]. Exoskeleton samples from these same crayfish showed the greatest differences in microbiome composition based on preservation method, with ethanol-preserved vs. frozen individuals differing in terms of dominant phyla, major ASVs, and alpha diversity indices. Looking at the differences, there is the possibility that ethanol-preservation decreased the percentages of dominant taxa making the exoskeleton bacterial community more even, although it is equally possible that freezing may have had the opposite effect. Sampling the microbiome from crayfish immediately after collection would be useful as a control for direct comparisons to preserved samples, but it is generally necessary to freeze crayfish prior to scrubbing the exoskeleton, and the humane way of euthanizing invertebrates typically entails freezing or ethanol immersion, making microbiome sampling from freshly collected individuals difficult.

Gut microbiomes from the field-collected crayfish *P. vioscai paynei* were similar between ethanol-preserved and frozen samples with Betaproteobacteria being the most prevalent phyla identified. This similarity in microbiome composition regardless of preservation method was further supported through alpha and beta diversity indices. Although Betaproteobacteria accounted for the greatest percentage of sequences in these samples, the most frequently detected ASVs in the gut microbiomes of both ethanol-preserved and frozen *P. vioscai paynei* were identified as belonging to Firmicutes phylum, taxa that have been regarded as common in the guts of other crayfish species [34,35]. Firmicutes were the most prevalent phylum in exoskeleton samples of the field-collected crayfish, regardless of preservation method, although the most commonly detected ASVs were identified as members of the phyla Verrucomicrobia and Proteobacteria. While there was variability in the most frequently identified ASVs in the exoskeleton microbiome of ethanol-preserved and frozen samples, alpha and beta diversity metrics suggested that preservation method had little impact on the overall microbiome associated with the exoskeleton of field-collected crayfish.

The gut microbiomes of all three species of dragonfly nymphs were dominated by 16S rRNA gene sequences classified within phylum Proteobacteria, which is consistent with previous studies analyzing the gut microbiome of dragonfly nymphs [38,39]. Those previous studies also found that host species had a significant effect on the gut microbial community of dragonflies, and the three species of nymphs examined in this study (*Libellula luctuosa, Pachydiplax longipennis, Erythemis simplicicollis*) were also found previously to have distinct gut microbiomes [38]. Preservation method had no effect on any of the microbiome community parameters that we examined, suggesting that future studies could be conducted to look at the gut microbiomes of dragonfly nymphs, as well as other aquatic insects, that are commonly stored in ethanol. That said, larger studies on the effects of sample preservation on the aquatic insect microbiome are needed, as the results of this portion of our study are potentially limited by a relatively low sample size.

Using a consistent method of sample preservation within a study is important to accurately assess ecological patterns in microbiome composition. This is evident from our finding that, while most types of samples yielded similar microbiome data regardless of whether samples were frozen or preserved in ethanol, the exoskeleton microbiome of *F. virilis* differed substantially with preservation method. Others have found significant differences between frozen and ethanol-preserved tadpole feces (*Nanorana parkeri*), although that was acknowledged, in part, as being due to thawing of frozen samples during transport to the laboratory [8]. Of the few studies that have analyzed the effect of perseveration method on the microbiome of other aquatic invertebrates, most have concluded that the microbiome of organisms is capable of being retained after specimen storage [13–15]. From the current study, it was determined that 95% ethanol is an acceptable method to conserve the internal microbiome and a potential way to conserve the external microbiome of aquatic invertebrates. The potential for ethanol to be used as quick and economical method of preserving specimens in the field shows promise and would reduce potential issues with the transportation of frozen specimens for later microbiome analysis.

Standardized protocols for preserving aquatic invertebrate samples gives the scientific community the opportunity to directly compare the effects of species, habitat, climate, nutrients, etc., on the microbiome of these aquatic organisms. Ethanol is one of the most frequently used preservation methods for storage of aquatic invertebrate specimens for study and in museum collections, because of its ability to fix specimen, morphologically and molecularly [30–33], and our study shows that it can also be used for preservation of the gut microbiome. One limitation of our study, however, could be the length of time that samples were stored (almost three months) and future work could examine how longer storage times relate to the reliability of recovering a representative microbiome community, especially if long-term ethanol-preserved specimens, such as in museum collections [30–33], are to be examined. Regardless, this initial study shows that ethanol-preservation was as successful as freezing in conserving the gut microbial community of a variety of aquatic invertebrates. Future work should further examine the impacts of sample preservation methods on the microbiome of other aquatic animals that are commonly preserved in ethanol, such as mollusks and even vertebrates.

**Author Contributions:** Conceptualization, S.N.V. and C.R.J.; validation, S.N.V. and C.R.J.; formal analysis: S.N.V.; investigation, S.N.V. and C.R.J.; resources, C.R.J.; data curation, S.N.V.; writing—original draft preparation, S.N.V.; writing—review and editing, S.N.V. and C.R.J.; visualization, S.N.V.; supervision, C.R.J.; project administration, C.R.J.; funding acquisition, C.R.J. All authors have read and agreed to the published version of the manuscript.

**Funding:** This research was funded by the National Science Foundation award DEB 1831531 to C.R.J.

**Institutional Review Board Statement:** Not applicable.

**Informed Consent Statement:** Not applicable.

**Data Availability Statement:** Raw sequences are deposited in the NCBI Sequence Reads Archive under BioProject ID PRJNA797466.

**Conflicts of Interest:** The authors declare no conflict of interest.

## References

1. Cani, P.D. Human Gut Microbiome: Hopes, threats and promises. *Gut* **2018**, *67*, 1716–1725. [CrossRef] [PubMed]
2. Clooney, A.G.; Fouhy, F.; Sleator, R.D.; O'Driscroll, A.; Stanton, C. Comparing Apples and Oranges?: Next Generation Sequencing and Its Impact on Microbiome Analysis. *PLoS ONE* **2016**, *11*, e0148028. [CrossRef]
3. Colston, T.J.; Jackson, C.R. Microbiome evolution along divergent branches of the vertebrate tree of life: What is known and unknown. *Mol. Ecol.* **2016**, *25*, 3776–3800. [CrossRef] [PubMed]
4. Ma, J.; Sheng, L.; Hong, Y.; Xi, C.; Gu, Y.; Zheng, N.; Li, M.; Chen, L.; Wu, G.; Li, Y.; et al. Variations of Gut Microbiome Profile Under Different Storage Conditions and Preservation Periods: A Multi-Dimensional Evaluation. *Front. Microbiol.* **2020**, *11*, 972. [CrossRef] [PubMed]
5. Greay, T.L.; Gofton, A.W.; Paparini, A.; Ryan, U.M.; Oskam, C.L.; Irwin, P.J. Recent insights into the tick microbiome gained through next-generation sequencing. *Parasites Vectors* **2018**, *11*, 12. [CrossRef] [PubMed]
6. Ghanbari, M.; Kneifel, W.; Domig, K.J. A new view of the fish gut microbiome: Advances from next-generation sequencing. *Aquaculture* **2015**, *448*, 464–475. [CrossRef]
7. Foster, J.A.; Bunge, J.; Gilbert, J.A.; Moore, J.H. Measuring the microbiome: Perspectives on advances in DNA-based techniques for exploring microbial life. *Brief. Bioinform.* **2012**, *13*, 420–429. [CrossRef]
8. Anslan, S.; Li, H.; Kunzel, S.; Vences, M. Microbiomes from feces vs. gut in tadpoles: Distinct community compositions between substrates and preservation methods. *Salamandra* **2021**, *57*, 96–104.
9. Majumder, R.; Sutcliffe, B.; Taylor, P.W.; Chapman, T.A. Next-Generation Sequencing reveals relationship between the larval microbiome and food substrate in the polyphagous Queensland fruit fly. *Sci. Rep.* **2019**, *9*, 14292. [CrossRef]
10. Song, S.J.; Amir, A.; Metcalf, J.L.; Amato, K.R.; Xu, Z.Z.; Humphrey, G.; Knight, R. Preservation methods differ in fecal microbiome stability, affecting suitability for field studies. *mSystems* **2016**, *1*, e00021-16. [CrossRef]
11. Horng, K.R.; Ganz, H.H.; Eisen, J.A.; Marks, S.L. Effects of preservation method on canine (*Canis lupus familiaris*) fecal microbiota. *PeerJ* **2018**, *6*, e4827. [CrossRef] [PubMed]
12. De Cock, M.; Virgilio, M.; Vandamme, P.; Augustinos, A.; Bourtzis, K.; Willems, A.; De Meyer, M. Impact of Sample Preservation and Manipulation on Insect Gut Microbiome Profiling. A Test Case with Fruit Flies (Diptera, Tephritidae). *Front. Microbiol.* **2019**, *10*, 2833. [CrossRef] [PubMed]
13. Hammer, T.J.; Dickerson, J.C.; Fierer, N. Evidence-based recommendations on storing and handling specimens for analyses of insect microbiota. *PeerJ* **2015**, *3*, e1190. [CrossRef]
14. Simister, R.; Schmitt, S.; Taylor, M.W. Evaluating methods for the preservation and extraction of DNA and RNA for analysis of microbial communities in marine sponges. *J. Exp. Mar. Biol. Ecol.* **2011**, *397*, 38–43. [CrossRef]
15. Rocha, J.; Coelho, F.; Peixe, L.; Gomes, N.C.M.; Calado, R. Optimization of preservation and processing of sea anemones for microbial community analysis using molecular tools. *Sci. Rep.* **2014**, *4*, 6986. [CrossRef] [PubMed]
16. Blekhman, R.; Tang, K.; Archie, E.A.; Barreiro, L.B.; Johnson, Z.P.; Wilson, M.E.; Kohn, J.; Yuan, M.L.; Gesquiere, L.; Grieneisen, L.E.; et al. Common methods for fecal sample storage in field studies yield consistent signatures of individual identity in microbiome sequencing data. *Sci. Rep.* **2016**, *6*, 31519. [CrossRef]
17. Lauber, C.L.; Zhou, N.; Gordon, J.I.; Knight, R.; Fierer, N. Effect of storage conditions on the assessment of bacterial community structure in soil and human-associated samples. *FEMS Microbiol. Lett.* **2010**, *307*, 80–86. [CrossRef] [PubMed]
18. Vaughn, C.C. Ecosystem services provided by freshwater mussels. *Hydrobiologia* **2018**, *810*, 15–27. [CrossRef]
19. Weingarten, E.A.; Atkinson, C.L.; Jackson, C.R. The gut microbiome of freshwater Unionidae mussels is determined by host species and is selectively retained from filtered seston. *PLoS ONE* **2019**, *14*, e0224796. [CrossRef]

20. Prather, C.M.; Pelini, S.L.; Laws, A.; Rivest, E.; Woltz, M.; Bloch, C.P.; Del Toro, I.; Ho, C.-K.; Kominoski, J.; Newbold, T.A.S.; et al. Invertebrates, ecosystem services and climate change. *Biol. Rev.* **2013**, *88*, 327–348. [CrossRef]
21. Kozich, J.J.; Westcott, S.L.; Baxter, N.T.; Highlander, S.K.; Schloss, P.D. Development of a dual-index sequencing strategy and curation pipeline for analyzing amplicon sequence data on the MiSeq Illumina sequencing platform. *Appl. Environ. Microbiol.* **2013**, *79*, 5112–5120. [CrossRef] [PubMed]
22. Jackson, C.R.; Stone, B.W.G.; Tyler, H.L. Emerging perspectives on the natural microbiome of fresh produce vegetables. *Agriculture* **2015**, *5*, 170–187. [CrossRef]
23. Stone, B.W.G.; Jackson, C.R. Biogeographic patterns between bacterial phyllosphere communities of the Southern Magnolia (*Magnolia grandiflora*) in a small forest. *Microb. Ecol.* **2016**, *71*, 954–961. [CrossRef] [PubMed]
24. Callahan, B.J.; McMurdie, P.J.; Rosen, M.J.; Han, A.W.; Johnson, A.J.A.; Holmes, S.P. DADA2: High-resolution sample inference from Illumina amplicon data. *Nat. Methods* **2016**, *13*, 581–583. [CrossRef] [PubMed]
25. RStudio Team. *RStudio: Integrated Development for R*; RStudio PBC: Boston, MA, USA, 2020. Available online: http://www.rstudio.com/ (accessed on 7 July 2020).
26. Cole, J.R.; Wang, Q.; Fish, J.A.; Chai, B.; McGarrell, D.M.; Sun, Y.; Brown, C.T.; Porras-Alfaro, A.; Kuske, C.R.; Tiedje, J.M. Ribosomal Database Project: Data and tools for high throughput rRNA analysis. *Nucleic Acids Res.* **2014**, *42*, D633–D642. [CrossRef] [PubMed]
27. Oksanen, J.; Kindt, R.; Legendre, P.; O'Hara, B.; Simpson, G.L.; Solymos, P.; Stevens, M.H.H.; Wagner, H. vegan: Community Ecology Package. *Version 2.5–7* **2008**, *10*, 631–637.
28. Lahti, L.; Sudarshan, S.; Blake, T.; Salojarvi, J. Tools for microbiome analysis in R. *Version* **2017**, *1*, 28.
29. Hale, V.L.; Tan, C.L.; Knight, R.; Amato, K.R. Effect of preservation method on spider monkey (*Ateles geoffroyi*) fecal microbiota over 8 weeks. *J. Microbiol. Methods* **2015**, *113*, 16–26. [CrossRef]
30. Krogmann, L.; Holstein, J. Preserving and Specimen Handling: Insects and other Invertebrates. In *Manual on Field Recording Techniques and Protocols for All Taxa Biodiversity Inventories*; Eymann, J., Degreef, J., Hauser, C., Monje, J.C., Samyn, Y., VandenSpiegel, D., Eds.; ABC Taxa: Brussels, Belgium, 2010; Volume 1, pp. 463–479.
31. Schiller, E.K.; Haring, E.; Daubl, B.; Gaub, L.; Szeiler, S.; Sattmann, H. Ethanol concentration and sample preservation considering diverse storage parameters: A survey of invertebrate wet collections of the Natural History Museum Vienna. *Ann. Naturhist. Mus. Wien* **2014**, *116*, 41–61.
32. Moreau, C.S.; Wray, B.D.; Czekanski-Moir, J.E.; Rubin, B.E.R. DNA preservation: A test of commonly used preservatives for insects. *Invert. Syst.* **2013**, *27*, 81–86. [CrossRef]
33. Nagy, Z.T. A hands-on overview of tissue preservation methods for molecular genetic analyses. *Org. Divers. Evol.* **2010**, *10*, 91–105. [CrossRef]
34. Xavier, R.; Soares, M.C.; Silva, S.M.; Banha, F.; Gama, M.; Ribeiro, L.; Anastacio, P.; Cardoso, S.C. Environment and host-related factors modulate gut and carapace bacterial diversity of the invasive red swamp crayfish (*Procambarus clarkii*). *Hydrobiologia* **2021**, *848*, 1045–4057. [CrossRef]
35. Chen, X.; Fan, L.; Qiu, L.; Dong, X.; Wang, Q.; Hu, G.; Meng, S.; Li, D.; Chen, J. Metagenomics Analysis Reveals Compositional and Functional Differences in the Gut Microbiota of Red Swamp Crayfish, *Procambarus clarkii*, Grown on Two Different Culture Environments. *Front. Microbiol.* **2021**, *12*, 3070. [CrossRef] [PubMed]
36. Skelton, J.; Geyer, K.M.; Lennon, J.T.; Creed, R.P.; Brown, B.L. Multi-scale ecological filters shape the crayfish microbiome. *Symbiosis* **2016**, *72*, 159–170. [CrossRef]
37. Dragicevic, P.; Bielen, A.; Petric, I.; Vuk, M.; Žucko, J.; Hudina, S. Microbiome of the successful freshwater invader, the signal crayfish, and its changes along the invasion range. *Microbiol. Spectr.* **2021**, *9*, e00389-21. [CrossRef] [PubMed]
38. Nobles, S.; Jackson, C.R. Effects of Life Stage, Site, and Species on the Dragonfly Gut Microbiome. *Microorganisms* **2020**, *8*, 183. [CrossRef] [PubMed]
39. Deb, R.; Nair, A.; Agashe, D. Host dietary specialization and neutral assembly shape gut bacterial communities of wild dragonflies. *PeerJ* **2019**, *7*, e8058. [CrossRef]

*Article*

# Gut Microbial Characterization of Melon-Headed Whales (*Peponocephala electra*) Stranded in China

**Shijie Bai [1], Peijun Zhang [1,*], Xianfeng Zhang [2], Zixin Yang [1] and Songhai Li [1,*]**

[1] Institute of Deep-Sea Science and Engineering, Chinese Academy of Sciences, Sanya 572000, China; baishijie@idsse.ac.cn (S.B.); yangzx@idsse.ac.cn (Z.Y.)
[2] Institute of Hydrobiology, Chinese Academy of Sciences, Wuhan 430072, China; zhangx@ihb.ac.cn
* Correspondence: pjzhang@idsse.ac.cn (P.Z.); lish@idsse.ac.cn (S.L.)

**Abstract:** Although gut microbes are regarded as a significant component of many mammals and play a very important role, there is a paucity of knowledge around marine mammal gut microbes, which may be due to sampling difficulties. Moreover, to date, there are very few, if any, reports on the gut microbes of melon-headed whales. In this study, we opportunistically collected fecal samples from eight stranded melon-headed whales (*Peponocephala electra*) in China. Using high-throughput sequencing technology of partial 16S rRNA gene sequences, we demonstrate that the main taxa of melon-headed whale gut microbes are Firmicutes, Fusobacteriota, Bacteroidota, and Proteobacteria (Gamma) at the phylum taxonomic level, and *Cetobacterium*, *Bacteroides*, *Clostridium sensu stricto*, and *Enterococcus* at the genus taxonomic level. Meanwhile, molecular ecological network analysis (MENA) shows that two modules (a set of nodes that have strong interactions) constitute the gut microbial community network of melon-headed whales. Module 1 is mainly composed of *Bacteroides*, while Module 2 comprises *Cetobacterium* and *Enterococcus*, and the network keystone genera are *Corynebacterium*, *Alcaligenes*, *Acinetobacter*, and *Flavobacterium*. Furthermore, by predicting the functions of the gut microbial community through PICRUSt2, we found that although there are differences in the composition of the gut microbial community in different individuals, the predicted functional profiles are similar. Our study gives a preliminary inside look into the composition of the gut microbiota of stranded melon-headed whales.

**Keywords:** melon-headed whale; gut; microbial communities; aquatic mammal

**Citation:** Bai, S.; Zhang, P.; Zhang, X.; Yang, Z.; Li, S. Gut Microbial Characterization of Melon-Headed Whales (*Peponocephala electra*) Stranded in China. *Microorganisms* **2022**, *10*, 572. https://doi.org/10.3390/microorganisms10030572

Academic Editor: Konstantinos Ar. Kormas

Received: 30 January 2022
Accepted: 5 March 2022
Published: 6 March 2022

**Publisher's Note:** MDPI stays neutral with regard to jurisdictional claims in published maps and institutional affiliations.



## 1. Introduction

The melon-headed whale (*Peponocephala electra*) is a member of the subfamily Globicephalinae, where it is most closely related to the larger pilot whales (*Globicephala melas* and *G. macrorhynchus*), and it is also not a well-known species [1]. This whale is mostly dark gray in color, with a faint dark gray cloak on its back and a narrow head that slopes downward below a tall sickle-shaped dorsal fin. This species is difficult to distinguish at sea from the pygmy killer whale (*Feresa attenuata*). However, in stranded specimens, the melon-headed whale can be identified from all other pygmy killer whales by its high tooth count, as the melon-headed whale has ~25 teeth per row, while the pygmy killer whale has only about ~15 teeth per row [2]. Melon-headed whales are found worldwide in tropical and warm–temperate waters [3]. They mainly feed on fish, squid, cuttlefish, and shrimp, foraging from the littoral zone down to the bathypelagic zone [2,4,5].

Microbes are exceedingly abundant and varied in the gut of mammals [6]. Interactions between microbes and their host are necessary for the regulation of health, survival, and physiological functions of the host [7–9]. The majority of microbes reside in the gut, and their associated phenotypes shape the immune system of the host and contribute to nutrient uptake and defense against infectious diseases [10,11]. Therefore, revealing the mammalian gut microbiota is essential to fully understand the physiology and health status of mammals themselves. To date, most studies have focused on human gut microbiota, and information on the gut microbial composition of other mammals, especially cetaceans, although there are some reports, remains relatively scarce due to sampling limitations.

According to previous reports, gut samples from cetaceans are mainly obtained from three approaches: (1) feces in the wild just post-defecation. For example, Sanders et al. [12] investigated the microbial diversity and function of gut microbiomes in baleen whales feces and found them harbored unique gut microbiomes whereas still kept a functional capacity similar to that of both carnivores and herbivores; (2) fecal samples from human cared animals, such as studies on belugas (*Delphinapterus leucas*), Pacific white-sided dolphins (*Lagenorhynchus obliquidens*) and common bottlenose dolphins (*Tursiops truncatus*) [13,14], and Yangtze finless porpoises (*Neophocaena phocaenoides asiaeorientalis*) [15]; and (3) from dead, stranded animals. A few of studies sequenced along the gastrointestinal tracts of stranded cetaceans to investigate the distribution of microorganisms in different gut regions [16–19].

In this study, we opportunistically collected fecal samples from eight melon-headed whales stranded in China. Through investigating this infrequently known cetacean species, we aim to address the gut microbial compositions and diversity and gut microbial community network and predict the potential function of gut microbes in melon-headed whales.

## 2. Materials and Methods

### 2.1. Sample Collection

A rare mass stranding of 12 melon-headed whales happened on 6 July 2021, Tumen Port, Linhai, Taizhou City, Zhejiang Province, China. In this group of melon-headed whales, three individuals were found dead, two were released back immediately during the rescue course, and the remaining seven individuals were temporarily kept for recovery and released back to the wild the next day. We thus collected seven fecal samples from the recovering melon-headed whales before their release.

Another melon-headed whale stranding case happened on 25 May 2021, in Houan Town, Wanning City, Hainan Province, China. The animal was rescued and kept in Fuli Oceanarium (Lingshui, Hainan Province, China) for recovery. We collected one fecal sample from this animal on June 10 during its recovery time, before its death on 20 June 2021.

All fecal samples were harvested by veterinarians using anal swabs, with a diameter of 12 mm, which were inserted 10–15 cm into the rectum. All fecal samples were collected when animals were lifted out of water, and frozen at −20 °C until DNA extraction. Detailed information of these sampling animals is shown in Table S1.

### 2.2. DNA Extraction and Sequencing

The DNA of all fecal samples and three extraction blank control samples were extracted using MoBio PowerSoil extraction kits (Mo Bio Laboratories, Carlsbad, CA, USA) in accordance with the manufacturer's instructions. The extracted DNA was quantified using a Qubit fluorometer (Invitrogen Inc. Manufacturer: Life Technologies Holdings Pte Ltd., Singapore) and primer pair 515f Modified and 806r Modified were used to amplify the V4 region of the 16S rRNA gene [20]. The PCR amplification was performed under the following conditions: denaturation at 95 °C for 3 min, followed by 27 cycles at 95 °C for 30 s, 55 °C for 30 s, and 72 °C for 45 s, and a final extension at 72 °C for 10 min. PCR amplification results in triplicate were combined after purification with a TaKaRa purification kit (TaKaRa, Kusatsu, Japan). PCR products were prepared for library construction using the TruSeq DNA sample preparation kit (Illumina, San Diego, CA, USA) in accordance with the manufacturer's instructions. The libraries were sequenced at MajorBio Co. Ltd. (Shanghai, China) using the HiSeq platform (Illumina, San Diego, CA, USA) with reads of 250 bp at the paired end [13].

*2.3. Microbial Community Analysis*

After sequencing and obtaining the raw data, barcodes were removed as well as forward and reverse primers (one mismatch each was allowed) to obtain clean data. The FLASH program version 1.2.8 [21] was used to obtain paired-end of sufficient length with at least a 30 bp overlap combined into full-length sequences, and the average fragment length was 253 bp. The high-quality sequences without Ns contained were recruited using the Btrim program (version 0.2.0), and the sequences of 245 bp to 260 bp were used for the next analysis [22]. UNOISE3 was applied to generate amplicon sequence variants (ASVs) with default settings [23]. A representative sequence from each ASV was selected for taxonomic annotation via comparison with the SILVA 132 database [24], which includes bacterial, archaeal, and eukaryotic sequences; the Chloroplast and mitochondrial reads were excluded. To take into count the different sequencing depths, ASVs were randomly resampled to normalize the reads for each sample. The diversity of the microbial communities from the fecal samples of different individuals was determined via statistical analysis of the α-diversity indices, such as the Shannon, Inverse Simpson, Chao1 indices [25], and observed richness. R language [26] and the Mothur program [27] were used to calculate these α-diversity indices.

Molecular ecological network analysis (MENA) was used to perform the structure of microbial community networks [28,29]. Only the ASVs that appeared in more than four of the eight fecal samples of melon-headed whales were included in the network analysis. Correlations were calculated using Spearman's coefficient and a random matrix theory (RMT)-based approach was employed to delimit the microbial network interactions between samples. The keystone taxa were allocated according to the within-module connectivity (Zi) and among-module connectivity (Pi) according to a previously used method [28]. Nodes (ASVs) can be divided into four categories: (1) peripherals, which includes the nodes with Zi $\leq$ 2.5 and Pi $\leq$ 0.62, indicating nodes interconnected by a few links within the modules; (2) connectors, which includes the nodes with Zi $\leq$ 2.5 and Pi $<$ 0.62, indicating nodes linking to various modules; (3) module hubs, which includes the nodes with Zi $<$ 2.5 and Pi $\leq$ 0.62, indicating nodes within the modules are highly connected; and (4) network hubs, which includes the nodes with Zi $<$ 2.5 and Pi $<$ 0.62, indicating nodes highly connected among modules. The Phylogenetic Investigation of Communities by Reconstruction of Unobserved States (PICRUSt2) was used to predict microbial community function based on the MetaCyc database [30,31]. The raw sequencing reads of all samples were deposited to the NCBI database (http://www.ncbi.nlm.nih.gov/, accessed on 29 January 2022) under BioProject accession number: PRJNA801934.

## 3. Results

*3.1. Sequencing Statistics and Microbial Diversity*

Originally, a total of 642,263 sequences were obtained from 8 fecal samples of 8 stranded melon-headed whales (assigned as PE1 to PE8, Table S1) after quality assessment. To obtain more accurate α-diversity results to analyze microbial diversity, composition, and structure, we rarefied the sequences of each sample to 34,224. The α-diversities of microbial communities from the gut of eight melon-headed whales were calculated. The results showed PE8 and PE2 had lower Shannon and Inverse Simpson indices, while PE1 and PE4 had lower Chao1 indices and observed richness (Figure 1).

The relative abundance of gut microbes was apparent at the phylum, family, and genus levels, with a similarity of 97% for ASV taxonomy, and provided detailed relative abundance information on gut microbial community composition (Figures 2–4). Furthermore, we also provided the datasets of ASV table and the information of classification (Table S2). Firmicutes, Fusobacteriota, and Bacteroidota were the dominant bacterial lineages in the fecal samples of melon-headed whales, while the majority of the fecal samples from the PE8 in this study were dominated by Proteobacteria (Gamma), accounting for 82%. At the family taxonomic level, Fusobacteriaceae, Enterococcaceae, and Bacteroidaceae, which are affiliated with Fusobacteriales, Lactobacillales, and Bacteroidales, respectively, were the dominant bacterial lineages in the fecal samples of PE1 to PE7. However, the respective compositions of different fecal samples were slightly different; for instance, the fecal sample of PE8 was dominated by Shewanellaceae (Enterobacterales, 72%). Furthermore, at the genus taxonomic level, the gut microbial communities of melon-headed whales were mainly composed of Cetobacterium, Bacteroides, Clostridium sensu stricto, and Enterococcus. Nevertheless, the distribution of these dominant bacterial lineages in different fecal samples is different. For instance, Cetobacterium was dominant in the fecal samples of PE4, PE5, and PE6; Bacteroides was dominant in the samples of PE1, PE4, and PE7; and Clostridium sensu stricto was dominant in the samples of PE1, PE6, and PE7. The fecal samples of PE2 and PE3 were dominated by Enterococcus, which accounted for 68% and 53%, respectively. Only one ASV was annotated with Shewanella, and this ASV was annotated at the level of species as Shewanella algae. This bacterium was distributed in all fecal samples, but in the sample of PE8, Shewanella algae was the overwhelmingly dominant bacterium, accounting for 72% (Figure 4).

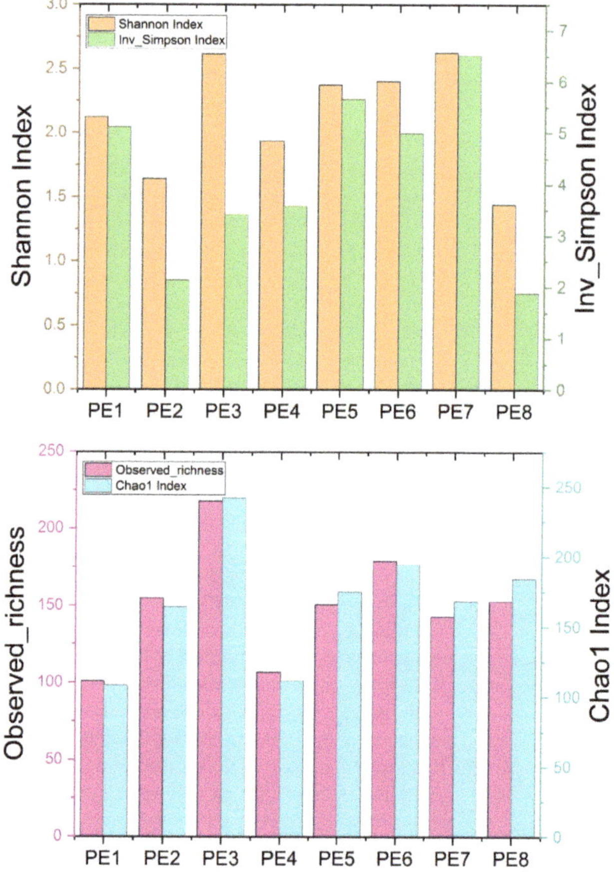


**Figure 1.** Four α-diversity indices—Shannon index, Inverse Simpson index, observed richness, and Chao1 index—of the eight fecal specimens from eight stranded melon-headed whales (PE1-8). The results are based on the ASV datasets.

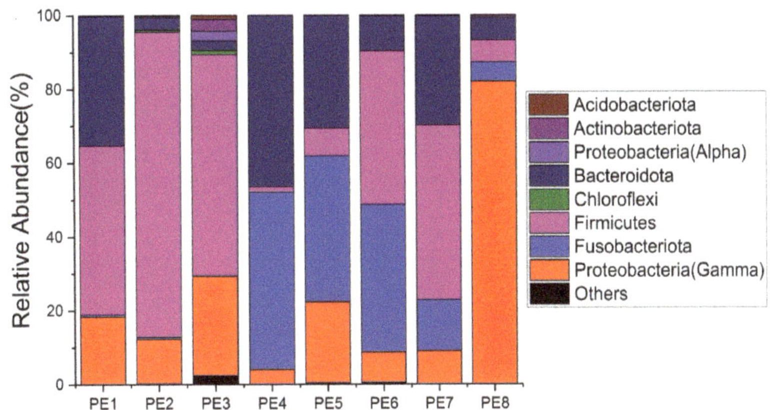


**Figure 2.** Gut microbial community members of eight stranded melon-headed whales (PE1-8) at the phylum level.

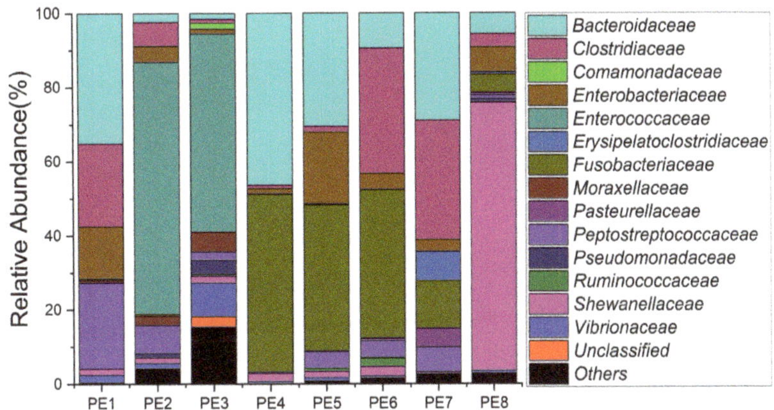


**Figure 3.** Gut microbial community members of eight stranded melon-headed whales (PE1-8) at the family level.

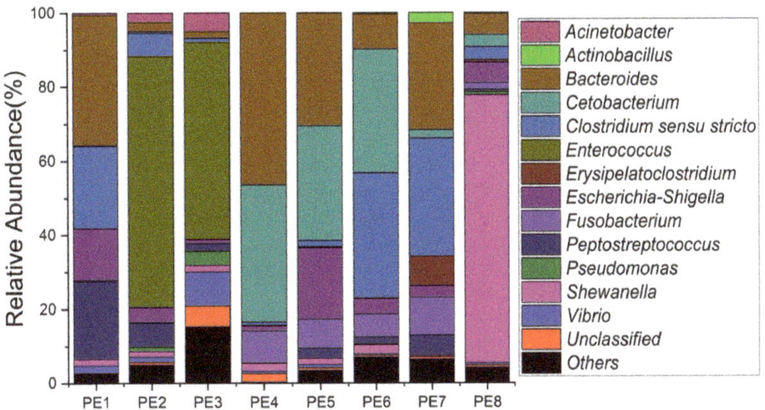


**Figure 4.** Gut microbial community members of eight stranded melon-headed whales (PE1-8) at the genus level.

### 3.2. Co-Occurrence Network and Functional Profile of Gut Microbial Communities

In order to reveal the gut microbial community interactions of melon-headed whales, the network was constructed through the MENA approach. The nodes and links of this network were 128 and 676, respectively. The average clustering coefficient (avgCC) was 0.337, and the average path distance (GD) was 2.513. This network formed a total of two modules (a set of nodes that have strong interactions): module one was mainly composed of *Bacteroides*, while module two was mainly composed of *Cetobacterium* and *Enterococcus*. Moreover, the keystone taxa belonged to module hubs, composed of those ASVs with $Zi < 2.5$, $Pi \leq 0.62$, in the microbial network of melon-headed whales; the keystone genera were *Acinetobacter*, *Alcaligenes*, *Corynebacterium*, and *Flavobacterium* (Figure 5).

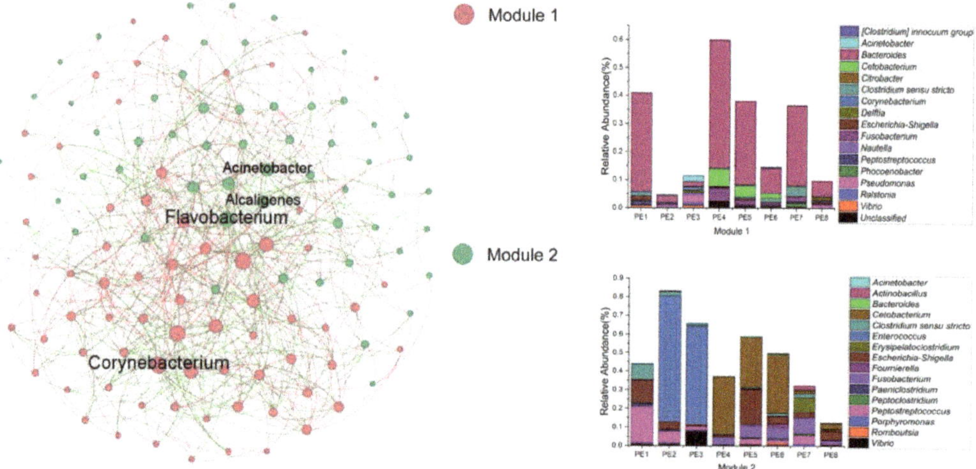


**Figure 5.** Co-occurrence networks of gut microbial communities. Stacked bar chart shows relative abundance of ASVs in Modules 1 and 2; a Module is a set of nodes that have strong interactions; these samples were collected from eight stranded melon-headed whales (PE1-8).

To better understand the potential functions of melon-headed whale gut bacteria, we explored the functional features of microbial communities using the newly updated PICRUSt2 software. No obvious functional difference was found between individuals. The main functions involved in the gut microbes of stranded melon-headed whales include the following: RNA processing and modification; energy production and conversion; cell cycle control, cell division, chromosome partitioning; amino acid transport and metabolism; nucleotide transport and metabolism; carbohydrate transport and metabolism; coenzyme transport and metabolism; lipid transport and metabolism; translation, ribosomal structure, and biogenesis; transcription; replication, recombination, and repair; cell wall/membrane/envelope biogenesis; cell motility; post-translational modification, protein turnover, chaperones; inorganic ion transport and metabolism; secondary metabolites biosynthesis, transport, and catabolism; signal transduction mechanisms; intracellular trafficking, secretion, and vesicular transport, and defense mechanisms (Figure 6). The detailed results of PICRUSt2 were provided in Table S3.

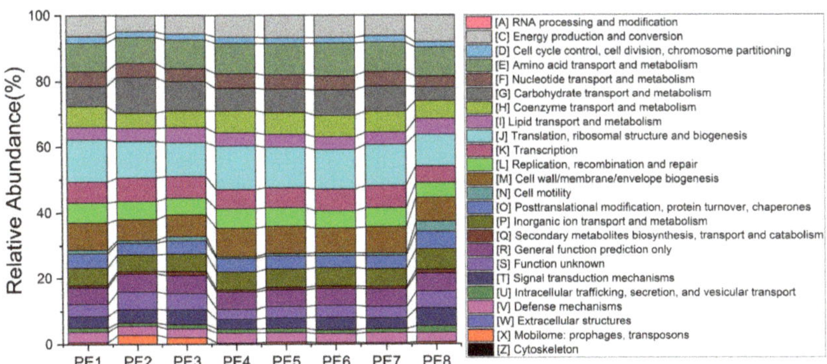


**Figure 6.** Functional profiles of gut microbial communities predicted by PICRUSt2; these samples were collected from eight stranded melon-headed whales (PE1-8).

## 4. Discussion

Due to the difficulty of sample collection, studies on cetacean gut microbes are usually from animals in zoos and oceanariums (e.g., [13,14,32]), or stranded cetaceans (e.g., [16,17,19]). To date, there are very few, if any, reports on the gut microbial communities of melon-headed whales. In this study, we obtained eight fecal samples from eight different stranded melon-headed whales. Through 16S rRNA gene sequencing, we revealed that members of *Cetobacterium*, *Bacteroides*, *Clostridium sensu stricto*, and *Enterococcus* constituted the vast majority of the gut microorganisms in melon-headed whales. We also found the distribution of gut microorganisms in different individuals was different; in spite of this, the functional profiles between individuals were similar. Thus, we propose that a functional-driven strategy may play an important role in the composition of the gut microbial community in melon-headed whales, rather than a species-driven strategy. However, further studies are warranted.

We also want to mention that PE8 in our study was not healthy, and was treated with antibiotics, i.e., penicillins and cephalosporin, for two weeks under human care before sample collection. Antibiotic treatment had a potential to affect the composition of gut microbial communities in PE8. A necropsy of PE8 showed it suffered from lung lesion, which might be the reason of its death. When we document the composition of gut microbial communities in melon-headed whales in our study, we always carefully consider the situation of PE8 first, and then make a cautious conclusion.

The genera *Cetobacterium*, from the phylum Fusobacteria, can be found in the gut of many cetacean species, such as short-finned pilot whales [16], toothed whales [12], and southern right whales *Eubalaena australis* [33]. Polysaccharides comprise the most abundant type of biopolymers, and therefore, the most abundant source of biological food. Carbohydrate fermentation by *Bacteroides* and other intestinal bacteria produces large amounts of volatile fatty acids, which are absorbed through the large intestine and utilized by the host as an energy source, providing a large portion of the host's daily energy needs [34]. Although most *Bacteroides* are symbiotic in the intestine, several species can also cause infections, including *Bacteroides fragilis*, *Bacteroides distasonis*, *Bacteroides ovatus*, *Bacteroides thetaiotaomicron*, *Bacteroides vulgatus*, and *Bacteroides uniformis*, with significant morbidity and mortality [35]. The genera *Clostridium sensu stricto* are other common microorganisms in the gut of cetaceans, for example, beluga whales *Delphinapterus leucas*, Pacific white-sided dolphins *Lagenorhynchus obliquidens*, common bottlenose dolphins *Tursiops truncatus*, and short-finned pilot whales [13,16]. *Clostridium* is one of the most common genera of cetacean gut microorganisms, while some studies suggest that members of *Clostridium* have low virulence and can pose a potential threat to unhealthy cetaceans [33,36]. The members of *Enterococcus* can also be found in the gut of some cetaceans, such as pygmy sperm whales *Kogia breviceps*, Pacific white-sided dolphins, and common bottlenose dolphins [13,16].

We detected an overwhelming dominance of *Shewanella algae* in PE8. However, the group of *Shewanella algae* was not found to be particularly common in the gut of cetaceans in this or in previous studies; we detected this bacterial linage in the gut of short-finned pilot whales[16] and melon-headed whales. Furthermore, the ASVs were all annotated as *Shewanella algae*. *Shewanella algae* is ubiquitous in the marine environments and has been identified as conditionally pathogenic bacteria that can cause serious infections, primarily associated with exposure to seawater and ingestion of raw seafood, and this group of bacteria can exhibited b-hemolytic activity, strong biofilm-adherence capabilities, and multiple antibiotic resistances [37–42]. We think that *Shewanella algae* should not be a dominant group (though it can be present) in melon-headed whales or short-finned pilot whales [16]; indeed, the overwhelming dominance of *Shewanella algae* in the gut of PE8 might have been a potential trigger of its death (Figure 3). A necropsy of PE8 showed that it likely died of lung lesion.

Functional profiles are characteristics that influence the adaptability of microbial communities under specific environmental conditions. However, because of the continuous exchange and transfer of horizontal genes between microorganisms and adaptive evolution, functional characteristics of microbial communities can be delinked from their taxonomic relevance [43]. In the present study, although there were differences in the microbial community structures between different samples, their predicted functional profiles were similar. The recently developed approach of molecular ecological networks can reveal the interrelations within a microbial community. We found two modules in the gut microbial community network of the eight stranded melon-headed whales. The microbial communities of Module 1 were dominated by *Bacteroides*, and the keystone genus was *Corynebacterium*. *Cetobacterium* and *Enterococcus* were the dominant bacterial lineages in Module 2, and *Alcaligenes*, *Acinetobacter*, and *Flavobacterium* were the keystone genera. The genus *Corynebacterium* represents a group of Gram-positive, rod-shaped, and typically club-shaped bacterial cells [44]. Some species of *Corynebacterium* are well-known pathogens of mammals and may occasionally cause infections, while some other species are normal microorganisms of microbial communities where it belongs [44]. In this study, ASV 56 was the keystone genus of Module 1 and could be annotated to the genus level. The keystone ASVs of Module 2, *Alcaligenes* and *Acinetobacter*, could also be annotated to the genus level, while another keystone ASV of Module 2, annotated as *Flavobacterium jumunjinense*, was isolated from lagoon water in Korea [45]. The genus *Alcaligenes* consists of motile Gram-negative rod-shaped bacteria that are chemoorganotrophic microbes. The members of *Alcaligenes* are common in water, soil, vertebrate intestinal tracts, and in clinical samples as a result of opportunistic infection [46]. Some *Alcaligenes* strains are able to be isolated from some contaminated environmental samples; therefore, they may show potential in the development of biodegradation processes or as biosensors. Moreover, some species of *Alcaligenes* are used in the food and healthcare industries, while some enzymes and polysaccharides produced by *Alcaligenes* have been used in the cosmetic industry and as food additives, showing potential for the treatment of certain immune diseases [46]. *Acinetobacter* spp. are Gram-negative coccobacilli; they are ubiquitous in the environment and are considered to be nonpathogenic to healthy individuals [47]. Although we detected both groups of bacteria (*Alcaligenes* and *Acinetobacter*) in the gut of melon-headed whales, their relative abundance was very low, and their roles are still unclear.

## 5. Conclusions

It is important to reveal the gut microbial communities of specific cetacean species, especially some poorly understood ones. In our study, the composition, functional profile, and interactions of gut microbial communities of eight stranded melon-headed whales were systematically studied. We conclude that the microbial community composition mainly consists of *Cetobacterium*, *Bacteroides*, *Clostridium sensu stricto*, and *Enterococcus*. Two modules constitute the network of the gut microbes of melon-headed whales; *Bacteroides* was the main microbial taxon in Module 1, while Module 2 mainly comprised *Cetobacterium* and *Enterococcus*. Moreover, based on network analysis, the keystone taxa (module hubs) were assigned to *Corynebacterium*, *Alcaligenes*, *Acinetobacter*, and *Flavobacterium*. Our study gives a preliminary inside look into the composition of the gut microbiota of stranded melon-headed whales. Furthermore, we also want to mention that we have very limited microbial information in melon-headed whales, as only one group of whales was studied. This may strongly affect the informational value of the obtained data. All whales may have had an exchange of the microbiota and may have been affected by the same environmental conditions. Other studies of whale feces microbiota studied samples collected from whales at different locations and different time points should be further conducted. In addition, metagenomics, transcriptomics, and proteomics should be used to better understand the functional information of the gut microbes in melon-headed whales.

**Supplementary Materials:** The following supporting information can be downloaded at: https://www.mdpi.com/article/10.3390/microorganisms10030572/s1, Table S1: Sampling information of the eight melon-headed whales; Table S2: ASV datasets and the classification information; Table S3: The detailed functional results predicted by PICRUSt2.

**Author Contributions:** Conceptualization, P.Z. and S.B.; methodology, S.B.; software, S.B.; formal analysis, S.B.; resources, P.Z., X.Z., and Z.Y.; data curation, S.B. and P.Z.; writing—original draft preparation, S.B. and P.Z.; writing—review and editing, S.L., X.Z., and Z.Y.; supervision, P.Z. and S.L.; project administration, S.L.; funding acquisition, P.Z. All authors have read and agreed to the published version of the manuscript.

**Funding:** This research and APC were all funded by the Youth Innovation Promotion Association of Chinese Academy of Sciences, grant number 2020363.

**Institutional Review Board Statement:** The study was approved by the Institutional Ethics Committee of Institute of Deep-sea Science and Engineering CAS (No. IDSSE-SYLL-MMMBL-01 approval on January 2017).

**Informed Consent Statement:** Not applicable.

**Data Availability Statement:** The raw sequencing reads of all samples were deposited to the NCBI database (http://www.ncbi.nlm.nih.gov/, accessed on 29 January 2022)) under BioProject accession number: PRJNA801934.

**Acknowledgments:** We want to thank all the staff who participated in the rescue work of these whales from official fishery administration departments and the aquaria in Zhejiang and Hainan provinces. We also want to thank Xi Wang and Huanshan Wang from Institute of Hydrobiology CAS for their kind help during this study.

**Conflicts of Interest:** The authors declare no conflict of interest. The funders had no role in the design of the study; in the collection, analyses, or interpretation of data; in the writing of the manuscript, or in the decision to publish the results.

## References

1. Vilstrup, J.T.; Ho, S.Y.; Foote, A.D.; A Morin, P.; Kreb, D.; Krützen, M.; Parra, G.J.; Robertson, K.M.; De Stephanis, R.; Verborgh, P.; et al. Mitogenomic phylogenetic analyses of the Delphinidae with an emphasis on the Globicephalinae. *BMC Evol. Biol.* **2011**, *11*, 65. [CrossRef] [PubMed]
2. Perryman, W.L.; Danil, K. Melon-Headed Whale. In *Encyclopedia of Marine Mammals*, 3rd ed.; Wursig, B., Thewissen, J.G.M., Kovacs, K., Eds.; Academy Press: Cambridge, MA, USA, 2018; pp. 593–595. [CrossRef]
3. Jefferson, T.A.; Webber, M.A.; Pitman, R. *Marine Mammals of the World: A Comprehensive Guide to their Identification*, 2nd ed.; Academic Press: San Diego, CA, 2015.
4. Jefferson, T.A.; Barros, N.B. Peponocephala electra. *Mammalian Species* **1997**, *553*, 1–6. [CrossRef]
5. Spitz, J.; Cherel, Y.; Bertin, S.; Kiszka, J.; Dewez, A.; Ridoux, V. Prey preferences among the community of deep-diving odontocetes from the Bay of Biscay, Northeast Atlantic. *Deep Sea Res. Part I Oceanogr. Res. Pap.* **2011**, *58*, 273–282. [CrossRef]
6. Gensollen, T.; Iyer, S.S.; Kasper, D.L.; Blumberg, R.S. How colonization by microbiota in early life shapes the immune system. *Science* **2016**, *352*, 539–544. [CrossRef]
7. Dang, A.T.; Marsland, B.J. Microbes, metabolites, and the gut–lung axis. *Mucosal Immunol.* **2019**, *12*, 843–850. [CrossRef]
8. Krajmalnik-Brown, R.; Ilhan, Z.-E.; Kang, D.-W.; DiBaise, J.K. Effects of Gut Microbes on Nutrient Absorption and Energy Regulation. *Nutr. Clin. Pr.* **2012**, *27*, 201–214. [CrossRef]

9. Woting, A.; Blaut, M. The Intestinal Microbiota in Metabolic Disease. *Nutrients* **2016**, *8*, 202. [CrossRef]
10. Dzutsev, A.; Badger, J.H.; Perez-Chanona, E.; Roy, S.; Salcedo, R.; Smith, C.K.; Trinchieri, G. Microbes and Cancer. *Annu. Rev. Immunol.* **2017**, *35*, 199–228. [CrossRef]
11. Quin, C.; Gibson, D.L. Human behavior, not race or geography, is the strongest predictor of microbial succession in the gut bacteriome of infants. *Gut Microbes* **2020**, *11*, 1143–1171. [CrossRef]
12. Sanders, J.G.; Beichman, A.C.; Roman, J.; Scott, J.; Emerson, D.; McCarthy, J.J.; Girguis, P.R. Baleen whales host a unique gut microbiome with similarities to both carnivores and herbivores. *Nat. Commun.* **2015**, *6*, 8285. [CrossRef]
13. Bai, S.; Zhang, P.; Zhang, C.; Du, J.; Du, X.; Zhu, C.; Liu, J.; Xie, P.; Li, S. Comparative Study of the Gut Microbiota Among Four Different Marine Mammals in an Aquarium. *Front. Microbiol.* **2021**, *12*. [CrossRef] [PubMed]
14. Suzuki, A.; Segawa, T.; Sawa, S.; Nishitani, C.; Ueda, K.; Itou, T.; Asahina, K.; Suzuki, M. Comparison of the gut microbiota of captive common bottlenose dolphins Tursiops truncatus in three aquaria. *J. Appl. Microbiol.* **2018**, *126*, 31–39. [CrossRef] [PubMed]
15. McLaughlin, R.W.; Chen, M.; Zheng, J.; Zhao, Q.; Wang, D. Analysis of the bacterial diversity in the fecal material of the endangered Yangtze finless porpoise, Neophocaena phocaenoides asiaeorientalis. *Mol. Biol. Rep.* **2011**, *39*, 5669–5676. [CrossRef]
16. Bai, S.; Zhang, P.; Lin, M.; Lin, W.; Yang, Z.; Li, S. Microbial diversity and structure in the gastrointestinal tracts of two stranded short-finned pilot whales (*Globicephala macrorhynchus*) and a pygmy sperm whale (*Kogia breviceps*). *Integr. Zool.* **2021**, *16*, 324–335. [CrossRef]
17. Tian, J.; Du, J.; Lu, Z.; Han, J.; Wang, Z.; Li, D.; Guan, X.; Wang, Z. Distribution of microbiota across different intestinal tract segments of a stranded dwarf minke whale, Balaenoptera acutorostrata. *MicrobiologyOpen* **2020**, *9*. [CrossRef]
18. Wan, X.; Li, J.; Cheng, Z.; Ao, M.; Tian, R.; Mclaughlin, R.W.; Zheng, J.; Wang, D. The intestinal microbiome of an Indo-Pacific humpback dolphin (*Sousa chinensis*) stranded near the Pearl River Estuary, China. *Integr. Zool.* **2021**, *16*, 287–299. [CrossRef] [PubMed]
19. Wan, X.-L.; McLaughlin, R.W.; Zheng, J.-S.; Hao, Y.-J.; Fan, F.; Tian, R.-M.; Wang, D. Microbial communities in different regions of the gastrointestinal tract in East Asian finless porpoises (*Neophocaena asiaeorientalis sunameri*). *Sci. Rep.* **2018**, *8*, 1–10. [CrossRef] [PubMed]
20. Walters, W.; Hyde, E.R.; Berg-Lyons, D.; Ackermann, G.; Humphrey, G.; Parada, A.; Gilbert, J.A.; Jansson, J.K.; Caporaso, J.G.; Fuhrman, J.; et al. Improved Bacterial 16S rRNA Gene (V4 and V4-5) and Fungal Internal Transcribed Spacer Marker Gene Primers for Microbial Community Surveys. *mSystems* **2016**, *1*, e00009-15. [CrossRef]
21. Magoč, T.; Salzberg, S.L. FLASH: Fast length adjustment of short reads to improve genome assemblies. *Bioinformatics* **2011**, *27*, 2957–2963. [CrossRef]
22. Kong, Y. Btrim: A fast, lightweight adapter and quality trimming program for next-generation sequencing technologies. *Genomics* **2011**, *98*, 152–153. [CrossRef]
23. Edgar, R.C. UNOISE2: Improved error-correction for Illumina 16S and ITS amplicon sequencing. *BioRxiv* **2016**, 081257. Available online: https://www.biorxiv.org/content/10.1101/081257.abstract (accessed on 29 January 2022).
24. Quast, C.; Pruesse, E.; Yilmaz, P.; Gerken, J.; Schweer, T.; Yarza, P.; Peplies, J.; Glöckner, F.O. The SILVA ribosomal RNA gene database project: Improved data processing and web-based tools. *Nucleic Acids Res.* **2013**, *41*, D590–D596. [CrossRef] [PubMed]
25. Chao, A. Nonparametric-estimation of the number of classes in a population. *Scand. J. Stat.* **1984**, *11*, 265–270.
26. R Core Team. R: A Language and Environment for Statistical Computing. Available online: https://www.R-project.org/ (accessed on 29 January 2022).
27. Schloss, P.D.; Westcott, S.L.; Ryabin, T.; Hall, J.R.; Hartmann, M.; Hollister, E.B.; Lesniewski, R.A.; Oakley, B.B.; Parks, D.H.; Robinson, C.J.; et al. Introducing mothur: Open-Source, Platform-Independent, Community-Supported Software for Describing and Comparing Microbial Communities. *Appl. Environ. Microbiol.* **2009**, *75*, 7537–7541. [CrossRef] [PubMed]
28. Deng, Y.; Jiang, Y.-H.; Yang, Y.; He, Z.; Luo, F.; Zhou, J. Molecular ecological network analyses. *BMC Bioinform.* **2012**, *13*, 113. [CrossRef] [PubMed]
29. Deng, Y.; Zhang, P.; Qin, Y.; Tu, Q.; Yang, Y.; He, Z.; Schadt, C.; Zhou, J. Network succession reveals the importance of competition in response to emulsified vegetable oil amendment for uranium bioremediation. *Environ. Microbiol.* **2015**, *18*, 205–218. [CrossRef] [PubMed]
30. Caspi, R.; Billington, R.; Keseler, I.M.; Kothari, A.; Krummenacker, M.; Midford, P.E.; Ong, W.K.; Paley, S.; Subhraveti, P.; Karp, P.D. The MetaCyc database of metabolic pathways and enzymes—A 2019 update. *Nucleic Acids Res.* **2020**, *48*, D445–D453. [CrossRef]
31. Douglas, G.M.; Maffei, V.J.; Zaneveld, J.R.; Yurgel, S.N.; Brown, J.R.; Taylor, C.M.; Huttenhower, C.; Langille, M.G.I. PICRUSt2 for prediction of metagenome functions. *Nat. Biotechnol.* **2020**, *38*, 685–688. [CrossRef]
32. Soverini, M.; Quercia, S.; Biancani, B.; Furlati, S.; Turroni, S.; Biagi, E.; Consolandi, C.; Peano, C.; Severgnini, M.; Rampelli, S.; et al. The bottlenose dolphin (Tursiops truncatus) faecal microbiota. *FEMS Microbiol. Ecol.* **2016**, *92*, fiw055. [CrossRef]
33. Marón, C.F.; Kohl, K.; Chirife, A.; Di Martino, M.; Fons, M.P.; Navarro, M.A.; Beingesser, J.; McAloose, D.; Uzal, F.A.; Dearing, M.D.; et al. Symbiotic microbes and potential pathogens in the intestine of dead southern right whale (*Eubalaena australis*) calves. *Anaerobe* **2019**, *57*, 107–114. [CrossRef]
34. Hooper, L.V.; Midtvedt, T.; Gordon, J.I. How host-microbial interactions shape the nutrient environment of the mammalian intestine. *Annu. Rev. Nutr.* **2002**, *22*, 283–307. [CrossRef] [PubMed]
35. Wexler, H.M. Bacteroides: The Good, the Bad, and the Nitty-Gritty. *Clin. Microbiol. Rev.* **2007**, *20*, 593–621. [CrossRef] [PubMed]

36. Erwin, P.M.; Rhodes, R.G.; Kiser, K.B.; Keenan-Bateman, T.F.; McLellan, W.A.; Pabst, D.A. High diversity and unique composition of gut microbiomes in pygmy (*Kogia breviceps*) and dwarf (*K. sima*) sperm whales. *Sci. Rep.* **2017**, *7*, 7205. [CrossRef] [PubMed]
37. Han, Z.; Sun, J.; Lv, A.; Sung, Y.; Shi, H.; Hu, X.; Xing, K. Isolation, identification and characterization of *Shewanella algae* from reared tongue sole, *Cynoglossus semilaevis* Günther. *Aquaculture* **2017**, *468*, 356–362. [CrossRef]
38. Holt, H.M.; Gahrn-Hansen, B.; Bruun, B. Shewanella algae and Shewanella putrefaciens: Clinical and microbiological characteristics. *Clin. Microbiol. Infect.* **2005**, *11*, 347–352. [CrossRef]
39. Ibrahim, N.N.N.; Nasir, N.M.; Sahrani, F.K.; Ahmad, A.; Sairi, F. Characterization of putative pathogenic Shewanella algae isolated from ballast water. *Veter-World* **2021**, *14*, 678–688. [CrossRef]
40. Lemaire, O.N.; Méjean, V.; Iobbi-Nivol, C. The Shewanella genus: Ubiquitous organisms sustaining and preserving aquatic ecosystems. *FEMS Microbiol. Rev.* **2020**, *44*, 155–170. [CrossRef]
41. Paccalin, M.; Grollier, G.; Le Moal, G.; Rayeh, F.; Camiade, C. Rupture of a Primary Aortic Aneurysm Infected with Shewanella Alga. *Scand. J. Infect. Dis.* **2001**, *33*, 774–775. [CrossRef]
42. Shanmuganathan, M.; Goh, B.L.; Lim, C.; NorFadhlina, Z.; Fairol, I. Shewanella algae Peritonitis in Patients on Peritoneal Dialysis. *Perit. Dial. Int. J. Int. Soc. Perit. Dial.* **2016**, *36*, 574–575. [CrossRef]
43. Yang, Y. Emerging Patterns of Microbial Functional Traits. *Trends Microbiol.* **2021**, *29*, 874–882. [CrossRef]
44. Bernard, K. The Genus Corynebacterium and Other Medically Relevant Coryneform-Like Bacteria. *J. Clin. Microbiol.* **2012**, *50*, 3152–3158. [CrossRef] [PubMed]
45. Joung, Y.; Kim, H.; Joh, K. Flavobacterium jumunjinense sp. nov., isolated from a lagoon, and emended descriptions of Flavobacterium cheniae, Flavobacterium dongtanense and Flavobacterium gelidilacus. *Int. J. Syst. Evol. Microbiol.* **2013**, *63*, 3937–3943. [CrossRef] [PubMed]
46. Batt, C. Alcaligenes. In *Encyclopedia of Food Microbiology*, 2nd ed.; Batt, C., Tortorello, M., Eds.; Academic Press: Cambridge, MA, USA, 2014; Volume 1, pp. 38–42. [CrossRef]
47. Paterson, D.L.; Peleg, A.Y. Acinetobacter. In *Antimicrobial Drug Resistance, Mayers, D. L, Eds.*; Humana Press: Totowa, NJ, USA, 2009; Volume 2, pp. 819–823. [CrossRef]

MDPI
St. Alban-Anlage 66
4052 Basel
Switzerland
Tel. +41 61 683 77 34
Fax +41 61 302 89 18
www.mdpi.com

*Microorganisms* Editorial Office
E-mail: microorganisms@mdpi.com
www.mdpi.com/journal/microorganisms

www.ingramcontent.com/pod-product-compliance
Ingram Content Group UK Ltd.
Pitfield, Milton Keynes, MK11 3LW, UK
UKHW062005290726
14090UKWH00022B/1399